Drug Delivery Nanoparticles Formulation and Characterization

DRUGS AND THE PHARMACEUTICAL SCIENCES

A Series of Textbooks and Monographs

For information on volumes 1–149 in the *Drugs and Pharmaceutical Science Series*, please visit www.informahealthcare.com

150. Laboratory Auditing for Quality and Regulatory Compliance, *Donald Singer, Raluca-Ioana Stefan, and Jacobus van Staden*
151. Active Pharmaceutical Ingredients: Development, Manufacturing, and Regulation, *edited by Stanley Nusim*
152. Preclinical Drug Development, *edited by Mark C. Rogge and David R. Taft*
153. Pharmaceutical Stress Testing: Predicting Drug Degradation, *edited by Steven W. Baertschi*
154. Handbook of Pharmaceutical Granulation Technology: Second Edition, *edited by Dilip M. Parikh*
155. Percutaneous Absorption: Drugs–Cosmetics–Mechanisms–Methodology, Fourth Edition, *edited by Robert L. Bronaugh and Howard I. Maibach*
156. Pharmacogenomics: Second Edition, *edited by Werner Kalow, Urs A. Meyer and Rachel F. Tyndale*
157. Pharmaceutical Process Scale-Up, Second Edition, *edited by Michael Levin*
158. Microencapsulation: Methods and Industrial Applications, Second Edition, *edited by Simon Benita*
159. Nanoparticle Technology for Drug Delivery, *edited by Ram B. Gupta and Uday B. Kompella*
160. Spectroscopy of Pharmaceutical Solids, *edited by Harry G. Brittain*
161. Dose Optimization in Drug Development, *edited by Rajesh Krishna*
162. Herbal Supplements-Drug Interactions: Scientific and Regulatory Perspectives, *edited by Y. W. Francis Lam, Shiew-Mei Huang, and Stephen D. Hall*
163. Pharmaceutical Photostability and Stabilization Technology, *edited by Joseph T. Piechocki and Karl Thoma*
164. Environmental Monitoring for Cleanrooms and Controlled Environments, *edited by Anne Marie Dixon*
165. Pharmaceutical Product Development: In Vitro-In Vivo Correlation, *edited by Dakshina Murthy Chilukuri, Gangadhar Sunkara, and David Young*
166. Nanoparticulate Drug Delivery Systems, *edited by Deepak Thassu, Michel Deleers, and Yashwant Pathak*
167. Endotoxins: Pyrogens, LAL Testing and Depyrogenation, Third Edition, *edited by Kevin L. Williams*
168. Good Laboratory Practice Regulations, Fourth Edition, *edited by Anne Sandy Weinberg*
169. Good Manufacturing Practices for Pharmaceuticals, Sixth Edition, *edited by Joseph D. Nally*
170. Oral-Lipid Based Formulations: Enhancing the Bioavailability of Poorly Water-soluble Drugs, *edited by David J. Hauss*
171. Handbook of Bioequivalence Testing, *edited by Sarfaraz K. Niazi*

172. Advanced Drug Formulation Design to Optimize Therapeutic Outcomes, *edited by Robert O. Williams III, David R. Taft, and Jason T. McConville*
173. Clean-in-Place for Biopharmaceutical Processes, *edited by Dale A. Seiberling*
174. Filtration and Purification in the Biopharmaceutical Industry, Second Edition, *edited by Maik W. Jornitz and Theodore H. Meltzer*
175. Protein Formulation and Delivery, Second Edition, *edited by Eugene J. McNally and Jayne E. Hastedt*
176. Aqueous Polymeric Coatings for Pharmaceutical Dosage Forms, Third Edition, *edited by James McGinity and Linda A. Felton*
177. Dermal Absorption and Toxicity Assessment, Second Edition, *edited by Michael S. Roberts and Kenneth A. Walters*
178. Preformulation Solid Dosage Form Development, *edited by Moji C. Adeyeye and Harry G. Brittain*
179. Drug-Drug Interactions, Second Edition, *edited by A. David Rodrigues*
180. Generic Drug Product Development: Bioequivalence Issues, *edited by Isadore Kanfer and Leon Shargel*
181. Pharmaceutical Pre-Approval Inspections: A Guide to Regulatory Success, Second Edition, *edited by Martin D. Hynes III*
182. Pharmaceutical Project Management, Second Edition, *edited by Anthony Kennedy*
183. Modified Release Drug Delivery Technology, Second Edition, Volume 1, *edited by Michael J. Rathbone, Jonathan Hadgraft, Michael S. Roberts, and Majella E. Lane*
184. Modified-Release Drug Delivery Technology, Second Edition, Volume 2, *edited by Michael J. Rathbone, Jonathan Hadgraft, Michael S. Roberts, and Majella E. Lane*
185. The Pharmaceutical Regulatory Process, Second Edition, *edited by Ira R. Berry and Robert P. Martin*
186. Handbook of Drug Metabolism, Second Edition, *edited by Paul G. Pearson and Larry C. Wienkers*
187. Preclinical Drug Development, Second Edition, *edited by Mark Rogge and David R. Taft*
188. Modern Pharmaceutics, Fifth Edition, Volume 1: Basic Principles and Systems, *edited by Alexander T. Florence and Jürgen Siepmann*
189. Modern Pharmaceutics, Fifth Edition, Volume 2: Applications and Advances, *edited by Alexander T. Florence and Jürgen Siepmann*
190. New Drug Approval Process, Fifth Edition, *edited by Richard A. Guarino*
191. Drug Delivery Nanoparticles Formulation and Characterization, *edited by Yashwant Pathak and Deepak Thassu*

Drug Delivery Nanoparticles Formulation and Characterization

edited by

Yashwant Pathak

Sullivan University College of Pharmacy

Louisville, Kentucky, USA

Deepak Thassu

PharmaNova, Inc.

Victor, New York, USA

informa

healthcare

New York London

Informa Healthcare USA, Inc.
52 Vanderbilt Avenue
New York, NY 10017

Printed in the United States of America on acid-free paper
10 9 8 7 6 5 4 3 2 1

International Standard Book Number-10: 1-4200-7804-6 (Hardcover)
International Standard Book Number-13: 978-1-4200-7804-6 (Hardcover)

Library of Congress Cataloging-in-Publication Data

Drug delivery nanoparticles formulation and characterization /
edited by Yashwant Pathak, Deepak Thassu.
p. ; cm. – (Drugs and the pharmaceutical sciences ; 191)
Includes bibliographical references and index.
ISBN-13: 978-1-4200-7804-6 (hardcover : alk. paper)
ISBN-10: 1-4200-7804-6 (hardcover : alk. paper) 1. Nanoparticles. 2. Drug delivery systems. I. Pathak, Yashwant. II. Thassu, Deepak. III. Series: Drugs and the pharmaceutical sciences ; v. 191.
[DNLM: 1. Drug Delivery Systems–methods. 2. Drug Carriers.
3. Nanoparticles. W1 DR893B v.191 2009 / QV 785 D79384 2009]
RS201.N35D78 2009
615′.6–dc22

2009007551

For Corporate Sales and Reprint Permissions call 212-520-2700 or write to: Sales Department, 52 Vanderbilt Avenue, 16th floor, New York, NY 10017.

Visit the Informa Web site at
www.informa.com

and the Informa Healthcare Web site at
www.informahealthcare.com

To the loving memories of my parents and Dr. Keshav Baliram Hedgewar, who gave a proper direction, my wife Seema, who gave a positive meaning, and my son Sarvadaman, who gave golden lining to my life.

Yashwant Pathak

I dedicate this book to my fellow scientists, my family – wife Anu, daughter Sakshi Zoya, and son Alex Om, and my parents, who taught me love, life, and compassion.

Deepak Thassu

Foreword

Drug delivery research is clearly moving from the micro- to the nanosize scale. Nanotechnology is therefore emerging as a field in medicine that is expected to elicit significant therapeutic benefits. The development of effective nanodelivery systems capable of carrying a drug specifically and safely to a desired site of action is one of the most challenging tasks of pharmaceutical formulation investigators. They are attempting to reformulate and add new indications to the existing blockbuster drugs to maintain positive scientific outcomes and therapeutic breakthroughs. The nanodelivery systems mainly include nanoemulsions, lipid or polymeric nanoparticles, and liposomes. Nanoemulsions are primarily used as vehicles of lipophilic drugs following intravenous administration. On the other hand, the ultimate objective of the other nanodelivery systems is to alter the normal biofate of potent drug molecules in the body following their intravenous administration to markedly improve their efficacy and reduce their potential intrinsic severe adverse effects.

Despite three decades of intensive research on liposomes as drug delivery systems, the number of systems that have undergone clinical trials and then reached the market has been quite modest. Furthermore, the scientific community has been skeptical that such goals could be achieved, because huge investments of funds and promising research studies have frequently ended in disappointing results or have been slow to yield successfully marketed therapeutic dosage forms based on lipid nanotechnology. Thus, the focus of the research activity has shifted to nanoparticulate drug delivery systems, as there are still significant unmet medical needs in target diseases such as cancer, autoimmune disorders, macular degeneration, and Alzheimer's disease. Most of the active ingredients used to treat these severe diseases can be administered only through the systemic route. Indeed, both molecular complexity associated with drugs and inaccessibility of most pharmacological targets are the major constraints and the main reasons behind the renewed curiosity and expanding research on nanodelivery systems, which can carry drugs directly to their site of action. Ongoing efforts are being made to develop polymeric nanocarriers capable of delivering active molecules specifically to the intended target organ. This approach involves modifying the pharmacokinetic profile of various therapeutic classes of drugs through their incorporation into nanodelivery systems. These site-specific delivery systems allow an effective drug concentration to be maintained for a longer interval in the target tissue and result in decreased adverse effects associated with lower plasma concentrations in the peripheral blood. Thus, drug targeting has evolved as the most desirable but elusive goal in the science of drug nanodelivery.

Drug targeting offers enormous advantages but is highly challenging and extremely complicated. Increased knowledge on the cellular internalization mechanisms of the nanocarriers is crucial for improving their efficacy, site-specific delivery, and intracellular targeting. Optimal pharmacological responses require both spatial placement of the drug molecules and temporal control at the site of action. Many hurdles still need to be overcome through intensive efforts and concentrated interdisciplinary scientific collaborations to reach the desired goals. However, in recent years, efforts have started to yield results with the approval by health authorities of nanoparticles containing paclitaxel (Abraxane®) for improved cancer therapy, which has rapidly become a commercial success. A large number of clinical trials are currently underway and are again raising the hopes and interest in drug nanodelivery systems.

There are various techniques to prepare drug-loaded nanoparticles, the selection of which depends on the physicochemical properties of the bioactive molecule and the polymer. The nanoparticulate drug delivery field is complex and requires considerable interdisciplinary knowledge. To facilitate the comprehension of this field, Dr. Deepak Thassu, Dr. Michele Deleers, and Dr. Yashwant Pathak co-edited their first book in a series on nanoparticulate drug delivery systems, which was published by Informa Healthcare in 2007. The book covered recent trends and emerging technologies in the field and was very well received by the scientific community. In this second volume on nanoparticulate drug delivery systems, Dr. Pathak and Dr. Thassu are covering various aspects of the field with a focus on formulations and characterization—two crucial but poorly understood issues in this technology.

The chapter authors come from a number of countries including the U.S.A., the U.K., India, Portugal, Canada, and South Korea, and represent many laboratories in the forefront of nanotechnology research. Chapters 1 to 11 cover various formulation aspects of nanoparticulate drug delivery systems. They embrace delivery of small molecules, macromolecules like therapeutic proteins, applications in gene therapy, and drug delivery systems for cancer, diabetes treatment, dermal applications, and many more. Chapters 12 to 15 cover the in vitro and in vivo evaluation as well as characterizations of the nanoparticulate drug delivery systems. The remaining chapters describe various analytical techniques used for the characterization of nanomaterials with special reference to nanomedicines. These sections highlight microscopic and spectroscopic characterization, SEM, TEM applications, structural fingerprinting of nanomaterials, mechanical characterization, and nanomaterial applications in bioimaging. Thus, a better understanding of physicochemical and physiological obstacles that a drug needs to overcome should provide the pharmaceutical scientist with information and tools needed to develop successful designs for drug targeting delivery systems. The book is therefore a timely publication that provides an opportunity for scientists to learn about the complex development issues of nanoparticulate drug delivery systems.

The book clearly and comprehensively presents recent advances and knowledge related to formulation and characterization of nanoparticulate drug delivery systems and is an excellent reference for researchers in the field of nanomedicine.

Dr. Yashwant Pathak and Dr. Deepak Thassu are to be complimented for both their judicial choice of topics in nanodelivery systems and their characterization techniques as well as for their selection of such respected and expert contributors from the field. Drs. Pathak and Thassu through their book will contribute to advancements in designing and successfully developing new generations of nanodelivery systems.

Simon Benita
The Institute of Drug Research
School of Pharmacy
The Hebrew University of Jerusalem
Jerusalem, Israel

Preface

Modern nanotechnology is an emerging and dynamic field. It is multidisciplinary in nature. It appears that *Mother Nature* was the first scientist offering nanoscale materials abundantly and they were used by the human beings from time immemorial. Several ancient practices have been developing nanoparticles through the traditional processes but these were not identified as nanosystems or nanoparticles. Ayurveda, the ancient traditional system of medicine in India, has described several "bhasmas," which have particles with sizes in nano range and have been used traditionally.

Nanotechnology employs knowledge from the fields of physics, chemistry, biology, materials science, health sciences, and engineering. It has immense applications in almost all the fields of science and human life. As generally acknowledged, the modern nanotechnology originated in 1959 as a talk given by Richard Feynman, "There's plenty of room at the bottom." However, the actual term "nanotechnology" was not coined until 1974 by Norio Taniguchi from Japan. The impetus for modern nanotechnology was provided by interest in interface and colloid science together with the development of analytical tools such as the scanning tunneling microscope (1981) and the atomic force microscope (1986).

People and scientists argue that nanotechnology is likely to have a horizontal impact across an entire range of industries and great implications on human health, environment, sustainability, and national security. The impact of nanotechnology is felt by everyone. The increasing amount of money governments are pouring worldwide in developing these technologies is an encouraging sign. It is observed that many facets of the science are impacted and people are revisiting many research areas with a nanoview to understand how the same thing can work at nano level. This phenomenon is revolutionizing pharmaceutical sciences, and many drugs are again being relooked for possibilities of delivering as a nanosystem.

We had our first volume entitled *Nanoparticulate Drug Delivery Systems: Recent Trends and Emerging Technologies,* which was edited by Drs. Deepak Thassu, Michele Deleers, and Yashwant Pathak. The book was published by Informa Healthcare in April 2007 and shares the status of nanotechnology worldwide. It has been very well received by the scientific and industrial community globally.

We are pleased to submit this second volume edited by Drs. Yashwant Pathak and Deepak Thassu. The objective of the book is to address formulation and characterization properties of nanoparticulate drug delivery systems (NPDDS) and also fulfill the void felt by the scientific community the last 2 years.

The volume comprises 20 chapters written by the leading scientists in this field. The first chapter covers the recent developments and features of NPDDS. This is followed by 8 chapters that address formulation aspects covering small molecules, macromolecules, gene delivery, protein-based nanodelivery systems, therapeutic and diagnostic applications of gold nanoparticles, and the application of NPDDS in cancer, diabetes, and dermal and transdermal deliveries.

The following group of chapters, which includes chapters 10 to 14, deals with in vitro and in vivo characterization, covering various methods used for characterizing the NPDDS in vitro, mathematical models used in analyzing the in vitro data, in vitro characterization of the interaction of nanoparticles with cell and blood constituents, pharmacological and toxicological characterization of nanosystems, and in vivo evaluation of solid lipid nanoparticle–based NPDDS.

The final group of chapters from 14 to 20 covers various analytical techniques used for characterizing the NPDDS and nanosystems. This includes various microscopic and spectroscopic characterizations. Various advanced techniques used to characterize the nanosystems include the scanning electron microscopy, transmission electron microscopy, structural fingerprinting of the nanocrystals, mechanical properties of the nanosystems, application of fullerene nanosystems for magnetic resonance imaging analysis, and the use of nanosystems for bioimaging. The chapters are authored by experts in these fields, and they have discussed their own researches as well as the developments in their fields of interest.

It is our sincere hope that this multiauthored book covering the formulation and characterization aspects of NPDDS will be welcomed by the scientific community and get a response similar to that received for our first volume. We sincerely hope that the book will assist and enrich readers to understand various aspects of formulation of NPDDS and their therapeutic applications in different disease conditions. The book also highlights the in vitro, in vivo, and analytical characterization in depth providing an insight to the readers in characterizing the NPDDS. We deem that this book will be equally relevant to scientists from varied backgrounds working in the field of drug delivery systems representing industry as well as academia. The text is organized in such a way that each chapter represents an independent area of research and can be easily followed without referring to other chapters.

We express our sincere thanks to Evelyn Kuhn and Jamie Hampton from Creative Communications of Sullivan University for their kind help in developing appropriate figures for publications. We express our sincere thanks to Ms. Allison Koch for her kind help in manuscript development, word processing, corrections, and punctuation.

Special thanks to Ms. Carolyn Honour, Sherri Niziolek, and Sandra Beberman from Informa Healthcare for their kind help and patience in seeing this book being completed.

We will be failing in our duty if we do not express our sincere thanks to all the authors who took trouble and time from their busy schedule to write chapters and provide them in time for publication. We are grateful to Dr. Simon Benita for the wonderful foreword to this book.

We appreciate our respective families for without their continuous support this work could not have been completed.

Yashwant Pathak
Deepak Thassu

Contents

Foreword Simon Benita .... *ix*
Preface .... *xiii*
Contributors .... *xvii*

1. **Recent Developments in Nanoparticulate Drug Delivery Systems** *1*
Yashwant Pathak

2. **Polymeric Nanoparticles for Small-Molecule Drugs: Biodegradation of Polymers and Fabrication of Nanoparticles** *16*
Sheetal R. D'Mello, Sudip K. Das, and Nandita G. Das

3. **Formulation of NPDDS for Macromolecules** *35*
Maria Eugénia Meirinhos Cruz, Sandra Isabel Simões, Maria Luísa Corvo, Maria Bárbara Figueira Martins, and Maria Manuela Gaspar

4. **Formulation of NPDDS for Gene Delivery** *51*
Ajoy Koomer

5. **NPDDS for Cancer Treatment: Targeting Nanoparticles, a Novel Approach** *57*
Karthikeyan Subramani

6. **Design and Formulation of Protein-Based NPDDS** *69*
Satheesh K. Podaralla, Radhey S. Kaushik, and Omathanu P. Perumal

7. **Gold Nanoparticles and Surfaces: Nanodevices for Diagnostics and Therapeutics** *92*
Hariharasudhan D. Chirra, Dipti Biswal, and Zach Hilt

8. **NPDDS for the Treatment of Diabetes** *117*
Karthikeyan Subramani

9. **Nanosystems for Dermal and Transdermal Drug Delivery** *126*
Venkata Vamsi Venuganti and Omathanu P. Perumal

10. **In Vitro Evaluation of NPDDS** *156*
R. S. R. Murthy

11. **In Vitro Characterization of Nanoparticle Cellular Interaction** *169*
R. S. R. Murthy and Yashwant Pathak

12. **In Vitro Blood Interaction and Pharmacological and Toxicological Characterization of Nanosystems** *190*
R. S. R. Murthy and Yashwant Pathak

13. **In Vivo Evaluations of Solid Lipid Nanoparticles and Microemulsions** *219*
Maria Rosa Gasco, Alessandro Mauro, and Gian Paolo Zara

14. **Microscopic and Spectroscopic Characterization of Nanoparticles** *239*
Jose E. Herrera and Nataphan Sakulchaicharoen

15. **Introduction to Analytical Scanning Transmission Electron Microscopy and Nanoparticle Characterization** *252*
Zhiqiang Chen, Jinsong Wu, and Yashwant Pathak

16. **Structural Fingerprinting of Nanocrystals in the Transmission Electron Microscope: Utilizing Information on Projected Reciprocal Lattice Geometry, 2D Symmetry, and Structure Factors** *270*
Peter Moeck and Sergei Rouvimov

17. **Mechanical Properties of Nanostructures** *314*
Vladimir Dobrokhotov

18. **Fullerene-Based Nanostructures: A Novel High-Performance Platform Technology for Magnetic Resonance Imaging (MRI)** *330*
Krishan Kumar, Darren K. MacFarland, Zhiguo Zhou, Christpher L. Kepley, Ken L. Walker, Stephen R. Wilson, and Robert P. Lenk

19. **Semiconducting Quantum Dots for Bioimaging** *349*
Debasis Bera, Lei Qian, and Paul H. Holloway

20. **Application of Near Infrared Fluorescence Bioimaging in Nanosystems** *367*
Eunah Kang, Ick Chan Kwon, and Kwangmeyung Kim

Index 387

Contributors

Debasis Bera Department of Materials Science and Engineering, University of Florida, Gainesville, Florida, U.S.A.

Dipti Biswal Department of Chemical and Materials Engineering, University of Kentucky, Lexington, Kentucky, U.S.A.

Zhiqiang Chen Institute for Advanced Materials and Renewable Energy, University of Louisville, Louisville, Kentucky, U.S.A.

Hariharasudhan D. Chirra Department of Chemical and Materials Engineering, University of Kentucky, Lexington, Kentucky, U.S.A.

Maria Luísa Corvo Unit New Forms of Bioactive Agents (UNFAB)/INETI and Nanomedicine & Drug Delivery Systems Group [iMed.UL], Lisbon, Portugal

Maria Eugénia Meirinhos Cruz Unit New Forms of Bioactive Agents (UNFAB)/INETI and Nanomedicine & Drug Delivery Systems Group [iMed.UL], Lisbon, Portugal

Nandita G. Das Department of Pharmaceutical Sciences, Butler University, Indianapolis, Indiana, U.S.A.

Sudip K. Das Department of Pharmaceutical Sciences, Butler University, Indianapolis, Indiana, U.S.A.

Sheetal R. D'Mello Department of Pharmaceutical Sciences, Butler University, Indianapolis, Indiana, U.S.A.

Vladimir Dobrokhotov Department of Physics and Astronomy, Western Kentucky University, Bowling Green, Kentucky, U.S.A.

Maria Rosa Gasco Nanovector s.r.l., Turin, Italy

Maria Manuela Gaspar Unit New Forms of Bioactive Agents (UNFAB)/INETI and Nanomedicine & Drug Delivery Systems Group [iMed.UL], Lisbon, Portugal

Jose E. Herrera Department of Civil and Environmental Engineering, University of Western Ontario, London, Ontario, Canada

Zach Hilt Department of Chemical and Materials Engineering, University of Kentucky, Lexington, Kentucky, U.S.A.

Paul H. Holloway Department of Materials Science and Engineering, University of Florida, Gainesville, Florida, U.S.A.

Eunah Kang Biomedical Research Center, Korea Institute of Science and Technology, Seoul, South Korea

Radhey S. Kaushik Department of Biology & Microbiology/Veterinary Science, South Dakota State University, Brookings, South Dakota, U.S.A.

Christpher L. Kepley Luna a nanoWorks (A Division of Luna Innovations, Inc.), Danville, Virginia, U.S.A.

Kwangmeyung Kim Biomedical Research Center, Korea Institute of Science and Technology, Seoul, South Korea

Ajoy Koomer Department of Pharmaceutical Sciences, Sullivan University College of Pharmacy, Louisville, Kentucky, U.S.A.

Krishan Kumar Luna a nanoWorks (A Division of Luna Innovations, Inc.), Danville, Virginia, U.S.A.

Ick Chan Kwon Biomedical Research Center, Korea Institute of Science and Technology, Seoul, South Korea

Robert P. Lenk Luna a nanoWorks (A Division of Luna Innovations, Inc.), Danville, Virginia, U.S.A.

Darren K. MacFarland Luna a nanoWorks (A Division of Luna Innovations, Inc.), Danville, Virginia, U.S.A.

Maria Bárbara Figueira Martins Unit New Forms of Bioactive Agents (UNFAB)/INETI and Nanomedicine & Drug Delivery Systems Group [iMed.UL], Lisbon, Portugal

Alessandro Mauro Department of Neurosciences, University of Turin, Turin, and IRCCS—Istituto Auxologico Italiano, Piancavallo (VB), Italy

Peter Moeck Laboratory for Structural Fingerprinting and Electron Crystallography, Department of Physics, Portland State University, Portland, Oregon, U.S.A.

R. S. R. Murthy Pharmacy Department, The M. S. University of Baroda, Vadodara, India

Yashwant Pathak Department of Pharmaceutical Sciences, Sullivan University College of Pharmacy, Louisville, Kentucky, U.S.A.

Omathanu P. Perumal Department of Pharmaceutical Sciences, South Dakota State University, Brookings, South Dakota, U.S.A.

Satheesh K. Podaralla Department of Pharmaceutical Sciences, South Dakota State University, Brookings, South Dakota, U.S.A.

Lei Qian Department of Materials Science and Engineering, University of Florida, Gainesville, Florida, U.S.A.

Sergei Rouvimov Laboratory for Structural Fingerprinting and Electron Crystallography, Department of Physics, Portland State University, Portland, Oregon, U.S.A.

Nataphan Sakulchaicharoen Department of Civil and Environmental Engineering, University of Western Ontario, London, Ontario, Canada

Sandra Isabel Simões Unit New Forms of Bioactive Agents (UNFAB)/INETI and Nanomedicine & Drug Delivery Systems Group [iMed.UL], Lisbon, Portugal

Karthikeyan Subramani Institute for Nanoscale Science and Technology (INSAT), University of Newcastle upon Tyne, Newcastle upon Tyne, U.K.

Venkata Vamsi Venuganti Department of Pharmaceutical Sciences, South Dakota State University, Brookings, South Dakota, U.S.A.

Ken L. Walker Luna a nanoWorks (A Division of Luna Innovations, Inc.), Danville, Virginia, U.S.A.

Stephen R. Wilson Luna a nanoWorks (A Division of Luna Innovations, Inc.), Danville, Virginia, U.S.A.

Jinsong Wu Department of Materials Science and Engineering, Northwestern University, Evanston, Illinois, U.S.A

Gian Paolo Zara Department of Anatomy, Pharmacology and Forensic Medicine, University of Turin, Turin, Italy

Zhiguo Zhou Luna a nanoWorks (A Division of Luna Innovations, Inc.), Danville, Virginia, U.S.A.

1 Recent Developments in Nanoparticulate Drug Delivery Systems

Yashwant Pathak

Department of Pharmaceutical Sciences, Sullivan University College of Pharmacy, Louisville, Kentucky, U.S.A.

INTRODUCTION

In the last 35 years, the growth of nanotechnology has opened several new vistas in medical sciences, especially in the field of drug delivery. New and new moieties are coming handy for treating diseases. The biotechnology has also produced several potent drugs, but many of these drugs encounter problems delivering them in biological systems. Their therapeutic efficacy is significantly marred owing to their incompatibilities and specific chemical structure. The input of today's nanotechnology is that it allows real progress to achieve temporal and spatial site-specific delivery. The market of nanotechnology and drug delivery systems based on this technology will be widely felt by the pharmaceutical industry. In recent years, the number of patents and products in this field is increasing significantly. The most straightforward application is in cancer treatment, with several products (Table 1) in market such as Caelyx®, Doxil®, Transdrug®, Abraxane®, etc. (1).

In 1904, Paul Ehrlich (1854–1915), one of the great architects of medical science, published three articles in the *Boston Medical and Surgical Journal*, the immediate predecessor of the *New England Journal of Medicine*. These articles, which concerned Ehrlich's work in immunology, were summaries of the Herter lectures he had given at Johns Hopkins University. They dealt with immunochemistry, the mechanism of immune hemolysis in vitro, and the side-chain theory of antibody formation. Whether such articles would appear in a clinical journal today is debatable. At the time of the Herter lectures, Ehrlich was at the peak of his intellectual powers and scientific influence. He was not only the father of hematology but also one of the founders of immunology. He made key contributions in the field of infectious diseases and, with his idea of the "magic bullet," initiated a new era of chemotherapy (4). The concept of Paul's magic bullet has turned out to be a reality with the approval of several forms of drug-targeting systems for the treatment of certain cancer and infectious diseases. Nanoparticulate drug delivery systems are going in this direction.

FEATURES OF NANOPARTICULATE DRUG DELIVERY SYSTEMS

Several terminologies have been used to describe nanoparticulate drug delivery systems. In most cases, either polymers or lipids are used as carriers for the drug, and the delivery systems have particle size distribution from few nanometers to few hundred nanometers (Table 2).

New and newer polymers have been tried to develop nanoparticles for their application as drug carriers. Craparo et al. described the preparation and

TABLE 1 Marketed or Scientifically Explored Nanosystems [Based on Information from Review Article, Ref. (2)]

Product/group	Company	Applications
Daunoxome®	Gilead Science, Cambridge, U.K.	Cancer
Doxil®	Johnson & Johnson, Bridgewater, NJ	Cancer
Myocet®	Sepherion Therapeutics, Princeton, NJ	Cancer
Amphotec®	Amphotec, Beverly, MA	Antifungal
Gendicine®	China approved this for Chinese market	Gene delivery
Accell	Powderject Vaccine, Inc. Madison, WI	Gene gun
Helios	Bio Rad Labs, Hercules, CA	Gene gun
Rexin-G™	Epeius Biotech Corp, Glendale, CA	Gene delivery
Vitravene™	ISIS Pharm, Carlsbad, CA	AIDS related
Medusa®	Flamel Technologies, Lyon, France	Generic amphiphilic polymer technology
Transdrug®	Bioalliance, Paris, France	Cancer
Caelyx®	Johnson & Johnson, Bridgewater, NJ	Cancer
Nanoedge®	Baxter Health Corp, Deerfield, IL	Generic delivery system
Emend®	Elan/Merck & Co, King of Prussia, PA	Antinausea drug
Rapamune	Elan/Wyeth, King of Prussia, PA	Immunosuppressant
Nanocrystal technology	Elan, Dublin, Ireland	Generic technology
Abraxane®/paclitaxel crystals	American Bioscience, Santa Monica, CA	Cancer
Nanobase	Yamanouchi Japan	Solid lipid generic carriers for drugs and cosmetics
Nanoxel	Dabur Pharma, India	Paclitaxel
Bioimaging/ diagnostics		
Endorem™	Guebert CCL, Bloomington, IN	Supermagnetic iron oxide MRI system
Multifunctional nanocarriers	Georgia Tech, Atlanta, GA	Breast cancer bio-imaging (3)

physicochemical and in vitro biological characterization of nanoparticles based on PEGylated, acryloylated polyaspartamide polymers (27). These systems were obtained by UV irradiation of poly(hydroxyethylaspartamide methacrylate) (PHM) and PHM/PEG-2000 as an inverse microemulsion by using the aqueous solution of the PHM/PHM PEG-2000 copolymer mixture as the internal phase and triacetin saturated with water as the external phase. Various parameters such as particle size distribution, dimensional analysis, and zeta-potential were used to characterize these systems. These nanosystems were evaluated for cell compatibility, and their ability to escape phagocytosis was also characterized. Rivastigmine was used as a drug model.

Protein-based nanoparticulate drug delivery systems are defined as proteins being biodegradable, biocompatible, very versatile molecules, and can be used as drug carriers (28). A protein-based nanoparticulate drug delivery system is already in the market (paclitaxel-loaded albumin nanoparticles, Abraxane®). Protein macromolecules offer many advantages over their synthetic counterparts (synthetic polymers that are commonly used as drug carriers). Owing to the presence of several synthetic functional groups in protein molecules, these molecules are more versatile and can offer covalent or noncovalent modifications of the molecules

TABLE 2 Types of Terminologies Used for Nanoparticulate Drug Delivery Systems (Based on Information from Various References Mentioned in the Last Column)

S. No.	Terminologies used	Particle size distribution (nm)	References
Polymeric systems			
1	Dendrimers	1–10	1,5
2	Polymer micelles	10–100	1
3	Niosomes	10–150	1
4	Nanoparticles	50–500	1,6–10
5	Nanocapsules	100–300	1,11,12
6	Nanogels	200–800	1
7	Polymer–drug nanoconjugates	1–15	13–16
8	Chitosan polymers	100–800	17,18
9	Methacrylate polymers	100–800	19
Lipid systems			
1	Solid lipid nanoparticles	50–400	20
2	Lipid nanostructured systems	200–800	21
3	Cubosomes	50–700	1
4	Liposomes	10–1000	22
5	Polymerosomes	100–300	13
6	Immunoliposomes	100–150	13
Protein/peptide nanotubes			
1	Peptide nanotubes	1–100	23
2	Fusion proteins and immunotoxins	3–15	13
Metal nanostructures			
1	Metal colloids	1–50	1,9
2	Carbon nanotubes	1–10 (diameter) and 1–1000 (length)	1
3	Fullerene	1–10	1
4	Gold nanoparticles	100–200	13,24
5	Gold nanoshells	10–130	13
6	Silicone nanoparticles		25
7	Magnetic colloids	100–600	26

when used as drug carriers. Owing to this property, these can be used for delivering different drug molecules. As these protein molecules are biocompatible and biodegradable, this is a distinct advantage over their synthetic counterparts. More detailed description of these protein carriers is described elsewhere (chap. 6). Some of the natural organic and protein molecules are also described as carriers for drug. These are fabricated as nanoparticles or nanofibers for delivering the drugs (29–31).

SOME MAJOR APPLICATIONS

Since the first nanoparticulate drug delivery systems as Liposome proposed by Dr. Gregory Gregoriadis in 1974 (32) lead to several breakthrough discoveries by using nanoparticles as drug carriers resulting from cutting-edge researches based on multidisciplinary approaches, many more applications have developed. We have discussed in detail about the nanoparticulate drug delivery systems in our first volume in chapters 1 and 13, which covered most of the development and technologies and applications till 2005. Hence, this chapter restricts to the developments mostly in 2007 and 2008. Several research reports have been published on the applications of nanoparticulate drug delivery systems using various drug entities and polymers and different forms of drug delivery systems. Table 3 provides details of several of these research reports.

NANOPARTICULATE DRUG DELIVERY SYSTEMS AND BLOOD–BRAIN BARRIER CANCER TREATMENT

Effectiveness of the chemotherapy of brain pathologies is often impeded by insufficient drug delivery across the blood–brain barrier (BBB). Galeperina from Russia has patented poly(butyl cyanoacrylate) nanoparticles coated with polysorbate 80, showing the efficient brain-targeting drug delivery system crossing the BBB (33). Doxorubicin in free form cannot pass the BBB. The employment of poly(butyl cyanoacrylate) nanoparticles showed high efficacy of nanoparticle-bound doxorubicin in intracranial glioblastoma in rats. Another interesting review, covering various techniques used for crossing the BBB, discusses the application of nanoparticulate drug delivery systems for this purpose (60). Kreuter et al. have described the application of covalently bonded apolipoprotein A-I and apolipoprotien B-100 to albumin nanoparticles, enabling these to deliver the drug in brain (61). An interesting review on the application of nanotechnology in breast cancer therapy is covered by Tanaka et al. (62). A PEGylated form of liposomally encapsulated doxorubicin is routinely used for the treatment of metastatic cancer, and albumin nanoparticulate chaperones of paclitaxel are approved for the locally recurrent and metastatic cancer tumors. More than 150 clinical trials are being conducted worldwide for the treatment of breast cancer by using nanotechnology-based products. This review covers different generations of nanotechnology tools used for drug delivery, especially in breast cancer.

Injectable drug delivery nanovectors are used for cancer therapy, especially when multiple-drug therapy is used. These vectors need to be large enough to evade the body defense but should be sufficiently small to avoid blockages in even the capillaries. The nanosize plays an important and helpful role in such capillary blockages. As these vectors are smaller than the diameters of the capillaries, the blockages can be effectively prevented (13). The anticancer drug or drugs can be incorporated in such nanovectors. These nanovectors can functionalize in order to actively bind to specific sites and cells after extravasation thorough ligand–receptor interactions. To maximize the specificity, a surface marker (receptor or antibody) should be overexpressed on target cells relative to normal ones. There is a need for increasing the mucoadhesion of nanocarriers. Another area that is being explored is to use the external energy or the environmental system to release cytotoxic drugs at the site of action by using metabolic markers or acidity levels that accompany inflammatory states, infections, and neoplastic processes (13). Nanosized vectors include fusion proteins and immunotoxins/polymers, dendrimers, polymer–drug conjugates, polymeric micelles, polymerosomes and liposomes, and metal nanoparticles such as gold nanoparticles or nanoshells. The major concern of nanovectors based on polymers is their biocompatibility, biodegradability, and release of drug from the polymer nanosystem in the body at the site of action. In case of lipid-based systems, the problems of biocompatibility and biodegradability are not

TABLE 3 Recent Research Reports Covering Various Applications of Nanoparticulate Drug Delivery Systems (96)

Drug used	Polymer used	Form of the drug delivery systems suggested	Therapeutic indication (references)
Doxorubicin	Poly(butyl cyanoacrylate)	Drug-loaded nanoparticles	Brain cancer (33)
Clarithromycin	Gliadin/pluronic F 68	Mucoadhesive nanoparticles	Broad-spectrum antibiotics (34)
Celecoxib	Lipid system	Nanostructured carriers	Inflammation and other allied conditions (21)
Acyclovir	Cholesteryl acyl didanosine and cholesteryl adipoyl didanosins	Self-assembled drug delivery systems	Anti-HIV drug (35)
Vaccine	Poly(γ-glutamic acid)	Nanoparticles	Vaccines against cancer (6)
Vaccine	Hydroxyapatite/ bovine serum albumin	Nanoparticles	Vaccine/antigen carrier (7)
Camptothecin	PLGA	Nanoparticles	Cancer (36)
Rifampicin, isoniazid	Alginates	Nanoparticles	Anti-TB drugs (37)
Psoralen	Solid lipid system	Lipid nanostructure	Topical treatment of psoriasis (20)
Docetaxel	Poly(D,L-lactide-co-glycolide)	Long-acting nanoparticles	Cancer tumors (38)
Antimycobacterial agents	Various polymers	Liposomes and nanoparticles	Anti-TB (22)
Rivastigmine	PEGylated acryloylated polyaspartamide	Nanoparticles	Dementia in Alzheimer's disease and Parkinson's disease
Plasmid IGF-1	Cationized gelatin	Nanoparticles	Gene delivery (39)
γ-Interferon	Albumin	Nano particles	Antibacterial against *Brucella abortus* (40)
Protein drugs	Poly(lactide)-tocopheryl polyethylene glycol succinate copolymers	Nano particles	Protein drug delivery (41,42)
Paclitaxel	PEO-PPO-PEO/PEG cross-linked polymer	Nano capsules	Anticancer (11)
	PLGA polymer	Nano capsules	Anticancer (12)

(*Continued*)

TABLE 3 Recent Research Reports Covering Various Applications of Nanoparticulate Drug Delivery Systems (*Continued*)

Drug used	Polymer used	Form of the drug delivery systems suggested	Therapeutic indication (references)
Oridonin	Poly(D,L-lactic acid)	Nanocapsules	Multiple myeloma (43)
Cisplatin	Phosphatidylethanolamine	Liposomes	Melanomas (44)
Aclarubicin	Albumin-conjugated PEGylated drug	Nanoparticles	Glioma (45)
Amphotericin B	Lipoproteins	Nanodisks	Antibiotic (46)
Docetaxel	PEG derivatives	Nanosized PEG drug assembly	Cancer (16)
Ceramide and paclitaxel	Lipid carriers	Nanoemulsions	Cancer (47)
Cisplatin	Hyaluronic acid–drug conjugate	Nanoparticles	Cancer (48)
Fludarabine and mitoxantrone	Lipid carriers	Liposomes	Lymphoproliferative disorder (49)
Estradiol	PLGA	Nanoparticles	Hormone (50)
Cyclosporine	Polymeric micelles delivery	Colloidal	Antibiotics (51)
Flurbiprofen	PLGA	Nanospheres	Ocular delivery (52)
Insulin	Sodium alginate and chitosan	Nanospheres	Diabetes mellitus (53)
Thymopentin	Poly(butyl cyanoacrylate) polymer	Nanoparticles	Protein peptide drug for cancer treatment (54)
Indinavir	Lipids	Lipid–drug conjugates	AIDS treatment (55)
Cyclosporin	Polysorbate 80	Nanodispersion	Pulmonary infections (56)
Ketoprofen and diflunisal	Polyamidoamine dendrimers	Drug–polymer complex suspension	Anti-inflammatory (57)
Halofantrine and procubol	Cremaphore RH 40	Self-emulsifying nanoemulsion system	Lipophilic drug models (58)
Etoposide	Poly(D,L-lactide) block copolymers	Nanoparticles	Cancer (59)

Abbreviations: PEG, poly(ethylene glycol); PEO, poly(ethylene oxide); PLGA, poly(lactic-glycolic acid); PPO, poly(phenylene oxide).

encountered. Liposomes, either single layered or multilayered, have shown significant potential as nanovectors for cancer treatment. They have shown preferential accumulation in tumor via enhanced permeability and retention effect. However, too long circulating liposomes may lead to extravasation of the drug into undesired sites. Long circulating half-life, soluble or colloidal behavior, high binding affinity, biocompatibility, easy functionalization, easy intracellular penetration, controlled pharmacokinetic, and high drug protection are all characteristics simultaneously required for an optimal nanocarrier design and efficient applications. Increasing the adhesion of nanovectors will be very useful. Pugna has shown in his article that controlling adhesion in highly flexible nanovectors can help in smartly delivering the drug (13). The high flexibility of nanovectors is used to release the drug only during adhesion by nanopumping, and, as a limit case, by the new concept of adhesion-induced nanovector implosion. He recommended that fast pumping and slow diffusion of drug could thus be separately controlled. See other similar studies reported by Decuzzi and Ferrari (8), Gentile et al. (63), Peng et al. (64), Sanhai et al. (65), Steinhauser et al. (66), and Silva (67).

Another interesting study was reported by Bae et al. in which they used PEO-PPO-PEO/PEG shell cross-linked nanocapsules for target-specific delivery system of the anticancer drug paclitaxel (11). They synthesized oil-encapsulating, composite, polymeric nanoshells by dissolving an oil (lipiodol) and an amine-reactive PEO-PPO-PEO derivative in dichloromethane, and subsequently dispersing in an aqueous solution containing amine-functionalized, six-armed, branched polyethylene glycol by ultrasonication. The resultant nanoshells were sized around 110 nm, and they incorporated paclitaxel in the oil phase. They have shown that such a nanoshell delivery system can be used for different hydrophobic oil-soluble drugs. Jin et al. reported an interesting application of paclitaxel-loaded nanocapsules with radiation on hypoxic MCF-7 human breast carcinoma cells (12). They reported that paclitaxel could be effectively released from biodegradable poly(lactic-glycolic acid) nanoparticle delivery system, while maintaining potent, combined, cytotoxic, and radio-sensitizing abilities for hypoxic human breast tumor cells.

Kim et al. explained that poly(methoxypolyethylene glycol cyanoacrylate-co-hexadecylcyanoacrylate) (PEGPHDCA) nanoparticles have the capacity to diffuse through the BBB after intravenous administration (68). However, they could not elucidate the mechanism of transport of these nanoparticles. They did some in vitro cellular uptake studies, which showed that nanoparticles preincubated with apolipoprotien E and blocked low-density lipoprotein (LDL), suggesting that LDL-mediated pathway may be involved in the endocytosis of PEGPHDCA nanoparticles by rat brain endothelial cells.

Kommareddy and Amiji studied PEG chain–modified thiolated gelatin nanoparticles and examined their long circulating and tumor-targeting properties in vivo in an orthotopic human breast adenocarcinoma xenograft model (69). They reported that upon modifications with PEG, the nanoparticles were having long circulating time with the plasma and tumor half-lives of 15.3 and 37.8 hours, respectively. Several other studies have shown the application of nanoparticulate drug delivery systems in cancer treatment (70–74).

ANTIBODY TARGETING OF NANOPARTICLES

Many studies have reported the antibody mediation of the nanoparticles to develop targeted drug delivery systems, especially in the application of cancer treatment

(36). Antibody targeting of drug substances can improve the therapeutic efficacy of the drug substance, as well as improve the distribution and concentration of the drug at the targeted site of drug action. McCarron et al. studied two novel approaches to create immunonanoparticles with improved therapeutic effect against colorectal tumor cells (36). They used poly(lactide) polymers and CD95/APO-1 antibody to target nanoparticles. Pan et al. used dendrimer–magnetic nanoparticles for efficient delivery of gene-targeted systems for cancer treatment (5). Olton et al. have described the use of nanostructured calcium nanophosphates for nonviral gene delivery and studied the influence of synthesis parameters on transfection efficiency (75).

NANOPARTICULATE DRUG DELIVERY SYSTEMS FOR VACCINE DELIVERY

Nanoscopic systems incorporating therapeutic agents with molecular-targeting and diagnostic imaging capabilities are emerging as the next generation of functional nanomedicines to improve the outcome of therapeutics (35). Yoshikawa et al. developed a technique to prepare uniform nanoparticles based on poly-γ-glutamic acid nanoparticles and used them successfully as carriers for vaccines in the treatment of cancer (6). The development of compounds that enhance immune responses to recombinant or synthetic epitopes is of considerable importance in vaccine research. An interesting approach for formulating aquasomes is described by Goyal et al. (7). These were prepared by self-assembling hydroxyapatite by coprecipitation, and then they were coated with polyhydroxyl oligomers (cellobiose and trehalose) and adsorbed on bovine serum albumin (BSA) as a model antigen. BSA-immobilized aquasomes were approximately 200 nm in diameter, and it was observed that these formulations elicit combined T-helper Th1 and Th2 immune responses (7).

LIPID NANOPARTICLES AND NANOSTRUCTURED LIPID CARRIERS

Lipid nanoparticles have been used for many years and are still showing lots of interest in delivering drugs, as well as nanostructured lipid carriers for drugs. Fang et al. recently conducted a study by using lipid nanoparticles for the delivery of topical psoralen delivery. This study compared lipid nanoparticles with nanostructured lipid carriers composed of precirol and squalene, a liquid lipid. They showed that the particle size was between 200 and 300 nm for both the carriers. It was used for the treatment of psoriasis. Their results showed that the entrapment of 8-methoxypsoralen in nanoparticulate systems could minimize the permeation differentiation between normal and hyperproliferative skin compared with that of free drug in aqueous control (20).

MUCOADHESIVE NANOPARTICULATE DRUG DELIVERY SYSTEMS AND IMPROVING THE GASTROINTESTINAL TRACT ABSORPTION

A novel nanoparticle system to overcome intestinal degradation and drug transport–limited absorption of P-glycoprotein substrate drugs is reported by Nassar et al. (76). Dr. Juliano has written a very good article about the challenges in macromolecular drug delivery and the use of various techniques including polymeric carriers for the macromolecular drugs (77). Zidan et al. had an interesting report on quality by design for understanding the product variability of a mucoadhesive, self-nanoemulsified drug delivery system (SNEDDS) of cyclosporine A (78). This is probably one of the first of its kind of research report on quality by design in the field of pharmaceutical nanotechnology. They used near infrared and chemometric analysis and several other well-known processes for the characterization of emulsions during processing. Their study demonstrated the ability to understand the impact of nanodroplets' size on the SNEDDS variability by different product-analyzing tools.

HYDROGEL NANOPARTICLES IN DRUG DELIVERY

Hamidi et al. have written a very good review on hydrogel nanoparticles and their applications in drug delivery as well as therapeutic applications in various disease conditions (79). Another polymeric group tried by Lee et al. was poly(lactide)-tocopheryl polyethylene glycol succinate (PLA-TPGS) copolymers, which they used to deliver protein and peptide drugs (41). They used double-emulsion technique for protein drug formulation, with BSA as the model protein drug. They used confocal laser scanning microscopy observations to demonstrate the intracellular uptake of the PLA-TPGS nanoparticles by fibroblast cells and Caco-2 cells, showing great potential of these polymeric carriers for protein and peptide drugs.

Cheng et al. showed that the size of the nanoparticles affects the biodistribution of targeted and nontargeted nanoparticles in an organ-specific manner (10). Nair et al. described the enhanced intratumoral uptake of quantum dots concealed within hydrogel nanoparticles (80). To develop a functional device for tumor imaging, they embedded quantum dots within hydrogel nanoparticles. Their results suggest that the derivatized quantum dots enhance tumor monitoring through quantum dot imaging and that they are useful in cancer monitoring and chemotherapy. An interesting work was reported by Vihola et al. (81). They have discussed the effect of cross-linking on the formation and properties of thermosensitive polymer particles of poly(*N*-vinyl caprolactum) (PVCL) and PVCL grafted with poly(ethylene oxide) macromonomer. They showed different levels of drug release profiles based on varying polymer cross-linking. Baroli wrote a review on hydrogels for tissue engineering, and it has lots of information on the formulation and characterization of hydrogels for various applications including NPDDS (82).

NANOPARTICLES IN DIAGNOSTIC MEDICINE

Lee et al. have reported very interesting study on the subject of nanoparticles in diagnostic medicine (83). They used antibody-conjugated, hydrophilic, magnetic nanocrystals as smart nanoprobes for the ultrasensitive detection of breast cancer via magnetic resonance imaging (MRI) (83). $MnFe_2O_4$ nanocrystals employed as MRI-contrast agents for MRI were synthesized by thermal decomposition. The surfaces were then modified with amphiphilic triblock copolymers. They showed clear advantage as a contrast medium to detect breast cancer tumors. Faure et al. have shown different methods to detect streptavidin by attaching a molecule to dielectric particles made of a rare earth oxide core and a polysiloxane shell containing fluoreschein for biodetection (84). A new, interesting class of magnetic nanoparticles, gadolinium hydroxide and dysprosium oxide, are characterized by different methods by using X-ray diffraction, NMR relaxometry, and magnetometry at multiple fields. These have very good applications in diagnostic purposes (85). Shao et al. have used nanotube antibody biosensor arrays for the detection of circulating breast cancer cells (86). This is the first report giving information on the new technique by using the nanotube and the antibody cancer cell detection system.

RECENT REVIEWS

Prokop and Davidson have recently published a wonderful review on nanovehicular intracellular delivery systems (2). This review extensively covers various aspects of nanodrug delivery systems and their uptake in biological system at cellular levels. They have discussed in detail the applications of various nanosystems and their interactions at cellular levels and the mechanism for the uptake of the nanosystems (2). Devapally et al. discussed in detail the role of nanotechnology in the product development of drugs, with several drugs and extensive references (87). With many examples, they have shown that nanoparticulate drug delivery systems show a promising approach to obtain desirable delivery properties by altering the biopharmaceutic and pharmacokinetic properties of the molecule. It summarizes the parameters and approaches used to evaluate NPDDS in early stages of formulation developments. A detailed description on micro (nano) emulsions has been recently discussed in a review by Gupta and Moulik (88). It covered the development and characterization of biocompatible micro (nano) emulsion systems and their evaluation as probable vehicle for encapsulation, stabilization, and delivery of bioactive natural products and prescription drugs. In a recently published review, Drummond et al. discussed the pharmacokinetics and in vivo drug release rates in liposomal nanocarrier development (89). They discussed the rationale for selecting optimal strategies of liposomal drug formulations with respect to drug encapsulation, retention, and release, and how these strategies can be applied to maximize the therapeutic benefit in vivo. An interesting review about the application of nanoparticulate drug delivery systems in nasal delivery is reported by Illum (90). The author discusses the possible benefits of NPDDS in nasal delivery over the commonly used simple delivery systems. The nasal delivery of protein and peptide drugs is facilitated when incorporated in NPDDS, but it did not show significant advantages; however, when it came to vaccine delivery, the immune responses were much enhanced when incorporated in NPDDS for nasal delivery of these drugs. There is a need for more studies in this area to prove the efficacy of NPDDS and its nasal applications.

Gene therapy is considered to be a promising therapeutic strategy to combat root causes of genetic or acquired diseases rather than just treating the symptoms (97). There is a need for nontoxic and efficient gene delivery vectors; an interesting review by Mozafari and Omri discusses important aspects of divalent cations in nanolipoplex gene delivery (91). They reviewed the role of divalent cations in nucleic acid delivery, particularly with respect to the potential improvement of transfection efficiency of nanolipoplexes.

Salonen et al. have provided a good review about the mesoporous silicon-based drug delivery systems (92). They have described the application of silicon in drug delivery. The size and surface chemistry of mesoporous silicon-based drug delivery systems can be useful in delivering many drugs, including protein and peptide drugs. The review covered the fabrication and chemical modification of mesoporous silicon-based drug delivery systems for biomedical applications and also discussed the potential advantages of these delivery systems. Cheng et al. have written a very good review on dendrimers as drug carriers and their applications in the development of different delivery systems used by delivery routes (93). The review covered potential applications of dendrimers as polymeric carriers for intravenous, oral, transdermal, and ocular delivery systems. They discussed the dendrimer–drug interactions and mechanisms, encapsulation, electrostatic interactions, and covalent conjugation of drug and dendrimer molecules. Two important reviews discuss the application of Raman and terahertz spectroscopies for the characterization of nanoscale drug delivery systems (94) and high-field resolution NMR spectroscopy as a tool to understand and assess the interaction of protein with small-molecule ligands with applications in NPDDS (95).

CONCLUSION

The industrial scene of nanotechnology developments is very promising. The application of nanotechnology to drug delivery is widely expected to create novel therapeutics, capable of changing the landscape of pharmaceutical and biotechnology industries. Various nanotechnology platforms are being investigated, either in development or in clinical stages, and many areas of interest where there will be effective and safer targeted therapeutics for a myriad of clinical applications. It will be evolving out very soon for the benefit of humanity at large.

REFERENCES

1. Couvreur P, Vauthier C. Nanotechnology: Intelligent design to treat complex disease. Pharm Res 2006; 23:1417–1450.
2. Prokop A, Davidson JM. Nanovehicular intracellular delivery systems. J Pharm Sci 2008; 97:3518–3590.
3. Karathanasis E, Chan L, Balusu SR, et al. Multifunctional nanocarriers for mammographic quantification of tumor dosing and prognosis of breast cancer therapy. Biomaterials 2008; 29:4815–4822.
4. Silverstein A. Paul Ehrlich's Receptor Immunology. New York, NY: Academic Press, 2001.
5. Pan B, Cui D, Sheng Y, et al. Dendrimer-modified magnetic nanoparticles enhance efficiency of gene delivery system. Cancer Res 2007; 67:8156–8163.
6. Yoshikawa T, Okada N, Oda A, et al. Development of amphiphilic gamma-PGA-nanoparticle based tumor vaccine: Potential of the nanoparticulate cytosolic protein delivery carrier. Biochem Biophys Res Commun 2008; 366(2):408–413.
7. Goyal AK, Khatri K, Mishra N, et al. Aquasomes: A nanoparticulate approach for the delivery of antigen. Drug Dev Ind Pharm 2008; 34:1297–1305.
8. Decuzzi P, Ferrari M. Design maps for nanoparticles targeting the diseased microvasculature. Biomaterials 2008; 29;377–384.
9. Ziv O, Avtalion RR, Margei S. Immunogenecity of bioactive magnetic nanoparticles: Natural and acquired antibodies. J Biomater Res 2008; 85:1011–1021.
10. Cheng J, Teply BJA, Sherifi I, et al. Formulation of functionalized PLGA-PEG nanoparticles for in vivo targeted drug delivery. Biomaterials 2007; 28:869–876.
11. Bae KH, Lee Y, Park TG. Oil encapsulating PEO-PPO-PEO/PEG shell cross-linked nanocapsules for target-specific delivery of paclitaxel. Biomacromolecules 2007; 8:650–656.
12. Jin C, Wu H, Liu J, et al. The effect of paclitaxel-loaded nanoparticles with the radiation on hypoxic MCF-7 cells. J Clin Pharm Ther 2007; 32:41–47.
13. Pugna NM. A new concept for smart drug delivery: Adhesion induced nanovector implosion. Open Med Chem J 2008; 2:62–65.
14. Farokhzad OC, Cheng J, Teply BA, et al. Targeted nanoparticle-aptamer bioconjugates for cancer chemotherapy in vivo. Proc Natl Acad Sci U S A 2006; 103;6315–6320.
15. Park JS, Koh YS, Bang JY, et al. Antitumor effect of all-trans retinoic acid-encapsulated nanoparticles of methoxy poly(ethylene glycol)-conjugated chitosan against CT-26 colon carcinoma in vitro. J Pharm Sci 2008; 97:4011–4019.

16. Liu J, Zahedi P, Zeng F, et al. Nanosized assemblies of a PEG docetaxel conjugates as a formulation strategy for docetaxel. J Pharm Sci 2008; 97:3274–3290.
17. Liu X, Howard KA, Dong M, et al. The influence of polymeric properties on chitosan/siRNA nanoparticle formulation and gene silencing. Biomaterials 2007; 28: 1280–1288.
18. Chan P, Kurisawa M, Chung JE, et al. Synthesis and characterization of chitosan-g-ploy(ethylene glycol)-folate as anon viral carrier for tumor targeted gene delivery. Biomaterials 2007; 28:540–549.
19. Feng M, Li P. Amine containing core shell nanoparticles as potential drug carriers for intracellular delivery. J Biomed Mater Res 2007; 80:184–193.
20. Fang JY, Fang CL, Liu CH, et al. Lipid nanoparticles as vehicles for topical psoralen delivery: Solid lipid nanoparticles (SLN) versus nanostructured lipid carriers (NLC). Eur J Pharm Biopharm 2008; 70:633–640.
21. Joshi M, Patravale V. Nanostructured lipid carrier (NLC) based gel of celecoxib. Int J Pharm 2007; 346:124–132.
22. Gasper MM, Cruz A, Farga AG, et al. Developments on drug delivery systems for the treatment of mycobacterial infections. Curr Top Med Chem 2008; 8:579–591.
23. Sppher NB, Abrams ZR, Reches M, et al. Integrating peptide nanotubes in microfabrication processes. J Micromech Microeng 2007; 17:2360–2365.
24. Anil Kumar S, Peter YA, Nadeau JL. Facile biosynthesis, separation and conjugation of gold nanoparticles to doxorubicin. Nanotechnology 2008; 19:495101 (10pp).
25. Tasciotti E, Liu XW, Bhavane R, et al. Mesoporous silicon particles as a multistage delivery system for imaging and therapeutic applications. Nat Nanotechnol 2008; 3: 151–157.
26. Duran JDG, Arias JL, Gallardo V, et al. Magnetic colloids as drug vehicles. J Pharm Sci 2008; 97:2948–2983.
27. Craparo EF, Pitarresi G, Bondi ML, et al. A nanoparticulate drug delivery system for rivastigmine: Physicochemical and in vitro biological characterization. Macromol Biosci 2008; 8:247–259.
28. Wang G, Uludag H. Recent developments in nanoparticle based drug delivery and targeting systems with emphasis on protein based nanoparticles. Expert Opin Drug Deliv 2008; 5:499–515.
29. Torres GS, Gimenrez E, Lagaron JM. Characterization of the morphology and thermal properties of Zein Prolamine nanostructures obtained by electrospinning. Food Hydrocolloids 2007; 22:601–614.
30. Zhang Y, Shen W, Xiang R, et al. Formation of silk fibroin nanoparticles in water miscible organic solvent and their characterization. J Nanopart Res 2007; 9:885–900.
31. Liu X, Sun Q, Wang H, et al. Microspheres of corn protein, zein for an invermectin drug delivery system. Biomaterials 2005; 26:109–115.
32. Gregoriadis G, Willis EJ, Swain CP, et al. Drug carrier potential of liposomes in cancer chemotherapy. Lancet 1974; 1:1313–1316.
33. Galeprina S. Nanoparticulate drug delivery systems for the non-invasive chemotherapy of brain tumors. Proceedings of the NSTI Conference, Boston, USA, May 7–11, 2006.
34. Ramteke S, Maheshwari RBU, Jain NK. Clarithromycin based oral sustained release nanoparticulate drug delivery system. Indian J Pharm Sci 2006; 68:479–484.
35. Jin Y, Ai P, Xin R, et al. Self assembled drug delivery systems; part I: In vitro in vivo studies of the self assembled nanoparticulates of cholesteryl acyl didanosine. Int J Pharm 2006; 309:199–207 [part III in Oct 2008 (in press)].
36. McCarron PA, Marouf WM, Quinn DJ, et al. Antibody targeting of camptothecin-loaded PLGA nanoparticles to tumor cells. Bioconjug Chem 2008; 19:1561–1569.
37. Ahmed Z, Pandey R, Sharma S, et al. Alginate nanoparticles as anti tuberculosis drug carriers: Formulation development, pharmacokinetics and therapeutic potential. Indian J Chest Dis Allied Sci 2006; 48:171–176.
38. Sethilkumar M, Mishra P, Jain NK. Long circulating PEGylated poly(D,L-lactide-co-glycolide nanoparticulate delivery of docetaxel to solid tumors. J Drug Target 2008; 16:424–435.

39. Xu X, Capito RM, Spector M. Delivery of plasmid IGF-1 to chondrocytes via cationized gelatin nanoparticles. J Biomed Mater Res 2008; 84:73–83.
40. Segura S, Gamazo C, Irache JM. Gamma interferon loaded onto albumin nanoparticles: In vitro and in vivo activities against *Brucella abortus*. Antimicro Agents Chemother 2007; 51:1310–1314.
41. Lee SH, Zhang Z, Feng SS. Nanoparticles of poly(lactide) tocopheryl polyethylene glycol succinate (PLA-TPGS) copolymers for protein drug delivery. Biomaterials 2007; 28:2041–2050.
42. Gao H, Fang YN, Fan YG, et al. Conjugates of poly(D,L-lactide-co-glycolide) on amino cyclodextrins and their nanoparticles as protein delivery system. J Biomed Mater Res 2007; 80:111–122.
43. Xing J, Zhang D, Tan T. Studies on the oridonin-loaded poly(D,L-lactic acid) nanoparticles in vitro and in vivo. Int J Biol Macromol 2007; 40:153–158.
44. Hwang TL, Lee WR, Hua SC, et al. Cisplatin encapsulated in phosphatidylethanolamine liposomes enhances the in vitro cytotoxicity and in vivo intratumor drug accumulation against melanomas. J Dermatol Sci 2007; 46:11–20.
45. Lu W, Wan J, Zhang Q, et al. Aclarubicin-loaded cationic albumin-conjugated pegylated nanoparticles for glioma chemotherapy in rats. Int J Cancer 2007; 120:420–431.
46. Tufteland M, Ren G, Ryan RO. Nanodisks derived from amphotericin B lipid complex. J Pharm Sci 2008; 97:4425–4432.
47. Desai A, Vyas T, Amiji M. Cytotoxicity and apoptosis enhancement in brain tumor cells upon coadministration of aclitaxel and ceramide in nanoemulsion formulations. J Pharm Sci 2008; 97:2745–2756.
48. Jeong Y, Kim ST, Jin SG, et al. Cisplatin incorporated hyaluronic acid nanoparticles based on ion complex formation. J Pharm Sci 2008; 97:1268–1276.
49. Zhao X, Wu J, Muthuswami N, et al. Liposomal coencapsulated fludarabine and mitoxantrone for lymphoproliferative disorder treatment. J Pharm Sci 2008; 97:1508–1518.
50. Sahana DK, Mittal G, Bharadwaj V, et al. PLGA nanoparticles for oral delivery of hydrophobic drugs: Influence of organic solvents on nanoparticle formation and release behavior in vitro and in vivo using estradiol as a model drug. J Pharm Sci 2008; 97:1530–1542.
51. Aliabadi HM, Elhasi S, Brocks DR, et al. Polymeric micelles delivery reduces kidney distribution and nephritic effects of cyclosporine A after multiple dosing. J Pharm Sci 2008; 97:1916–1926.
52. Vega E, Gamisans F, Garcia ML, et al. PLGA nanospheres for the ocular delivery of flurbiporfen: Drug release and interactions. J Pharm Sci 2008; 97:5306–5317.
53. Reis CP, Veiga FJ, Ribeiro AJ, et al. Nanoparticulate biopolymers deliver insulin orally eliciting pharmacological response. J Pharm Sci 2008; 97:5291–5305.
54. He W, Jiang X, Zghang ZR. Preparation and evaluation of poly-butylcyanoacrylate nanoparticles for oral delivery of thymopentin. J Pharm Sci 2008; 97:2250–2259.
55. Choi SU, Bui T, Ho RJY. pH dependent interactions of indinavir and lipids in nanoparticles and their ability to entrap a solute. J Pharm Sci 2008; 97:931–943.
56. Tam JM, McConville T, Williams RO III, et al. Amorphous cyclosporine nanodispersions for enhanced pulmonary deposition and dissolution. J Pharm Sci 2008; 97:4915–4933.
57. Yiyun C, Na M, Tongwen X, et al. Transdermal delivery of nonsteroidal anti-inflammatory drugs mediated by polyamidoamine (PAMAM) dendrimers. J Pharm Sci 2007; 96:595–602.
58. Nielsen FS, Gibault E, Wahren HL, et al. Characterization of prototype self-nanoemulsifying formulations of lipophilic compounds. J Pharm Sci 2007; 96:876–892.
59. Gaucher G, Poreba M, Ravelelle F, et al. Poly(*N*-vinyl-pyrrolidone)-block-poly(D,L-lactide) as polymeric emulsifier for the preparation of biodegradable nanoparticles. J Pharm Sci 2007; 96:1763–1775.
60. Juillerat JL. The targeted delivery of cancer drugs across the blood brain barrier: Chemical modifications of drugs or drug nanoparticles. Drug Discov Today 2008; 13: 1099–1106.

61. Kreuter J, Hekmatara T, Dreis S, et al. Covalent attachment of apolipoprotein AI and apolipoprotein B-100 to albumin nanoparticles enables drug transport to brain. J Control Rel 2007; 118:54–58.
62. Takemi T, Paolo D, Massimo C, et al. Nanotechnology for breast cancer. Biomed Microdevices 2009; 11:49–63.
63. Gentile F, Ferrari M, Decuzzi P. The transport of nanoparticles in blood vessels: The effect of vessel permeability and blood rheology. Ann Biomed Eng 2008; 36:254–261
64. Peng J, He X, Wang K, et al. An antisense oligonucleotide carrier based on amino silica nanoparticles for antisense inhibition of cancer cells. Nanomedicine 2006; 2:113–120.
65. Sanhai WR, Sakamoto JH, Canady R, et al. Seven challenges for nanomedicine. Nat Nanotechnol 2008; 3:242–244.
66. Steinhauser B, Spankuch K, Strebhardt S, et al. Trastuzumab-modified nanoparticles: Optimization of preparation and uptake in cancer cells. Biomaterials 2006; 27;4975–4983.
67. Silva GA. Nanotechnology approaches for drug and small molecule delivery across the blood brain barrier. Surg Neurol 2007; 67:113–116.
68. Kim HR, Gil S, Andrieux K, et al. Low density lipoprotein receptor mediated endocytosis of PEGylated nanoparticles in rat brain endothelial cells. Cell Mol Life Sci 2007; 64: 356–364.
69. Kommareddy S, Amiji M. Biodistribution and pharmacokinetic analysis of long-circulating thiolated gelatin nanoparticles following systemic administration in breast cancer-bearing mice. J Pharm Sci 2007; 96:397–407.
70. Lu H, Li B, Kang Y, et al. Paclitaxel nanoparticle inhibits growth of ovarian cancer xenografts and enhances lymphatic targeting. Cancer Chemother Pharmacol 2007; 59:175–181.
71. Hatakeyama H, Akita H, Kogure K, et al. Development of a novel systemic gene delivery system for cancer therapy with a tumor specific cleavable PEG lipid. Gene Ther 2007; 14:68–77.
72. Myc A, Majoros IJ, Thomas TP, et al. Dendrimer-based targeted delivery of an apoptotic sensor in cancer cells. Biomacromolecules 2007; 8:13–18.
73. Nikanjam M, Blakely EA, Bjornstad KA, et al. Synthetic nano-low density lipoprotein as targeted drug delivery vehicle for glioblastoma multiforme. Int J Pharm 2007; 328: 86–94.
74. Wang ZY, Zhao Y, Ren L, et al. Novel gelatin–siloxane nanoparticles decorated by Tat peptide as vectors for gene therapy. Nanotechnology 2008; 19:445103.
75. Olton D, Li J, Wilson ME, et al. Nanostructured nanophosphates for non viral gene delivery: Influence of the synthesis parameters on transfection efficiency. Biomaterials 2007; 28:1267–1279.
76. Nassar T, Rom A, Nyska A, et al. A novel nanocapsule delivery system to overcome intestinal degradation and drug transport limited absorption of P glycoprotein substrate drug. Pharm Res 2008; 25:2019–2029.
77. Juliano R. Challenges to macromolecular drug delivery. Biochem Soc Trans 2007; 35: 41–43.
78. Zidan AS, Sammour OA, Hammad MA, et al. Quality by design: Understanding the product variability of a self-nanoemulsifying drug delivery system of cyclosporine A. J Pharm Sci 2007; 96:2409–2423.
79. Hamidi M, Azadi A, Rafiei P. Hydrogel nanoparticles in drug delivery. Adv Drug Deliv Rev 2008; 60:1638–1649.
80. Nair A, Shen J, Thevonot P, et al. Enhanced intratumoral uptake of quantum dots concealed within hydrogel nanoparticles. Nanotechnology 2008; 19:485102.
81. Vihola H, Laukknen A, Tenhu H, et al. Drug release characteristics of physically cross-linked thermosensitive poly(*N*-vinyl caprolactum) hydrogel particles. J Pharm Sci 2008; 97:4783–4793.
82. Baroli B. Hydrogels for tissue engineering and delivery of tissue inducing substances. J Pharm Sci 2007; 96:2197–2223.

83. Lee J, Yang J, Seo SB, et al. Smart nanoprobes for ultrasensitive detection of breast cancer via magnetic resonance imaging. Nanotechnology 2008; 19:485101.
84. Faure AC, Barbillon G, Ou M, et al. Core/shell nanoparticles for multiple biological detection with enhanced sensitivity and kinetics. Nanotechnology 2008; 19:485103.
85. Gossuin Y, Hocq A, Vuong QL, et al. Physico-chemical and NMR relaxometry characterization of gadolinium hydroxide and dysprosium oxide nanoparticles. Nanotechnology 2008; 19:475102.
86. Shao N, Wickstrom E, Panchapakesan B. Nanotube antibody biosensor arrays for the detection of circulating breast cancer cells. Nanotechnology 2008; 19:465101.
87. Devapally H, Chakilam A, Amiji MM. Role of nanotechnology in pharmaceutical development. J Pharm Sci 2007; 96:2547–2565.
88. Gupta S, Moulik SP. Biocompatible microemulsions and their prospective uses in drug delivery. J Pharm Sci 2008; 97:22–45.
89. Drummond DC, Noble CO, Hayes ME, et al. Pharmacokinetics and in vivo drug release rates in liposomal nanocarrier development. J Pharm Sci 2008; 97:4696–4740.
90. Illum L. Nanoparticulate systems for nasal delivery of drugs: A real improvement over simple systems. J Pharm Sci 2007; 96:473–483.
91. Mozafari MR, Omri A. Importance of divalent cations in nanolipoplex gene delivery. J Pharm Sci 2007; 96:1955–1966.
92. Salonen J, Kaukonen AM, Hirvonen J, et al. Mesoporous silicon in drug delivery applications. J Pharm Sci 2008; 97:632–653.
93. Cheng Y, Xu Z, Ma M, et al. Dendrimers as drug carriers: Applications in different routes of drug administration. J Pharm Sci 2008; 97:123–143.
94. McGoverin CM, Rades T, Gordon KC. Recent pharmaceutical applications of Raman and terahertz spectroscopies. J Pharm Sci 2008; 97:4598–4621.
95. Skinner AL, Laurence JS. High-field NMR resolution spectroscopy as a tool for assessing protein interactions with small molecules ligands. J Pharm Sci 2008; 97:4670–4695.
96. Nahar M, Mishra D, Dubey V, et al. Development, characterization and toxicity evaluation of amphotericin B-loaded gelatin nanoparticles. Nanomedicine 2008; 4:252–261.
97. Green JJ, Chiu E, Leshchiner ES, et al. Electrostatic ligand coatings of nanoparticles enable ligand-specific gene delivery to human primary cells. Nano Lett 2007; 7:874–879.

2 Polymeric Nanoparticles for Small-Molecule Drugs: Biodegradation of Polymers and Fabrication of Nanoparticles

Sheetal R. D'Mello, Sudip K. Das, and Nandita G. Das

Department of Pharmaceutical Sciences, Butler University, Indianapolis, Indiana, U.S.A.

INTRODUCTION

The three important parameters on which the selection of the most suitable drug delivery system is based are the drug, the disease state, and the latter's location in the body. Currently, small-molecule drugs continue to dominate the pharmaceutical market despite biotech drugs making a distinct niche for themselves, because the former enjoys the advantages of small molecular size, solubility, and permeability, which are favorable for passive membrane diffusion. However, the perspective of a drug as a chemical compound used for the prevention, diagnosis, and treatment of a disease state has changed drastically over the past couple of decades as we learn that the mode of delivery of a drug could radically change the therapeutic outcomes of a disease state. For instance, the entrapment of a molecule in a nanoparticulate system could pave the way for better cellular uptake or drug targeting to specific tissues for those drugs with poor bioavailability, in addition to providing prolonged drug release effects.

Polymeric nanoparticles are defined as colloidal particles ranging between 10 and 1000 nm in size and composed of natural or artificial polymers (1). Since the diameter of the smallest capillaries in the human body is about 4 μm, in order for the nanoparticles to access all locations in the body by the intravenous, intramuscular, or subcutaneous route, the solid particles should preferably have a small diameter (2). The small particle size also reduces potential irritant reactions at the injection site (3). The utility of a nanoparticle delivery system is dependent upon the bioacceptability of the carrier polymer, which, in turn, is affected by the particle size and physicochemical properties of the polymer. Ultimately, the bioacceptability of the polymer, physicochemical properties of the drug, and the therapeutic goal will influence the final choice of the appropriate polymer, particle size, and the manufacturing method. Based on the manufacturing method for the nanoparticles, drug molecules can be either dissolved in a liquid core or dispersed within a dense polymeric matrix, resulting in nanocapsules or nanospheres. Liposomes, niosomes, and microemulsions are similar to polymeric nanoparticles with respect to their shape, size, and mode of administration, and thus serve as alternative modes of novel colloidal drug carrier systems. However, nanoparticles offer additional advantages when compared with the other colloidal carriers, such as higher stability when in contact with biological fluids, high drug-loading capacities, and protection by the solid matrix of the incorporated drug against degradation, thus leading to increased intracellular concentration of the drug (4). Also, because of their polymeric nature, controlled drug release can be obtained with nanoparticles. The surface of the polymeric nanoparticles can be covalently conjugated to folic acid, monoclonal antibodies, and aptamers to achieve targeted delivery and cell-specific uptake. For injectable nanoparticles, surface modification with poly(ethylene glycol) (PEG) can help evade phagocytosis by the macrophages of the mononuclear phagocyte system and improve the chances of the nanoparticles reaching the site of action.

BIOPHARMACEUTIC CLASSIFICATION AND DRUG MOLECULES

Amidon et al. (5) first proposed the classification of drugs based on their solubility and permeability, which opened up various avenues and strategies for formulation development based on those physical properties. In 2000, the US Food and Drug Administration (FDA) issued the biowaiver guidelines based on the Biopharmaceutic Classification System (BCS), which offers biowaivers to class I (highly soluble and highly permeable) drugs. While many reports have demonstrated the increase in solubility of poorly soluble small molecules by modifying the shape, size, and functional groups present on the molecule, as well as the increase in permeability by the incorporation of lipid components into the drug, the entrapment of drugs in nanoparticles offers opportunities for the modulation of both solubility and permeability. Since the nanoparticle could itself provide a carrier-driven cellular entry mechanism in certain situations that would be irrespective of the solubility or permeability of the entrapped drug, the unique properties of these carrier systems could be used for the drugs that belong to class II (low solubility-high permeability), class III (high solubility-low permeability), or class IV (low solubility-low permeability) drugs. With an increase in the use of computational and combinatorial processes in drug design, often based on receptor morphology, many of the new drugs approved fall into the category of class II or IV, with significant solubility problems, and nanoparticulate delivery approaches could play a major role in the intracellular delivery of such drugs. Nifedipine, a class II drug, when encapsulated in poly (ε-caprolactone) (PCL) and Eudragit®, shows significantly increased bioavailability (6). Giannavola et al. (7) reported that acyclovir (a class III drug), when loaded in poly(D,L-lactic acid) (PLA) nanospheres, showed improved ocular pharmacokinetics compared with the free drug. In a recent study, it was shown that paclitaxel, a BCS class IV drug, when loaded in PEG–poly(lactide-co-glycolide) nanoparticles showed greater tumor growth inhibition than free paclitaxel (8).

BIODEGRADABLE POLYMERS USED IN THE FABRICATION OF NANOPARTICLES

Biodegradable polymers are advantageous in many ways over other materials for use in drug delivery systems such as nanoparticles. They can be fabricated into various shapes and sizes, with tailored pore morphologies, mechanical properties, and degradation kinetics to suit a variety of applications. By selecting the appropriate polymer type, molecular weight, and copolymer blend ratio, the degradation/erosion rate of the nanoparticles can be controlled to achieve the desired type and rate of release of the encapsulated drug. The common biodegradable polymers used in drug delivery include (*i*) polyesters, such as lactide and glycolide copolymers, polycaprolactones, poly(β-hydroxybutyrates), (*ii*) polyamides, which includes natural polymers such as collagen, gelatin, and albumin, and semisynthetic pseudo-poly(amino acids) such as poly(*N*-palmitoyl hydroxyproline ester), (*iii*) polyurethanes, (*iv*) polyphosphazenes, (*v*) polyorthoesters, (*vi*) polyanhydrides, and (*vii*) poly(alkyl cyanoacrylates).

DEGRADATION AND BIOCOMPATIBILITY OF BIODEGRADABLE POLYMERS

Lactide and Glycolide Copolymers

Biodegradation

One of the most popular biodegradable polymers used in drug delivery are aliphatic polyester copolymers based on lactic and glycolic acids. Poly(D,L-lactic-co-glycolic acid) (PLGA) is used for the manufacture of implants and internal sutures and is known to be biocompatible, degrading to produce the natural products lactic acid and glycolic acid (9). PLGA nanoparticles undergo homogenous hydrolytic degradation, which is modulated by various factors such as chemical composition, porosity, hydrophilicity/hydrophobicity, morphology (crystalline/amorphous), and molecular weight and molecular weight distribution (10). Owing to the presence of methyl groups in the lactide polymers, they are more hydrophobic than the glycolide polymers. Also, the water uptake increases as the glycolide ratio in the copolymer increases (11). Of the homopolymers, PLA is highly crystalline compared with PGA and erodes slowly since it is more resistant to hydrolysis, whereas the PLGA copolymers with an increasing ratio of PGA tend to be less crystalline and thus have a faster rate of biodegradation. The transition glass temperatures of the copolymers range from 36°C to about 67°C.

PLGA polymers undergo bulk hydrolysis/erosion of the ester bonds, wherein the molecular weight decreases whereas the mass remains unchanged while they are metabolized to monomeric acids and undergoes elimination through Krebs cycle. Pitt et al. demonstrated that molecular weight of the polymer decreases in the first stage of degradation owing to the random hydrolytic cleavage of the ester linkage, followed by the onset of weight loss and a change in the rate of chain scission in the second stage (12). Furthermore, hydrolysis is enhanced by the accumulation of acidic products and the reduction of pH facilitated by the carboxylic acid end groups, which is an autocatalytic degradation process (13–15). The degradation of these polymers differs in vivo and in vitro, mainly because, although in vivo there is no major influence of enzymes during the glassy state of the polymer, these enzymes can play a significant role when the polymer becomes rubbery (15). Normally, 50:50 lactide/glycolide copolymers have the fastest half-life of degradation, around 50 to 60 days, whereas 65:35, 75:25, and 85:15 lactide/glycolide copolymers have progressively longer degradation half-lives in vivo. Jalil et al. (17–19) demonstrated that although physical properties of the microparticles were not seriously affected by the molecular weight of poly(D,L-lactide), swelling properties (which are a function of hydrophilicity of the polymer) could be affected owing to the variations in the molecular weight and the core loading.

The half-life of these linear polyesters can be increased by coblending with more hydrophobic comonomers such as polycaprolactone. Visscher et al. performed the biodegradation studies of poly(D,L-lactide) and 50:50 poly(D,L-lactide-co-glycolide) in the rat gastrocnemius muscles (9,20). The complete breakdown of the poly(D,L-lactide) nanoparticles was achieved within 480 days, whereas the PLGA nanoparticles degraded in 63 days, the reason being hydrophilic and semicrystalline nature of the glycolide part. The 50:50 ratio of PLGA is thus advantageous as

TABLE 1 Responses to Polymer Material

Phase	Duration	Response
I	1–2 wk	Acute or chronic inflammatory responses that are independent of the degradation rate and the polymer composition
II	0–3 wk	Response depends on the rate of polymer degradation and includes granular tissue development, foreign-body reaction, and fibrosis
III	3 wk	Phagocytosis by macrophages and foreign-body giant cells

compared with other polymers due to its fastest degradation rate, and as a result, fastest drug release from the nanoparticles.

Biocompatibility

The evaluation of the biocompatibility of biodegradable polymers takes into consideration the incidence of the inflammatory and healing responses of the injected and implanted materials. The particles after an intramuscular or a subcutaneous injection usually have a high surface area/low volume ratio within a given tissue volume.

Table 1 outlines the tissue responses to the polymer materials that are divided into three time phases (10).

Nanoparticles, when given intravenously, can modulate the inflammatory and healing responses in their presence. These responses are usually lesser in magnitude because the particles are present as single, isolated particles and not in groups and also because the cellular injury at the site of the particle is minimal (21). The 50:50 PLGA nanoparticles have a phase II response of 50 to 60 days, whereas for the PLA microspheres, it takes around 350 to 400+ days, thereby indicating its dependence on the rate of biodegradation of the nanoparticles (9,20). By modifying the polymer type, the copolymer composition, the polymer molecular weight, and the porosity of the microspheres, their degradation rate can be varied from days to months.

Poly(ε-Caprolactones)

Biodegradation

PCL is a water-permeable polymer with hydrophobic and high crystalline properties. It undergoes bulk erosion by random hydrolytic chain cleavage in the first phase, resulting in a decrease in the molecular weight of the polymer. This is followed by the second phase, in which these low molecular weight fragments undergo phagocytosis or solubilization in the body fluids. It may require 2 to 4 years for the complete degradation and elimination of PCL homopolymers in vivo. Their degradation rate can be enhanced by the addition of additives such as oleic acid and tertiary amines, which act as catalysts in the chain hydrolysis process. Also, copolymerization with lactide and glycolide decreases crystallinity, and thus hastens the polymer degradation rate due to its higher water uptake (12,22,23).

Biocompatibility

The biocompatibility study of PCL was reported in rats by the subcutaneous injection of bupivacaine-loaded PCL microspheres (24). They could be considered safe because it was observed that (*i*) there were multinucleate, foreign-body giant cells, which are macrophagic cells present in normal processes of polymer degradation and reaction of the organism; (*ii*) mast cells were absent, and therefore histamine granules among microspheres were not observed, indicating that there were no inflammatory and immunogenic processes; (*iii*) lymphocytes were not present in the implants, which indicates the lack of rejection of the implanted microspheres; (*iv*) the implant was surrounded by conjunctive tissue, with tissue infiltration within the particle aggregate – this characteristic is indicative of the integration of the implant in the body of the animal, and thus the absence of rejection; and (*v*) the accumulation of extravascular liquids in the implant area was not detected, which indicates the lack of an acute inflammatory response.

Polyanhydrides

Biodegradation

These hydrophobic and crystalline materials have been shown to undergo erosion by surface hydrolysis, minimizing water diffusion into the bulk of the delivery device (25,26). The monomeric anhydride bonds have extreme reactivity toward water and undergo hydrolysis to generate the dicarboxylic acids (27). Although hydrolysis is catalyzed by both acid and base, an increase in pH enhances the rate of hydrolytic degradation. At low pH, oligomeric products formed at the surface of the matrix have poor solubility; this hinders the degradation of the core. The degradation rate of these polymers can be either accelerated by the incorporation of sebacic acid, a relatively more water-soluble aliphatic comonomer than carboxyphenoxy propane, into the polymer or reduced by increasing the methylene groups or long-chain fatty acid terminal such as stearic acid into the polymer backbone, thereby increasing the monomeric chain length, its hydrophobicity, and the erosion rate (28,29). The branching of poly(sebacic acid) with either 1,3,5-tricarboxylic acid or low molecular weight poly(acrylic acid) results in an increased erosion rate (30). It is also known that aliphatic anhydrides and their copolymers undergo a first-order, self-depolymerization reaction, under anhydrous conditions both in solid state and in solution (31). The rate of depolymerization is found to increase with temperature and the polarity of the solvent. Studies on copolymers of several polyanhydride families have shown that varying comonomer ratios produce erosion profiles ranging from days to years (32). The polyanhydride 20:80 poly[(1,3,bis-*p*-carboxyphenoxy propane)-co-(sebacic anhydride)] has been approved by the FDA for use in biomedical devices (33) and has been successful in delivering carmustine (BCNU) for the treatment of brain cancer (34,35).

Biocompatibility

During biocompatibility testing of linear aliphatic polyanhydrides in rats, histopathological examination of tissue specimens that were in direct contact with the polymer device showed mild inflammation, but macroscopically, no swelling or pathological signs were observed (27). In another set of compatibility studies, these polyanhydrides were shown to be nontoxic, nonmutagenic, and nonteratogenic (36). A rabbit cornea bioassay indicated the absence of an inflammatory response with implanted polyanhydrides (37). In a rabbit animal model, blank polyanhydride particles did not elicit any inflammatory response; however, when a tumor angiogenic factor was incorporated within the polymer matrix, it resulted in a significant vascularization response, further proving the innocence of the polymer by itself (38,39). When tested in rats, polyanhydrides based on ricinoleic acid did not show any signs of tissue necrosis 21 days postimplantation, while only minimal subacute inflammation and mild fibrosis were noted (40). Clinical trials in humans with a polyanhydride dosage form, Gliadel®, produced no systemic or central toxicity, thus demonstrating its biocompatibility and acceptability for human use (41).

Polyorthoesters

Biodegradation

Although polyorthoesters are hydrophobic in nature, their orthoester linkage is acid sensitive and highly unstable in the presence of water. The primary mechanism by which these polymers degrade is hydrolysis, and, depending on the reactants used during the polymer synthesis, the degradation products are a diol, a triol, or a mixture of diols and carboxylic acid. This in situ production of acid further catalyzes the breakdown of these orthoester linkages, thus resulting in the bulk erosion of the matrix. The rate of this acid-catalyzed hydrolysis of the pH-sensitive linkage can be controlled by incorporating either acidic or basic salts into the polymer matrix (42). This was demonstrated in experiments with 5-fluorouracil-embedded polyorthoester nanoparticles – when suberic acid was incorporated as an additive, the acidic excipient accelerated the rate of hydrolysis and caused significantly faster release of the drug (43). Alternatively, when the interior of the matrix is buffered with basic salts, the generated acid is neutralized and hydrolysis can be retarded. In this way, they stabilize the bulk of the matrix but allow the drug to escape from the surface region, thus converting the system into a surface-eroding polymer type. For example, the release of tetracycline from a polyorthoester matrix was found to be extremely rapid; however, the addition of 0.5 wt% $Mg(OH)_2$ as an excipient resulted in a sustained release over 10 days, with 1 wt% $Mg(OH)_2$ the release period extended to about 25 days, and with 2 wt% up to 75 days (44). Certain polyorthoesters containing glycolide sequences exist that undergo hydrolytic degradation by autocatalysis without the use of any excipients (45). The control over the erosion rate can also be extended by altering the amount of catalyst, phthalic anhydride, present in the polymer (46).

Biocompatibility

Biocompatibility studies conducted by Alza Corp. (Mountain View, CA) on some prototype polyorthoesters demonstrated local tissue irritation in human clinical trials (47). They, however, lacked any acute cytotoxicity or abnormal inflammatory response.

PRODUCTION OF POLYMER NANOPARTICLES

The use of a particular manufacturing technique in the preparation of nanoparticles depends on the nature of the polymer employed, nature of the drug to be encapsulated, intended use of the system, and intended duration of the therapy. The various parameters that can be externally controlled to yield nanoparticles of desired physicochemical characteristics, drug entrapment efficiency, and drug release rate properties include the nature and solubility of the drug to be encapsulated, polymer type and concentration, its molecular weight, composition of the copolymers, drug-loading concentrations, type and volume of the organic solvent, the water phase volume, pH, temperature, concentration, types of surfactants, and the mechanical speed of agitation. In vitro and in vivo responses from the nanoparticles are influenced by their various properties, such as the particle size and size distribution, surface morphology, porosity, surface chemistry, surface adhesion, zeta-potential, drug stability, drug-encapsulation efficiency, surface/bulk erosion/degradation, diffusion of the drug, kinetics of drug release, and the hemodynamic properties of the nanoparticles. Conventionally, nanoparticles can be prepared either by dispersion of the preformed polymers or by the in situ polymerization of the monomers.

Laboratory-Scale Production of Nanoparticles

Phase Separation in Aqueous System

The use of coacervation technique to develop polyester microspheres was first reported by Fong in 1979 (48) and modifications of the same are used today for the production of nanoparticles. This technique depends on the precipitation of the drug-entrapping polymer either by the addition of a third compound to the polymer solution or by some other physical means. The point has to be reached where two liquid phases are formed, the polymer-rich coacervate and the supernatant liquid phase, which is depleted in the polymer. Briefly, two steps are involved in the process: (*i*) the formation of liquid droplets of the polymer from the complete solution phase, which depends on the solubility parameters of the polymer, and (*ii*) subsequent hardening of the polymer droplets due to extraction or evaporation of the polymer solvent. A number of organic solvents, such as dichloromethane, isopropanol, and heptanes, have been used as solvent, coacervating agent, and hardening agent. If a drug is initially dispersed in the polymer solution, it can be coated by the coacervate. Phase separation could occur as a result of changes in pH (49) or counterions (50), or as a result of the aqueous phase acting as a nonsolvent for the polymer. Both hydrophilic and hydrophobic drugs can be entrapped by this principle, albeit with different drug-entrapment efficiencies. For example, hydrophilic drugs can be solubilized in water and this aqueous phase can be added to an organic solution of the polymer (w/o emulsion) (51), whereas lipophilic drugs can be dissolved/dispersed in the polymer solution. Hydrophilic drug–entrapment efficiency decreases significantly if a large volume of water is used in the process, or water is used as a coacervating agent. Various process variables such as the aqueous phase/organic phase volume ratio, stirring rate, addition rate of the nonsolvent, polymer concentration, polymer solvent/nonsolvent ratio, and viscosity of the nonsolvent affect the characteristics of the nanoparticles such as morphology, internal porosity, and the size distribution (52,53). The surface porosity of particles normally depends on the solvent extraction process, whereas the shape is normally spherical. The main advantage of phase-separation method is that it protects active drugs from partitioning out into the dispersed phase. However, the residual solvent content is a major concern, especially when organic solvents are used as the hardening agent (54).

Emulsion-Solvent Evaporation/Extraction

In this method, the polymer is first dissolved in a water-immiscible, volatile, organic solvent such as chloroform, dichloromethane, or ethyl acetate (55). The drug is added to this polymer solution and the mixture is emulsified into an outer water phase containing an emulsifier, such as poly(vinyl alcohol) (PVA), gelatin, polysorbate 80, or polaxamer-188 to yield an o/w emulsion. To harden the nanoemulsion droplets into solid nanoparticles, the organic solvent is evaporated or extracted from the system after it diffuses into the external aqueous phase. Emulsification is facilitated by high-speed homogenization or sonication. For the removal of solvent, the stirring process may be continued for several hours at high-temperature/low-pressure conditions; a quicker option to harden the particles may be to pour the emulsion into water, causing the solvent to phase toward the surfactants in the interface and eventually diffuse out into the aqueous phase. Normally, the rate of solvent extraction or evaporation has significant effects on the porosity of the nanoparticles, which, in turn, significantly affects the drug release from the nanoparticles. Since the solvent extraction is normally faster than the evaporation rate (the latter depends on the boiling point of the solvent), the resultant porosity of the nanoparticle matrix prepared by the solvent extraction method is usually greater than the nanoparticles prepared by using the evaporation process (56). Nanoparticles may be harvested by centrifugation or filtration, washed, and freeze-dried to produce free-flowing nanoparticles. One of the challenges encountered in this method is the poor entrapment and burst release effect of moderately – water-soluble and hydrophilic drugs. The encapsulation efficiencies of the water-soluble drugs can be increased by using a w/o emulsification method in which the solution of the drug and polymer of interest are dissolved in a water-miscible organic solvent, such as acetonitrile or acetone, and emulsified in an oil, such as light mineral oil containing an oil-soluble surfactant. Finally, the emulsion is subjected to solvent removal processes and the oil is removed from the particles by washing with hexane (56,57). A diagrammatic representation of o/w single emulsion solvent evaporation method is depicted in Figure 1.

A modification of the single-emulsion method is made by the preparation of a water-in-oil-in-water (w/o/w) type multiple emulsion, which allows for the better incorporation of hydrophilic drugs; this process is termed as the double- or multiple-emulsion method. The process consists of adding the aqueous solution of the drug to the polymer solution in an organic solvent with vigorous stirring to form the first o/w emulsion. This emulsion system is then added gently with stirring to a large quantity of water containing PVA, resulting in a w/o/w double emulsion. The

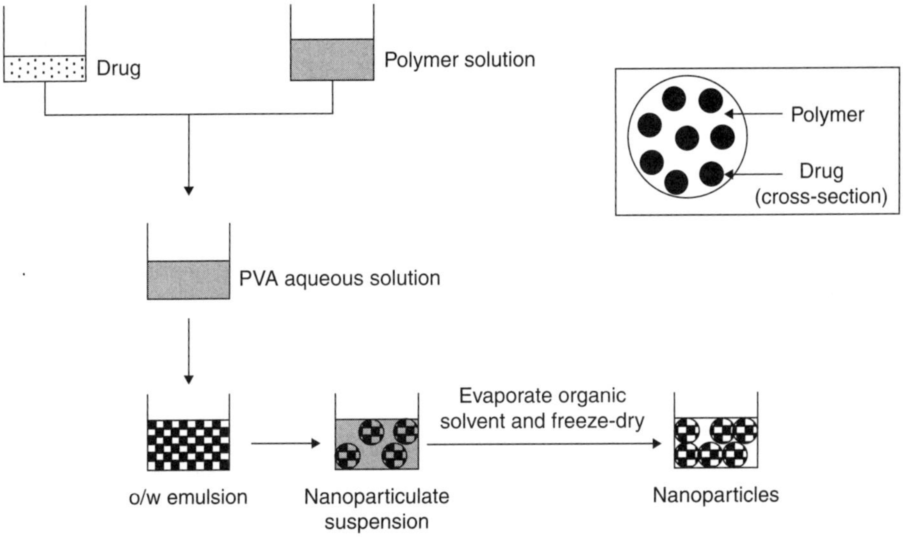

FIGURE 1 Schematic for the preparation of drug-loaded poly(D,L-lactic-co-glycolic acid) nanoparticles by using o/w single-emulsion, solvent evaporation method. *Abbreviation*: PVA, poly(vinyl alcohol).

solvent is then evaporated or extracted from the emulsion as previously described. The extraction of the polymer solvent (e.g., acetone or acetonitrile) from the polymer droplets predominantly takes place when the external phase (e.g., water) is miscible with the polymer solvent. On the other hand, the evaporation process assumes the predominant step if the polymer solvent (e.g., dichloromethane) is not miscible with the external phase (e.g., water). Early reports on the multiple emulsion (w/o/w) solvent evaporation method for the preparation of poly(D,L-lactide)- and poly(lactide-co-glycolide)–biodegradable nanoparticles by Bodmeier and McGinity (58,59) and Ogawa et al. (60) started appearing in literature in the 1980s. This method was subsequently modified and applied toward the delivery of proteins and other small-molecule drugs by a number of different research groups (61,62). The major existing challenges of this method for the production of nanoparticles are the parameters that control the particle size and the outcome of uniform size distribution for small particles. Moreover, the common solvent used to solubilize the polymer, dichloromethane, is a class 2 solvent that poses problems in use in pharmaceutical preparations due to its potential toxicity (63). The common class 3 solvent, acetone, produces highly porous particles that eventually adversely facilitate the drug release, especially for hydrophilic small-molecule drugs (64). Moreover, processing with acetone must be done very carefully because of its high flammability. A recent study by Sani et al. reported the use of ethyl acetate (a class 3 solvent) for PEG-PLGA that produced uniform small size range of nanoparticles (65).

In another modification of the solvent evaporation method (66), the oil phase consists of water-miscible organic solvents such as methanol or acetone together with water-immiscible chlorinated organic solvents. During the formation of an o/w emulsion, acetone/methanol rapidly diffuses into the outer water phase and causes an interfacial turbulence between the two phases, thus resulting in the formation of smaller particles.

Salting Out

The salting-out method and emulsification solvent diffusion techniques for the production of nanoparticles have been developed to meet the US FDA specification on the residual amount of organic solvents in injectable colloidal systems. Polymeric nanoparticles can be prepared by using an emulsion technique that avoids surfactants and chlorinated solvents and involves a salting-out process between two miscible solvents to separate the phases (67). The preparation method consists of adding an electrolyte-saturated (usually magnesium chloride hexahydrate) or a non–electrolyte-saturated aqueous solution containing PVA as a viscosity-increasing agent as well as a stabilizer to an oil phase composed of the polymer and the drug dissolved in acetone under continuous mechanical stirring at room temperature. The saturated aqueous solution prevents complete miscibility of both the phases by virtue of the high salt content. After the preparation of the initial water-in-oil emulsion (w/o), water is immediately added in sufficient quantity to cause a phase inversion from water-in-oil (w/o) to oil-in-water (o/w) type emulsion; this induces complete diffusion of acetone from the internal nonaqueous phase into the continuous external aqueous phase, thus leading to the formation of nanoparticles. The final emulsion is then stirred overnight at room temperature to allow for the complete removal of acetone. Centrifugation may also be used to remove the organic solvent, free PVA, and electrolytes from the raw nanoparticle suspension, after which the nanoparticles can be purified by cross-flow microfiltration and subsequently freeze-dried.

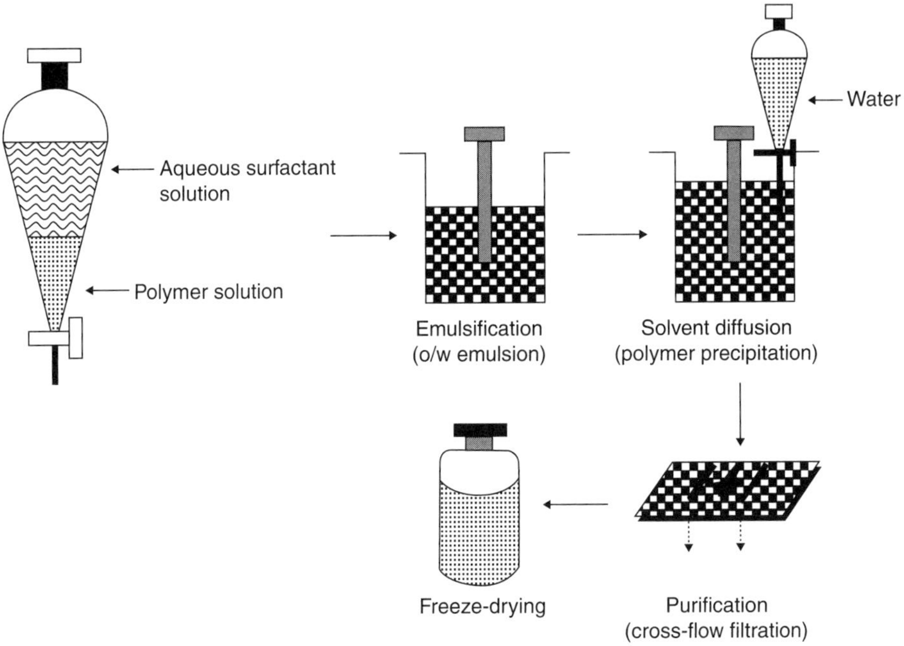

FIGURE 2 Schematic of the emulsification solvent diffusion method. *Source*: From Ref. 68.

Emulsification Solvent Diffusion Method

In the technique developed by Quintanar-Guerrero et al. (68), the solvent and water are mutually saturated at room temperature before use to ensure the initial thermodynamic equilibrium of both liquids. Later, the organic solvent containing the dissolved polymer and the drug is emulsified in an aqueous surfactant solution (usually with PVA as a stabilizing agent) by using a high-speed homogenizer. Water is subsequently added under constant stirring to the o/w emulsion system, thus causing phase transformation and outward diffusion of the solvent from the internal phase, leading to the nanoprecipitation of the polymer and the formation of colloidal nanoparticles. Finally, the solvent can be eliminated by vacuum steam distillation or evaporation. A schematic diagram of the emulsification-solvent diffusion method is presented in Figure 2.

Emulsion Polymerization

This method has been used to prepare poly(alkyl cyanoacrylate) nanoparticles with an approximate diameter of 200 nm (69). A schematic diagram for preparation of Poly(alkyl-cyanoacrylate) nanoparticles by anionic polymerization is presented in Figure 3. The alkyl cyanoacrylate monomer is dispersed in an aqueous acidic medium containing stabilizers such as dextrans and poloxamers (70). Surfactants such as polysorbates can be used as well. The low pH favors the formation of stable and high molecular mass nanoparticles. Under vigorous mechanical stirring, polymerization follows the anionic mechanism since it is initiated usually by nucleophilic initiators such as OH^-, CH_3O^-, and CH_3COO^- and proceeds at ambient temperature. The nonpolar ends within the interior of the surfactant micelles help solubilize the monomer. In the presence of water-soluble initiators, chain growth

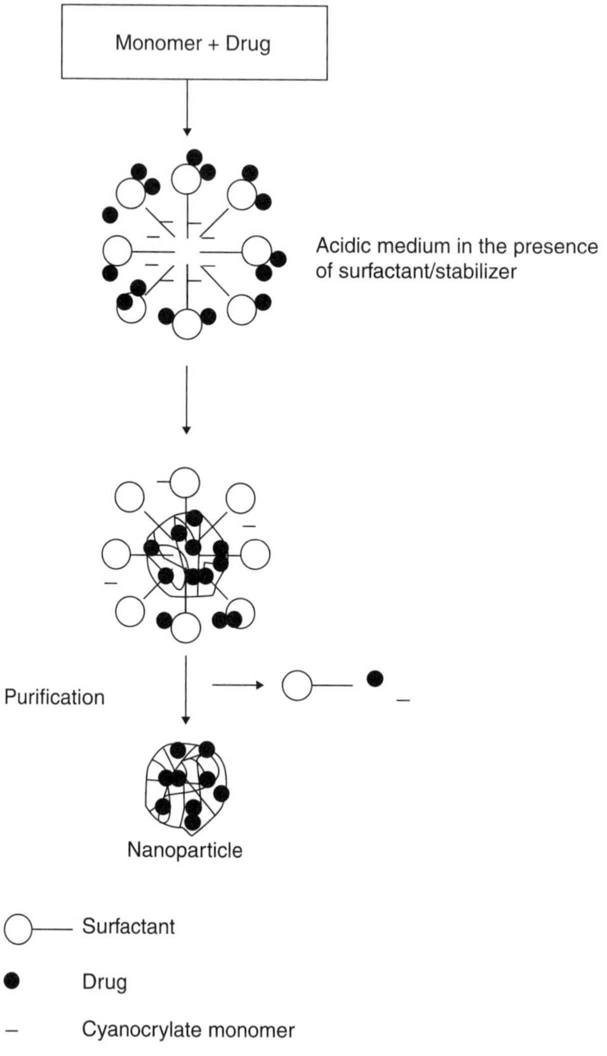

FIGURE 3 Schematic for the preparation of poly(alkyl cyanoacrylate) nanoparticles by anionic polymerization.

commences at the hydrophilic surface of the micelle. When the monomer in the interior of the micelle gets depleted, more monomer droplets from the exterior aqueous phase enter inside; thus, the polymerization reaction proceeds inward and continues until it is terminated by the free radicals. The drug can be solubilized in the polymerization medium either before the monomer is added or later when the reaction has ended. Finally, the nanoparticulate suspension is purified either by ultracentrifugation or by redispersing the nanoparticles in an isotonic medium. The various factors affecting the formation of particles, their size, and molecular mass include monomer concentration, stirring speed, surfactant/stabilizer type and concentration, and the pH of the polymerization medium (71).

Phase Separation in Nonaqueous System

Unlike the single o/w and double w/o/w emulsion techniques, this process can be used to encapsulate both hydrophilic and lipophilic drugs, offering distinct advantages in terms of the entrapment efficiency over the application of predominantly aqueous systems that wash away highly hydrophilic drugs. In this method, hydrophilic drugs are solubilized in water and added to an organic solution of the polymer (w/o emulsion), whereas lipophilic drugs can be dissolved/dispersed in the polymer solution (51). Subsequently, an organic nonsolvent (e.g., silicone oil), which is miscible with the organic solvent (e.g., dichloromethane) but does not dissolve either the drug or the polymer, is added to the emulsion system with stirring; this gradually extracts the organic polymer solvent. With the loss of the solvent, there is a reduction in the polymer solubility, and the coating polymer in the solution undergoes phase separation, with the coacervate phase containing the polymer coacervate droplets. The polymer coacervate adsorbs on to the drug particle surface, resulting in the encapsulation of the drug by the precipitated polymer. Kim et al. (72) compared the loading efficiency of aqueous and nonaqueous phase-separation solvent evaporation systems in terms of the size of particles produced and loading efficiency of felodipine, a poorly water-soluble drug. It was noted that the o/w emulsion template produced smaller particle size and higher entrapment efficiency for the hydrophobic drug than the nonaqueous o/o template used for the same drug. On the other hand, the entrapment of a highly water-soluble agent, a Bowman-Birk inhibitor, was significantly increased by using the nonaqueous o/o template (73).

Large-Scale Pilot Production of Drug-Loaded Nanoparticles

Spray Drying

Some of the challenges faced by this technique include the production of small-sized nanoparticles and the need for innovative methods to increase the drug-entrapment efficiency. However, when compared with other methods, it provides a relatively rapid and convenient production technique that is easy to scale up, involves mild processing conditions, and has relatively less dependence on the solubility characteristics of the drug and the polymer. In this method, a solution or dispersion (w/o) of a drug in an organic solvent containing the polymer is sprayed from the sonicating nozzle of a spray dryer and subsequently dried to yield nanoparticles. The process parameters that can be varied include the inlet and outlet air temperatures, spray flow, and compressed spray air flow (represented as the volume of the air input). In a novel, low-temperature, freeze–spray-drying method (74), the solution or dispersion of the drug in an organic solvent containing the dissolved polymer is sprayed or atomized through an ultrasonic nozzle into a vessel containing liquid nitrogen overlaying frozen ethanol and frozen at −80°C and lyophilized. The liquid nitrogen is evaporated, whereupon the melting liquefied ethanol extracts the organic solvent from the frozen droplets causing the particles to harden. The nanoparticles are filtered and dried under vacuum. A schematic diagram for production of nanoparticles by spray-drying is presented in Figure 4.

Higher encapsulation efficiency for hydrophilic drugs can be achieved with the spray-drying method using aqueous solutions. With an aim to avoid significant product loss due to nanoparticulate adhesion on to the interior wall of the spray dryer, as well as to prevent the aggregation of the nanoparticles, a double-nozzle spray-drying method has been developed together with the use of mannitol as an antiadherent (75). In this technique, drug solution or suspension in the

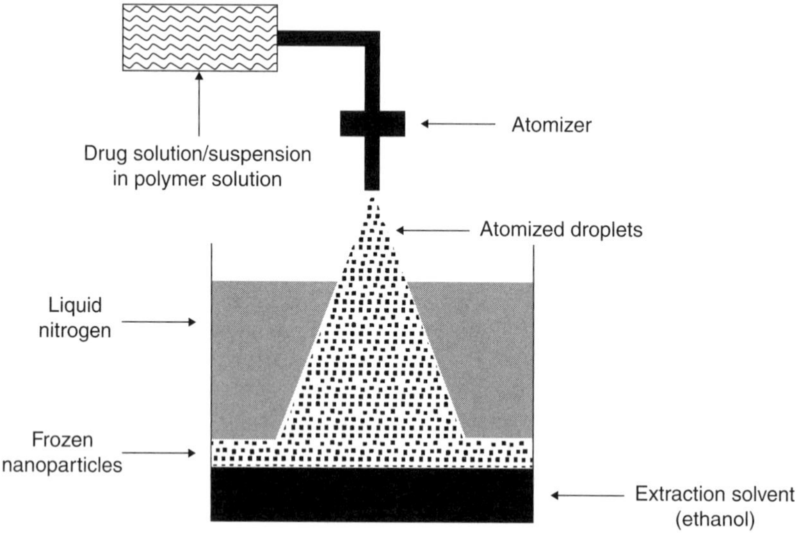

FIGURE 4 Schematic for the production of nanoparticles by spray-drying. *Source*: From Ref. 74.

polymer solution is sprayed from one nozzle, with the aqueous mannitol solution being simultaneously sprayed from another nozzle. In this process, the surface of the spray-dried nanoparticles gets coated with mannitol and the degree of agglomeration is reduced.

Supercritical Fluid Spraying

This technology is advantageous in that the use of an organic solvent/surfactant can be avoided or minimized, thus producing nanoparticles that are free from toxic impurities. Carbon dioxide is nontoxic, nonflammable, and environmentally acceptable, and supercritical CO_2 can be easily obtained by pressurizing and heating the CO_2 system to a minimum of 73.8 bars and 31.05°C, respectively.

In the supercritical antisolvent method (76–78), both the drug and the polymer are dissolved in a suitable organic solvent and are atomized through a nozzle into supercritical CO_2. The dispersed organic solvent phase and the antisolvent CO_2 phase diffuse into each other and since CO_2 is miscible only with the solvent, the solvent gets extracted causing the supercritical fluid–insoluble solid to precipitate as nanoparticles. The rates of two-way mass transfer are much faster than those for conventional organic antisolvents. When the density of CO_2 decreases, the atomization of the spray is intensified, resulting in faster mass transfer rates associated with high surface area of the associated droplets, thus rapid nucleation and smaller particle sizes (79). The dry, micronized powder is then collected following the depressurization of CO_2.

In the gas antisolvent method (80), antisolvent CO_2 is introduced into the organic solution containing the solutes of interest. Supercritical CO_2 is miscible with the solvent but does not solubilize the solutes. This causes the solvent concentration to be significantly lowered, resulting in the precipitation of the drug inside the polymer matrix. Later, the solid product is flushed with fresh CO_2 to strip the residual solvent. The rate of addition of CO_2 to the organic solution affects the final particle

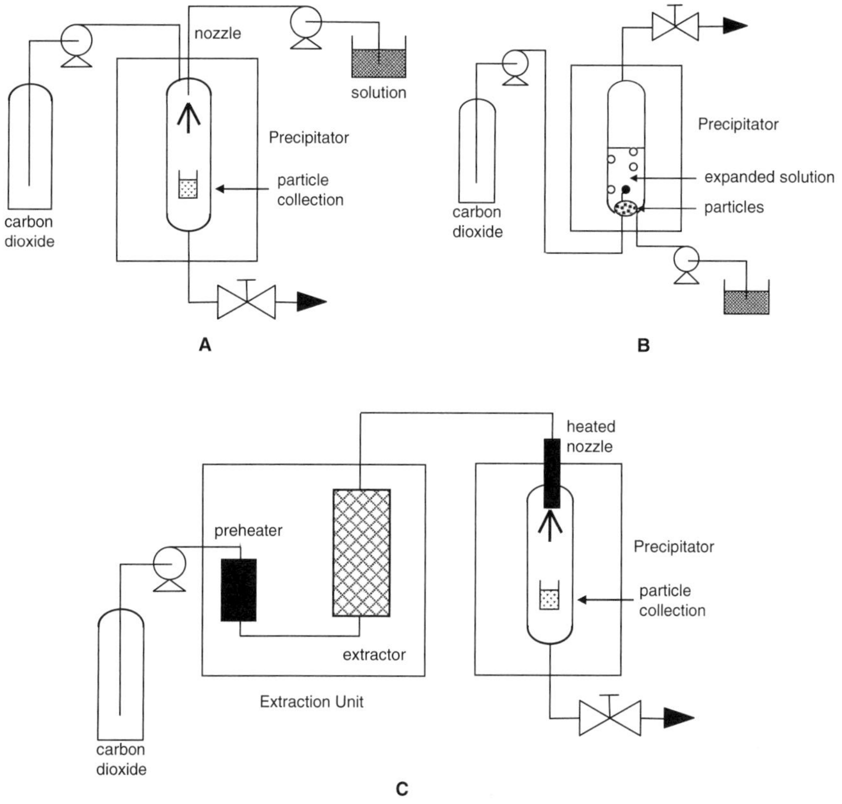

FIGURE 5 Schematic diagram of the (a) SAS, (b) GAS and (c) the RESS processes. *Source*: Reprinted with permission of John Wiley & Sons, Inc. Ref. 77.

size. A major challenge of this process is the need to filter the precipitate from the organic solvent solution without particle growth and aggregation. Schematic diagrams representing SAS, GAS and RESS processes are presented in Figure 5.

In the rapid expansion of supercritical solutions technique (81), the solute is dissolved in supercritical CO_2 and this solution is atomized through a nozzle into a collection chamber at atmospheric conditions. When expanded, CO_2 immediately evaporates and the solute precipitates as a coprecipitate of the drug embedded in the polymer matrix. Various parameters that affect the resulting particle size and morphology are the pre- and postexpansion temperature and pressure, nozzle geometry, and solution concentration (77,82). The disadvantages of this method include the use of higher temperatures to form homogenous precipitates (thus degrading thermally labile drugs) and the limited solubility of the polymers and drugs that result in low drug loading (83).

CONCLUSION AND FUTURE DIRECTIONS

Colloidal drug carriers (nanoparticles, nanoemulsions, nanocapsules, liposomes, nanosuspensions, mixed micelles, microemulsions, and cubosomes) are generally considered to have sizes below 1 μm (84). In general, they possess several potential advantages, such as better oral bioavailability for poorly water-soluble drugs, formulating intravenous injections, and targeting of drugs to specific tissues, thus reducing general toxicity. The nanoparticulate mode of drug delivery using biodegradable polymers is viewed as one of the most promising approaches for (*i*) improving the bioavailability of the drug with the possibility of reduction of the effective dose, thus reducing the chance of potential toxicity and the adverse effects of the drug, (*ii*) passive drug targeting to specific tissues, and (*iii*) effective stabilization of the drug in the polymer matrix, protecting from enzymes and other normal defense mechanisms of the body. Besides nanoparticles, other colloidal carriers such as emulsions for the administration of drugs and parenteral nutrition offer the advantage of reduction of adverse effects such as pain and inflammation at the injection site. Successful commercial products include Diazemuls®, Diazepam-Lipuro®, Etomidate-Lipuro®, and Diprivan®. However, a major disadvantage of this system is the critical, physical instability caused by the incorporated drug, which leads to a decrease in the zeta-potential and thus promotes agglomeration, drug expulsion, and, finally, breaking of the emulsion (85). Also, the FDA generally regarded as safe oils such as soybean oil, medium- and large-chain triglycerides, and their mixtures tend to have limited solubility for the possible drugs of interest to be formulated into emulsions. The expensive toxicity studies associated with the search for new oils with improved solubility properties present a challenge to the further development of this delivery system (86). Another particulate carrier system, liposomes, have been introduced to reduce the toxic adverse effects of the highly potent drugs and thereby enhance the efficacy of the system. Marketed products include Doxil®, DaunoXome®, and Ambisome®. However, low physical stability, drug leakage, nonspecific tumor targeting, nonspecific phagocytosis, problems in upscaling and their high cost limit the total number of products in the market (87,88). The polymeric nanoparticulate carrier system consisting of either biodegradable or nonbiodegradable polymers are thus advantageous in terms of site-specific targeting and controlled release of the encapsulated drug molecules (89). While both formulation stability and in vivo stability are big advantages of nanoparticles, their disadvantage arises from the cytotoxicity of polymers after being internalized by the cells such as macrophages and their subsequent degradation as in the case of polyester polymers (90). Thus far, the lack of a suitable large-scale production method that would be cost-effective and lead to an acceptable product by the regulatory authorities has lead to very few marketed nanoparticle preparations. The major challenges faced by the pharmaceutical industry in the manufacturing of nanoparticles are controlling batch-to-batch variation of the particle size and drug loading. The formation of uniform sized microdrops of solvated polymer by the use of piezoelectric transducers have been reported. Although the size distribution of the particles were narrow, the average size was larger than 10 μm. In addition, the freeze-drying of the nanoparticles with bioactive cryoprotectants and the processing of sterile products offer major challenges. The aggregation of nanoparticles in the biological medium poses another challenge, as the aggregate size and not the individual particle size determines the transport of the drug and the cellular uptake. Regardless of these challenges, given the potential of nanoparticulate polymeric drug delivery systems in improving dug therapy, it appears to be a promising strategy for the drug delivery industry to allocate R&D initiatives in this area. Moreover, a number of drugs that were previously removed from the pipeline owing to unfavorable biopharmaceutic properties could now be potentially revisited by using the nanoparticle carrier systems.

ACKNOWLEDGMENT

The authors thank Lilly Endowment Inc. for providing a graduate stipend to Sheetal R. D'Mello.

REFERENCES

1. Kreuter J. Nanoparticles. In: Swarbrick J, and Boylan JC, eds. Encyclopedia of Pharma. Tech. New York: Marcel Dekker, 1994:165–190.
2. Thews G, Mutschler E, Vaupel P. Anatomie, Physiologie, Pathophysiologie des Menschen. Stuttgart: Wissenschaftliche Verlagsgesellschaft, 1980:229.
3. Little K, Parkhouse J. Tissue reactions to polymers. Lancet 1962; 2:857.
4. Roney C, Kulkarni P, Arora V, et al. Targeted nanoparticles for drug delivery through the blood-brain barrier for Alzheimer's disease. J Control Release 2005; 108(2–3):193–214.
5. Amidon GL, Lennernäs H, Shah VP, et al. A theoretical basis for a biopharmaceutic drug classification: the correlation of in vitro drug product dissolution and in vivo bioavailability. Pharm Res 1995; 12:413–420.
6. Kim YI, Fluckiger L, Hoffman M, et al. The antihypertensive effect of orally administered nifedipine-loaded nanoparticles in spontaneously hypertensive rats. Br J Pharmacol 1997; 120:399–404.
7. Giannavola C, Bucolo C, Maltese A, et al. Influence of preparation conditions on acyclovir-loaded poly-D,L-lactic acid nanospheres and effect of PEG coating on ocular drug bioavailability. Pharm Res 2003; 20:584–590.
8. Danhier F, Lecouturier N, Vroman B, et al. Paclitaxel-loaded PEGylated PLGA-based nanoparticles: In vitro and in vivo evaluation. J Control Release 2009; 133:11–17.
9. Visscher GE, Robison RL, Mauling HV, et al. Biodegradation of and tissue reaction to 50:50 poly(D,L-lactide-co-glycolide) microcapsules. J Biomed Mater Res 1985; 19:349–365.
10. Anderson JM, Shive MS. Biodegradation and biocompatibility of PLA and PLGA microspheres. Adv Drug Deliv Rev 1997; 28:5–24.
11. Gilding DK, Reed AM. Biodegradable polymers for use in surgery: Poly(glycolic)/poly(lactic acid) homo- and copolymers; part 1. Polymer 1979; 20:1459.
12. Pitt CG, Gratzel MM, Kimmel GL, et al. Aliphatic polyesters; part 2: The degradation of poly(D,L-lactide), poly(ε-caprolactone) and their copolymers in vivo. Biomaterials 1981; 2:215.
13. Pistner H, Bendix DR, Mühling J, et al. Poly(L-lactide): A long-term degradation study in vivo; part III: Analytical characterization. Biomaterials 1993; 14:291–304.
14. Li SM, Garreau H, Vert M. Structure–property relationships in the case of the degradation of massive aliphatic poly(α-hydroxy acids) in aqueous media; part 1. J Mater Sci: Mater Med 1990; 1:123–130.
15. Li SM, Garreau H, Vert M. Structure–property relationships in the case of the degradation of massive aliphatic poly(α-hydroxy acids) in aqueous media; part 3. J Mater Sci: Mater Med 1990; 1:198–206.
16. Holland SJ, Tighe BJ, Gould PL. Polymers for biodegradable medical devices; part I: The potential of polyesters as controlled macromolecular release system. J Control Release 1986; 4:155.
17. Jalil R, Nixon JR. Microencapsulation using poly(L-lactic acid); part I: Microcapsule properties affected by the preparative techniques. J Microencapsul 1989; 6:473–484.
18. Jalil R, Nixon JR. Microencapsulation using poly(D,L-lactic acid); part II: Effect of polymer molecular weight on the microcapsule properties. J Microencapsul 1990; 7:245–254.
19. Jalil R, Nixon JR. Microencapsulation using poly(D,L-lactic acid); part III: Effect of polymer molecular weight on the release kinetics. J Microencapsul 1990; 7:357–374.
20. Visscher GE, Robison RL, Maulding HV, et al. Biodegradation of and tissue reaction to poly(DL-lactide) microcapsules. J Biomed Mater Res 1986; 20:667–676.

21. Spenlehauer G, Vert M, Benoit JP, et al. In vitro and in vivo degradation of poly(DL-lactide/glycolide) type microspheres made by solvent evaporation method. Biomaterials 1989; 10:557–563.
22. Pitt G, Chasalow FI, Hibionada YM, et al. Aliphatic polyesters; part I: The degradation of poly(ε-lactone) in vivo. J Appl Polym Sci 1981; 26:3779–3787.
23. Woodward SC, Brewer PS, Moatamed F, et al. The intracellular degradation of poly (ε-caprolactone). J Biomed Mater Res 1985; 19:437–444.
24. Blanco MD, Bernardo MV, Sastre RL, et al. Preparation of bupivacaine-loaded poly (ε-caprolactone) microspheres by spray drying: Drug release studies and biocompatibility. Eur J Pharm Biopharm 2003; 55(2):229–236.
25. Goepferich A, Langer R. The influence of microstructure and monomer properties on the erosion mechanism of a class of polyanhydrides. J Polym Sci Part A: Polym Chem 1993; 31:2445–2458.
26. Goepferich A, Shieh L, Langer R. Aspects of polymer erosion. Mater Res Soc Symp Proc 1995; 394:155–160.
27. Domb AJ, Nudelman R. In vivo and in vitro elimination of aliphatic polyanhydrides. Biomaterials 1995; 16(4):319–323.
28. Domb A, Gallardo CF, Langer R. Poly(anhydrides); part 3: Poly(anhydrides) based on aliphatic-aromatic diacids. Macromolecules 1989; 22:3200–3204.
29. Teomim D, Domb AJ. Fatty acid terminated polyanhydrides. J Polym Sci Part A: Polym Chem 1999; 37:3337–3344.
30. Maniar M, Xie X, Domb A. Poly(anhydrides); part V: Branched poly(anhydrides). Biomaterials 1990; 11:690.
31. Domb A, Langer R. Solid state and solution stability of poly(anhydrides) and poly(esters). Macromolecules 1989; 22:2117–2122.
32. Heller J. Controlled release of biologically active compounds from bioerodible polymers. Biomaterials 1980; 1:51–57.
33. Langer R. New methods of drug delivery. Science 1990; 249:1527–1533.
34. Tamargo R, Epstein J, Reinhard C, et al. Brain biocompatibility of a biodegradable, controlled-release polymer in rats. J Biomed Mater Res 1989; 23:253–266.
35. Tamargo RJ, Myseros JS, Epstein JI, et al. Interstitial chemotherapy of the 9L gliosarcoma: Controlled release polymers for drug delivery in the brain. Cancer Res 1993; 53: 329–333.
36. Leong KW, D'Amore P, Marletta M, et al. Bioerodible polyanydrides as drug-carrier matrices: biocompatibility and chemical reactivity. J Biomed Mater Res 1986; 20:51–64.
37. Rock M, Green M, Fait C, et al. Evaluation and comparison of biocompatibility of various classes of anhydrides. Polym Preprints 1991; 32:221–222.
38. Langer R, Brem H, Tapper D. Biocompatibility of polymeric delivery systems for macromolecules. J Biomed Mater Res 1981; 15:267–277.
39. Langer R, Lund D, Leong K, et al. Controlled release of macromolecules: Biological studies. J Control Release 1985; 2:331–341.
40. Teomim D, Nyska A, Domb AJ. Ricinoleic acid based biopolymers. J Biomed Mater Res 1999; 45:258–267.
41. Domb AJ, Maniar M, Bogdansky S, et al. Drug delivery to the brain using polymers. Crit Rev Therap Drug Carrier Syst 1991; 8:1–17.
42. Heller J. Poly(ortho esters). Adv Polym Sci 1993; 107:41–92.
43. Seymour LW, Duncan R, Duffy J, et al. Poly(ortho ester) matrices for controlled release of the antitumor agent 5-fluorouracil. J Control Release 1994; 31:201–206.
44. Roskos KV, Fritzinger BK, Rao SS, et al. Development of a drug delivery system for the treatment of periodontal disease based on bioerodible poly(ortho esters). Biomaterials 1995; 16:313–317.
45. Ng SY, Vandamme T, Taylor MS, et al. Synthesis and erosion studies of self-catalyzed poly(ortho ester)s. Macromolecules 1997; 30:770.
46. Sparer RV, Shih C, Ringeisen CD, et al. Controlled release from erodible poly(orthoester) drug delivery systems. J Control Release 1984; 1(1):23–32.
47. Gabelnick HL. Biodegradable implants: alternative approaches. In: Mishell DR Jr, ed. Long-acting steroid contraception. New York: Raven Press, 1983:149–173.
48. Fong JW. Processes for preparation of microspheres. US patent 4,166,800. 1979.

49. El-Samaligy MS, Rohdewald P. Reconstituted collagen nanoparticles, a novel drug carrier delivery system, J Pharm Pharmacol 1983; 35:537–539.
50. Rajaonarivony M, Vauthier C, Couarraze G, et al. Development of a new drug carrier made from alginate. J Pharm Sci 1993; 82:912–917.
51. Ruiz JM, Tissier B, Benoit JP. Microencapsulation of peptide: A study of the phase separation of poly(D,L-lactic acid-co-glycolic acid) copolymers 50/50 by silicone oil. Int J Pharm 1989; 49:69–77.
52. Nihant N, Stassen S, Grandfils C, et al. Microencapsulation by coacervation of poly(lactide-co-glycolide); part III: Characterization of the final microspheres. Polym Int 1994; 34:289–299.
53. Nihant N, Grandfils C, Jérôme R, et al. Microencapsulation by coacervation of poly(lactide-co-glycolide); part IV: Effect of the processing parameters on coacervation and encapsulation. J Control Release 1995; 35(2–3):117–125.
54. Muller BW, Bleich J, Wagenaar B. Microparticle production without organic solvent. Proceedings of the 9th International Symposium on Microencapsulation, 13–15 September 1993.
55. Wu XS. Preparation, characterization, and drug delivery applications of microspheres based on biodegradable lactic/glycolic acid polymers. In: Wise DL, Trantolo DJ, Altobelli DE, et al., eds. Encyclopedic Handbook of Biomaterials and Bioengineering. New York: Marcel Dekker, 1995:1151–1200.
56. Arshady R. Preparation of biodegradable microspheres and microcapsules; part 2: Polylactides and related polyesters. J Control Release 1991; 17:1–22.
57. Jalil R, Nixon JR. Biodegradable poly(lactic acid) and poly(lactide-co-glycolide) microcapsules: problems associated with preparative techniques and release properties. J Microencapsul 1990; 7:297–325.
58. Bodmeier R, McGinity JW. The preparation and evaluation of drug-containing poly(DL-lactide) microspheres formed by the solvent evaporation method. Pharm Res 1987; 4:465–471.
59. Bodmeier R, McGinity JW. Solvent selection in the preparation of poly(lactide) microspheres prepared by the solvent evaporation method. Int J Pharm 1988; 43:179–186.
60. Ogawa Y, Yamamoto M, Okada H, et al. A new technique to efficiently entrap leuprolide acetate into microcapsules of polylactic acid or copoly(lactic/glycolic) acid. Chem Pharm Bull (Tokyo) 1988; 36:1095–1103.
61. Jeffery H, Davis SS, O'Hagan DT. The preparation and characterization of poly(lactide-co-glycolide) microparticles; part II: The entrapment of a model protein using a (water-in-oil)-in-water emulsion solvent evaporation technique. Pharm Res 1993; 10:362–368.
62. Chattaraj SC, Rathinavelu A, Das SK. Biodegradable microparticles of influenza viral vaccine: comparison of the effects of routes of administration on the in vivo immune response in mice. J Control Release 1999; 58:223–232.
63. ICH Harmonized Tripartite Guideline. Impurities: Guideline for Residual Solvents, Q3C(R3). Available at: http://www.ich.org/LOB/media/MEDIA423.pdf. Accessed October 15, 2007.
64. Ruan G, Feng SS. Preparation and characterization of poly(lactic acid)-poly(ethylene glycol)-poly(lactic acid) (PLA-PEG-PLA) microspheres for controlled release of paclitaxel. Biomaterials 2003; 24:5037–5044.
65. Sani SN, Das NG, Das SK. Effect of microfluidization parameters on the physical properties of PEG-PLGA nanoparticles prepared using high pressure microfluidization. J Microencapsul 2008. doi:10.1080/02652040802500655.
66. Niwa T, Takeuchi H, Hino T, et al. Preparations of biodegradable nanospheres of water-soluble and insoluble drugs with D,L-lactide/glycolide copolymer by a novel spontaneous emulsification solvent diffusion method and the drug release behavior. J Control Release 1993; 25:89–98.
67. Allémann E, Gurny R, Doelker E. Preparation of aqueous polymeric nanodispersions by a reversible salting-out process: Influence of process parameters on particle size. Int J Pharm 1992; 87(1–3):247–253.
68. Quintanar-Guerrero D, Fessi H, Allémann E, et al. Influence of stabilizing agents and preparative variables on the formation of poly(D,L-lactic acid) nanoparticles by an emulsification-diffusion technique. Int J Pharm 1996; 143(2):133–141.

69. Couvreur P, Kante B, Roland M, et al. Polycyanoacrylate nanocapsules as potential lysosomotropic carriers: Preparation, morphology and sorptive properties. J Pharm Pharmacol 1979; 31:331–332.
70. Das SK, Tucker IG, Hill DJ, et al. Evaluation of poly(isobutylcyanoacrylate) nanoparticles for mucoadhesive ocular drug delivery; part I: Effect of formulation variables on physicochemical characteristics of nanoparticles. Pharm Res 1995; 12:534–540.
71. Behan N, Birkinshaw C, Clarke N. A study of the factors affecting the formation of poly(*n*-butylcyanoacrylate) nanoparticles. Proc Int Symp Control Rel Bioact Mater 1999; 26:1134–1135.
72. Kim BK, Hwang SJ, Park JB, et al. Characteristics of felodipine-located poly(epsilon-caprolactone) microspheres. J Microencapsul 2005; 22:193–203.
73. Morgan AW, Das NG, Das SK. Development of poly(lactide-co-glycolide) nanoparticles of Bowman-Birk inhibitor using non-aqueous solvent evaporation method. Proceedings of the annual meeting of the American Association of Pharmaceutical Scientists; Baltimore, MD; November 7–11, 2004.
74. Johnson OL, Jaworowicz W, Cleland JL, et al. The stabilization and encapsulation of human growth hormone into biodegradable microspheres. Pharm Res 1997; 14(6):730–735.
75. Takada S, Uda Y, Toguchi H, et al. Application of a spray drying technique in the production of TRH-containing injectable sustained-release microparticles of biodegradable polymers. PDA J Pharm Sci Technol 1995; 49(4):180–184.
76. Dixon DJ, Johnston JP, Bodmeier RA. Polymeric materials formed by precipitation with a compressed fluid antisolvent. AIChE J 1993; 39:127–139.
77. Subramaniam B, Rajewski RA, Snavely K. Pharmaceutical processing with supercritical carbon dioxide. J Pharm Sci 1997; 86:885–890.
78. Bodmeier R, Wang J, Dixon DJ, et al. Polymeric microspheres prepared by spraying into compressed carbon dioxide. Pharm Res 1995; 12:1211–1217.
79. Sassiat PR, Mourier P, Caude MH, et al. Measurement of diffusion coefficients in supercritical carbon dioxide and correlation with the equation of Wilke and Chang. Anal Chem 1987; 59:1164–1170.
80. Randolph TW, Randolph AD, Mebes M, et al. Sub-micron sized biodegradable particles of poly(L-lactic acid) via the gas antisolvent spray precipitation process. Biotechnol Prog 1993; 9:429–435.
81. Matson DW, Peterson RC, Smith RD. Production of fine powders by the rapid expansion of supercritical fluid solutions. Adv Ceram 1987; 21:109.
82. Philips EM, Stella VJ. Rapid expansion from supercritical solutions: Application to pharmaceutical processes. Int J Pharm 1993; 94:1–10.
83. Ting SST, MacNaughton SJ, Tomasko DL, et al. Solubility of naproxen in supercritical carbon dioxide with and without cosolvents. Ind Eng Chem Res 1993; 32:1471–1481.
84. Westesen K. Novel lipid-based colloidal dispersions as potential drug administration systems – expectations and reality. Coll Polym Sci 2000; 278(7):608.
85. Collins-Gold L, Feichtinger N, Warnheim T. Are lipid emulsions the drug delivery solution? Modern Drug Discov 2000; 3(3):44–48.
86. Müller RH, Mäder K, Gohla S. Solid lipid nanoparticles (SLN) for controlled drug delivery – a review of the state of the art. Eur J Pharm Biopharm 2000; 50(1):161–177.
87. Lasic DD. Novel applications of liposomes. Trends Biotechnol 1998; 16:307–321.
88. Heath TD. Liposome dependent drugs. In: Gregoriadis G, ed. Liposomes as Drug Carriers: Recent Trends and Progress. John Wiley & Sons, Chichester, U.K.: 1988; 709–718.
89. Allemann E, Gurny R, Dorlkar E. Drug loaded nanoparticles – preparation methods and drug targeting issues. Eur J Pharm Biopharm 1993; 39:173–191.
90. Smith A, Hunneyball IM. Evaluation of polylactid as a biodegradable drug delivery system for parenteral administration. Int J Pharm 1986; 30:215–230.
91. Yeo Y, Chen AU, Basaran OA, Park K. Solvent exchange method: A novel miroencapsulation technique using dual microdispensers. Pharm. Res., 2004; 21:1419–1427.
92. Berkland C, Kyekyoon K, Pack DW. Fabrication of PLG microspheres with precisely controlled and monodisperse size distributions. J. Control. Rel., 2001; 73:59–74.

3 Formulation of NPDDS for Macromolecules

Maria Eugénia Meirinhos Cruz, Sandra Isabel Simões, Maria Luísa Corvo, Maria Bárbara Figueira Martins, and Maria Manuela Gaspar

Unit New Forms of Bioactive Agents (UNFAB)/INETI and Nanomedicine & Drug Delivery Systems Group [iMed.UL], Lisbon, Portugal

INTRODUCTION

Novel therapeutic macromolecules, such as proteins or nucleic acids, are being introduced progressively into the pharmaceutical research field as a potent therapeutic promise. Biologically active macromolecules, namely, proteins, have generally low oral bioavailability and short biological half-times (1,2). The physicochemical characteristics of proteins are responsible for deficient systemic delivery, thus requiring frequent injections to maintain blood concentration within therapeutic levels. This results in oscillating protein concentration in the blood and poor patient compliance. Proteins are sensitive molecules and their three-dimensional structure can be disrupted by a number of factors such as hydrophobic environments, high shear, change in temperature or pH, and absence of water. A change in their structure could affect the therapeutic effect of proteins and also trigger adverse immune reactions. Formulation of proteins is not comparable with those of conventional low molecular weight (MW) drugs, as it is mandatory to maintain the protein's natural structure. Also, in the case of enzymes, the catalytic activity is dependent on the free accessibility of the active center. One way to circumvent these problems is, instead of using "naked" proteins, to promote the association with delivery systems that are able to maintain protein structure and activity, change their pharmacokinetics, and deliver them to the target tissues, thus improving safety and efficacy of proteins as drugs. The most common systems to deliver proteins in vivo are nanoparticulate drug delivery systems (NPDDS), either of lipid or polymeric nature (2–4). An alternative, or complementary, strategy for the association of proteins with NPDDS as pharmaceutical nanocarriers is their administration by noninvasive routes, for example, pulmonary or transdermal delivery (5,6).

The global market of therapeutic proteins was around $47 million in 2007 and is expected to increase up to 170% by 2013 (7). Taking into account these numbers, nonconventional and more sophisticated formulations of proteins than those available nowadays are urgently needed. Without such a strategy, there is a real risk of losing the enormous potential of proteins as therapeutic agents. Among therapeutic proteins, enzymes represent a special class due to their specificity and selectivity of action, representing a global market of $3 million, with increasing projections (7).

This chapter focuses mainly on the formulation of therapeutic enzymes by using NPDDS. The methodologies for the formulation of enzymes in NPDDS, parameters for their characterization, and methods for in vitro and in vivo evaluation can be easily extrapolated to other proteins.

ENZYMES WITH THERAPEUTIC ACTIVITY

The development of recombinant DNA technology together with the capacity of large-scale production of pure enzymes has increased the number of enzymes with therapeutic applications (8,9). Table 1 gives some examples of therapeutic enzymes and the indicated treatment application. Enzyme therapy can be used for a broad range of diseases. However, so far, enzyme NPDDS have not been approved for clinical use.

PHARMACEUTICAL NANOCARRIERS FOR ENZYMES

Pharmaceutical nanocarriers, herein designated as NPDDS, can be classified in different ways, which are according to the raw materials, physicochemical characteristics (size, charge, number of lamellae, permeability), preparation methods, in vivo behavior, etc. In a classification according to the materials used in their preparation, NPDDS can be of lipidic nature as liposomes, micelles, Transfersomes, and solid lipid nanoparticles, or of polymeric nature as nanoparticles, micelles, niosomes (4,14,15).

LIPIDIC NANOPARTICULATE DRUG DELIVERY SYSTEMS

Liposomes

Liposomes are colloidal particles made of phospholipids, organized in bilayers that can incorporate different kind of substances, independent of MW, solubility, or electric charge. Liposomes, first prepared by Bangham et al. (16), are the most commonly studied NPDDS used for the incorporation of different kind of bioactive agents in general (low MW drugs and macromolecules). In the last decades, they gave rise to several approved and commercial formulations (14,17). In particular, liposomes are good systems for the stabilization of enzymes of different characteristics and solubility. They have inner aqueous spaces, in which hydrophilic enzymes can be solubilized, and lipid bilayers, in which hydrophobic enzymes can be accommodated while preserving their structure and conformation (3,18,19). There are different kinds of liposomes with respect to lipid composition, number of bilayers, size, charge, and preparation methods (20). The size of liposomes can range from few nanometers to several microns. This characteristic, together with lipid composition, is crucial to determine their in vivo behavior (14,20). In general, it can be said that stability and circulation time of conventional liposomes increase with size reduction (21). Liposomes with long-circulating characteristics can be obtained by the inclusion of certain lipids and polymers in their composition (22). Long-circulation and small-sized liposomes play an important role in the delivery of liposomes to certain tissues, namely, sites of infection, inflammation, and solid tumors (14,22). Conventional liposomes of larger size can reach tissues of the mononuclear phagocyte system, primarily in the liver and the spleen (4,17,23).

Liposomes can be tailor made according to the kind of bioactive agent to be incorporated (hydrophilic, hydrophobic, and amphipathic) so as to achieve the final goal of the liposomal formulation (therapy or diagnosis) and to the target to be reached. The designs of liposomes appropriate for the incorporation of enzymes have slight differences from those of other bioactive agents, which can be crucial to preserve the complex structure of the macromolecules, provided that nonaggressive or destructive solvents or methodologies are used (3,18).

TABLE 1 Some Examples of Therapeutic Enzymes and the Indicated Treatment Application (8–13)

Enzyme	Treatment	Enzyme	Treatment
Asparaginase[a]	Acute lymphoblastic leukemia	Prolylendopeptidase	Celiac disease/gluten intolerance
Collagenase	Dermal ulcers	Streptokinase[a]	Thrombolytic agent/acute myocardial infarction/pulmonary embolism/deep venous thrombosis
Factor VII (proconvertin)	Coagulation/uncontrollable bleeding in hemophilic patients	Hyaluronidase	Stroke/tissue penetrating adjuvant
Uroquinase[a]	Thrombolytic agent/blood clots	Uricase[a]	Gout
Rhodanase	Detoxifies cyanide	Ribonuclease	Antiviral
β-Lactamase	Antibiotic (allergy to penicillin)		
Trypsin (pancreatic form)	Inflammation	Trypsin (microbial form)	Thrombolytic agent/blood clots
Superoxide dismutase	Inflammation/oxidative stress	Catalase	Inflammation/oxidative stress
β-Glucocerebrosidase[a]	Gaucher disease	Galactosidase[a]	Fabry disease
DNAzymes	Ablation of gene target: Egr-1: balloon injury, restenosis, tumor growth (breast carcinoma) TGF-β: glomerulonephritis c-jun: neovascularization (cornea), tumor growth (melanoma, squamous cell carcinoma), inflammatory disease VEGFR-2: tumor growth (breast cancer) TNF-α: myocardial infarction VDUP1: myocardial ischemia PAI-1: myocardial infarction	Enzybiotics (lysins: endo-β-*N* acetylglucosaminidases, acetylmuramidases (lysozyme), amidases)	Bacterial infections – catalyzes hydrolysis of bacterial cell

[a] Approved for clinical use.

Native Hydrophilic Enzymes and Liposomes

Although several proteins, such as interleukins, insulin, and albumins, were successfully encapsulated in liposomes, only few attempts were reported concerning the enzyme encapsulation. Examples are therapeutic thrombolytic enzymes such as streptokinase and β-glucuronidase. The encapsulation of streptokinase offers a potential method of improved fibrinolytic treatment of clot-based disorders (24). For β-glucuronidase, used as a model protein, the feasibility of preparing an enzyme-loaded liposomal formulation for pulmonary delivery was successfully evaluated (25). The antioxidant enzyme catalase (CAT) was successfully encapsulated in different liposomal formulations but with limited therapeutic activity when used individually (26,27). Examples of enzymes extensively studied in terms of encapsulation parameters, pharmacokinetics, biodistribution, and therapeutic activity are L-asparaginase (21,28) and superoxide dismutase (SOD) (29,30). L-Asparaginase was approved for clinical use (Table 1) both in free form and in covalently bound to poly(ethylene glycol) (PEG) form, whereas SOD is no longer approved owing to its bovine origin.

There are several procedures regarding the association of enzymes in liposomes. Multilamellar vesicles may be obtained by thin-film hydration, dehydration–rehydration method, reverse-phase evaporation, detergent dialysis, and ethanol injection. The achievement of unilamellar vesicles may result from sonication, extrusion, and high-pressure homogenization (HPH) of a multilamellar dispersion. When dealing with enzymes, several precautions must be taken. These include use of organic solvents, type of phospholipids, ultrasound, and heat production, which may result in the loss of enzymatic activity and consequently in the reduction of the biological therapeutic effect.

Reverse-phase evaporation, which consists in emulsifying a lipid-containing water-immiscible solvent (chloroform, ether) with an aqueous phase, has the drawback of requiring organic solvents, which is not appropriate when the incorporation of enzymes is involved (31).

Detergent dialysis consists in mixing a solubilized dispersion of phospholipids with a solution containing the enzyme to be incorporated. The solubilization is achieved with detergents, leading to the formation of mixed micelles. The encapsulation efficiency of hydrophilic molecules is very low (20).

The dehydration–rehydration vesicle (DRV) appears as the method of choice to obtain high encapsulation efficiencies and preservation of enzymatic activity. In this method, dried lipids are homogenized with an aqueous solution containing the enzyme to be encapsulated, frozen, and lyophilized. The lyophilized powder is then hydrated in one-tenth of the starting volume of the liposome dispersion, gently stirred, and completed with the rest of the volume after a hydration step (21,32). This method has been used with success for the encapsulation of L-asparaginase, SOD (21,29,30), and hydrophobic forms of these enzymes with slight modifications (19,33,34).

The fate of liposomes in vivo after intravenous administration is dependent on several factors, namely, lipid composition, surface charge, steric effect, fluidity of the lipid bilayer, and mean size of liposomes. The size plays an important role in the in vivo profile of liposomes (i.e. size reduction being related to long residence time). Several techniques were assayed to reduce the size of liposomes, namely, sonication, HPH, and extrusion. Sonication is a mechanical method in which liposomal suspension is subjected to ultrasound by using either sonication probe or sonicator bath. The use of sonicator probes may present some drawbacks such as heat production, degradation of lipids, aerosol generation, and the presence of titanium particles originating from the probe leading to the contamination of the liposomal formulation (20). This method is not appropriate for the encapsulation of enzymes in liposomes, as it results in low loadings (around 2%) and a 50% loss of the catalytic activity (35). The French press cell reduces the particle size of liposomes by forcing them to pass through a small orifice under high pressure (36). Although the reproducibility, lower leakage of vesicle contents, and ease of preparing liposomes compare favorably with the sonication technique, the temperature of this process must be carefully controlled, as heat resulting from the extrusions may damage lipids or associated enzymes (31). The best method to reduce the size of liposomes while preserving enzymatic activity is achieved by filtering the suspension through polycarbonate membranes with defined pore sizes (range, 5–0.03 μm). It yields the best vesicles with respect to size homogeneity and is suitable for the preparation of liposomes in a scale ranging from one to hundreds of milliliters. Different enzymes incorporated in liposomes were sized by this procedure without loss of enzymatic activity (6,19,28,30).

Chemically Modified Enzymes and Liposomes

The formulation of hydrophilic therapeutic enzymes in liposomes is not restricted to the encapsulation or retention of the macromolecules into the inner aqueous space of the vesicles (Fig. 1) (19,34,37). Enzymes can be bound to the liposome surface, building an enzymosome [Fig. 1(B) and 1(C)] (a liposome that expresses catalytic activity in the intact form, which means before the disruption of the vesicle). The binding of enzymes to liposomes outer surface can be done by two main approaches: 1) by linking the enzyme with functional hydrophobic anchors, such as long-chain fatty acids, or 2) by directly linking the enzyme to some of the phospholipids of the liposome bilayer (11). In the former, the enzyme conjugate is incorporated into the liposomal membrane during liposome formation. In the latter, the anchor is included in the liposome bilayer and the coupling reaction occurs on the liposome surface. In both cases, owing to the complexity and structural diversity of the enzyme molecules, each process must be optimized to both preserve the enzyme function and get an appropriate enzyme load into the liposomal bilayer. The main differences between the two approaches are as follows: the number of enzyme molecules exposed to the outer bilayer of the enzymosome, the stability of the enzyme–liposome conjugation, the accessibility to the active site, and the characteristics of the modified enzyme, as the molecules bound to the enzyme are considerably different, namely, long-chain fatty acids, phospholipids, or polymer chains linked to phospholipids. The selection of the approach to be used has to be performed according to each case of therapeutic enzyme delivery mediated by enzymosomes.

Acylation of Enzymes to Promote Hydrophobic Interaction with Liposomes

The conjugation of a hydrophilic enzyme to acyl chains (Ac-enzyme) switches the affinity of the enzyme from hydrophilic to hydrophobic microenvironments (11,38). The level of hydrophobicity of the Ac-enzyme is modulated by the number and/or the length of fatty chains linked to the enzyme surface. The preservation of other properties of the modified enzyme is dependent on suitable strategies during conjugation. An example is the case of Ac-L-asparaginase, which preserves 100% of the catalytic activity if the active site is blocked with the substrate during conjugation (38,39). To maximize the load of such an Ac-enzyme into a liposome structure,

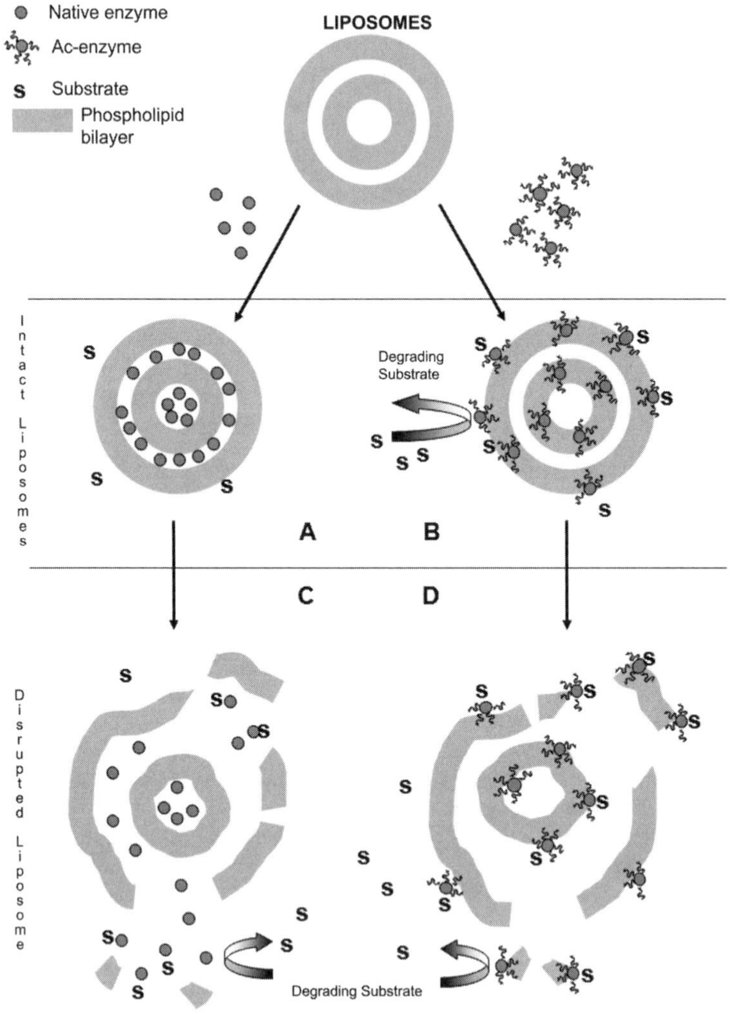

FIGURE 1 Schematic of representation of enzyme and Ac-enzyme. Shown here is the localization both in the internal aqueous space (A) and in the lipid bilayer (B) of liposomes and the corresponding release (C and D). In case A, the enzyme is not available for substrate degradation when the liposome is intact; in case B, the enzyme is available even before liposome disruption.

appropriate strategies are needed. The Ac-enzyme can be partially inserted into the liposome bilayer or buried into the hydrophobic lipid matrix of the vesicles, which depends on the number and localization of the hydrophobic chains linked to the enzyme surface. Consequently, to incorporate efficiently an Ac-enzyme into the bilayer of DRV liposomes, an additional step was used to improve the contact between the fragments of the dehydrated lipid bilayers and the Ac-enzyme. The new procedure was developed to incorporate the bioconjugate Ac-L-asparaginase into liposomes (40). In brief, a process in which empty DRV liposomes are formed and the Ac-L-asparaginase is added as a dry solid (lyophilized) to the liposomes before the lyophilization stage. After the extrusion step used to reduce the size of the enzymosomes, any Ac-enzyme not incorporated is removed by gradient centrifugation.

These methodologies of conjugation were also successfully used to convert the hydrophilic enzyme SOD into Ac-SOD. The decrease of affinity of this bioconjugate for water as a function of the acyl chain length and the effect of the chemical modification on the charge and hydrophobicity of Ac-SOD were evaluated in comparison with Ac-L-asparaginase (33). The Ac-enzyme–liposomal bilayer association depends on the overall electrostatic interactions between the enzyme-associated charges and on the hydrophobic interactions. The incorporation of Ac-enzymes into the bilayer of liposomes is efficiently evaluated by the ratio between the catalytic activity quantified in intact versus disrupted enzymosomes. Significant enzyme activity was found in intact enzymosomes [Fig. 1(B)]. No significant activity was found in intact liposomes loaded with the native enzyme [Fig. 1(A)] (41). The characteristics of liposomes, such as ionic charge, vesicle size, composition, and PEG-coated vesicle bilayers, play an important role in the incorporation of Ac-SOD into liposomes for Ac-SOD enzymosome construction (34). Relevant points for the preparation of either SOD or Ac-SOD enzymosomes are described in detail elsewhere (18).

Chemical Link of Enzymes Directly to Liposome Surface

As mentioned before, the other approach to build enzymosomes is by directly linking the hydrophilic enzyme to lipids of the liposome bilayer. The direct conjugation of therapeutic enzymes to the outer surface of lipid vesicles remains a challenge, as few publications report the construction of liposomes with surface-attached enzymes. In contrast, many publications report the attachment of antibodies to the liposome surface, a concept widely used for the active targeting of liposomes (14,37). SOD enzymosomes were also obtained by covalent linkage of the therapeutic enzyme directly to the outer surface of the phospholipids bilayer. Two different processes used to link proteins to liposomes were optimized to minimize alterations of the activity of SOD. The conjugation of the thiolated enzyme, SOD-AT, directly to *N*-[4-(*p*-maleimidophenyl)butyrate]phosphatidylethanolamine–reactive groups located at the liposome outer surface was performed by using different liposome compositions and for several degrees of SOD thiolation. The conjugation of SOD to the outer surface of liposomes was also performed by coupling the exposed ε-NH_2 groups of SOD by means of a carbodiimide to the linker lipid *N*-glutaryl phosphatidylethanolamine previously incorporated into liposomes. The efficiency of both enzyme–liposome link procedures was evaluated (42). Long-circulation time SOD enzymosomes were also obtained by the direct conjugation of the thioacetylated enzyme, SOD-ATA, with the maleimide-reactive group located at the terminus of the polymer phosphoethanolamine-*N*-[maleimide (polyethylene glycol)-2000] (maleimide-PEG-PE). A study on the effect of the percentage of maleimide-PEG-PE in the conjugation parameters was conducted. The total PEG chains (reactive and nonreactive) were constant for all the cases. The percentage of maleimide-PEG-PE was corrected by an equivalent percentage of PEG-2000 [methoxy(polyethylene glycol)-2000] (PEG-PE) in order to keep the total number of PEGylated chains constant. A suitable enzyme load, keeping the vesicle structural integrity and preserving the enzyme activity, was achieved (43).

Solid Lipid Nanoparticles

Solid lipid nanoparticles (SLNs) are particles made from solid lipids, with a mean diameter in the range from 50 to 1000 nm, and described in the mid 1990s (44). They combine the advantages of other carrier systems, especially regarding lipophilic drug incorporation and parenteral administration. They represent an alternative to polymeric particulate systems and are considered alternative carriers for peptides, proteins, and antigens. For macromolecules with hydrophilic nature, it is not expected to obtain high encapsulation efficiency into the hydrophobic matrix of SLNs. SLNs are relatively recent NPDDS; nevertheless, a recent publication presents an exhaustive list of peptide and protein molecules incorporated in lipid microparticles and nanoparticles (44), such as peptides (calcitonin, cyclosporine A, insulin, luteinizing hormone-releasing hormone, somatostatin), protein antigens (HB and malaria), and model proteins (bovine serum albumin, lysozyme). However, no results on therapeutic enzyme incorporation in SLNs have been reported so far. SLNs exhibit physical stability, protection of incorporated labile drugs from degradation, controlled release, excellent tolerability, and site-specific targeting (45). These colloidal systems are made from solid lipids (highly purified triglycerides, complex glyceride mixtures, or waxes) and stabilized by surfactant(s). There is no need for potentially toxic organic solvents for their production, which is important in protein formulation. The solid lipid core of SLNs should increase the chemical stability of the incorporation of macromolecules and protect them from degradation. Recently, SLNs based on a mixture of solid lipids and liquid lipids, high amounts of lecithins, amphipathic cyclodextrins, and *p*-acyl-calixarenes have been introduced and studied (46).

Proteins and other macromolecules can be incorporated in SLNs by two main production techniques: HPH (at elevated temperature—hot HPH technique, or at or below room temperature—cold HPH technique) (47,48) and microemulsion (49). Other procedures were described: solvent emulsification/evaporation method (50) or emulsification/diffusion technique (51), water/oil/water double-emulsion method, or high-speed stirring and/or ultrasonication technique. Supercritical fluid technology has recently been used to prepare lipid particles. Among these, loading onto preformed lipid nanoparticles by sorption procedures has also been introduced. As in other NPDDS, each protein should be considered as a special case, and thus the lipid mixture and the technique employed should be carefully studied (44). In spite of lack of release mechanism knowledge and kinetic characterization, the prolonged in vitro release, and subsequent in vivo sustained effect of various proteins are described (46). Controlled drug release for peptides and protein-loaded SLNs, especially for oral drug delivery (which is not the first choice of administration route for these types of drugs), has been shown. Garcia-Fuentes et al. studied the interactions of surface-modified lipid nanoparticles loaded with salmon calcitonin with Caco-2 cells, as well as evaluated the potential of these nanostructures as oral delivery systems for salmon calcitonin and suggested their potential as carriers for oral peptide delivery (52). To increase the cyclosporine oral bioavailability, SLNs have been used (53) and a low variation in bioavailability of the drug was achieved with better blood profile compared with the commercial formulation (54). Insulin-loaded SLNs were developed with both high encapsulation efficiency and good stability characteristics, providing interesting possibilities as delivery systems for oral administration (55).

POLYMERIC NPDDS

Synthetic biodegradable polymers have been used for the preparation of polymeric nanoparticles for drug delivery, in particular for enzymes (56,57).

Polymeric nanoparticles are obtained by different processes based on two main approaches: polymerization reactions and the use of preformed polymers (56,57). The term "polymeric nanoparticle" encompasses nanospheres and nanocapsules. Nanospheres are defined as a polymeric matrix in which the drug is uniformly dispersed and nanocapsules are described as a polymeric membrane that surrounds the drug in the matrix core (58).

Some of the most used polymers for drug delivery systems approved for human use are the poly(lactic acid) (59), poly(lactic-co-glycolic acid) (PLGA) (60,61) and poly(ε-caprolactone) (62,63), poly(alkyl cyanocrylate) (PACA) (64,65). In addition, to avoid toxicological problems associated with synthetic polymers (66), there is a growing list of natural polymers, including chitosan (67), gelatin (68) and sodium alginate (69), for the preparation of NPDDS.

PACAs are synthetic, biologically degradable polymers (70). Their nanoparticles are easily obtained by an emulsion polymerization process developed by Couvreur (64). Only a few macromolecules, antibodies and enzymes, were incorporated in PACA nanoparticles. Owing to the structural complexity of enzymes, for their incorporation in nanoparticles, both the interaction of the enzyme with the components of the emulsion polymerization system and the effect of the process of polymerization on the characteristics of the enzyme must be taken into account. Mild conditions are required, and each process must be optimized for each enzyme to maximize the enzyme load and minimize the loss of catalytic activity. The more obvious advantage of the emulsion polymerization is the absence of organic solvents. Limitative parameters are the low pH and high reactivity of the monomer. Conformational changes of the enzyme with consequent partial inactivation or strong modification of the kinetics are the main drawbacks. The effect of increasing the pH after initialization of polymerization both on the characteristics of the enzyme SOD-loaded poly(isobutyl cyanocrylate) (PIBCA) nanoparticles and on the reduction of enzyme activity was reported (71). The effect of process parameters of the isobutyl cyanoacrylate (IBCA) emulsion polymerization on the characteristics of L-asparaginase and SOD-loaded PIBCA nanoparticles were reported (72). The incorporation of enzymes in PIBCA nanoparticles seems to be strongly dependent on the enzyme structure according to results obtained with L-asparaginase and SOD, macromolecules with very different three-dimensional structures. In brief, the monomer was added under stirring to the polymerization medium in which an amount of enzyme was added. In some cases, enzyme substrate was also added to the polymerization medium. The incorporation of the L-asparaginase in PIBCA nanoparticles has low efficiency in comparison with other nanocarriers. However, the incorporation of SOD in PIBCA nanoparticles can be compared with the efficiencies of incorporation observed for this enzyme incorporated in liposomes (30). In conclusion, the incorporation of enzymes in PIBCA nanoparticles imposes a number of constraints in process parameters, such as enzyme MW, structure, concentration, pH, presence of other molecules such as enzyme substrate, monomer concentration, stabilizers, or ionic strength, all of which may affect physicochemical properties of the nanoparticles formed.

When using preformed polymers such as PLGA for the preparation of nanoparticles, mainly three methods are used: water-oil-water (w/o/w) double-emulsion technique, phase-separation method, and spray drying (73). In the double-emulsion method, enzymes in the aqueous solvent were emulsified with nonmiscible organic solution of the polymer to form a w/o emulsion. The organic solvent dichloromethane was mainly used and the homogenization step was carried out by using either high-speed homogenizers or sonicators. This emulsion was then rapidly transferred to an excess of an aqueous medium, containing a stabilizer, usually PVA. A homogenization step or intensive stirring is necessary to form a double emulsion of w/o/w. Then, the removal of organic solvent by heating and vacuum evaporation is done by either extracting organic solvent or adding a nonsolvent (i.e., silicone oil), thereby inducing coacervation. The first process is designated as w/o/w, whereas the second is known as the phase-separation technique. In both cases, nanoparticles occur in the liquid phase. In the spray-drying technique, particle formation is achieved by atomizing the emulsion into a stream of hot air under vigorous solvent evaporation.

Enzymes encapsulated into nanoparticles by w/o or w/o/w techniques are susceptible to denaturation, aggregation, oxidation, and cleavage, especially at the aqueous phase–solvent interface. Enzyme denaturation may also result in a loss of biological activity. Improved enzymatic activity has been achieved by the addition of stabilizers such as carrier proteins (e.g., albumin), surfactants during the primary emulsion phase, or molecules such as trehalose and mannitol to the protein phase.

L-Asparaginase was efficiently encapsulated within PLGA nanoparticles by using a double-emulsion technique. The nanospheres obtained could continuously release the enzyme while preserving the enzymatic activity (74). The encapsulation efficiency, kinetics, and activity of the enzyme released were dependent on the MW and hydrophilicity of the copolymer used. Best results were achieved for nanoparticles based on PLGA with carboxyl end groups. These results were attributed to a favorable interaction of the enzyme with this specific copolymer (74,75).

DEFORMABLE NPDDS: DERMAL AND TRANSDERMAL DELIVERY

Currently, the parental route is considered the major route for the clinical administration of therapeutic macromolecules formulated in NPDDS. Transdermal drug delivery has been approved and has become widely accepted for the systemic administration of drugs. This noninvasive approach avoids the hepatic "first-pass" metabolism, maintains a steady drug concentration (extremely important both in the case of drugs with a short half-life and in the case of chronic therapy), allows the use of drugs with a low therapeutic index, and improves patient compliance. However, the outermost layer of the skin, stratum corneum (SC), prevents transdermal permeation of most drugs at clinically useful rates. Currently, the market for transdermal delivery comprises only a few drugs that have low MW, solubility in lipids, and low therapeutic doses. For charged and polar molecules or macromolecules, skin delivery is difficult and has advanced substantially within the last few years. To facilitate the delivery of such entities, a number of strategies were developed. These include the use of chemical or physical techniques to enhance molecular diffusion through the SC. Lipid-based carriers have been investigated for the dermal and transdermal delivery of many drugs, but only a few NPDDS have found drug delivery enhancement using incorporated macromolecules (76). In recent years, specially designed carriers have claimed the ability to cross the skin intact and deliver the loaded drugs into the systemic circulation, being at the same time responsible for the percutaneous absorption of the drug within the skin. To differentiate them from the more conventional carriers, liposomes are named as deformable vesicles. Deformability is the characteristic that enables these carriers to penetrate the narrow gaps between the cells of the SC and ensure the delivery of loaded or associated material. Concerning the enzymes, only mixed lipid vesicles, so-called Transfersomes (a trademark of IDEA, AG, Munich, Germany), have proved an effective transport into the skin. Transfersomes are composed of highly flexible membranes obtained by combining into single-structure phospholipids (which give structure and stability to the bilayers) and an edge-active component (to increase the bilayer flexibility) that gives them the capacity to move spontaneously against water concentration gradient in the skin. It has now been proven that intact Transfersomes, in contrast to liposomes, penetrate the skin without disruption (77). These carriers comprise at least phosphatidylcholine and an edge-active molecule acting as membrane softener. The vesicles have a typical diameter size of 150 ± 50 nm. In structural terms, Transfersomes are related to liposomes and many of the techniques for their preparation and characterization are common. For Transfersomes, a properly defined composition is responsible for membrane flexibility and consequently for vesicle deformability necessary for through-the-skin passagework. Transfersomes are much more flexible and deformable than liposomes, which are assessed by using membrane penetration assays (78).

Among the many drugs that can be incorporated in Transfersomes (79,80), including polypeptides and proteins (81–85), enzymes were also reported to be transferred into the body through the skin after incorporation in these systems. Simões et al. developed and characterized SOD- and CAT-loaded Transfersomes (78) in terms of carrier structure, the efficiency of protein association, and/or the incorporation and retention of catalytic activity after association (86). In vitro penetrability of deformable vesicles was characterized and was not affected by the incorporation of the studied enzymes (78). Successful enzyme incorporation was obtained by using other membrane-softening agents such as Tween 80, without compromising the vesicles deformability (87). Using a mouse ear edema model of inflammation, it was observed that antioxidant enzymes, such as SOD and CAT, delivered by means of ultradeformable lipid vesicles can serve as a novel, region-specific treatment option for inflammation (88). Moreover, it was shown for the first time that SOD incorporated into Transfersomes and applied onto the skin that is not necessarily close to the inflamed tissue can promote noninvasive treatment of induced arthritis. This study on transdermal transport of antioxidant enzymes contributed to an innovative approach in the field of the protein transdermal delivery (6).

Ethosomes are a special kind of unusually deformable vesicles in which the abundant ethanol makes lipid bilayers very fluid, and thus by inference soft (89). This reportedly improves the delivery of various molecules into deep skin layers (90). The results showed the preferential incorporation of hydrophobic and low MW drugs. No reports on transdermal or dermal region-specific delivery of enzymes mediated by ethosomes are available to date.

Other so-called "elastic vesicles" were found to be responsible for major morphological changes in the intercellular lipid bilayer structure in comparison with rigid vesicles (91). The structures are liquid-state vesicles made of L-595/PEG-8-L/sulfosuccinate (50/20/5) and have an average size of 100 to 120 nm. These high elastic structures served to demonstrate the presence of channel-like penetration pathways in the SC, which could be seen containing vesicular structures (92).

No results on the transdermal delivery of enzymes by using these systems were reported.

Liposomal recombinant SOD was developed for topical application for the treatment of Peyronie's disease (93). This study is one of the few reporting topical application of enzymes, while using nondeformable liposomes.

CONCLUSIONS

Owing to their selectivity and broad field of applications, enzymes are very attractive as potential therapeutic agents, providing that correct formulations are achieved. A number of NPDDS, either of lipid or polymeric nature, could fulfill this purpose. Examples in literature demonstrated that it is possible to modulate the properties of NPDDS to efficiently incorporate therapeutic enzymes without disrupting their activity and/or to reach different targets, according to each particular aim. Although proteins in general and enzymes in particular are relatively new as therapeutic agents, it is envisaged that they will play an important role in the battery of nonconventional formulations of this millennium.

ACKNOWLEDGMENT

The authors thank Dr. M. Costa Ferreira for revising the manuscript.

REFERENCES

1. Hu FQ, Hong Y, Yuan H. Preparation and characterization of solid lipid nanoparticles containing peptide. Int J Pharm 2004; 273:29–35.
2. Storm G, Koppenhagen F, Heeremans A, et al. Novel developments in liposomal delivery of peptides and proteins. J Control Release 1995; 36:19–24.
3. Walde P, Ichikawa S. Enzymes inside lipid vesicles: Preparation, reactivity and applications. Biomol Eng 2001; 18:143–177.
4. Couvreur P, Vauthier C. Nanotechnology: Intelligent design to treat complex disease. Pharm Res 2006; 23:1417–1450.
5. Gaspar MM, Bakowsky U, Ehrhardt C. Inhaled liposomes - current strategies and future challenges. J Biomed Nanotechnol 2008; 4:245–257.
6. Simoes SI, Delgado TC, Lopes RM, et al. Developments in the rat adjuvant arthritis model and its use in therapeutic evaluation of novel non-invasive treatment by SOD in Transfersomes. J Control Release 2005; 103:419–434.
7. Available at: http://www.bccresearch com/report/BIO021C.html. Accessed October 2008. Protein drugs: Global markets and manufacturing technologies.
8. Walsh G. Pharmaceutical biotechnology products approved within the European Union. Eur J Pharm Biopharm 2003; 55:3–10.
9. Marshall SA, Lazar GA, Chirino AJ, et al. Rational design and engineering of therapeutic proteins. Drug Discov Today 2003; 8:212–221.
10. Cerf-Bensussan N, Matysiak-Budnik T, Cellier C, et al. Oral proteases: A new approach to managing coeliac disease. Gut 2007; 56:157–160.
11. Verma N, Kumar K, Kaur G, et al. L-Asparaginase: A promising chemotherapeutic agent. Crit Rev Biotechnol 2007; 27:45–62.
12. Isaka Y. DNAzymes as potential therapeutic molecules. Curr Opin Mol Ther 2007; 9:132–136.
13. Veiga-Crespo P, Ageitos JM, Poza M, et al. Enzybiotics: A look to the future, recalling the past. J Pharm Sci 2007; 96:1917–1924.
14. Torchilin VP. Multifunctional nanocarriers. Adv Drug Deliv Rev 2006; 58:1532–1555.
15. Crommelin DJA, Storm G, Jiskoot W, et al. Nanotechnological approaches for the delivery of macromolecules. J Control Release 2003; 87:81–88.

16. Bangham AD, Standish MM, Watkins JC. Diffusion of univalent ions across lamellae of swollen phospholipids. J Mol Biol 1965; 13:238–252
17. Torchilin VP. Recent advances with liposomes as pharmaceutical carriers. Nat Rev Drug Discov 2005; 4:145–160.
18. Cruz MEM, Gaspar MM, Martins MBF, et al. Liposomal superoxide dismutases and their use in the treatment of experimental arthritis. Methods Enzymol 2005; 391:395–413.
19. Jorge JCS, PerezSoler R, Morais JG, et al. Liposomal palmitoyl-L-asparaginase – characterization and biological-activity. Cancer Chemother Pharmacol 1994; 34:230–234.
20. Crommelin DJA, Schreier H. Liposomes. In: Kreuter J, ed. Colloidal Drug Delivery Systems. New York: Marcel Dekker, Inc., 1994:73–190.
21. Cruz MEM, Gaspar MM, Lopes F, et al. Liposomal L-asparaginase – in-vitro evaluation. Int J Pharm 1993; 96:67–77.
22. Storm G, Woodle MC. Long circulating liposome therapeutics: From concept to clinical reality. In: Woodle MC, Storm G, eds. Long Circulating Liposomes: Old Drugs, New Therapeutics. New York: Springer-Verlag, Landes Bioscience, 1998:1–12.
23. Gaspar MM, Neves S, Portaels F, et al. Therapeutic efficacy of liposomal rifabutin in a *Mycobacterium avium* model of infection. Antimicrob Agents Chemother 2000; 44:2424–2430.
24. Leach JK, O'Rear EA, Patterson E, et al. Accelerated thrombolysis in a rabbit model of carotid artery thrombosis with liposome-encapsulated and microencapsulated streptokinase. Thromb Haemost 2003; 90:64–70.
25. Lu D, Hickey AJ. Liposomal dry powders as aerosols for pulmonary delivery of proteins. AAPS PharmSciTech 2005; 6:E641–E648.
26. Jubeh TT, Nadler-Milbauer M, Barenholz Y, et al. Local treatment of experimental colitis in the rat by negatively charged liposomes of catalase, TMN and SOD. J Drug Target 2006; 14:155–163.
27. Ledwozyw A. Protective effect of liposome-entrapped superoxide dismutase and catalase on bleomycin-induced lung injury in rats; part I: Antioxidant enzyme activities and lipid peroxidation. Acta Vet Hung 1991; 39:3–4.
28. Gaspar MM, PerezSoler R, Cruz MEM. Biological characterization of L-asparaginase liposomal formulations. Cancer Chemother Pharmacol 1996; 38:373–377.
29. Corvo ML, Jorge JCS, van't Hof R, et al. Superoxide dismutase entrapped in long-circulating liposomes: Formulation design and therapeutic activity in rat adjuvant arthritis. Biochim Biophys Acta 2002; 1564:227–236.
30. Corvo ML, Martins MB, Francisco AP, et al. Liposomal formulations of Cu,Zn-superoxide dismutase: Physicochemical characterization and activity assessment in an inflammation model. J Control Release 1997; 43:1–8.
31. Betageri GV, Jenkins SA, Parsons DL. Preparation of liposomes. In: Betageri GV, Jenkins SA, Parsons DL, eds. Liposome Drug Delivery Systems. Basel, Switzerland: Technomic Publishing Company, Inc., 1993:1–26.
32. Deamer DW, Barchfeld GL. Encapsulation of macromolecules by lipid vesicles under simulated prebiotic conditions. J Mol Evol 1982; 18:203–206.
33. Martins MBF, Cruz MEM. Characterization of bioconjugates of L-asparaginase and Cu,Zn-superoxide dismutase. Proceedings of the Third European Symposium on Controlled Drug Delivery; University of Twente, Noodwijk aan Zee, The Netherlands; April 6–8, 1994.
34. Gaspar MM, Martins MB, Corvo ML, et al. Design and characterization of enzymosomes with surface-exposed superoxide dismutase. Biochim Biophys Acta 2003; 1609: 211–217.
35. Cruz MEM, Corvo ML, Jorge JS, et al. Liposomes as carrier systems for proteins: Factors affecting protein encapsulation. In: Lopez-Berestein G, Fidler I, eds. Liposomes in the Therapy of Infectious Diseases and Cancer. New York: Alan R. Liss, Inc., 1989: 417–426.
36. Lelkes PI. The use of French pressed vesicles for efficient incorporation of bioactive macromolecules and as drug carriers in vitro and in vivo. In: Gregoriadis G, ed. Liposome Technology. Boca Raton, FL: CRC Press, 1984:51–65.

37. Torchilin VP, Weissig V. Liposomes a Practical Approach. New York: Oxford University Press Inc., 2003.
38. Martins MBF, Jorge JCS, Cruz MEM. Acylation of L-asparaginase with total retention of enzymatic activity. Biochimie 1990; 72:671–675.
39. Martins MBAF, Goncalves APV, Cruz MEM. Biochemical characterization of an L-asparaginase bioconjugate. Bioconjug Chem 1996; 7:430–435.
40. Cruz MEM, Jorge JC, Martins MBF, et al. Liposomal compositions and processes for their production. European patent 0485143A1, 1991.
41. Martins MBF, Corvo ML, Cruz MEM. Lipophilic derivatives of SOD: characterization and immobilization in liposomes. Proceedings of the 19th International Symposium on Controlled Release of Bioactive Materials; Controlled Release Society, Orlando, FL; July 26–31, 1992.
42. Vale CA, Corvo ML, Martins LCD, et al. Construction of Enzymosomes: Optimization of coupling parameters, NANOTECH\'06; Nano Science and Technology Institute, Boston, MA; May 7–11, 2006.
43. Vale CA, Corvo ML, Martins LCD, et al. Enzymosomes: An innovative approach for therapeutic enzyme delivery. Proceedings of PEGS2008: The Protein Engineering Summit; Cambridge Healthtech Institute, Boston, MA; April 28–May 02, 2008.
44. Almeida AJ, Souto E. Solid lipid nanoparticles as a drug delivery system for peptides and proteins. Adv Drug Deliv Rev 2007; 59:478–490.
45. Rawat M, Singh D, Saraf S, et al. Lipid carriers: A versatile delivery vehicle for proteins and peptides. J Pharm Soc Jpn 2008; 128:269–280.
46. Joshi MD, Muller RH. Lipid nanoparticles for parenteral delivery of actives. Eur J Pharmaceut Biopharmaceut 2009; 71:161–172.
47. Jahnke S. The theory of high pressure homogenization. In: Muller RH, Benita S, Bohm B, eds. Emulsions and Nanosuspensions for the Formulation of Poorly Soluble Drugs. Stuttgart: Medpharm Scientific Publishers, 1998:177–200.
48. Mehnert W, Mader K. Solid lipid nanoparticles – Production, characterization and applications. Adv Drug Deliv Rev 2001; 47:165–196.
49. Gasco MR. Method for producing solid lipid microspheres having a narrow size distribution. Patent 5250236, 1993.
50. Sjostrom B, Bergenstahl B. Preparation of submicron drug particles in lecithin-stabilized o/w emulsions; part 1: Model studies of the precipitation of cholesteryl acetate. Int J Pharm 1992; 84:107–116.
51. Trotta M, Debernardi F, Caputo O. Preparation of solid lipid nanoparticles by a solvent emulsification-diffusion technique. Int J Pharm 2003; 257:153–160.
52. Garcia-Fuentes M, Prego C, Torres D, et al. A comparative study of the potential of solid triglyceride nanostructures coated with chitosan or poly(ethylene glycol) as carriers for oral calcitonin delivery. Eur J Pharm Sci 2005; 25:133–143.
53. Bekerman T, Golenser J, Domb A. Cyclosporin nanoparticulate lipospheres for oral administration. J Pharm Sci 2004; 93:1264–1270.
54. Muller RH, Runge S, Ravelli V, et al. Oral bioavailability of cyclosporine: Solid lipid nanoparticles (SLN (R)) versus drug nanocrystals. Int J Pharm 2006; 317:82–89.
55. Battaglia L, Trotta M, Gallarate M, et al. Solid lipid nanoparticles formed by solvent-in-water emulsion-diffusion technique: Development and influence on insulin stability. J Microencapsul 2007; 24:672–684.
56. Soppimath KS, Aminabhavi TM, Kulkarni AR, et al. Biodegradable polymeric nanoparticles as drug delivery devices. J Control Release 2001; 70:1–20.
57. Kreuter J. Nanoparticles. In: Kreuter J, ed. Colloidal Drug Delivery Systems. New York: Marcel Dekker, Inc., 1994:219–342.
58. Anton N, Benoit JP, Saulnier P. Design and production of nanoparticles formulated from nano-emulsion templates – A review. J Control Release 2008; 128:185–199.
59. Guiziou B, Armstrong DJ, Elliott PNC, et al. Investigation of in-vitro release characteristics of NSAID-loaded polylactic acid microspheres. J Microencapsul 1996; 13:701–708.
60. Aguiar MMG, Rodrigues JM, Cunha AS. Encapsulation of insulin-cyclodextrin complex in PLGA microspheres: A new approach for prolonged pulmonary insulin delivery. J Microencapsul 2004; 21:553–564.

61. Bilati U, Allemann E, Doelker E. Poly(D,L-lactide-co-glycolide) protein-loaded nanoparticles prepared by the double emulsion method-processing and formulation issues for enhanced entrapment efficiency. J Microencapsul 2005; 22:205–214.
62. Le Ray AM, Chiffoleau S, Iooss P, et al. Vancomycin encapsulation in biodegradable poly(epsilon-caprolactone) microparticles for bone implantation. Influence of the formulation process on size, drug loading, in vitro release and cytocompatibility. Biomaterials 2003; 24:443–449.
63. Kim BK, Hwang SJ, Park JB, et al. Characteristics of felodipine-located poly(epsilon-caprolactone) microspheres. J Microencapsul 2005; 22:193–203.
64. Couvreur P. Polyalkylcyanoacrylates as colloidal drug carriers. CRC Crit Rev Ther Drug Carrier Syst 1988; 5:1–20.
65. Fattal E, Vauthier C, Aynie I, et al. Biodegradable polyalkylcyanoacrylate nanoparticles for the delivery of oligonucleotides. J Control Release 1998; 53:137–143.
66. Feng SS, Mu L, Win KY, et al. Nanoparticles of biodegradable polymers for clinical administration of paclitaxel. Curr Med Chem 2004; 11:413–424.
67. Fernandez-Urrusuno R, Romani D, Calvo P, et al. Development of a freeze-dried formulation of insulin-loaded chitosan nanoparticles intended for nasal administration. STP Pharm Sci 1999; 9:429–436.
68. Farrugia CA, Groves MJ. Gelatin behaviour in dilute aqueous solution: Designing a nanoparticulate formulation. J Pharm Pharmacol 1999; 51:643–649.
69. Tonnesen HH, Karlsen J. Alginate in drug delivery systems. Drug Dev Ind Pharm 2002; 28:621–630.
70. Toub N, Malvy C, Fattal E, et al. Innovative nanotechnologies for the delivery of oligonucleotides and siRNA. Biomed Pharmacother 2006; 60:607–620.
71. Martins MBF, Simoes SID, Cruz MEM, et al. Development of enzyme-loaded nanoparticles: Effect of pH. J Mater Sci Mat Med 1996; 7:413–414.
72. Martins MBF, Simoes SID, Supico A, et al. Enzyme-loaded PIBCA nanoparticles (SOD and L-ASNase): Optimization and characterization. Int J Pharm 1996; 142:75–84.
73. Freitas S, Merkle HP, Gander B. Microencapsulation by solvent extraction/evaporation: Reviewing the state of the art of microsphere preparation process technology. J Control Release 2005; 102:313–332.
74. Gaspar MM, Blanco D, Cruz MEM, et al. Formulation of L-asparaginase-loaded poly(lactide-co-glycolide) nanoparticles: Influence of polymer properties on enzyme loading, activity and in vitro release. J Control Release 1998; 52:53–62.
75. Blanco MD, Alonso MJ. Development and characterization of protein-loaded poly(lactide-co-glycolide) nanospheres. Eur J Pharm Biopharm 1997; 43:287–294.
76. Cevc G. Lipid vesicles and other colloids as drug carriers on the skin. Adv Drug Deliv Rev 2004; 56:675–711.
77. Cevc G, Schatzlein A, Richardsen H. Ultradeformable lipid vesicles can penetrate the skin and other semi-permeable barriers unfragmented. Evidence from double label CLSM experiments and direct size measurements. Biochim Biophys Acta 2002; 1564:21–30.
78. Simões SID. Transdermal delivery of theraputic enzymes. Ph.D. Thesis, 2005 (Chapter 4), University of Lisbon, Portugal.
79. Cevc G, Blume G. Biological activity and characteristics of triamcinolone-acetonide formulated with the self-regulating drug carriers, Transfersomes (R). Biochim Biophys Acta 2003; 1614:156–164.
80. Rother M, Lavins BJ, Kneer W, et al. Efficacy and safety of epicutaneous ketoprofen in transfersome (IDEA-033) versus oral celecoxib and placebo in osteoarthritis of the knee: Multicentre randomised controlled trial. Ann Rheum Dis 2007; 66:1178–1183.
81. Cevc G, Gebauer D, Stieber J, et al. Ultraflexible vesicles, Transfersomes, have an extremely low pore penetration resistance and transport therapeutic amounts of insulin across the intact mammalian skin. Biochim Biophys Acta 1998; 1368:201–215.
82. Paul A, Cevc G. Non-invasive administration of protein antigens. Epicutaneous immunization with the bovine serum albumin. Vaccine Res 1995; 4:145–164.
83. Paul A, Cevc G, Bachhawat BK. Transdermal immunization with large proteins by means of ultradeformable drug carriers. Eur J Immunol 1995; 25(12):3521–3524.

84. Paul A, Cevc G, Bachhawat BK. Transdermal immunisation with an integral membrane component, gap junction protein, by means of ultradeformable drug carriers, Transfersomes. Vaccine 1998; 16:188–195.
85. Hofer C, Gobel R, Deering P, et al. Formulation of interleukin-2 and interferon-alpha containing ultradeformable carriers for potential transdermal application. Anticancer Res 1999; 19:1505–1507.
86. Simoes SI, Marques CM, Cruz MEM, et al. The effect of cholate on solubilisation and permeability of simple and protein-loaded phosphatidylcholine/sodium cholate mixed aggregates designed to mediate transdermal delivery of macromolecules. Eur J Pharm Biopharm 2004; 58:509–519.
87. Simoes SI, Tapadas JM, Marques CM, et al. Permeabilisation and solubilisation of soybean phosphatidylcholine bilayer vesicles, as membrane models, by polysorbate, Tween 80. Eur J Pharm Sci 2005; 26:307–317.
88. Simoes S, Marques C, Cruz ME, et al. Anti-inflammatory effects of locally applied enzyme-loaded ultradeformable vesicles on an acute cutaneous model. J Microencapsulation 2008; 17:1–10.
89. Touitou E, Dayan N, Bergelson L, et al. Ethosomes – Novel vesicular carriers for enhanced delivery: Characterization and skin penetration properties. J Control Release 2000; 65:403–418.
90. Horwitz E, Pisanty S, Czerninski R, et al. A clinical evaluation of a novel liposomal carrier for acyclovir in the topical treatment of recurrent herpes labialis. Oral Surg Oral Med Oral Pathol Oral Radiol Endod 1999; 87:700–705.
91. van den Bergh BAI, Vroom J, Gerritsen H, et al. Interactions of elastic and rigid vesicles with human skin in vitro: Electron microscopy and two-photon excitation microscopy. Biochim Biophys Acta 1999; 1461:155–173.
92. Honeywell-Nguyen PL, de Graaff AM, Groenink HWW, et al. The in vivo and in vitro interactions of elastic and rigid Vesicles with human skin. Biochim Biophys Acta 2002; 1573:130–140.
93. Riedl CR, Sternig P, Galle G, et al. Liposomal recombinant human superoxide dismutase for the treatment of Peyronie's disease: A randomized placebo-controlled double-blind prospective clinical study. Eur Urol 2005; 48:656–661.

4 Formulation of NPDDS for Gene Delivery

Ajoy Koomer

Department of Pharmaceutical Sciences, Sullivan University College of Pharmacy, Louisville, Kentucky, U.S.A.

INTRODUCTION

The full potential of gene therapy in medicine is being realized today. It is defined as the use of genes or genetic material (DNA, RNA, oligonucleotides) to treat a disease state, generally a genetic-based disease (1–3). Genes are introduced into cells or tissues either to inhibit undesirable gene expression or to express therapeutic proteins (3). Target disease states for gene delivery can be broadly categorized into two major classes: inherited and acquired. Inherited disease states are limited to sickle cell anemia, hemophilia, cystic fibrosis, Huntington's disease, and errors of metabolism; acquired diseases include cancer, HIV infection, and diabetes. The two standard procedures used in gene delivery are addition/replacement and ablation (4). While performing the former, a normal gene is introduced into the cell type to replace activity of the defective gene (4). On the contrary, ablation deals with destroying undesired cells, as in cancer.

Whether one performs ex vivo or in vivo gene therapy, important focal points are duration of expression of the gene or therapeutic protein and specificity in delivering the gene to the site of action with minimal adverse effects (1–3). Currently, genes packaged in viral vectors, such as retrovirus, adenovirus, adeno-associated virus, and herpes simplex virus, remain the leading therapeutic candidates for gene therapy, as they have produced functional improvements in several animal models of previously mentioned genetic diseased states. However, because of the risk factors (pathogenicity, immunogenicity) associated with viral vectors, a major emphasis has been placed on the formulation of nanoparticulate drug delivery vehicles for gene delivery (3).

The term "nanometer" in the metric scale of linear measurement refers to one-billionth of a meter. According to National Nanotechnology Initiative, nanotechnology is defined as research and technology resulting in "the controlled creation and usage" of unique small particles, varying from 1 to 100 nm in length. Looking at the biological systems, it is evident that they are composed of inherent "nanoblocks." While the width of a DNA molecule is 2.5 nm, the dimension of most proteins fall in the range 1 from 15 nm and the width of a typical mammalian cell membrane is around 6 nm. Thus, operating in a "nanoscale domain" at the biomolecular range, nanomaterials offer a "wide range of molecular biology related applications, including fluorescent biological labels, drug and gene delivery, probing of DNA structures and tissue engineering" (1,3,5,6). This chapter focuses on the formulation of nanoparticulate drug delivery systems for gene delivery.

The most common nanoparticulate-based drug delivery vehicles that can be used for gene delivery include gene gun or ballistic particle–mediated gene delivery, nanoparticle-mediated transfer of small transfer RNA, self-assembling gene delivery systems, polymeric micelles, block ionomer complexes, nanogels, and nanospheres/nanocapsules/aquasomes (7,8).

GENE GUN OR BALLISTIC PARTICLE–MEDIATED GENE DELIVERY

The gene gun, which was originally developed by John Sanford and his colleagues at Cornell University in the late 1980s, is based on the concept of a bullet-like projectile coated with gold particles acting as gene carriers to "transfect the target cells" (9). The gene-loaded gold nanoparticles target the cells at a critical velocity that can puncture the cell membrane, and ultimately release the genes into the cell nucleus (9). This method, which involves physical transfer of genes, has the potential to replace the traditional transfection techniques characterized by dismal efficiency rates and immunotoxicity (9). It is to be noted, however, that original gene gun suffers from lack of precision and can crush the cells due to "sticking of gold particles (pit damage)". To overcome these pitfalls, Pui and Chen's laboratory, at University of Minnesota, devised a similar gene gun by using the patented technique called "continuous gene transfection" (9). As reported by Pui and Chen, the gold particle–coated gene composite is loaded into a capillary with the help of a syringe. The applied electric field forces the gene suspension or spray out of the capillary at a constant velocity. The suspension is a complex mixture of "highly charged and dispersed gene-coated particles" (9). As mentioned before, the unusually high repelling velocity of similarly charged particles tear the cell membrane and "unload the genes into the cells" (9). Approximately 0.5 to 5 μg of DNA can be carried per milligram of gold (10). The pros include the fact that, unlike traditional transfection techniques, gene gun, or ballistic particle–mediated gene delivery, it is not restricted to any cell type or by the size of DNA that can be incorporated (10). Also, there is reduced or no risk of immunotoxicity and the cells can be transfected with plasmids as often as desired. Another added advantage is the possibility of incorporating multiple genes encoded by different plasmids (10). The potential problem of nonselectivity can be addressed by tagging the gold particles with specific antibodies.

It has been reported by Johnson-Saliba and Jans that the gene gun has been extensively used for the development of DNA vaccines, particularly in the treatment of cancer. Recently, however, researchers have reported an in situ application of this technique in introducing DNA into heart, liver, and cornea (10).

NANOPARTICLE-MEDIATED TRANSPORT OF SHORT INTERFERING RNA

Short interfering RNA (siRNA) offers huge potential for controlling gene expression with a large number of applications (11). Researchers have been intrigued by the ability of siRNA to inhibit tumor-promoting genes (12). However, translating this potential into reality is difficult, as it is extremely tricky to deliver these short nucleotides to the site of action without degradation. Recently, scientists at Children's Mercy Hospital at Los Angeles and University of Southern California have developed a nanoparticle-based siRNA technique that targets Ewing's sarcoma in mouse animal model without RNA degradation (12). In this model, a sugar-encased polymer, developed by California Institute of Technology, traps the engineered siRNA, forming nanoparticles that offer a protective shield around the nucleotides. The siRNA is attached to transferrin, a protein that supplies iron to bloodstream (12). After binding to transferring receptors, the protein is endocytosized, releasing siRNA nanoparticles into the cytosol, targeting EWS-FL11, a specific tumor-promoting gene that is active in Ewing's sarcoma (12). Since transferrin production is upregulated in Ewing's sarcoma, more copies of siRNA nanoparticles are delivered to the site of action (12). In spite of achieving moderate success with the delivery of siRNA nanoparticles, tracking their delivery and monitoring their transfection efficiency are challenging in the absence of a suitable tracking agent or marker (11). Recently, Tan et al. synthesized quantum dot–entrapped, chitosan-based nanoparticles and used them to deliver human epidermal growth factor receptor-2 (HER2/neu) siRNA (11). This unique nanocarrier aided in the monitoring of siRNA ex vivo/in vivo owing to the presence of fluorescent quantum dots in the nanoparticles (11).

SELF-ASSEMBLING GENE DELIVERY SYSTEMS

Cationic liposome–based gene delivery vehicles approximate 12% of the clinical trials in Europe and America (7,8). Although, in nascent stage, polycation-based gene delivery shows promise in vitro and in vivo studies. Polycation–DNA complexes lead to a better control of charge, size, and hydrophilic–lipophillic balance of the transfecting species (8). However, these systems usually suffer from low solubility and poor bioavailability (8). To circumvent these problems, scientists have developed a new class of nanoparticulate-based drug delivery systems known as nanocochleates (13). Originally developed by Papahadjopoulos in 1974 as an intermediate in the preparation of large unilamellar vesicles, the modified versions of nanocochleates (diameter range, 30–100 nm) are stable drug delivery vehicles for gene and drug delivery whose structure and properties differ enormously from those of liposomes (13).

It comprises a purified calcium (or any other divalent cation, such as zinc, magnesium, or barium)–soy-based phospholipid, with lipids accounting for three-fourths of the weight. Different lipids that make up the nanocochleates include phosphotidyl serine, dioleoylphosphatidylserine, phosphatidic acid, phosphatidylinositol, phosphatidyl glycerol, phosphatidyl choline, phosphatidylethanolamine, diphosphotidylglycerol, dioleoyl phosphatidic acid, distearoyl phosphatidylserine, and dimyristoyl phosphatidylserine, dipalmitoyl phosphatidylgycerol, or a mixture of one or more of these (13). Scanning electron microscopy reveals that nanocochleates have a unique solid lipid bilayer structure folded into a sheet and devoid of any aqueous internal space unlike a typical phospholipid (13). The divalent cations maintain the sheet structure by electrostatic interaction of its positive charge with the negatively charged lipid head groups in the bilayer (13). Nanocochleates can be formulated by any of the following techniques: hydrogel method, trapping method, liposomes before cochleates dialysis method, direct calcium dialysis method, or binary aqueous–aqueous emulsion system (13). Nanocochleates offer many advantages as a drug or gene delivery vehicle. The unique structure (which is extremely stable) protects the associated or encochleated drug or nucleotides from harsh conditions, enzymes, and digestion in the stomach (13). This feature also makes them an ideal vehicle for the oral and systemic delivery of drugs and polynucleotides, with the possibility of increasing oral bioavailibity of the delivered species such as drugs or genes (13).

POLYMERIC MICELLES

Gene-loaded nanoparticles, such as gene gun, are gaining prominence as gene delivery vehicles in the tissues owing to the absence of immunotoxicity associated with viral vectors. However, a major impediment in using them in vivo stems from their tendency to agglomerate or dissociate when challenged with salt and serum (14). Using biocompatible and biodegradable polymeric micelles as drug or gene delivery vehicles can solve this problem. Amphiphilic block copolymers organize into "micelles of mesogenic size in aqueous milieu owing to differences in solubility between hydrophobic and hydrophilic segments" (15). These copolymer micelles can be differentiated from surfactant micelles in that they have low critical micelle concentration and low dissociation constants (15). These features enhance the retention time of drugs or genes in polymeric micelles, ultimately "loading a higher concentration of genes into the target sites" (15). Recently, Kataoka used polyionic complex micelles to prolong the circulation time for plasmid DNA in the blood and reported reporter gene expression in the liver (15). Mao developed a series of biodegradable and biocompatible polymers that condensed with DNA to produce nanoparticulate polymeric micelles [14]. These micelles, which are composed of a DNA–polyphosphoramidate complex core and a poly(ethylene glycol) carrier, are used to deliver siRNA into target sites (15).

BLOCK INONOMER COMPLEXES

Block inonomers are unique complexes between polymers and surfactants. The polymeric entity is a copolymer containing two hydrophilic groups, one neutral and other charged. The charge of the surfactant usually counteracts that of the polymer (8). Block ionomer complex between poly(ethylene oxide)-[*b*]-polymethacrylate anions and cetylpyridinium cations produce nanoparticles in the size range from 30 to 40 nm (8). Despite neutralization of the charges of the polyion and the surfactant, this complex is soluble and stable, unlike the regular polyelectrolyte–surfactant complexes that are usually water insoluble. Block ionomer copolymers complex with nucleic acids and stabilize DNA through the neutralization of DNA charge and condensation (8). Researchers have demonstrated increased stability, transport, and efficiency of antisense oligonucleotides both in vitro and vivo, using cationic copolymers as gene-delivering vehicles (8). For example, Professor Sayon Roy (Boston University) demonstrated the reduction of fibronectin expression by intravitreal administration of antisense oligonucleotides, using block ionomer complexes (8).

NANOGELS

Hydrogels are hydrophilic, three-dimensional, polymeric networks composed of either homopolymers or copolymers and can entrap large amounts of fluids (8,16). Nanogels represent miniature hydrogel particles that were formulated by using an emulsification/solvent evaporation technique by chemically cross-linking polyethyleneimine with double-end–activated poly(ethylene oxide) (7,8,16). Polynuceotides can be easily entrapped in this system by mixing with nanogel suspensions. Oligonucleotide-loaded nanogel particles are small (<100 nm in diameter), stable in aqueous dispersions, show no agglomeration with time, cross the intestinal cell layers, and affect gene transcription in a sequence-specific manner (8,16). The nanogels form a protective coating around the oligonucleotides and prevent their degradation. Vinogradov et al. have reported polyplex nanogel formulations "for drug delivery of cytotoxic nucleoside analogs" (17).]

DENDRIMERS

Dendrimers are nanometer-sized, highly branched, monodispersed, supramolecular complexes with symmetrical architecture (16). They are composed of three functional units: the inner core, the internal shell containing the repetitive units, and the terminal functional groups (18). They can be synthesized by divergent approach (starting from the central core and proceeding toward the outermost periphery), a reverse convergent approach, or by covalent attachment or self-assembly of dendrons (18,19). Apparent similarities of dendrimer architecture with "rigidified micelles" make them attractive candidates for drug and gene delivery. Smaller drug moieties can be encapsulated in the inner core, whereas oligonucleotides can form complex with cationic surface groups (19). Cationic dendrons form ionic complexes with DNA called dendriplexes and have the potential to lend themselves as nonviral vectors for gene delivery (19). Using EGFP-C2 as a marker gene, it has been shown that polyamidoamine dendrimers, which have positively charged amine groups on their surface, can deliver the gene to various organs in the body after intravenous injection and have elevated expression in the lungs, liver, kidney, and spleen, with minimum or no cytotoxicity (20).

NANOSPHERES/NANOCAPSULES/AQUASOMES

These are solid-state nanoparticles, ranging from 10 to 200 nm in size, and can be either crystalline or amorphous (16). The drug is dissolved or encapsulated or attached to the nanoparticles and, depending on the methods used for preparation, one can get nanospheres, nanocapsules, or aquasomes (16). Nanospheres are spherical particles composed of natural polymers such as gum, chitosan, gelatin, albumin, or collagen and the drug or gene is uniformly dispersed in it (7,16). Nanocapsules are vesicular materials in which the drug or gene is encased in a cavity surrounded by a polymeric material (16). Aquasomes are spherical particles composed of calcium phosphate or ceramic diamond covered with a polyhydroxyl oligomeric film (7). Recently, biodegradable polymeric nanoparticles, consisting of poly(glycolide) or poly(lactide-co-glycolide), are attracting considerable attention as potential gene delivery vehicles, as they are able to deliver peptides and genes through a peroral route of administration (16).

REFERENCES

1. Ehdaie B. Application of nanotechnology in cancer research: Review of progress in the National Cancer Institute's Alliance for Nanotechnology. Int J Biol Sci 2007; 3(2):108–110.
2. Park K. Nanotechnology: What can it do for drug delivery? J Control Release 2007; 120(1–2):1–3.
3. Bergen JM. Nonviral approaches for neuronal delivery of nucleic acids. Pharm Res 2008; 25(5):983–995.
4. Medicinenet. Medicinenet. Available at: http://www.medicinenet.com/ablation_therapy_for_arrhythmias/page2.htm, 2008. Accessed October 2008.
5. Solata O. Applications of nanoparticles in biology and medicine. J Nanobiotechnol 2004; 2(3):1–6.
6. Gao X. Nonviral gene delivery: What we know and what is next. AAPS J 2007; 9(1): E92–E104.

7. Moghimi SM, Hunter AC, Murray JC. Nanomedicine: Current status and future prospects. FASEB J 2005; 19:311–330.
8. Nanomedicine Center, University of Nebraska Medical Center. Nanomedicine Group. Available at: http://nanomedicine.unmc.edu/template_view.cfm?PageID=22, 2008.
9. Pui D. Gene gun. Available at: http://www.it.umn.edu/news/inventing/2000_Fall/nano_genegun.html, 2000. Accessed October 2008.
10. Johnson-Saliba M, Jans DA. Gene therapy: Optimising DNA delivery to the nucleus. Curr Drug Targets 2001; 2(4):371–399.
11. Suri SS. Novel gene-silencing nanoparticles shown to inhibit Ewing's sarcoma. Available at: http://www.physorg.com/news3800.html, 2005.
12. Hu S. Novel gene-silencing nanoparticles shown to inhibit Ewing's sarcoma. Available at: http://www.physorg.com/news3800.html, 2005. Accessed October 2008.
13. Bhinge JR. Nanocochleates: A novel drug delivery technology. Available at: http://www.pharmainfo.net/reviews/recent-trends-novel-drug-delivery-system, 2008. Accessed October 2008.
14. Mao h-Q. Polymeric biodegradable micelles for gene delivery. Available at: http://www.mse.uiuc.edu/downloads/seminars/Mao.pdf, 2008. Accessed October 2008.
15. Kataoka K. Smart polymeric micelles as nanocarriers for drug and gene delivery. Available at: http://www.ibn.a-star.edu.sg/news_seminar_details.php?eventid=52#sem_top#sem_top, 2008. Accessed October 2008.
16. Patel TB. Recent trends in novel drug delivery system. Available at: http://www.pharmainfo.net/reviews/recent-trends-novel-drug-delivery-system, 2008. Accessed October 2008.
17. Vinogradov SV. Polyplex nanogel formulations for drug delivery of cytotoxic nucleoside analogs. J Control Release 2005; 107(1):143–157.
18. Gardikis K. An overview of dendrons and its biomedical applications. PHARMAKEFTIKI 2006; 19(4):88–92.
19. Florence A. Dendrimers & dendrons: Facets of pharmaceutical nanotechnology. Drug Deliv Technol 2003; 3(5). Available at: http://www.drugdeliverytech.com. Accessed October 2008
20. Zhong H, Li Z, Zheng L, et al. Studies on polyamidoamine dendrimers as efficient gene delivery vectors. J Biomater Appl 2008; 22(6):527–544.

5 NPDDS for Cancer Treatment: Targeting Nanoparticles, a Novel Approach

Karthikeyan Subramani

Institute for Nanoscale Science and Technology (INSAT), University of Newcastle upon Tyne, Newcastle upon Tyne, U.K.

INTRODUCTION

Nanotechnology is evolving rapidly with much more potential impacts in the treatment of chronic diseases such as cancer and diabetes, respiratory diseases such as asthma, and ocular diseases, and in gene therapy. Scientific community worldwide has been working toward discovering "nanoscale" solutions to treat these diseases by using nanoparticle-based drug delivery systems. The applications of such systems for cancer treatment are discussed in the following sections.

CANCER TREATMENT TECHNIQUES

Currently, there are numerous techniques that are used for cancer treatment. But each technique has its own limitations and adverse effects (1,2). Surgical treatment (excision of the tumor) is usually the first choice of treatment preferred by physicians. However, surgical excision is not effective when the cancer cells have infiltrated the nearby vital organs or have spread to distant parts of the body (metastasis). Surgical excision is preferred for the removal of larger tumors. Cryosurgery is another surgical technique that is used for freezing and killing the tumor cells. It is an alternative to surgical excision and is used to treat tumors that have not spread to distant organs and for the treatment of precancerous or noncancerous lesions. Chemotherapy is the treatment of cancer with anticancer drugs. These drugs are used as pills, intravenous injections, or topical applications. Chemotherapeutic drugs may destroy healthy tissue along with cancer cells and carcinomatous tissue (cytotoxicity). The cytotoxic effect of chemotherapeutic drugs is highest in bone marrow, gonads, hair follicles, and digestive tract, all of which contain rapidly proliferating cells. The adverse effects of chemotherapy include fatigue, nausea, vomiting, alopecia (loss of hair), gastrointestinal disturbance, impaired fertility, impaired ovarian function, and bone marrow suppression resulting in anemia, leucopenia, and thrombocytopenia (3,4). Another technique of cancer treatment is radiation therapy, which uses radiation energy to destroy cancer cells and reduce the size of tumors. Bone marrow transplantation and peripheral blood stem cell transplantation are done to restore stem cells that are destroyed by high doses of radiation or chemotherapy. Immunotherapy, sometimes called biological therapy, biotherapy, or biological response modifier (BRM) therapy, is a treatment technique that utilizes human body's immune system to destroy cancer cells (5). The immune system is stimulated by an outside source such as an antibody, synthetic immune system proteins, or BRMs. BRMs include interferons, interleukins, colony-stimulating factors, monoclonal antibodies, vaccines, gene therapy, and nonspecific immunomodulating agents. Recent research work has been focused on studying gene therapy for cancer treatment. Gene therapy is an experimental treatment that involves introducing genetic material into the cancer cells to destroy the cells (6). Angiogenesis inhibitors are also currently being evaluated in clinical trials. These are chemicals that inhibit the formation of blood vessels (angiogenesis). Angiogenesis plays an important role in the growth and spread of cancer cells (7). New blood vessels act as a source of oxygen and nutrients to the cancer cells, allowing these cells to grow, invade nearby tissue, spread to other parts of human body, and form new colonies of cancer cells. Angiogenesis inhibitors are used to prevent the formation of blood vessels, thereby depleting the cancer cells of oxygen and nutrients. Hyperthermia (also called thermal therapy or thermotherapy) is a type of cancer treatment technique in which the cancer cells are exposed to high temperatures (up to 113°F). Research has shown that high temperatures can damage and kill cancer cells with minimal injury to normal tissues (8). By damaging proteins and functional structures within cells, hyperthermia destroys cancer cells (9). Hyperthermia may make some cancer cells more sensitive to radiation or harm other cancer cells that radiation cannot damage. It can also enhance the anticarcinogenic effect of certain anticancer drugs. Thus, it is almost used with other forms of cancer therapy, such as radiation and chemotherapy (10). Hyperthermia is at clinical trial stage currently. Laser therapy uses high-intensity laser to treat cancer (11). Laser can be used to shrink or destroy tumors. Laser therapy is most commonly used to treat superficial tumors on the surface of the body or the lining of internal organs. Photodynamic therapy is a type of cancer treatment that uses a drug called a photosensitizer or photosensitizing agent (12). Photosensitizer is activated by light of a specific wavelength. When photosensitizers are exposed to this specific wavelength, they produce singlet oxygen, which destroys cancer cells. Targeted cancer therapy uses target-specific drugs that invade cancer cells and block the growth and metastasis of cancer cells by interfering with specific molecules involved in carcinogenesis and tumor growth (13). To overcome the disadvantages of current cancer treatment techniques, the scientific community has turned toward nanotechnology to develop newer and more effective drug carrier systems to safely shepherd the anticancer drugs to the cancer cells.

MECHANISM OF ACTION OF CHEMOTHERAPEUTIC AGENTS

In general, anticarcinogenic chemotherapeutic agents can be divided into three main categories based on their mechanism of action (14).

Prevention of Synthesis of Pre-DNA Molecule Building Blocks

DNA building blocks are folic acid, heterocyclic bases, and nucleotides, which are formed naturally within cells. All of these agents work to block some steps in the formation of nucleotides or deoxyribonucleotides (necessary for making DNA). When these steps are blocked, the nucleotides, which are the building blocks of DNA and RNA, cannot be synthesized. Thus, the cells cannot replicate due to impaired DNA synthesis. Examples of drugs in this class include methotrexate, fluorouracil, hydroxyurea, and mercaptopurine.

Chemical Damage of DNA in the Cell Nuclei

Some chemotherapeutic agents destroy DNA and RNA of cancer cells. They disrupt the replication of DNA and totally halt the replication of DNA or RNA, which may stimulate cancer cell formation. A few examples of drugs in this class include cisplatin and antibiotics such as daunorubicin, doxorubicin, and etoposide.

Disruption of Synthesis or Breakdown of Mitotic Spindles

Mitotic spindles serve as molecular railroads with "north and south poles" in the cell when it starts to divide. These spindles are very important because they help split the newly copied DNA such that a copy goes to each of the two new cells during cell division. These drugs disrupt the formation of these spindles and therefore interrupt cell division. Classic examples of drugs in this class of mitotic disrupters include vinblastine, vincristine, and paclitaxel.

The applications of nanoparticles as carriers for these anticancer drugs are discussed in the following sections.

NANOPARTICULATE-BASED DRUG DELIVERY IN CANCER TREATMENT

The critical step in cancer treatment is the detection of cancer at its initial stage of carcinogenesis. Results of numerous scientific research studies done in nanotechnology and nanomedicine are inspiring the scientific community to discover new, innovative, noninvasive tools at the nanoscale level for such purposes. Nanoscale cantilevers (15) and quantum dots (16,17) are being studied as cancer detection tools at the cellular level. If the tumor has not been detected in its early stage, treatment methods should be devised to eradicate the fully developed cancer cells without harming the normal, healthy cells of human body. The various types of nanoparticles that are currently studied for their use as drug delivery systems are polymeric micelles, magnetic nanoparticles, colloidal gold nanoparticles, and ceramic nanoparticles (18–20). These nanoparticulate-based drug delivery systems can be characterized for their localization in tumor cells by coating them with tumor-specific antibodies, peptides, sugars, hormones, and anticarcinogenic drugs. These nanoparticles have been effectively coupled with the abovementioned anticarcinogenic chemotherapeutic agents and have been tested for their target specificity. These nanoparticles are superior over conventionally available drug delivery systems, as the chemotherapeutic agents can be targeted to a specified area of the human body by adding nanoscale surface receptors. These receptors specifically recognize the target tissue and bind to it and release the drug molecules (21). Thus, healthy cells are spared from cytotoxic effects of the drug. Drugs can also be protected from degradation by encapsulating them with nanoparticle coatings (22). As nanoparticles are extremely small, they can penetrate through smaller capillaries and are easily taken up by cancer cells. This causes efficient drug accumulation at the target site. The use of biodegradable nanoparticles allows sustained drug release over a period of time (23). Thus, nanoparticles as drug delivery systems, with enhanced target specificity, can overcome the limitations of conventional cancer treatment techniques. There are numerous other nanobiotechnology-based approaches being developed to formulate nanoparticles as carriers of anticarcinogenic agents. These include dendrimers, chitosan nanoparticles, low-density lipoproteins, nanoemulsions, nanolipospheres, nanoparticle composites, polymeric nanocapsules, nanospheres, and nanovesicles. Their applications in nanoencapsulation and targeted drug delivery of anticancer drugs, in combination with radiotherapy, laser therapy, thermotherapy, photodynamic therapy, ultrasound therapy, and nanoparticle-mediated gene therapy, have been extensively reviewed in the literature (24). The following sections discuss the most promising groups of nanoparticles and their applications in drug delivery for cancer treatment.

Gold Nanoparticles for Anticarcinogenic Drug Delivery

Colloidal gold nanoparticles are the most commonly used nanoparticles for anticarcinogenic drug delivery. Colloidal gold nanoparticles are more biocompatible than other nanoparticles. The physical and chemical properties of colloidal gold nanoparticles allow more than one protein molecule to bind to a single particle of colloidal gold. The use of colloidal gold nanoparticles as drug delivery vectors of tumor necrosis factor (TNF) has been tested in a growing tumor in mice (25). Although TNF has been evaluated in cancer treatment, it causes adverse effects such as hypotension and, in some cases, organ failure resulting in death. But recent researches have shown that when coupled with colloid gold particles, therapeutic amounts of TNF can be successfully delivered to destroy the tumor cells in animals (26). The use of laser to destroy the tumor cells in human breast cancer tissue has been described by a technique using selective nanothermolysis of self-assembling gold nanoparticles (27) (Fig. 1). These gold nanoparticles were coated with secondary Ab goat anti-mouse IgG. This structural configuration showed specific localization in the adenocarcinomatous breast cells targeted with primary Ab. Colloidal gold nanoparticles can also function as safe and efficient gene delivery vehicles in gene therapy and immunotherapy of cancer. Plasmid DNA encoding for murine interleukin-2 was complexed with gold nanoparticles (28). Gold nanoparticles showed significantly higher cellular delivery and transfection efficiency than other gene delivery vehicles. Gold nanoparticles conjugated with antiepithelial growth factor receptor (EGFR) antibody have been used to treat epithelial carcinoma by using selective laser photothermal therapy (29). Benign epithelial cell lines

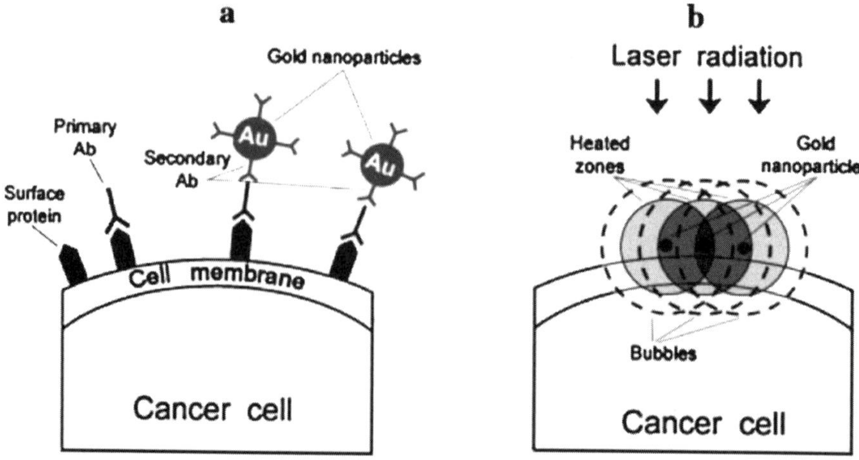

FIGURE 1 Principle of selective nanophotothermolysis of cancer cells targeted with gold nanoclusters. (a) The human breast adenocarcinoma cell targeted with primary antibodies (Abs), which are selectively attached to surface proteins; secondary Ab-goat antimouse immunoglobulin G, conjugated with 40-nm gold nanoparticles, is selectively attached to the primary Abs. (b) The schematics of laser-induced overlapping of heated zones and bubbles from the nanoparticles attached to cell membrane. *Source:* From Ref. 27.

incubated with anti-EGFR antibody–conjugated gold nanoparticles were exposed to continuous visible argon ion laser at 514 nm. It was found that the malignant cells required less than half the laser energy to be destroyed than the benign cells after incubation with anti-EGFR antibody–conjugated gold nanoparticles. A recent study showed that glucose-bound gold nanoparticles enhanced radiation sensitivity and toxicity in prostate cancer cells (30). Such nanoparticles can be effectively used to kill cancer cells at a lower radiation level.

Liposomes in Cancer Treatment

Liposomes are small artificial spherical vesicles made from naturally occurring, nontoxic phospholipids and cholesterol. There are four major types of liposomes (31). Conventional liposomes are either neutral or negatively charged. Sterically stabilized "stealth" liposomes carry polymer coatings to obtain prolonged circulatory duration. Immunoliposomes have specific antibodies or antibody fragments on their surface to enhance target specific binding. Cationic liposomes interact with negatively charged molecules and condense them to finer structure, thereby carrying them externally rather than encapsulating the molecules within. Owing to their size, biocompatibility, hydrophobicity, and ease of preparation, liposomes serve as promising systems for drug delivery. Their surfaces can be modified by attaching poly(ethylene glycol) (PEG) units to enhance the circulation time in bloodstream. Liposomes can also be conjugated with ligands or antibodies to improve their target specificity (Fig. 2). Antiestrogens solubilized within the oily core of liposomes was found to incorporate high amounts of 4-hydroxy tamoxifen (4-HT) or RU58668 (32). This combination is used in the treatment of multiple myeloma, as the cancer cells express estrogen receptors, and is of particular interest in the treatment of estrogen-dependant breast cancer. In 20% to 30% of breast cancer cells, there is a high amount of human epidermal growth factor receptor-2 (HER2) expression. Anti-HER2 antibody–conjugated immunoliposomes with magnetic nanoparticles were used to treat breast cancer cells in combination with hyperthermia (33). Such studies demonstrate the potential of liposomes as a drug delivery system in breast cancer treatment. Over recent years, liposomes have been evaluated as carriers for several anticancer drugs in the form of PEGylated liposomal cisplatin (SPI-77) and lurtotecan (OSI-211) and non-PEGylated liposomal daunorubicin (DaunoXome®), vincristine (Onco-TCS), and lipoplexes (Allovectin and LErafAON), in which cationic liposomes are used as carriers (33). Temperature-sensitive liposomal carriers (liposomal doxorubicin, Thermodox) have recently entered clinical trials (34). Recent preclinical studies have been done with different liposome–drug combinations (35) to improve the efficacy of these liposomal formulations by altering the outer coatings (36).

Magnetic Nanoparticles in Cancer Treatment

Magnetic nanoparticles are iron oxide particles sheathed with sugar molecules (Fig. 2). Therefore, these are not recognized by the immune system (37). When these particles are brought under the influence of an external magnetic field, they heat up the tumor cells and destroy them without affecting the surrounding healthy tissues. A group of researchers have synthesized biodegradable magnetic nanoparticles by using organic polymers and nanosized magnetites (38). After the characterization studies, an external magnetic field was used as a guidance system to direct the magnetic nanoparticles to the specified part of the experimental setup.

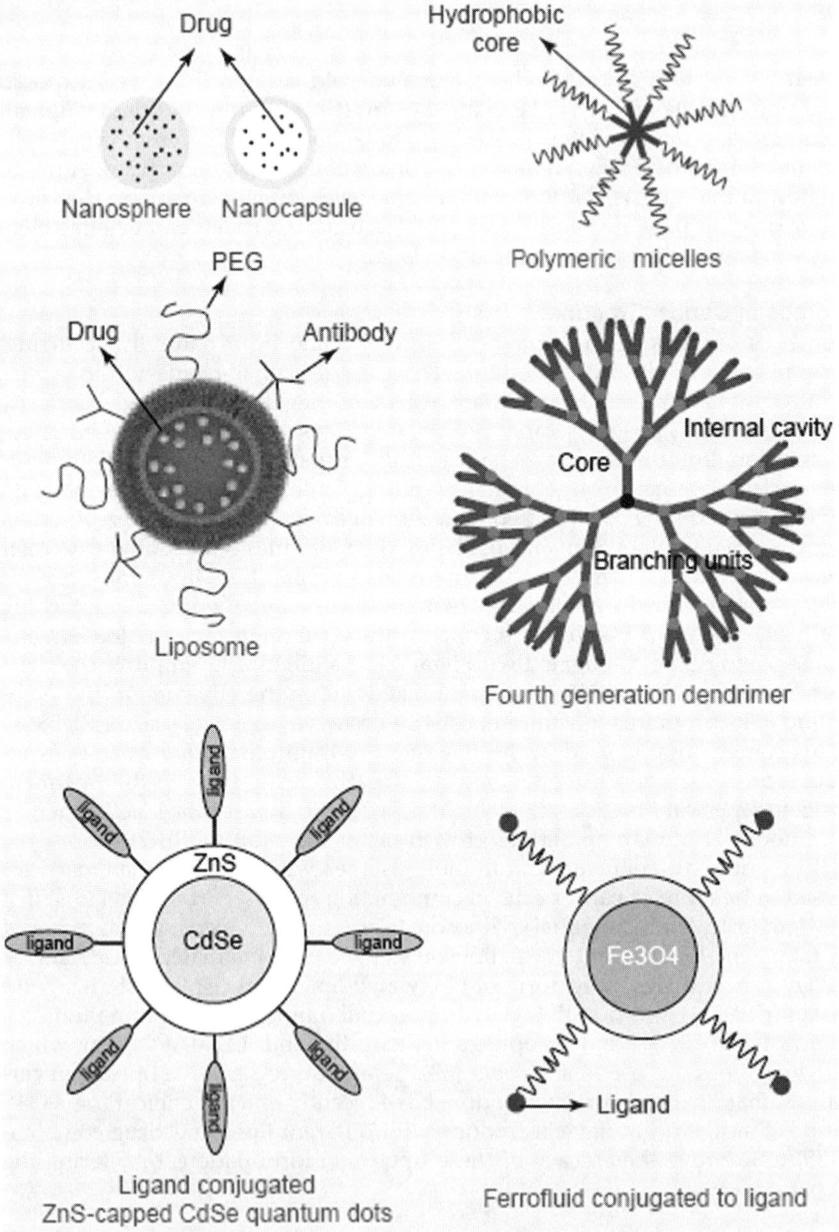

FIGURE 2 Schematics of different nanotechnology-based drug delivery systems. Nanoparticles are small polymeric colloidal particles with a therapeutic agent either dispersed in polymer matrix or encapsulated in polymer. Polymeric micelles are self-assembled block co-polymers, which in aqueous solution arrange to form an outer hydrophilic layer and an inner hydrophobic core. The miceller core can be loaded with a water insoluble therapeutic agent. Liposomes are lipid structures that can be made 'stealth' by PEGylation and further conjugated to antibodies for targeting. Dendrimers are monodispersed symmetric macromolecules built around a small molecule with an internal cavity surrounded by a large number of reactive end groups. Quantum dots are fluorescent nanocrystals that can be conjugated to a ligand and thus can be used for imaging purposes. Ferrofluids are colloidal solutions of iron oxide magnetic nanoparticles surrounded by a polymeric layer, which can be further coated with affinity molecules such as antibodies. *Source:* From Ref. 72.

The results of such studies substantiate the theory of targeting magnetic nanoparticles to specific areas of the human body by using an external magnetic field. Iron oxide nanoparticles can also be coated with amino groups to achieve cell-specific delivery of therapeutic agents, for example, to carcinomatous brain cells without unselectively invading the whole brain. This concept has been demonstrated in a study about functionalized superparamagnetic iron oxide nanoparticles (SPION) and the interaction with the brain cells (39). A recent in vitro study has tested the pH-dependant release of doxorubicin from SPION, with specific cytotoxicity toward cancer cells (40). An approach of localizing the iron oxide nanoparticles to specific cell receptors has been studied by functionalizing them with glycoproteins such as lactoferrin and ceruloplasmin (41). In breast cancer tissue, luteinizing hormone-releasing hormone (LHRH) receptors are expressed predominantly. Thus, to localize the iron oxide nanoparticles to the cancerous breast tissue, the magnetic nanoparticles can be conjugated with LHRH. Such an approach has demonstrated the target specificity of the iron oxide nanoparticles in breast cancer treatment (42). These approaches prove that magnetic nanoparticles can be functionalized with suitable moieties to localize them specifically to tumor cells. Carbon magnetic nanoparticles (CMNPs) are made up of spherical particles of 40- to 50-nm diameter, with iron oxide particles dispersed in a carbon-based host structure (43). Doxorubicin molecules immobilized on activated CMNP formed CMNP- doxorubicin conjugates, which were demonstrated to be effective in cancer cell cytotoxicity assays (44). This showed that CMNPs can be used as effective drug delivery systems. Recent studies have investigated the synthesis and release characteristics of poly(ethyl-2-cyanoacrylate)–coated magnetite nanoparticles containing anticancer drugs such as cisplatin and gemcitabine (45). The hydrophobic cisplatin showed sustained release pattern when compared with the hydrophobic gemcitabine.

Polymers as Drug Delivery Systems

Polymeric micelles serve as a novel drug delivery system due to their target specificity and controlled release of hydrophobic anticancer drugs (46). PEG-based micelles are biocompatible and biodegradable. Effective drug delivery of cytotoxic drugs to cancer cells by using PEG polymeric micelles has been demonstrated by conjugating doxorubicin with poly(ethylene glycol)-poly(α,β-aspartic acid) block copolymer and paclitaxel with poly(lactic-co-glycolic acid) (PLGA) (47). Doxorubicin (also known by its trade name Adriamycin) was physically entrapped and chemically bound to the core of the polymeric micelle. Owing to reduced uptake by the reticuloendothelial system (RES), this drug carrier had a prolonged circulation time in the bloodstream. Localization to the cancer cells can be achieved by linking monoclonal antibodies, sugars, and biotin or tumor-specific peptides to the polymeric micelles (48). PEG-coated biodegradable nanoparticles can be coupled with folic acid to target folate-binding protein, which is a soluble form of folate protein and is overexpressed on the surface of many tumor cells. These folate-linked nanoparticles have been tested and confirmed for their selective target binding (49). In 1994, PK1 was the first nanoparticle-based polymeric prodrug to enter clinical trials. PK1 has doxorubicin conjugated to poly[*N*-(2-hydroxypropyl)methacrylamide] (34). PK1 has been shown to improve the therapeutic index of doxorubicin by diminishing its cardiotoxic effect and has progressed to phase II evaluation (50). Following these proof of principles, combinations such as polymer–drug, polymer–protein, and protein–drug conjugates (e.g., paclitaxel, neocarcinostatin) have been evaluated clinically (34). These demonstrate the potential of polymers as a novel drug delivery system in cancer treatment.

Ceramic Nanoparticles in Photodynamic Therapy

Ceramic nanoparticles are made from calcium phosphate, silica, alumina, or titanium. These ceramic nanoparticles have certain advantages such as easier manufacturing techniques, high biocompatibility, ultralow size (<50 nm), and good dimensional stability (51). These particles effectively protect the doped drug molecules against denaturation caused by changes in external pH and temperature. Their surfaces can be easily modified with different functional groups as well as can be conjugated with a variety of ligands or monoclonal antibodies in order to target them to desired site (52). These nanoparticles can be manufactured with the desired size, shape, and porosity. A ceramic nanoparticle does not undergo swelling or porosity changes caused by fluctuations in temperature or pH and are small enough to evade the RES. The application of ultrafine, silica-based nanoparticles, which are ~35 nm in diameter with photosensitive anticarcinogenic drugs encapsulated within, has been described (53). These ceramic nanoparticles have been used to destroy cancer cells by photodynamic therapy. The photosensitizer encapsulated within the ceramic nanoparticles was [2-(1-hexyloxyethyl)-2-devinyl pyropheophorbide-a]. When activated by light of suitable wavelength (650 nm), the drug produces singlet oxygen, which necroses the tumor cells. This has entered clinical trials for esophageal cancer therapy (54). This concept of using silica nanoparticle platforms (Fig. 3), which can attach to the external surface of tumor cells and deliver photosensitizer such as meta-tetra(hydroxyphenyl)chlorin (m-THPC), and singlet oxygen–induced cancer cell apoptosis has also been demonstrated (55).

Nanoparticle-Based Delivery of Specific Anticarcinogenic Drugs

Methotrexate, a potent anticancer drug, has been coupled with poly(butyl cyanoacrylate) nanoparticles of different sizes (range, 70–345 nm) and tested for

FIGURE 3 SEM image of m-THP–C embedded silica nanoparticles. The average diameter of the particles is 180 nm. Scale bar = 1 μm. *Source:* From Ref. 55.

their ability to overcome blood–brain barrier in the treatment of brain cancer. This study showed that polysorbate 80–coated poly(butyl cyanoacrylate) nanoparticles of diameter below 100 nm can effectively overcome the blood–brain barrier (56). Research studies done on experimental rats have demonstrated that the nanoparticle formulation consisting of poly(amidoamine) modified with PEG-500 had the ability of sustained release of 5-fluorouracil and was target specific (57). PLGA-mPEG nanoparticles were used as a carrier for cisplatin, and this study showed that PLGA-mPEG effectively delivered the drug to human prostate cancer cells (58). PLGA nanoparticles have been shown as effective carriers of doxorubicin (59). Positively charged polysaccharide chitosan nanoparticles have also been tested for their target specificity to deliver doxorubicin (60). PLGA nanoparticles containing vitamin E have also been tested for their drug-carrying potential for paclitaxel, an anticarcinogenic drug that interferes with mitotic spindles and therefore inhibiting cell division (61). Nitrocamptothecin, an alkaloid drug belonging to a class of anticancer agents called topoisomerase inhibitors, have also been target specifically delivered to the cancer cells by PLGA nanoparticles (62). Trastuzumab (more commonly known under the trade name Herceptin) is a humanized monoclonal antibody that acts on HER2/neu (erbB2), which are overexpressed in breast cancer cells. Human serum albumin nanoparticles were used as drug delivery systems, and this study showed that a stable and biologically active system such as albumin can be used in cancer treatment (63). Another drug belonging to the same category and used in breast cancer treatment is tamoxifen. Poly(ethylene oxide)–modified poly(ε-caprolactone) polymeric nanoparticles have been tested for their target specificity as a carrier for tamoxifen (64). These research studies prove that a wide variety of nanoparticles can effectively function as drug delivery systems for anticancer drugs and thereby eliminating the adverse effects of these pharmaceutical agents.

CONCLUSIONS

Discussed in this chapter are the research works done in the past decade in targeting novel nanoparticles toward the treatment of cancer. This field of study has expanded tremendously in the past few years, as new nanoparticle carrier systems and anticancer drugs are being discovered. The use of nanoparticles for early diagnosis of cancer and in gene therapy has been extensively reviewed in the literature (65–71). A few of these innovative treatment techniques have made their way into clinical trials. There is a lot more to be done to treat or perhaps prevent advanced cancer by treating it in an early stage. This will require superior detection and targeting methods that many of the researchers are pursuing on nanoparticle-based drug delivery systems. These research studies in nanotechnology will definitely pave the way for early detection and prevention of cancer, thereby improving the life and quality of cancer patients. The limitations of nanoparticles as drug delivery systems and their applications in diabetes treatment are discussed in chapter 8.

REFERENCES

1. Burish TG, Tope DM. Psychological techniques for controlling the adverse side effects of cancer chemotherapy: Findings from a decade of research. J Pain Symptom Manage 1992; 7:287–301.
2. Schwartz CL. Late effects of treatment in long-term survivors of cancer. Cancer Treat Rev 1995; 21:355–366.

3. Tormey DC, Gray R, Taylor SG. Postoperative chemotherapy and chemohormonal therapy in women with node-positive breast cancer. Natl Cancer Inst Monogr 1986; 1:75–80.
4. Barton C, Waxman J. Effects of chemotherapy on fertility. Blood Rev 1990; 4: 187–195.
5. Durrant LG, Scholefield JH. Principles of cancer treatment by immunotherapy. Surgery 2003; 21:277–279.
6. El-Aneed A. Current strategies in cancer gene therapy. Eur J Pharm 2004; 498:1–8.
7. Fayette J, Soria JC, Armand JP. Use of angiogenesis inhibitors in tumour treatment. Eur J Cancer 2005; 41:1109–1116.
8. Van der Zee J. Heating the patient: A promising approach? Ann Oncol 2002; 13:1173–1184.
9. Hildebrandt B, Wust P, Ahlers O. The cellular and molecular basis of hyperthermia. Crit Rev Oncol Hematol 2002; 43:33–56.
10. Wust P, Hildebrandt B, Sreenivasa G. Hyperthermia in combined treatment of cancer. Lancet Oncol 2002; 3:487–497.
11. Goldstein NS. Laser therapy for small breast cancers. Am J Surg 2004; 187:149–150.
12. Schuitmaker JJ, Baas P, van Leengoed HLLM, et al. Photodynamic therapy: A promising new modality for the treatment of cancer. J Photochem Photobiol B 1996; 34:3–12.
13. Kim JA. Targeted therapies for the treatment of cancer. Am J Surg 2003; 186:264–268.
14. DiPiro JT, Talbert RL, Yee GC, et al. Cancer treatment and chemotherapy. In: Pharmacotherapy: A Pathophysiologic Approach, 5th ed. New York: McGraw Hill Publications, 2002:2175–2222.
15. Wee KW, Kang GY, Park J, et al. Novel electrical detection of label-free disease marker proteins using piezoresistive self-sensing micro-cantilevers. Biosens Bioelectron 2005; 20:1932–1938.
16. Voura EB, Jaiswal JK, Mattoussi H, et al. Tracking metastatic tumor cell extravasation with quantum dot nanocrystals and fluorescence emission-scanning microscopy. Nat Med 2004; 10:993–998.
17. Kaul Z, Yaguchi T, Kaul SC, et al. Mortalin imaging in normal and cancer cells with quantum dot immuno-conjugates. Cell Res 2003; 13:503–507.
18. Neuberger T, Schöpf B, Hofmann H, et al. Superparamagnetic nanoparticles for biomedical applications: Possibilities and limitations of a new drug delivery system. J Magn Magn Mater 2005; 293:483–496.
19. El-Sayed IH, Huang X, El-Sayed MA. Surface plasmon resonance scattering and absorption of anti-EGFR antibody conjugated gold nanoparticles in cancer diagnostics: Applications in oral cancer. Nano Lett 2005; 5:829–834.
20. Wang J, Mongayt D, Torchilin VP. Polymeric micelles for delivery of poorly soluble drugs: Preparation and anticancer activity in vitro of paclitaxel incorporated into mixed micelles based on poly(ethylene glycol)-lipid conjugate and positively charged lipids. J Drug Target 2005; 13:73–80.
21. Dinauer NB, Sabine W, Carolin K, et al. Selective targeting of antibody-conjugated nanoparticles to leukemic cells and primary T-lymphocytes. Biomaterials 2005; 26:5898–5906.
22. Pandey R, Ahmad Z, Sharma S, et al. Nano-encapsulation of azole antifungals: Potential applications to improve oral drug delivery. Int J Pharm 2005; 301:268–276.
23. Yoo HS, Oh JE, Lee KH, et al. Biodegradable nanoparticles containing doxorubicin-PLGA conjugate for sustained release. Pharm Res 1999; 16:1114–1118.
24. Jain KK. Nanotechnology-based drug delivery for cancer. Technol Cancer Res Treat 2005; 4:407–416.
25. Paciotti GF, Myer L, Weinreich D, et al. Colloidal gold: A novel nanoparticle vector for tumour directed drug delivery. Drug Deliv 2004; 11:169–183.
26. Miller G. Colloid gold nanoparticles deliver cancer-fighting drugs. Pharm Technol 2003; 7:17–19.
27. Zharov VP, Galitovskaya EN, Johnson C, et al. Synergistic enhancement of selective nanophotothermolysis with gold nanoclusters: Potential cancer therapy. Lasers Surg Med 2005; 39:219–226.

28. Noh SM, Kim WK, Kim SJ, et al. Enhanced cellular delivery and transfection efficiency of plasmid DNA using positively charged biocompatible colloidal gold nanoparticles. Biochim Biophys Acta 2007; 1770:747–752.
29. El-Sayed IH, Huang X, El-Sayed MA. Selective laser photo-thermal therapy of epithelial carcinoma using anti-EGFR antibody conjugated gold nanoparticles. Cancer Lett 2006; 239(1):129–135.
30. Zhang X, Xing JZ, Chen J, et al. Enhanced radiation sensitivity in prostate cancer by gold-nanoparticles. Clin Invest Med 2008; 31(3):E160–E167.
31. Storm G, Crommelin DJA. Liposomes: Quo vadis? Pharm Sci Technol Today 1998; 1:19–31.
32. Maillard S, Ameller T, Gauduchon J, et al. Innovative drug delivery nanosystems improve the anti-tumor activity in vitro and in vivo of anti-estrogens in human breast cancer and multiple myeloma. J Steroid Biochem Mol Biol 2005; 94:111–121.
33. Ito A, Kuga Y, Honda H, et al. Magnetite nanoparticle-loaded anti-HER2 immunoliposomes for combination of antibody therapy with hyperthermia. Cancer Lett 2004; 212:167–175.
34. Lammers T, Hennink WE, Storm G. Tumour-targeted nanomedicines: Principles and practice. Br J Cancer 2008; 99(3):392–397.
35. Banciu M, Metselaar JM, Schiffelers RM, et al. Liposomal glucocorticoids as tumor-targeted anti-angiogenic nanomedicine in B16 melanoma-bearing mice. J Steroid Biochem Mol Biol 2008; 111(1–2):101–110.
36. Romberg B, Hennink WE, Storm G. Sheddable coatings for long-circulating nanoparticles. Pharm Res 2008; 25(1):55–71.
37. Pankhurst QA, Connolly J, Jones SK, et al. Applications of magnetic nanoparticles in biomedicine. J Phys D Appl Phys 2003; 36:R167–R181.
38. Asmatulu R, Zalich MA, Claus RO, et al. Synthesis, characterization and targeting of biodegradable magnetic nanocomposite particles by external magnetic fields. J Magn Magn Mater 2005; 292:108–119.
39. Cengelli F, Maysinger D, Tschudi-Monnet F, et al. Interaction of functionalized superparamagnetic iron oxide nanoparticles with brain structures. J Pharmacol Exp Ther 2006; 318(3):1388.
40. Munnier E, Cohen-Jonathan S, Linassier C, et al. Novel method of doxorubicin-SPION reversible association for magnetic drug targeting. Int J Pharm 2008; 363(1–2): 170–176.
41. Gupta AK, Curtis ASG. Lactoferrin and ceruloplasmin derivatized superparamagnetic iron oxide nanoparticles for targeting cell surface receptors. Biomaterials 2004; 25:3029–3040.
42. Zhou J, Leuschner C, Kumar C, et al. Sub-cellular accumulation of magnetic nanoparticles in breast tumors and metastases. Biomaterials 2006; 27:2001–2008.
43. Qiu J, Li Y, Wang Y, et al. Preparation of carbon-coated magnetic iron nanoparticles from composite rods made from coal and iron powders. Fuel Process Technol 2004; 86:267–274.
44. Ma Y, Manolache S, Denes FS, et al. Plasma synthesis of carbon magnetic nanoparticles and immobilization of doxorubicin for targeted drug delivery. J Biomater Sci Polym Ed 2004; 15(8):1033–1049.
45. Yang J, Lee H, Hyung W, et al. Magnetic PECA nanoparticles as drug carriers for targeted delivery: Synthesis and release characteristics. J Microencapsul 2006; 23:203–212.
46. Kwon GS, Okano T. Polymeric micelles as new drug carriers. Adv Drug Deliv Rev 1996; 21:107–116.
47. Yokoyama M, Miyauchi M, Yamada N, et al. Characterization and anticancer activity of the micelle-forming polymeric anticancer drug Adriamycin-conjugated poly(ethylene glycol)-poly(aspartic acid) block copolymer. Cancer Res 1990; 50(6):1693–1700.
48. Gregoriadis G, Florence AT. Recent advances in drug targeting. Trends Biotechnol 1993; 11(11):440–442.
49. Stella B, Arpicco S, Peracchia MT. Design of folic acid-conjugated nanoparticles for drug targeting. J Pharm Sci 2000; 89:1452–1464.

50. Duncan R. Polymer conjugates as anticancer nanomedicines. Nat Rev Cancer 2006; 6(9):688–701.
51. Vollath DD, Szabo V, Haubelt J. Synthesis and properties of ceramic nanoparticles and nanocomposites. J Euro Ceram Soc 1997; 17(11):1317–1324.
52. Douglas SJ, Davis SS, Illum L. Nanoparticles in drug delivery. Rev Ther Drug Carr Syst 1987; 3(3):233–261.
53. Roy I, Ohulchanskyy T, Pudavar H, et al. Ceramic-based nanoparticles entrapping water-insoluble photosensitizing anticancer drugs: A novel drug-carrier system for photodynamic therapy. J Am Chem Soc 2003; 125:7860–7865.
54. Bechet D, Couleaud P, Frochot C, et al. Nanoparticles as vehicles for delivery of photodynamic therapy agents. Trends Biotechnol 2008; 26(11):612–621.
55. Yan F, Kopelman R. The embedding of meta-tetra(hydroxyphenyl)chlorin into silica nanoparticle platforms for photodynamic therapy and their singlet oxygen production and pH-dependent optical properties. Photochem Photobiol 2003; 78:587–591.
56. Gao K, Jiang X. Influence of particle size on transport of methotrexate across blood brain barrier by polysorbate 80-coated polybutylcyanoacrylate nanoparticles. Int J Pharm 2006; 310:213–219.
57. Bhadra D, Bhadra S, Jain P, et al. PEGylated dendritic nanoparticulate carrier of fluorouracil. Int J Pharm 2003; 257:111–124.
58. Avgoustakis K, Beletsi A, Panagi Z, et al. PLGA-mPEG nanoparticles of cisplatin: In vitro nanoparticle degradation, in vitro drug release and in vivo drug residence in blood properties. J Control Release 2002; 79:123–135.
59. Yoo HS, Park TG. Biodegradable polymeric micelles composed of doxorubicin conjugated PLGA-PEG block copolymer. J Control Release 2001; 70(1–2):63–70.
60. Janes KA, Fresneau MP, Marazuela A, et al. Chitosan nanoparticles as delivery systems for doxorubicin. J Control Release 2001; 73(2–3):255–267.
61. Mu L, Feng SS. A novel controlled release formulation for the anticancer drug paclitaxel (Taxol): PLGA nanoparticles containing vitamin E TPGS. J Control Release 2003; 86:33–48.
62. Derakhshandeh K, Erfan M, Dadashzadeh S. Encapsulation of 9-nitrocamptothecin, a novel anticancer drug, in biodegradable nanoparticles: Factorial design, characterization and release kinetics. Eur J Pharm Biopharm 2007; 66:34–41.
63. Steinhauser I, Spankuch B, Strebhardt K, et al. Trastuzumab-modified nanoparticles: Optimisation of preparation and uptake in cancer cells. Biomaterials 2006; 27:4975–4983.
64. Shenoy DB, Amiji MM. Poly(ethylene oxide)-modified poly(epsilon-caprolactone) nanoparticles for targeted delivery of tamoxifen in breast cancer. Int J Pharm 2005; 293(1–2):261–270.
65. Kayser O, Lemker A, Trejo NH. The impact of nanobiotechnology on the development of new drug delivery systems. Curr Pharm Biotechnol 2005; 6:3–5.
66. Sonavane G, Tomoda K, Makino K. Biodistribution of colloidal gold nanoparticles after intravenous administration: Effect of particle size. Coll Surf B Biointerfaces 2008; 66(2):274–280.
67. De Jong WH, Borm PJ. Drug delivery and nanoparticles: Applications and hazards. Int J Nanomed 2008;3(2):133–149.
68. Yezhelyev MV, Gao X, Xing Y, et al. Emerging use of nanoparticles in diagnosis and treatment of breast cancer. Lancet Oncol 2006; 7:657–667.
69. Wagner E. Programmed drug delivery: Nanosystems for tumor targeting. Expert Opin Biol Ther 2007; 7:587–593.
70. Brannon-Peppas L, Blanchette JO. Nanoparticle and targeted systems for cancer therapy. Adv Drug Deliv Rev 2004; 56:1649–1659.
71. Liu Y, Miyoshi H, Nakamura M. Nanomedicine for drug delivery and imaging: A promising avenue for cancer therapy and diagnosis using targeted functional nanoparticles. Int J Cancer 2007; 120:2527–2537.
72. Sahoo SK, Labhasetwar V. Nanotech approaches to drug delivery and imaging. Drug disc today 2003; 8:1112–1120.

6 Design and Formulation of Protein-Based NPDDS

Satheesh K. Podaralla and Omathanu P. Perumal

Department of Pharmaceutical Sciences, South Dakota State University, Brookings, South Dakota, U.S.A.

Radhey S. Kaushik

Department of Biology & Microbiology/Veterinary Science, South Dakota State University, Brookings, South Dakota, U.S.A.

INTRODUCTION

Proteins are a versatile class of biopolymers whose functional properties are dictated by their amino acid composition. The biodegradation of proteins into simple amino acids makes them attractive polymers for drug delivery applications. As a result, protein polymers are being increasingly investigated for various nano-enabled drug delivery systems (1). In fact, the first protein-based nanoparticulate system is already in the market. Albumin-bound paclitaxel nanoparticles (Abraxane™) was approved by the U.S. Food and Drug Administration (FDA) in 2005. In general, nanosystems used in drug delivery range in size from 100 to 1000 nm and are prepared using natural or synthetic polymers or lipids. The drug is encapsulated inside the nanosystem (nanocapsule or nanospheres) or is entrapped in the matrix of the nanosystem (nanoparticles). Alternatively, drugs or other agents of interest are adsorbed, complexed, or conjugated to the surface of the nanosystems. Proteins offer a number of advantages over synthetic polymers. Unlike synthetic polymers, which usually have a single type of functional groups, proteins have numerous functional groups ($-NH_2$, $-COOH$, and $-SH$) for covalent and noncovalent modifications of the nanosystems. A distinct advantage over synthetic polymers is the proven biocompatibility of proteins. Furthermore, they are biodegradable and are broken down into nontoxic by-products. Protein polymers are derived from animals or plants and are, therefore, devoid of monomers or initiators that are found in synthetic polymers. However, the composition and purity of protein biopolymers are difficult to control. Similarly, it is important to protect the protein from premature proteolytic degradation in the body. Although most of the proteins are generally safe, nonautologus proteins can be immunogenic (2). Recombinant DNA (rDNA) technology or combination of synthetic and protein polymers is used to address these limitations of protein polymers (3,4).

Proteins can be classified into different types based on their source or structural features. This chapter is mainly focused on natural protein polymers derived from animal or plant sources. The properties of different protein polymers used in drug delivery are listed in Table 1. Majority of the published literature on protein nanoparticles is on animal proteins such as gelatin and albumin. On the other hand, there are limited reports on plant protein–based nanoparticles. Similarly,

TABLE 1 Physicochemical Properties of Protein Polymers

Protein	Source	Composition	Molecular Weight	Isoelectric point
Gelatin	Collagen	4-Hydroxylysine, hydroxyproline, glycine, alanine, and proline	15–250 kDa	7.0–9.0 (type A) 4–5 (type B)
Albumin	Plasma protein	Single polypeptide with 585 amino acids	66 kDa	4.7
Whey protein	Milk protein By-product of cheese industry	α- and β-lactoglobulins	β-Lactoglobulin: 18 kDa α-Lactoglobulin: 14 kDa	3.5–5.2
Casein	Milk protein	α_{s1}, α_{s2}, β, and κ subunits	α_{s1}: 23 kDa α_{s2}: 25 kDa β: 24kDa κ: 19 kDa	4.6
Silk fibroin	*Bombyx mori* *Nephila clavipes*	Repeating units of alanine and glycine β-pleated sheets	60–150 kDa	3–5
Zein	*Zea mays* L.	High proportions of glutamine and proline	α-zein: 22–24 kDa β-zein: 44 kDa γ-zein: 14 kDa	5.0–9.0
Gliadin	Wheat flour	High proportions of glutamine and proline	28–55 kDa	6.8
Legumin	*Pisum sativum* L.	Higher proportion of acidic amino acids	300–400 kDa; 20- to 40-kDa subunits	4.8

most of the proteins used for nanoparticles are globular proteins and there is very little work with fibrous proteins. Animal proteins carry the risk of infection from pathogen contamination with animal tissues. However, animal proteins, such as gelatin, can be easily sterilized due to their thermal stability. In contrast, plant proteins are devoid of these issues, and further plant proteins are universally accepted as they can be used even by people who do not consume animal proteins for personal or religious reasons. This chapter discusses the properties of selected animal and plant proteins with an emphasis on factors that influence the preparation of nanoparticles, methods for preparing protein nanoparticles, biological interactions of protein-based nanoparticles, and their drug delivery applications.

CHARACTERISTICS OF PROTEIN POLYMERS

Animal/Insect Proteins

Gelatin

Gelatin is one of the most widely studied animal proteins and is obtained by the hydrolysis of collagen. It is obtained by heat-induced alkaline (type B) or acid (type A) hydrolysis of collagen from animal (bovine or porcine) skin, bones, and tendons (5). Gelatins obtained by these two processes differ in their isoelectric point (pI), molecular weight, amino acid composition, and viscosity (6). For example, type A gelatin has a pI of 7 to 9, whereas type B gelatin has a pI of 4 to 5. Gelatin is a weak immunogen (2) and is a generally regarded as safe (GRAS) excipient approved by the USFDA for use in pharmaceutical preparations. Gelatin capsules are a popular oral dosage form. Apart from its biomedical applications, gelatin is also widely used in the food industry as a stabilizer and a protective coating material. Gelatin is soluble in hot water, glycerin, acids, strong acids, and alkalis. It is practically insoluble in acetone, chloroform, ethanol, ether, and methanol (7–9). Two-thirds of gelatin is composed of 4-hydroxylysine, hydroxyproline, glycine, alanine, and proline, while methionine, tyrosine, and histidine are present in low quantities. Gelatin has many ionizable groups such as carboxyl, phenol, ε-amino, α-amino, guanidine, and imidazole groups, which are potential sites for conjugation or chemical modification. Gelatin consists of four molecular fractions, which include (*i*) α-chain (80–125 kDa), (*ii*) β-chain (125–130 kDa), comprising of dimers of tropocollagen, (*iii*) γ-chain, which is a trimer (230–340 kDa) of tropocollagen, and (*iv*) fraction delta, which is a tetramer with a high degree of cross-linking (10). The distribution of charge in gelatin depends on the frequency of side groups and their dissociation constants. The pH value of the solution influences the physical properties of gelatin (6). Gelatin can be made resistant to enzymatic degradation by covalent cross-linking using acid halides, anhydrides, aziridines, epoxides, polyisocyanates, aldehydes, ketones, methanesulfonate biesters, and carbodiimides (7,11,12). Noncovalent cross-linking can be achieved through electrovalent and coordinate interactions (11). Functional groups present on gelatin, such as amino, carboxyl, and hydroxyl side chains, form attractive sites for chemical conjugation with therapeutic agents, targeting ligands or poly(ethylene glycol) (PEG). Temperature and pH influence the various functional properties of gelatin. The hydrolytic degradation of gelatin is least at neutral pH and highest at acidic pH (13–15).

Albumin

Albumin is obtained from a variety of sources, including egg white (ovalbumin), bovine serum albumin (BSA), and human serum albumin (HSA). Albumin is the most abundant plasma protein (35–50 mg/mL), with a half-life of 19 days (16). Endogenous albumin is involved in the maintenance of osmotic pressure, binding and transport of nutrients to the cell. Many drugs and endogenous substances are known to bind to albumin, and the albumin serves as a depot and transporter protein (16). Human albumin has a molecular weight of about 66 kDa and has a single polypeptide chain consisting of 585 amino acids (16). Albumin contains 1 tryptophan residue, 6 methionine residues, 17 cysteine residues, 36 aspartic acid residues, 61 glutamic acid residues, 59 lysine residues, and 23 arginine residues (16). It contains seven disulfide bridges. Human albumin has a secondary structure that is about 55% α-helix (17). The remaining 45% is believed to be β-turn structures and disordered β-pleated structures. Seven disulfide bridges of albumin contribute to its chemical and spatial conformation (18). Albumin is freely soluble in dilute salt solution and water. The high solubility of albumin (up to 40% w/v) at pH 7.4 makes it an attractive carrier, capable of accommodating a wide variety of drugs. At physiological pH, it carries a net charge of −17 mV (19). It is stable in the pH range of 4 to 9 and can be heated at 60°C up to 10 hours without any deleterious effects (19).

It is devoid of toxicity and immunogenicity. HSA is obtained by the fractionation of human plasma and carry the risk of contamination with blood-borne pathogens. Recombinant HSA produced using bacteria and yeast has been used to address this limitation (20).

Milk Proteins

The two milk proteins that have been investigated for drug delivery applications include β-lactoglobulin (BLG) and casein. Bovine whey protein contains 53% of BLG, which has a molecular weight of 18.3 kDa. This protein contains two disulphide bonds and one free thiol group, which confers an unusual stability at low pH (below 2) (21). The ability to preserve its native, stable conformation at acidic pH makes it resistant to peptic and chymotryptic digestion (22). BLG has good gelling property, which can be useful for some drug delivery applications (23). Due to its abundance and low cost, BLG is a promising polymer for drug delivery applications (24–26).

Another potential milk protein for drug delivery applications is casein, which exists as micelles in the size range of 100 to 200 nm (27). It is a milk transport system that delivers calcium phosphate and amino acids from the mother to the offspring. The casein micelle contains small aggregates of 10 to 100 casein molecules (submicelles). They are composed of four phsophoproteins (Table 1), namely, α_{s1}-casein, α_{s2}-casein, β-caesin, and κ-caesin, at a molar ratio of about 4:1:4:1.3 (27). The casein micelles are held together by hydrophobic interactions and through calcium phosphate nanoclusters in the core (27). The surface is covered by κ-caesin, resulting in a hydrophilic charged surface that stabilizes the casein micelles by electrostatic and steric repulsions (27). Casein micelles per se (28) and particulate systems formed using cross-linked casein molecules (25) have been used as drug carriers.

Silk Fibroin

Silk is a fibrous protein that is produced by spiders (*Nephlia clavipes*) and silkworms (*Bombyx mori*). The amino acid composition and the mechanical properties of silk depends on the organism from which it is produced. Silk consists of both crystalline and amorphous regions (29). Crystalline regions of silk are responsible for its mechanical properties, which consist of β-pleated sheets of repetitive units of alanine and glycine residues that are interconnected with adjacent chains via hydrogen bonds and hydrophobic interactions (29). The noncrystalline domains are poorly oriented and randomly coiled protein chains, which are amorphous or semicrystalline. This domain is mainly composed of phenylalanine and tyrosine residues (29). The high degree of hydrogen bonding in silk makes it insoluble in most solvents, including water, dilute acids, and alkalis (30). However, silk can be solubilized by 9 M of lithium bromide and concentrated formic acid (30). To address its poor aqueous solubility, recombinant silk proteins have been synthesized, such as the phosphorylated silk, which has a higher aqueous solubility than the unphosphorylated form (31). Silk fibroin has remarkable mechanical properties, including extensibility, toughness, and thermal stability, at both low and high temperatures (31). Due to its superior mechanical properties and biocompatibility, silk fibroin is widely used as a surgical suture material. Liquid silk fibroin has been used as a biomaterial in various forms, including powder, film, gel, porous matrix, microparticles, and nanoparticles (31,32).

Plant Proteins

Zein

Zein belongs to a family of prolamine proteins, which contains a high proportion of hydrophobic amino acids proline and glutamine (33). As a result, the prolamine proteins are water insoluble. Zein is extracted from the endosperm of corn by using hydroalcoholic solvent system (33). It is widely used in the food and packaging industry for its film-forming properties and its ability to provide a moisture-impervious barrier (34). There are four different subunits of zein, namely, α, β, γ-zein, and δ-zein, which are characterized based on their solubility behavior and molecular weight (33). The commercially available zein contains mainly α-zein (75–85%). The α-zein subunit consists of two major protein subunits of about 210 to 245 amino acid residues, each with a molecular weight of 23 and 27 kDa, respectively (35). Zein is a GRAS polymer approved by the USFDA for human applications. It has been used to prepare particulate systems for drug delivery and food applications (34,36).

Gliadin

Gliadin is a gluten protein found in wheat and also belongs to the prolamine protein family. Most of the gliadins exist as monomers and are classified into α (25–35 kDa), β (30–35 kDa), γ (35–40 kDa), and ω (55–70 kDa) fractions in the order of decreasing electrophoretic mobility (37). ω-Gliadins are characterized by high proportions of glutamine, proline, and phenylalanine, which account for 80% of the total composition of gliadin. Gliadins have low water solubility, except at extreme pH levels (38). This is attributed to the presence of interpolypeptide disulphide bonds and to the cooperative hydrophobic interactions (38). Gliadins exhibit bioadhesive property and have been explored for oral and topical drug delivery applications (39,40).

Legumin

Legumin is a storage protein found in pea seeds (*Pisum sativum*). It belongs to the group of 11S globulins, with sedimentation coefficients between 11S and 14S (molecular weight = 300–400 kDa). Legumin has a complex globular structure that is made up of six pairs of subunits, with each subunit consisting of disulphide-linked acid (molecular weight = 40 kDa) and basic (molecular weight = 20 kDa) polypeptides (41). Legumin has a pI of 4.8 and has been used to prepare nanoparticles (41).

FACTORS THAT INFLUENCE NANOPARTICLE CHARACTERISTICS

Protein Composition

The physicochemical properties of the protein and the drug properties influence the preparation and characteristics of nanoparticles (Fig. 1). The composition of the protein depends on the source from which it is derived and can have a significant influence on the preparation of nanoparticles. Usually, the proteins are composed of different molecular weight fractions. Therefore, batch-to-batch variation can have a significant influence on the nanoparticle characteristics. Free thiol groups in HSA undergo oxidation, resulting in dimers (42). Langer et al. studied (42) the influence of dimers and higher aggregates of HSA on the preparation of albumin nanoparticles. The batches with higher molecular weight fractions resulted in larger particle size and higher polydispersity. Similarly, the reproducible particle

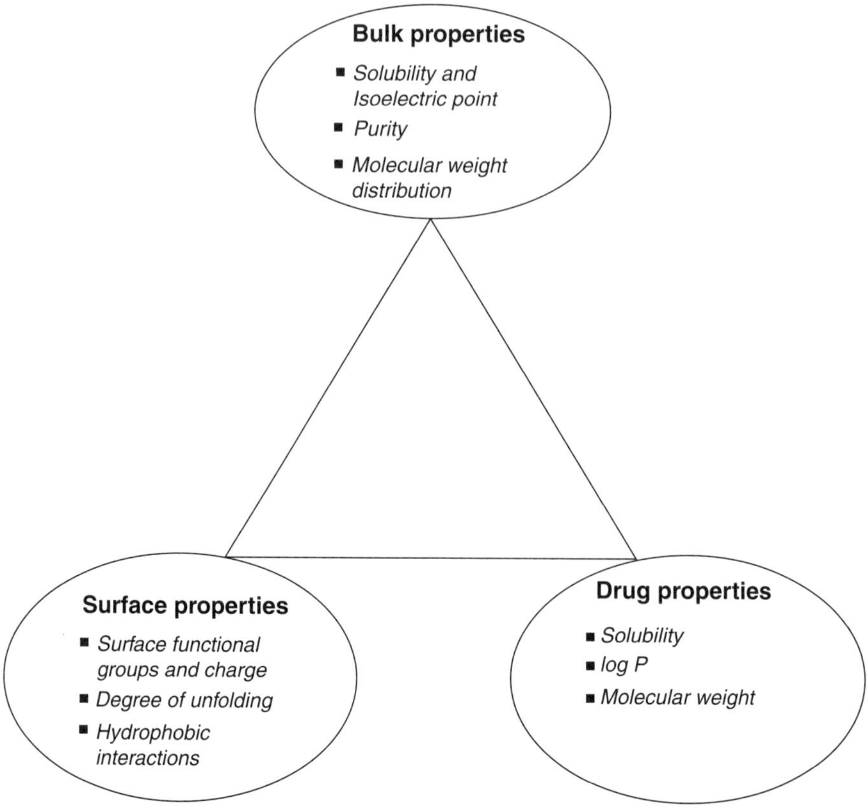

FIGURE 1 Factors that influence the preparation and performance of protein nanoparticles.

characteristics of gelatin nanoparticles are affected by the broad molecular weight distribution of gelatin (43). This can be overcome by using a two-step desolvation process to remove the large molecular weight aggregates (43). Proteins of human or animal origin also carry the risk of pathogen contamination. These issues can be addressed using rDNA technology, and further it can be used to produce proteins of consistent quality in large quantities (3,20). Similar to animal/human proteins, the plant proteins also contains various molecular weight fractions. In addition, the presence of pigments can also interfere with the preparation of nanoparticles. This necessitates a purification of the protein polymers before the preparation of nanoparticles (37).

Protein Solubility

The solubility of the protein in aqueous or organic solvent dictates the choice of the method and the nanoparticle characteristics. Protein nanoparticles are prepared by utilizing the differential solubility of the protein in aqueous and nonaqueous solvents, as proteins can fold or unfold depending on the polarity of the solvent (1). In the case of gliadin, a hydrophobic protein with low water solubility, it was found that the use of a solvent with a solubility parameter close to that of gliadin resulted in smaller sized nanoparticles (38,44). Proteins exhibit pH-dependent water solubility based on their pI. The pH value of the aqueous solution was found to have a significant influence on the size of the albumin nanoparticles (45). Particle size decreased with increasing pH above the pI of albumin (pI = 5.05) (45). At a pH away from the pI, the hydrophobic interactions in the protein are reduced, resulting in lesser aggregation (26). Type A gelatin has a pI of 7 to 9, whereas type B gelatin has a pI of 4 to 7 (Table 1). Hence, the source of gelatin can influence the size, zeta-potential, and drug loading in gelatin nanoparticles (46, 47). For type A gelatin and type B gelatin, pH 7 and pH 3, respectively, were found to be critical for the preparation of nanoparticles (48). At these pH values, the strengths of electrostatic interactions are maximized for obtaining stable particles with optimal drug loading. Furthermore, in the case of gelatin, due to its high viscosity, a preheating (40°C) step was used to prepare smaller size nanoparticles (48). However, a higher temperature (50°C) was found to increase the particle size, due to the excessive unfolding of the protein (48). In a comparative study between different strengths of gelatin (75, 175, and 300 bloom strength), it was found that 300 bloom strength was optimal in terms of particle size, polydispersity index, and drug encapsulation (48).

Surface Properties

One of the significant advantages of protein polymers is the presence of numerous surface functional groups that can be used for modifying the surface of the protein nanoparticles to alter their biodistribution or biocompatibility or drug loading and/or improve their enzymatic stability (Fig. 2). Particle size and surface properties of protein nanoparticles depend on the number of disulfide bonds, number of thiol groups, degree of unfolding, electrostatic repulsion among protein molecules, pH, and ionic strength. The conformational changes in a protein (i.e., unfolding of protein structure) expose its active interactive sites such as amine, carboxyl, and thiol groups. The surface amine groups can be cross-linked using cross-linkers such as glutaraldehyde (Fig. 2). The protonation or deprotonation of the surface amine groups can influence the degree of cross-linking (46). An increase in cross-linker concentration generally decreases the particle size of protein nanoparticles due to the formation of denser particles (48). Cross-linking helps in controlling the drug release from protein nanoparticles, as well as stabilizing the particles against proteolytic breakdown. The surface functional groups can also be used to load drugs by electrostatic interactions (47). Surface amino groups in proteins can be used to attach hydrophilic polymers such as PEG to avoid phagocytic uptake and increase the circulation half-life of the protein nanoparticles (4). Protein nanoparticles can be functionalized to respond to various stimuli. Thermoresponsive albumin nanoparticles were prepared by conjugating poly(*N*-isopropylacrylamide-co-acrylamide)-block-polyallylamine (PAN) on the surface carboxyl groups in albumin (49). Kommareddy and Amiji (50) thiolated the surface amine groups to form disulfide bonds in gelatin nanoparticles. These nanoparticles can release their cargo under the highly reducing environment in the tumor cells (50). Similarly, ligands have been attached to the surface of protein nanoparticles for drug targeting to specific tissues in the body (51,52). The surface functional groups in the protein can also directly interact with the biological membrane. Gliadin nanoparticles have been reported to show bioadhesive property in the intestinal membrane, where the surface amino acids adhered to the intestinal membrane through hydrogen bonding and hydrophobic interactions (39).

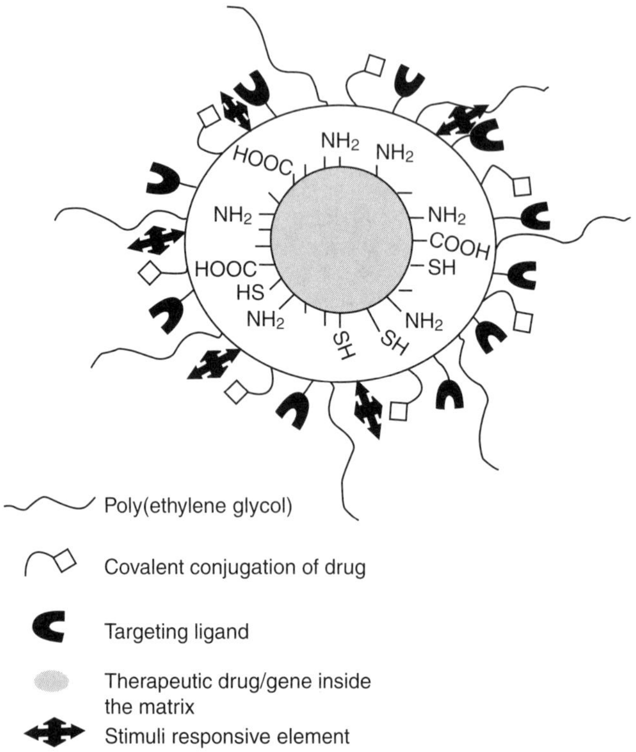

FIGURE 2 Surface modifications of protein nanoparticles for drug delivery applications.

Drug Properties

The physicochemical properties of the drug, such as solubility, log P, and molecular weight, influence the drug loading in protein nanoparticles. The drugs can be loaded through encapsulation in the nanoparticle or by the interaction of drug with the protein through covalent or noncovalent interactions. Highly hydrophobic drugs have been found to interact with cysteine residues in the albumin through hydrophobic interactions (19). For example, paclitaxel a highly hydrophobic drug is loaded in albumin nanoparticles by mixing albumin and paclitaxel in a high pressure homogenizer (53). A large number of drugs are known to bind to serum albumin, and hence albumin appears to be a promising carrier (19). In the case of gelatin nanoparticles, higher encapsulation efficiency was reported for hydrophilic drugs than for hydrophobic drugs (46). Doxorubicin was adsorbed onto gelatin-coated iron oxide nanoparticles for targeting using magnetic field (47). It was found that the adsorption of cationic doxorubicin onto gelatin nanoparticles increased with increasing pH due to the negative charge of gelatin at higher pH. On the other hand, the encapsulation of doxorubicin in gelatin-coated iron oxide nanoparticles showed a slower drug release than surface-adsorbed nanoparticles (47). The release of hydrophilic drugs from gelatin nanoparticles was found to occur by zero-order kinetics, whereas hydrophobic drugs were released by pseudo zero-order kinetics (46). Duclairoir et al. (54) studied the release of drugs of varying polarity from gliadin nanoparticles. It was found that hydrophobic drugs are slowly released because of their higher affinity to hydrophobic gliadin. On the other hand, hydrophilic drugs showed a burst release followed by slower drug diffusion from the nanoparticle matrix. Protein nanoparticles have also been used to encapsulate protein drugs and DNA (50,55). High encapsulation efficiency (90%) of bone morphogenetic protein was reported with albumin nanoparticles (55). Similarly, high encapsulation efficiency (100%) has been reported for DNA with gelatin nanoparticles (50). Unlike synthetic polymers, macromolecular drugs can be loaded in protein nanoparticles under milder conditions.

PREPARATION METHODS

The preparation of protein nanoparticles is based on balancing the attractive and repulsive forces in the protein. It is generally recognized that increasing protein unfolding and decreasing intramolecular hydrophobic interactions are critical determinants for nanoparticle formation (26). During nanoparticle formation, the protein undergoes conformational changes depending on its composition, concentration, preparation conditions (such as pH, ionic strength, solvent), and cross-linking methods. Usually, surfactants are required to stabilize the nanoparticles of water-insoluble proteins such as gliadin (40). The unfolding of the proteins during the preparation process exposes interactive groups such as disulfides and thiols. Subsequent thermal or chemical cross-linking leads to the formation of cross-linked nanoparticles with entrapped drug molecules. Coacervation/desolvation and emulsion-based methods are most commonly used for the preparation of protein nanoparticles.

Coacervation/Desolvation

Coacervation or desolvation is based on the differential solubility of proteins in solvents as a function of the solvent polarity, pH, ionic strength, or presence of electrolytes. The coacervation process reduces the solubility of the protein, leading to phase separation (Fig. 3). The addition of a desolvating agent leads to conformation change in the protein structure, resulting in precipitate or coacervate formation. By controlling the processing variables, the size of the particles in the coacervate phase can be controlled (45). After the nanoparticles are formed, they are cross-linked using glutaraldehyde or glyoxal (4,45). Organic solvents, including acetone or ethanol, have been used as antisolvents for the preparation of protein

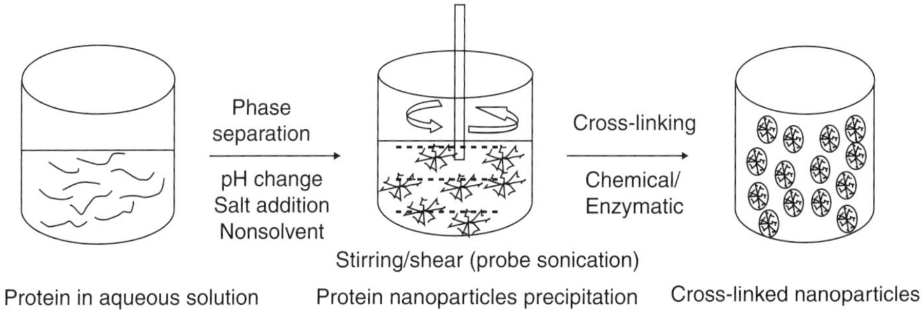

FIGURE 3 Preparation of protein nanoparticles by using coacervation or phase separation method.

nanoparticles (55). Gelatin nanoparticles are prepared by dissolving gelatin in an aqueous solution (pH 7), followed by changing the solvent composition from 100% water to 75 vol% of hydroalcoholic solution, upon gradual addition of ethanol with stirring (4). In the case of albumin nanoparticles, it was found that the use of acetone as an antisolvent produced smaller nanoparticles when compared to those obtained by the use of ethanol (55). An increase in the antisolvent/solvent ratio decreases the particle size. This is due to the rapid extraction, or diffusion, of the solvent into the antisolvent phase, thus limiting the particle growth (55). Langer et al. (45) studied the various process parameters that influenced HSA nanoparticles. The pH value prior to the desolvation step was found to be a critical factor governing the size of HSA nanoparticles. Higher pH values led to smaller nanoparticles in the size range of 100 to 300 nm. In this regard, it is essential to maintain the pH away from the pI to keep the protein in the deaggregated state and thus obtain smaller nanoparticles (26). A higher salt concentration can neutralize or mask the surface charge and promote particle agglomeration (45). On the other hand, in the case of legumin, a relatively more hydrophobic protein, an increase in ionic strength helped to increase the solubility of legumin in the aqueous phase and produce smaller nanoparticles (41).

The hydrophobicity of the protein can also influence the size of the nanoparticles. Smaller nanoparticles (~130 nm) were obtained using BLG, which has a similar pI value but lower hydrophobicity than BSA (26). Denaturation of BLG by heat treatment prior to phase separation further reduced the particle size of the nanoparticles to approximately 60 nm. Orecchioni et al. (44) studied the gliadin (a hydrophobic protein) nanoparticle formation using various ethanol/water ratios. Smaller nanoparticles were obtained at an ethanol/water ratio that matched the solubility parameter of gliadin and in which the protein was in the expanded conformation. For BSA, an increase in protein concentration decreased the particle size due to the increased nucleation of protein nanoparticles on addition of antisolvent (55). Protein nanoparticles can be rigidized by cross-linking using glutaraldehyde or glyoxal, and an increase in cross-linker generally decreases particle size due to the formation of denser particles (48). Lysine residues in the protein are generally involved in the cross-linking. In case of albumin, it was found that the non–cross-linked protein nanoparticles coalesced to form a separate phase (45). Therefore, cross-linking stabilizes the protein nanoparticles and, in addition, reduces enzymatic degradation and drug release from the protein nanoparticles (42,48). However, it is essential to remove the cross-linkers as completely as possible due to their known cytotoxic properties (1). Furthermore, the cross-linkers can also affect the stability of drugs, particularly protein drugs during entrapment in the nanoparticles. Glutaraldehyde cross-links the amino groups in interferon (IFN)-γ with the BSA matrix, leading to the loss of biological activity of IFN-γ (1). Bone morphogenetic protein–loaded BSA nanoparticles were prepared without using cross-linkers (55). Instead, the BSA nanoparticles were stabilized by surface coating with cationic polymers (polyetheyleneimine and polylysine).

Nanoparticles prepared using hydrophobic proteins generally require surfactants to stabilize the nanoparticles (41). Poloxamer was used to improve the solubility of legumin in the aqueous phase and stabilize the nanoparticles during phase separation (41). An increase in surfactant concentration increased the yield without appreciably altering the particle size (41). Drugs can be loaded by either surface adsorption onto preformed nanoparticles or entrapping the drug during the preparation of nanoparticles. In case of ganciclovir, a higher drug loading was achieved by entrapping the drug in HSA nanoparticles compared to that achieved by surface adsorption on the nanoparticles (56). However, the encapsulation efficiency depends on the drug properties and the drug/polymer ratio, among other factors.

Emulsion/Solvent Extraction

In this method, an aqueous solution of the protein is emulsified in oil by using a high-speed homogenizer or ultrasonic shear and the nanoparticles are formed at the w/o interface (Fig. 4). Surfactants, such as phosphatidyl choline or Span 80, are included as stabilizers to produce nanoparticles (57,58). Finally, the oil phase is removed using organic solvents. Mishra et al. (57) reported an emulsion-based method for preparing nanoparticles using HSA. Olive oil was used as an oil phase and was slowly added to the aqueous protein solution under mechanical stirring, followed by ultrasonication. Phosphatidyl choline was used as a surfactant in this method. The cross-linker glutaraldehyde was added to the emulsion to obtain nanoparticles in the size range of 100 to 800 nm. The protein concentration and the aqueous-phase volume (w/o) ratio influenced the particle size. An increase in w/o phase volume ratio increased the particle size of nanoparticles. Similarly, an increase in protein concentration increased the particle size. Yang et al. (58) used a w/o emulsion to prepare BSA nanoparticles for entrapping hydrophobic drug by using thermal cross-linking. In this method, the aqueous protein solution was emulsified with castor oil by using Span 80. The resultant emulsion was then added dropwise to heated (120°C–140°C) castor oil with stirring to evaporate the aqueous phase. Albumin concentration, emulsification time, and the rate of emulsion addition to castor oil affected the particle size. In general, the particle size of

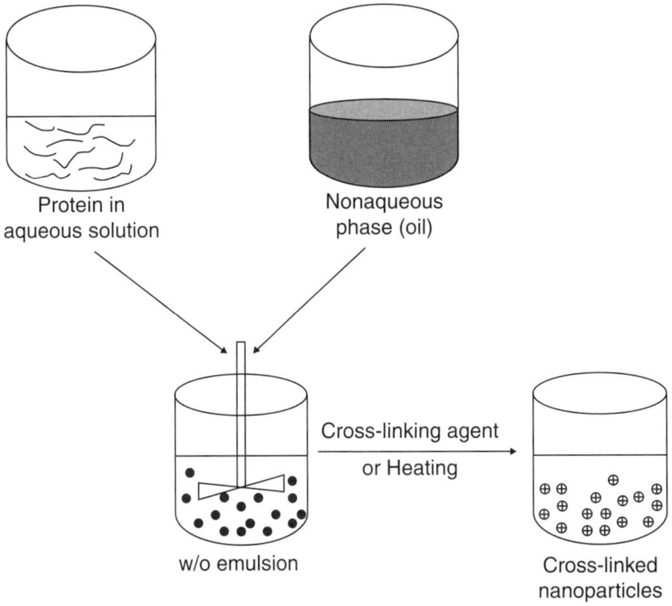

FIGURE 4 Preparation of protein nanoparticles by using the emulsion/solvent extraction method.

nanoparticles prepared by the emulsion method is larger than the nanoparticles prepared by coacervation (59). Furthermore, it is difficult to completely remove the oil phase or the organic solvent from the final formulation.

Complex Coacervation

Since proteins are amphoteric with a large number of charged functional groups, they can be made cationic or anionic by adjusting the pH below or above the pI of the protein, respectively. The charged protein can then undergo electrostatic interactions with other polyelectrolytes (Fig. 5). The electrostatic interaction is used to entrap or adsorb DNA or oligonucleotides in the protein nanoparticles by coacervation. Salt-induced complex coacervation has been reported for entrapping DNA in gelatin nanoparticles (60). At pH 5, gelatin is positively charged and therefore can form a complex coacervate with DNA. Salts such as sodium sulfate can be used to induce desolvation of the local water environment in the polyelectrolyte complex (200–750 nm). Furthermore, the nanoparticles can be stabilized by cross-linking with glutaraldehyde. The DNA is physically entrapped in the protein matrix. Endolysomotropic agents and other drugs can also be coencapsulated during complex coacervation. Rhaese et al. (61) prepared HSA–polyethyleneimine (PEI)–DNA nanoparticles by inducing complex coacervation through charge neutralization. The HSA solution (pH 4) was mixed with PEI and desolvation was achieved by adding sodium sulfate solution containing DNA. The nanoparticles were stabilized using 1-ethyl-3[3-dimethylaminopropyl]carbodiimide (EDC) as a cross-linking agent. The resultant nanoparticles were in the size range of 300 to 700 nm. The ratios of HSA, PEI, and DNA were found to be critical parameters that affected the nanoparticle characteristics and transfection. Smaller nanoparticles (30–300 nm) were prepared using a combination of HSA, oligonulceotides, and protamine (62). Alternatively, cationized proteins have been used to form complex coacervates with DNA. Zwiorek et al. (63) prepared cationized gelatin by covalent attachment of cholamine (a quartenary amine) to the free carboxyl groups in gelatin by using EDC as a coupling agent. In the first step, gelatin nanoparticles were prepared by coacervation by using acetone as a desolvating agent. This was followed by cross-linking with glutaraldehyde. Cholamine was conjugated to the surface of the gelatin nanoparticles at pH 4.5. The resultant cationized gelatin nanoparticles were used to adsorb the DNA at pH 7.4. These nanoparticles were in the size range of 183 to 288 nm, with a neutral or slightly positive zeta-potential. An alternative

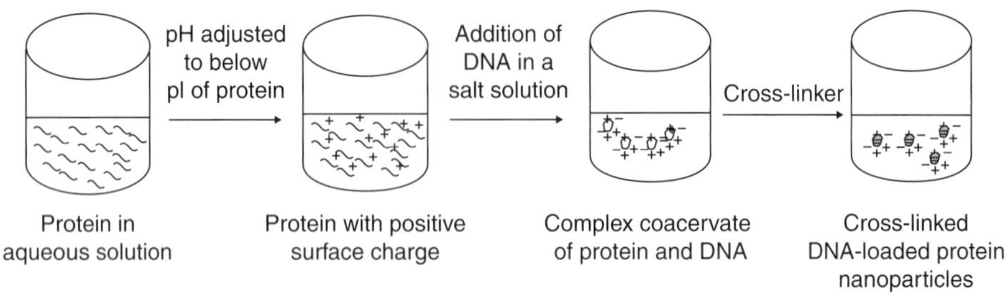

FIGURE 5 Preparation of DNA-loaded protein nanoparticles by using the complex coacervation method. *Abbreviation*: pI, isoelectric point.

approach is to prepare cationized gelatin and then form nanoparticles by salt-induced complex coacervation with DNA (64).

BIOLOGICAL CHARACTERISTICS OF PROTEIN NANOPARTICLES

Biocompatibility/Immunogenicity

An important determinant of in vivo performance of nanoparticulate systems is the opsonization and clearance of particles by the mononuclear phagocytic system or the reticuloendothelial system (RES) (65). Various steps involved in the opsonization and clearance of particles from the blood are shown in Figure 6. Opsonization is the process by which a foreign organism or particle becomes covered with the so-called opsonin proteins (65). There are more than 3700 plasma proteins which compete to bind to the nanoparticle surface (66). Initially, albumin and fibrinogen dominate the particle surface but are subsequently displaced by higher affinity proteins including immunoglobulins (IgM and IgG), laminin, fibronectin, C-reactive protein, and type I collagen (65,66). These bound proteins determine the subsequent particle uptake by various cells of the immune system and their interaction with other blood components. Opsonization can take a few seconds to many days to complete (65). Van der walls, electrostatic, and hydrophobic/hydrophilic forces are involved in the binding of opsonins to nanoparticles (65). The surface characteristics of nanoparticles, such as hydrophobicity, charge, composition, and method of preparation, primarily influence the adsorption of opsonins to their

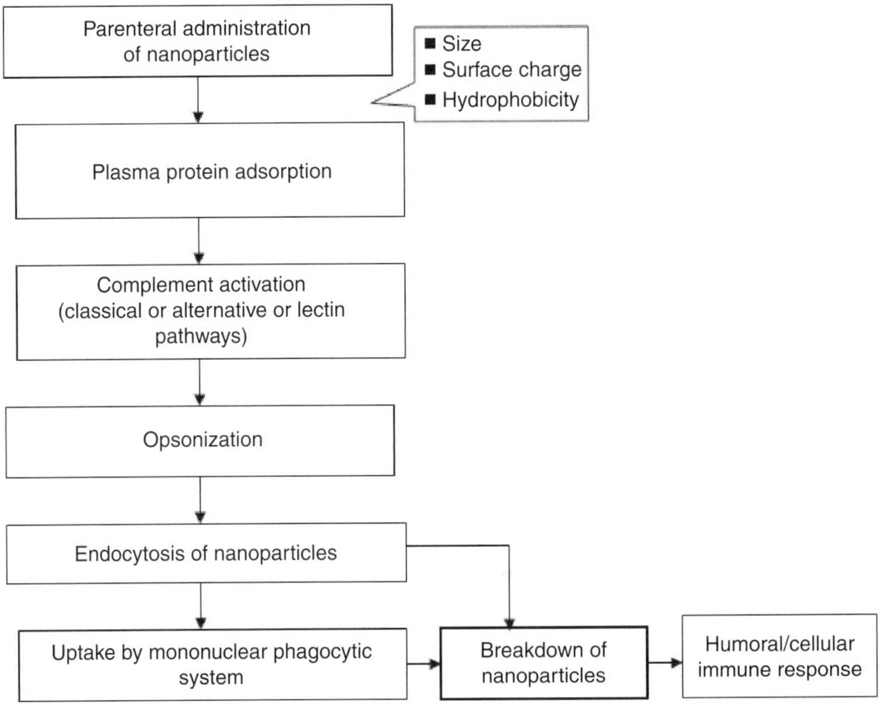

FIGURE 6 In vivo fate of systemically administered protein nanoparticles.

surface (65,67,68). Generally, neutral and hydrophilic particles undergo much lower opsonization than do charged and hydrophobic particles (68). Furthermore, the complement proteins involved in opsonization are specific to the surface functional groups in the nanoparticles (69). The particle size strongly influences the quality and quantity of proteins adsorbed on the nanoparticle surface. Lower protein adsorption is seen with smaller nanoparticles (50–70 nm) than with larger nanoparticles ($\geq$200 nm) (66,67). After opsonization, the proteins adsorbed on the particle surface interact with the monocytes and various subsets of macrophages and induce their cell uptake by phagocytosis. Alternatively, the complement proteins bound to the particle surface can activate one of the several pathways, including the classical pathway mediated by antigen–antibody binding, antibody independent, alternative pathway, or the lectin pathway, depending on the nature of the nanoparticle (65,69). Regardless of the pathway, finally the particles are taken up by the phagocytes and the degraded particles are presented to the immune system (65). Depending on the antigenicity of the protein fragments, it may induce a humoral or cellular immune response (68).

To avoid the rapid clearance of nanoparticles from the blood, it is critical to block the initial opsonization process. This is achieved by coating or by covalent attachment of shielding groups on the particle surface to block the electrostatic or hydrophobic interactions with the opsonins. PEGylation is one of the widely used strategies to prepare stealth particles that reduce or avoid opsonization (65). PEGylation refers to adsorption, grafting, or covalent attachment of hydrophilic PEG chains on the nanoparticle surface. Covalent attachment or grafting is more effective than simple adsorption of PEG to the particle surface for preventing opsonization (65). Since PEG is hydrophilic and has a neutral charge, it prevents hydrophobic and electrostatic interactions with the plasma proteins, thus avoiding opsonization. Furthermore, PEGylation reduces the immunogenicity and enzymatic breakdown of proteins. Generally, the PEG with a molecular weight of 2 kDa or higher provides better steric hindrance in avoiding opsonization (65). In addition to PEG chemistry, the size of the nanoparticles also influences the blood clearance. Larger PEGylated nanoparticles are cleared more rapidly than smaller PEGlyated nanoparticles (65). PEGylated gelatin nanoparticles were found to be twofold higher in the blood than non-PEGylated gelatin nanoparticles (4). As opposed to avoiding phagocytic uptake, for some applications, it is desirable to target the particles to macrophages. Particularly, this is true for immunostimulation of macrophages against microbial infections or immunosuppression of macrophages in autoimmune diseases. Segura et al. (70) used larger albumin nanoparticles (200–300 nm) to target IFN-γ to macrophages infected with *Brucella abortus* in mice. Similarly, gelatin nanoparticles ($\sim$200 nm) were used to target clodronate to macrophages for autoimmune diseases (71).

Immunogenicity is one of the major concerns with protein polymers, although most of the proteins used for drug delivery applications are known to be weak immunogens (2). This is because most of the animal or plant proteins used for drug delivery applications are endogenous or food substances. As a general rule, the body's immune system recognizes any foreign material or infectious agent as an antigen (immunogen) and mounts innate and/or adaptive immune response. Natural killer cells and phagocytes such as neutrophils, monocytes, macrophages, and dendritic cells mainly mediate the innate immune responses. Adaptive immune responses are mediated by antibodies produced by B cells (humoral immune response) or specifically activated T cells (cellular immune response). Almost all biologically important molecules such as proteins, sugars, lipids, and nucleic acids can act as antigens and induce an immune response (68). Humoral immune responses to these molecules can be T-cell dependent or T-cell independent. Most proteins are T-cell dependent antigens that are processed and presented by antigen-presenting cells through major histocompatibility complex molecules to T cells (68). These antigen-activated T cells help antigen-specific B cells to produce antibodies to protein antigens. However, a variety of factors can influence the immunostimulation or immunosuppression properties of protein nanoparticles (68). Gelatin and albumin are generally nonimmunogenic (2). It has been reported that natural antibodies are present in the body against these proteins, probably to clear the metabolites of collagen and damaged autologus albumin (72,73). Gelatin–magnetite nanoparticles did not produce any significant increase in antibodies over and above the natural antibodies in mice. However, a small percentage of the population is allergic to certain proteins (such as gelatin), where an immediate hypersensitivity reaction mediated by IgE and in some cases delayed hypersensitivity reactions have been observed (74).

Legumin nanoparticles were found to be nonimmunogenic, whereas legumin was found to produce antilegumin antibodies (75). The nanoparticle preparation method and the use of cross-linkers can alter the conformation of the protein, thus making it nonimmunogenic. Furthermore, the cross-linking may prevent the complete degradation of the protein for presenting it to the lymphocytes for immune activation (75). When nanoparticles are used as vaccine delivery carriers, the particulate is intended to act as an adjuvant to enhance the immune response to the vaccine. However, the carrier per se should not be immunogenic. Hurtado-Lopez and Murdan (76) found that when zein microspheres were used as a vaccine carrier, anti-zein antibodies were produced, albeit only slightly more than the control. Therefore, systematic studies are required to understand the influence of various factors on the immunological properties of protein polymers in the particulate form.

Biodegradation

One of the advantages of protein polymers is their biodegradability and is influenced by the physicochemical properties of the protein. Thus, it becomes essential to characterize the influence of proteolytic enzymes on drug release from protein nanoparticles. Albumin nanospheres were found to be stable at pH 7.4 and in rat serum, whereas the nanoparticles degraded in the presence of trypsin (77). Langer et al. (42) studied the degradation of HSA nanoparticles in the presence of various enzymes under appropriate pH conditions. The degree of cross-linking influenced the stability of the HSA nanoparticles towards various enzymes. In general, an increase in cross-linking reduced the enzymatic degradation of HSA nanoparticles, as it is difficult for the enzymes to penetrate into cross-linked particles (42). In the presence of cathepsin, a major lysosomal enzyme, HSA nanoparticles degrade slowly, which implies that the drug can be released in lysosomes after uptake by the cells. In a comparative study, Ko et al. (26) found that BLG nanoparticles were more stable to proteolytic enzymes than BSA nanoparticles. The difference was attributed to the smaller number of trypsin-susceptible lysine and arginine residues in BLG. Attachment of PEG chains on the surface of gelatin nanoparticles afforded greater proteolytic stability and slowed the drug release (78). PEGylation provides steric repulsion for the protein matrix from proteolytic degradation. Furthermore, both gelatin and PEGylated gelatin nanoparticles protected the encapsulated DNA from nuclease. Kommareddy and Amiji (79) prepared thiolated gelatin particles to specifically release the encapsulated contents under the highly reducing environment (i.e., by glutathione) inside the tumor cell. The in vitro release of the DNA from the thiolated gelatin nanoparticles increased with increasing concentrations of glutathione. Shen et al. (49) developed a thermoresponsive system by attaching PAN to the surface of the albumin nanoparticles. The drug release from the PAN–albumin nanoparticles was higher at 43°C than at 37°C due to the shrinkage of the PAN chains at higher temperature, thus facilitating enzymatic attack of albumin. The enzymatic release decreased with increasing molecular weight of PAN due to the steric hydrophilic barrier provided by the PAN chains.

Biodistribution and Applications

For systemic applications, it is generally accepted that the smaller particles (<500 nm) can avoid RES and result in a longer circulation half-life (79). Particularly, in case of solid tumors, the smaller nanoparticles can easily extravasate through the leaky tumor vasculature, whereas they are excluded from intact blood vessels in normal tissues. It has been estimated that the pore size of tumor vasculature varies from 200 to 600 nm, and this has been exploited for passive targeting of nanoparticles to tumors (80). Furthermore, once they reach the tumor interstitium, larger molecules are retained due to the poor lymphatic drainage. Figure 7 schematically shows the combination of enhanced permeability and retention results in the passive accumulation of nanoparticles in tumor tissues (81). A further level of targeting can be achieved by attaching targeting ligands to the nanoparticles. Table 2 provides a representative list of various protein-based nanocarriers used for different drug delivery applications. Albumin-bound paclitaxel nanoparticles (Abraxane) was

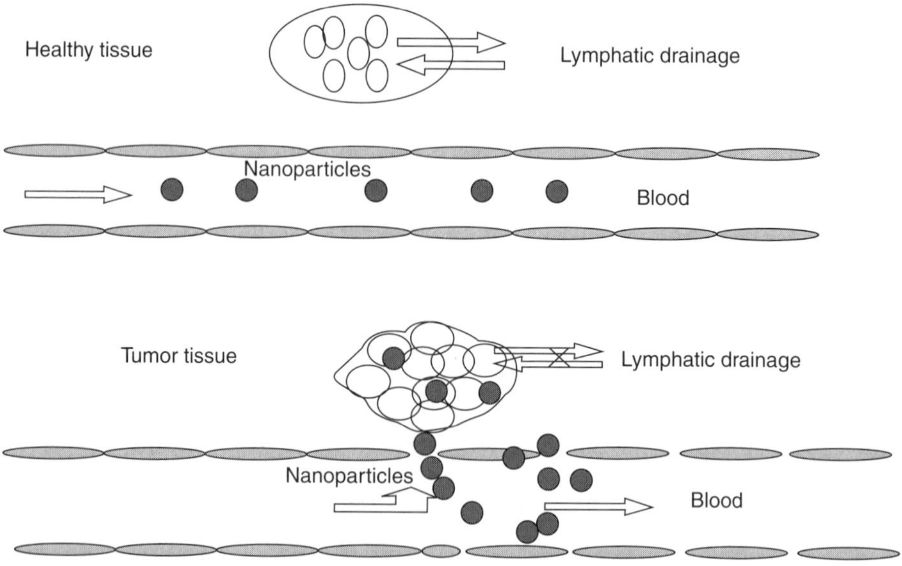

FIGURE 7 Schematic representation of passive targeting of nanoparticles by enhanced permeability and retention effect. *Source:* From Ref. 87.

TABLE 2 Representative Examples of Protein Polymers Used for Nanoparticulate Drug Delivery Systems

Protein polymer	Particle size (nm)	Drug/chemical	Application	Reference
Human serum albumin	200–300	Ganciclovir	Intravitreal delivery	56
Human serum albumin	130	Paclitaxel	Metastatic breast cancer	53
Human serum albumin	218–240	Loperamide: encapsulated Targeting ligands: apolipoprotein A-I, B-1000, and E-3	Brain targeting	52
Bovine Serum Albumin	200–400	Bone morphogenetic protein (BMP-2)	Bone tissue regeneration applications	55
Gelatin	301	Doxorubicin	Magnetic targeting to tumor	47
Gelatin	135–250	Doxorubicin	Passive targeting to squamous cell carcinoma –bearing mice	82
Casein	<200	Curcumin	Improved solubility and intracellular delivery of curcumin to tumor cells	28
β-Lactoglobulin	170 ± 17		Drug delivery system	26
Silk	35–125	Blank particles	Drug and cosmetic delivery applications	32
Gliadin	587 ± 35	Lectin conjugation	Lectin-mediated bioadhesive delivery system	39
	900	Vitamin E (α-tocopherol)	Cosmetic/antioxidant delivery system	40
	412 ± 35	Acetohydroxasamic acid	Treatment of *Helicobacter pylori* infections	83
	900–950	Benzalkonium chloride, linalool, linalyl acetate	Development of controlled release system for drugs of varying physicochemical properties	54
Legumin	250	Blank	Skin drug delivery	41
Zein	100	Essential oils	Controlled release of flavoring agents	36

approved by the USFDA in 2005 for breast cancer treatment (53). The nanoparticles are prepared by simply mixing paclitaxel with HSA in an aqueous solvent and the mixture is passed through a high-pressure homogenizer to form drug-loaded albumin nanoparticles in the size range of 100 to 200 nm. The use of HSA is based on the fact that albumin serves as a carrier for various endogenous and exogenous substances in the body (19). Furthermore, albumin is known to accumulate in tumor tissues due to the leaky tumor vasculature (19). It is also believed that the albumin nanoparticles are taken up inside the cells by a glycoprotein receptor via caveolae (53). Abraxane is a water-soluble paclitaxel formulation and has been found to be superior in various clinical studies compared with the conventional formulation, which contains Cremophor EL to solubilize paclitaxel (53). The albumin-based paclitaxel obviates the need for premedication with antihistamines and corticosteroids, which is the case with the Cremophor EL formulation. Most importantly, the patients can tolerate a higher paclitaxel dose with Abraxane. Also the patients showed higher response rates and longer time to tumor progression without increased toxicity compared to Cremophor EL formulation (53). Based on the huge commercial success of Abraxane ($300 million in 2 years), albumin-based nanoparticulate system is being explored for other drugs (84).

Surface modification with PEG has been used to prepare long-circulating gelatin nanoparticles (4). PEGylated gelatin nanoparticles achieved twofold higher blood levels than did the non-PEGylated gelatin nanoparticles. PEGylation also significantly enhanced the tumor accumulation of gelatin nanoparticles, and the tumor half-life of PEGylated gelatin nanoparticles was sixfold higher than that of the non-PEGylated gelatin nanoparticles (4). Lee et al. (82) showed that doxorubicin-loaded PEGylated nanoparticles significantly inhibited the tumor growth compared with the free drug and drug-loaded gelatin nanoparticles. Ligands have been attached to the surface of protein nanoparticles to target them to specific cells or tissues (51,52,57). Wartlick et al. (51) modified the surface of albumin nanoparticles by covalently attaching avidin, which was then utilized to attach biotinylated HER2-specific antibody. This was used to target the albumin nanoparticles to breast cancer cells, which overexpress HER2. Mishra et al. (57) modified the surface of HSA nanoparticles with PEG to attach transferrin. This nanoparticulate system was found to increase the delivery of azidothymidine by two- to threefold over PEGylated albumin nanoparticles. Alternatively, apolipoprotein has been attached to the surface of HSA nanoparticles through PEG-[*N*-hydroxyl succinimide] linker for brain targeting of loperamide (52). The pharmacological effect was measured using a tail-flick method in mice. A significantly higher antinoceptive response was achieved when animals are treated with drug-loaded, apolipoprotein-conjugated albumin nanoparticles.

Gliadin nanoparticles can be used as a bioadhesive delivery system for oral drug administration (39). The neutral amino acids in gliadin are believed to interact with the intestinal mucosa through hydrogen bonding, whereas the lipophilic amino acids in gliadin can interact with the mucus through hydrophobic interactions (39). Attachment of lectin to gliadin nanoparticles resulted in greater binding specificity to colonic mucosa. These properties can be used to develop colon-targeted drug delivery systems. Bhavsar and Amiji (85) developed a novel nanoparticle-in-microsphere oral system (NiMOS) for gene delivery to specific regions of the gastrointestinal (GI) tract (Fig. 8). Plasmid DNA was encapsulated in gelatin nanoparticles, which was further encapsulated in polycaprolactone (PCL) microspheres (<5 μm) by using an o/w emulsion preparation method. The PCL

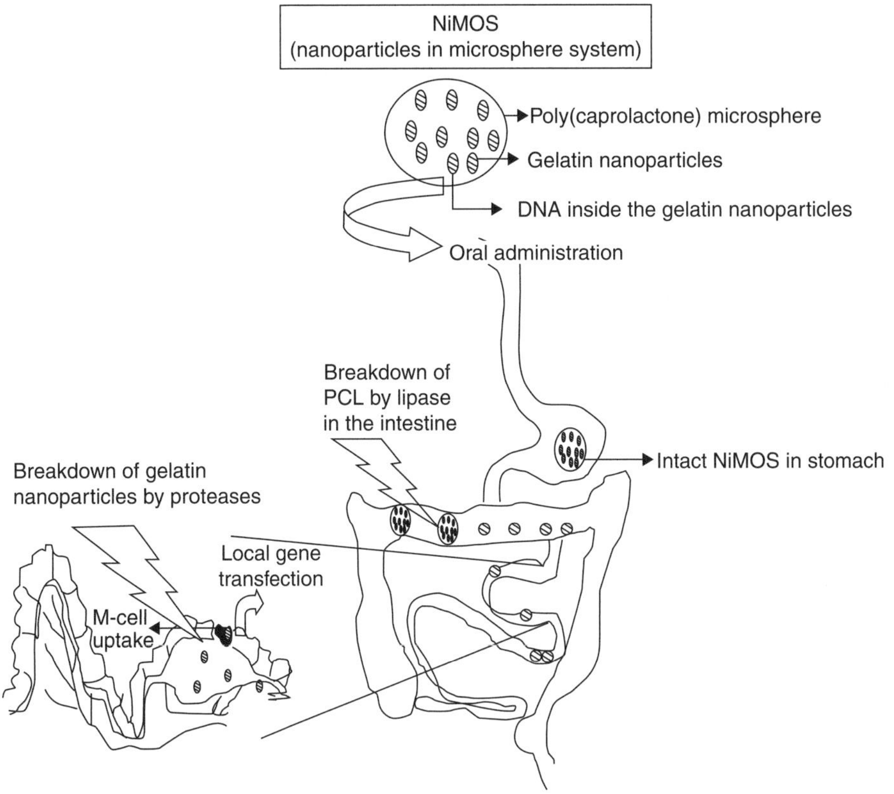

FIGURE 8 Schematic representation of localized gene delivery by using nanoparticle-in-microsphere oral system (NiMOS) through oral route.

layer was used to protect the gelatin nanoparticles from the acidic and enzymatic degradation in the stomach but is acted upon by lipases in the intestine to release the gelatin nanoparticles. The gelatin nanoparticles are then acted upon by proteases to release the DNA for localized delivery (Fig. 8). Following oral administration in rats, the gelatin nanoparticles traversed the GI tract fairly quickly, whereas NiMOS resided in the stomach and intestine for relatively longer duration. Only NiMOS was able to achieve transfection, and significant transgene expression was observed 5 days postadministration in small and large intestines. Mo et al. (86) used albumin nanoparticles for ocular delivery of Cu,Zn superoxide dismutase (*SOD1*) gene. Albumin nanoparticles were found to protect the DNA from degradation in vitreous humor and were taken up by the human retinal pigment cells in vitro through receptor-mediated endocytosis via caveolae. Intravitreal injection of this albumin nanoparticles achieved transgene expression after 48 hours. On the other hand, no expression was seen with the naked DNA.

SUMMARY

With an increasing interest in nanoparticulate delivery systems, there is a greater need to identify biomaterials that are biocompatible and safe for human applications. To this end, protein polymers from animal and plant sources are promising materials for designing nanocarriers. However, it is essential to ensure that there is batch-to-batch consistency with respect to purity and composition. This can be addressed using recombinant technology, in which the composition of the protein can be precisely defined and tailored for specific drug delivery applications such as drug release, targeting, and stimuli responsive drug release. Alternatively, the protein polymers can also be combined with other synthetic polymers to suit specific drug delivery applications. An important issue in protein polymers is the possibility of inducing immune or inflammatory response, particularly with nonautologous proteins. The protein may behave differently in a particulate form as opposed to the protein in the soluble form. Furthermore, the nanoparticle characteristics such as size, charge, and hydrophobicity may play a significant role in phagocytic uptake and initiating a subsequent immune response. This remains to be investigated systematically. Although protein polymers are biodegradable, it is essential to ensure that there is no premature enzymatic breakdown of the protein nanoparticles in the systemic circulation. Surface modification of the protein nanoparticles can be used to address this issue. Of the various proteins, gelatin and albumin have been widely studied for drug delivery applications. Plant proteins are yet to be investigated widely for drug delivery applications. The commercial success of albumin-based nanoparticles has created an interest in other proteins. An increased understanding of the physicochemical properties coupled with the developments in rDNA technology will open up new opportunities for protein-based nanoparticulate systems.

REFERENCES

1. Wang G, Uludag H. Recent developments in nanoparticle-based drug delivery and targeting systems with emphasis on protein-based nanoparticles. Expert Opin Drug Deliv 2008; 5:499–515.
2. Ziv O, Avtalion RR, Margel S. Immunogenicity of bioactive magnetic nanoparticles: Natural and acquired antibodies. J Biomater Res A 2008; 85:1011–1021.
3. Haider M, Megeed Z, Ghandehari H. Genetically engineered polymers: Status and prospects for controlled release. Adv Drug Deliv Rev 2004; 95:1–26.
4. Kaul G, Amiji MM. Biodistribution and targeting potential of poly(ethylene glycol)-modified gelatin nanoparticles in subcutaneous murine tumor model. J Drug Target 2004; 12:585–591.
5. Jones R. Gelatin: Structure and manufacture. In: Ridgway K, ed. Hard Capsules: Development and Technology. London, England: The Pharmaceutical Press, 1987:31–48.
6. Young S, Wong M, Tabata Y, et al. Gelatin as a delivery vehicle for the controlled release of bioactive molecules. J Control Release 2005; 109:256–274.
7. Perzon I, Djabourov M, Leblond J. Conformation of gelatin chains in aqueous solutions; part 1: A light and small angle neutron scattering study. Polymer 1991; 32:3201–3210.
8. Gouinlock E, Flory P, Scheraga H. Molecular confirmation of gelatin. J Polym Sci 1955; 16:383–395.
9. Herning T, Djabourov M, Leblond J, et al. Confirmation of gelatin chains in aqueous solutions: A quasi elastic light scattering study. Polymer 1991; 32:3211–3217.
10. Piez K, Weiss E, Lewis M. The separation and characterization of the alpha and beta components of calf skin collagen. J Biol Chem 1960; 235:1987–1991.
11. Digenis G, Gold T, Shah V. Cross-linking of gelatin capsules and its relevance to their in vitro-in vivo performance. J Pharm Sci 1994; 83:915–921.
12. Ames W. Heat degradation of gelatin. J Soc Chem Ind 1947; 66:279–284.
13. Courts A. The N-terminal amino acid residues of gelatin; part 1; Intact gelatins. Biochem J 1954; 58:70–74.

14. Courts A. The N-terminal amino acid residues of gelatin; part 2: Thermal degradation. Biochem J 1954; 58:74–79.
15. Courts A. The N-terminal amino acid residues of gelatin; part 3: Enzymatic degradation. Biochem J 1955; 58:382–386.
16. Peters T. Serum albumin. Adv Protein Chem 1985; 37:161–245.
17. Carter DC, Ho JX. Structure of serum albumin. Adv Protein Chem 1994; 45:153–203.
18. Gyathri VP. Biopolymer albumin for diagnosis and in drug delivery. Drug Dev Res 2003; 58:219–247.
19. Felix K. Albumin as a drug carrier: Design of prodrugs, drug conjugates and nanoparticles. J Control Release 2008; 132:171–181.
20. Chuang VTG, Kragh-Hansen U, Otagiri M. Pharmaceutical strategies utilizing recombinant human serum albumin. Pharm Res 2002; 19:569–577.
21. Papiz MZ, Sawyer L, Elipoulous EE, et al. The structure of beta-globulin and its similarity to plasma retinol binding protein. Nature 1986; 324:383–385.
22. Reddy IM, Kella NKP, Kinseller JE. Structural and conformational basis of the resistance of D-lactoglobulin to peptic and chymotrypsin digestion. J Agric Food Chem 1988; 36:737–741.
23. Chaitali B, Das KP. Characterization of microcapsules of β-lactoglobulin formed by chemical cross linking and heat setting. J Disp Sci Technol 2001; 22:71–78.
24. Lee SJ, Rosenberg M. Preparation and properties of glutaraldehyde cross-linked whey protein-based microcapsules containing theophylline. J Control Release 1999; 61:123–136.
25. Geraldine H, Stephaine B, Eric B, et al. Use of whey protein beads as a new carrier for recombinant yeasts in human digestive tract. J Biotechnol 2006; 127:151–160.
26. Ko S, Gunasekaran S. Preparation of sub-100-nm β-lactoglobulin (BLG) nanoparticles. J Microencapsul 2006; 23:887–898.
27. De Keuif CG, Holl C. Caesin micelle structure, function and interactions. In: Fox PE, McSweeney PLH, eds. Advanced Dairy Chemistry: Proteins, Part A, Vol. 1. New York: Kluwer Academic/Plenum, 2003:233–276.
28. Sahu A, Kasoju A, Bora U. Fluorescence study of the curcumin-casein micelle complexation and its application as a drug nanocarrier to cancer cells. Biomacromolecules 2008; 9:2905–2912.
29. Marsh RE, Corey RB, Pauling L. An investigation of the structure of silk fibroin. Biochim Biophys Acta 1955; 161:1–34.
30. Mello CM, Senecal K, Yeung B, et al. In: Kaplan DL, Vidams WW, Farmer B, Kiney C, eds. Silk Polymers Material Science and Biotechnology, ACS Symposium series vol 54. Washington, D.C.: American Chemical Society, 1994:67–79.
31. Hofmann S, Foo CT, Rossetti F, et al. Silk fibroin as an organic polymer for controlled drug delivery. J Control Release 2006; 111:219–227.
32. Zhang Y, Shen W, Xiang R, et al. Formation of silk fibroin nanoparticles in water-miscible organic solvent and their characterization. J Nanopart Res 2007; 9:885–900.
33. Shukla R, Cheriyan M. Zein: The industrial protein from corn. Ind Crops Prod 2001; 13:171–192.
34. Liu X, Sun Q, Wang H, et al. Microspheres of corn protein, zein, for an ivermectin drug delivery system. Biomaterials 2005; 26:109–115.
35. Torres-Giner S, Gimenerz E, Lagaron JM. Characterization of the morphology and thermal properties of zein prolamine nanostructures obtained by electrospinning. Food Hydrocolloids 2007; 22:601–614.
36. Parris N, Cooke PH, Hicks KB. Encapsulation of essential oils in zein nanospherical particles. J Agric Food Chem. 2005; 53:4788–4792.
37. Arangoa MA, Campanero MA, Popineau Y, et al. Electrophoretic separation and characterization of gliadin fractions from isolates and nanoparticulate drug delivery systems. Chromatographia 1999; 50:243–246.
38. Duclairoir C, Nakache E, Marchias H, et al. Formation of gliadin nanoparticles: Influence of the solubility parameter of the protein solvent. Colloid Polym Sci 1998; 276:321–327.
39. Arangoa MA, Ponchel G, Orecchioni AM, et al. Bioadhesive potential of gliadin nanoparticulate systems. Eur J Pharm Sci 2000; 11:333–341.

40. Duclairoir C, Orecchioni AM, Depraetere P, et al. Alpha tocopherol encapsulation and in vitro release from wheat gliadin nanoparticles. J Microencapsul 2002; 19:53–60.
41. Irache JM, Bergougnoux L, Ezpeleta I, et al. Optimization and in vitro stability of legumin nanoparticles obtained by a coacervation method. Int J Pharm 1995; 126:103–109.
42. Langer K, Anhorn MG, Steinhauser I, et al. Human serum albumin (HSA) nanoparticles: Reproducibility of preparation process and kinetics of enzymatic degradation. Int J Pharm 2008; 347:109–117.
43. Coester CJ, Langer K, van Briesen H, et al. Gelatin nanoparticles by two step desolvation – A new preparation method, surface modifications and cell uptake. J Microencapsul 2000; 17:187–193.
44. Orecchioni AM, Duclairoir C, Renard D, et al. Gliadin characterization by sans and gliadin nanoparticle growth modelization. J Nanosci Nanotechnol 2006; 6:3171–3178.
45. Langer K, Balthasar S, Vogel V, et al. Optimization of the preparation process for human serum albumin (HSA) nanoparticles. Int J Pharm 2003; 257:169–180.
46. Vandervoort J, Ludwig A. Preparation and evaluation of drug loaded gelatin nanoparticles for topical ophthalmic use. Eur J Pharm Biopharm 2004; 57:251–261.
47. Gaihra B, Khil MS, Lee DR, et al. Gelatin-coated magnetic iron oxide nanoparticles as carrier system: Drug loading and in vitro drug release study. Int J Pharm 2008; 365(1–2):180–189. .
48. Nahar M, Mishra D, Dubey V, et al. Development, characterization and toxicity evaluation of amphotericin B loaded gelatin nanoparticles. Nanomedicine 2008; 4:252–261.
49. Shen Z, Wei W, Zhao Y, et al. Thermosensitive polymer-conjugated albumin nanospheres as thermal targeting anti-cancer drug carrier. Eur J Pharm Sci 2008; 35(4):271–282.
50. Kommareddy S,Amiji M. Preparation and evaluation of thiol-modified gelatin nanoparticles for intracellular DNA delivery in response to glutathione. Bioconjug Chem 2005; 6:1423–1432.
51. Wartlick H, Michaelis K, Balthasar S, et al. Highly specific HER2-mediated cellular uptake of antibody-modified nanoparticles in tumor cells. J Drug Target 2004; 12:461–471.
52. Kreuter J, Hekmatara T, Dreis S, et al. Covalent attachment of apolipoprotein A-I and apolipoprotein B-100 to albumin nanoparticles enables drug transport into the brain. J Control Release 2007; 118:54–58.
53. Hawkins MJ, Soon-Shiong P, Desai N. Protein nanoparticles as drug carriers in clinical medicine. Adv Drug Deliv Rev 2008; 60:876–885.
54. Duclairoir C, Orecchioni AM, Depraetere P, et al. Evaluation of gliadins nanoparticles as drug delivery systems: A study of three different drugs. Int J Pharm 2003; 253:133–144.
55. Wang G, Siggers K, Zhang S, et al. Preparation of BMP-2 containing bovine serum albumin (BSA) nanoparticles stabilized by polymer coating. Pharm Res 2008; 25(12):2896–2909.
56. Merodio M, Arnedo A, Renedo MJ, et al. Ganciclovir-loaded albumin nanoparticles: Characterization and in vitro release properties. Eur J Pharm Sci 2001; 3:251–259.
57. Mishra V, Mahor S, Rawat A, et al. Targeted brain delivery of AZT via transferrin anchored pegylated albumin nanoparticles. J Drug Target 2006; 14:45–53.
58. Yang L, Cui F, Cun D, et al. Preparation, characterization and biodistribution of the lactone form of 10-hydroxycamptothecin (HCPT)-loaded bovine serum albumin (BSA) nanoparticles. Int J Pharm 2007; 340:163–172.
59. Müller BG, Leuenberger H, Kissel T. Albumin nanospheres as carriers for passive drug targeting: An optimized manufacturing technique. Pharm Res 1996; 13:32–37.
60. Truong-Le VL, August JT, Leong KW. Controlled gene delivery by DNA-gelatin nanospheres. Hum Gene Ther 1998; 9:1709–1717.
61. Rhaese S, von Briesen H, Rübsamen-Waigmann H, et al. Human serum albumin-polyethylenimine nanoparticles for gene delivery. J Control Release 2003; 92:199– 208.
62. Mayer G, Vogel V, Weyermann J, et al. Oligonucleotide-protamine-albumin nanoparticles: Protamine sulfate causes drastic size reduction. J Control Release 2005; 106:181–187.
63. Zwiorek K, Kloeckner J, Wagner E, et al. Gelatin nanoparticles as a new and simple gene delivery system. J Pharm Pharm Sci 2005; 7:22–28.

64. Xu X, Capito RM, Spector M. Delivery of plasmid IGF-1 to chondrocytes via cationized gelatin nanoparticles. J Biomed Mater Res A 2008; 84:73–83.
65. Owens DE, Peppas NA. Opsonization, biodistribution, and pharmacokinetics of polymeric nanoparticles. Int J Pharm 2006; 307:93–102.
66. Cedervall T, Lynch I, Lindman S, et al. Understanding the nanoparticle-protein corona using methods to quantify exchange rates and affinities of proteins for nanoparticles. Proc Natl Acad Sci U S A 2007; 104:2050–2055.
67. Lundqvist M, Stigler J, Elia G, et al. Nanoparticle size and surface properties determine the protein corona with possible implications for biological impacts. Proc Natl Acad Sci U S A 2008; 105:14265–14270.
68. Marina AD, Scott EM. Immunological properties of engineered nanomaterials. Nat Nanotechnol 2007; 2:469–478.
69. Vonarbourg A, Passirani C, Saulnier P, et al. Parameters influencing the stealthiness of colloidal drug delivery systems. Biomaterials 2006; 27:4356–4373.
70. Segura S, Gamazo C, Irache JM, et al. Gamma interferon loaded onto albumin nanoparticles: In vitro and in vivo activities against *Brucella abortus*. Antimicrob Agents Chemother 2007; 51:1310–1314.
71. Li P, Tan Z, Zhu Y, et al. Targeting study of gelatin adsorbed clodronate in reticuloendothelial system and its potential application in immune thrombocytopenic purpura of rat model. J Control Release 2006; 114:202–208.
72. Schwick H, Heide K. Immunohistochemistry and immunology of collagen and gelatin. Bibl Haematol 1969; 33:111–125.
73. Sansonno DE, DeTomaso P, Papanice MA, et al. An enzyme-linked immunosorbent assay for the detection of autoantibodies to albumin. J Immunol Methods 1986; 90:131–136.
74. Kumagai T, Yamanaka T, Wataya Y, et al. Gelatin-specific humoral and cellular immune responses in children with immediate- and nonimmediate-type reactions to live measles, mumps, rubella, and varicella vaccines. J Allergy Clin Immunol 1997; 100:130–134.
75. Mirshahi T, Irache JM, Nicolas C, et al. Adaptive immune responses of legumin nanoparticles. J Drug Target 2002; 10:625–631.
76. Hurtado-Lopez P, Murdan S. Formulation and characterization of zein microspheres as delivery vehicles. J Drug Deliv Sci Technol 2005; 15:267–272.
77. Lin W, Coombes AGA, Davies SS, et al. Preparation of sub-100 nm human serum albumin nanospheres using a pH-coacervation method. J Drug Target 1993; 1:237–243.
78. Kaul G, Amiji M. Cellular interactions and in vitro DNA transfection studies with poly (ethylene glycol)-modified gelatin nanoparticles. J Pharm Sci 2005; 94:184–198.
79. Kommareddy S, Amiji M. Poly(ethylene glycol)-modified thiolated gelatin nanoparticles for glutathione-responsive intracellular DNA delivery. Nanomedicine 2007; 3:32–42.
80. Yuan F, Dellian M, Fukumura D, et al. Vascular permeability in a human tumor xenograft: Molecular size dependence and cutoff size. Cancer Res 1995; 55:3752–3756.
81. Maeda H, Wu J, Sawa T, et al. Tumor vascular permeability and the EPR effect in macromolecular therapeutics: A review. J Control Release 2000; 65:271–284.
82. Lee GY, Park K, Nam JH, et al. Anti-tumor and anti-metastatic effects of gelatin-doxorubicin and PEGylated gelatin-doxorubicin nanoparticles in SCC7 bearing mice. J Drug Target 2006; 10:707–716.
83. Umamaheswari RB, Jain NK. Receptor mediated targeting of lectin conjugated gliadin nanoparticles in the treatment of *Helicobacter pylori*. J Drug Target 2003; 11:415–424.
84. Abraxis BioScience. Available at: www.abraxisbio.com.
85. Bhavsar MD, Amiji MM. Gastrointestinal distribution and in vivo gene transfection studies with nanoparticles-in-microsphere oral system (NiMOS). J Control Release 2007; 1198:339–348.
86. Mo Y, Barnett ME, Takemoto D, et al. Human serum albumin nanoparticles for efficient delivery of Cu, Zn superoxide dismutase gene. Mol Vis 2007; 13:746–757.
87. Kannan RM, Perumal O, Kannan S. Dendrimers and hyperbranched polymers for drug delivery. In: Labhasetwar V, Lesli-Pelecky DL. Biomedical Applications of Nanotechnology. New York: John Wiley & Co., 2007:105–129.

7 Gold Nanoparticles and Surfaces: Nanodevices for Diagnostics and Therapeutics

Hariharasudhan D. Chirra, Dipti Biswal, and Zach Hilt

Department of Chemical and Materials Engineering, University of Kentucky, Lexington, Kentucky, U.S.A.

INTRODUCTION

Gold has received much interest in the field of biomedical engineering. The use of gold as a key component in biodiagnostic and therapeutic fields has emerged primarily over a period of three decades, though it has been used for centuries for artistic purposes. Gold is known to be the main ingredient for the preparation of an ancient Roman elixir of life. An example of a historical use of gold was in the coloring of glass during the 17th century to produce intense shades of yellow, red, or brown depending on the concentration of gold. In 1842, colloidal gold was used in chrysotype, a photographic process to record images on paper. During the 19th century, the pure form of gold called activated gold, due to its inert behavior to harsh environments, was prominently employed for catalysis (1). With the advent of numerous tools, techniques, and concepts related to nanotechnology, in combination with the inherent property of gold to form functionalized bioconjugates via simple chemistry, gold has found importance in various biodiagnostic and therapeutic applications (2–6). Herein, we detail the progress made in the functionalization of gold surfaces, both planar and particulates, at the nanoscale for diagnostic and therapeutic applications.

FUNCTIONALIZATION OF GOLD

The unique chemical and physical properties of gold render it as effective sensing and delivery systems for pharmaceutical applications (7). The various properties of gold nanoparticles (GNPs) are mostly size dependent and surface characterized; therefore, the controlled synthesis of GNPs is important for bionanotechnological applications. Gold is mostly considered inert and nontoxic. Although gold can be directly used for biomedical applications, unique applications of this inert metal require functionalization with other biomolecules or biocompatible polymeric systems.

Monodispersed GNPs are relatively easy to form, with core sizes ranging from 1 to 150 nm. GNPs with varying core sizes are usually prepared by the reduction of gold salts in aqueous, organic phase, or two phases. However, the high surface energy of GNPs makes them highly reactive, and as a result, they undergo aggregation. The presence of an appropriate stabilizing agent prevents particle agglomeration by binding to the particle surface to impart high stability and also rich linking chemistry if it acts as a functional group for bioconjugation (8–10).

The functionalization of gold surfaces can be achieved by using either "grafting to" or "grafting from" methods (Fig. 1) (11). The "grafting to" method involves

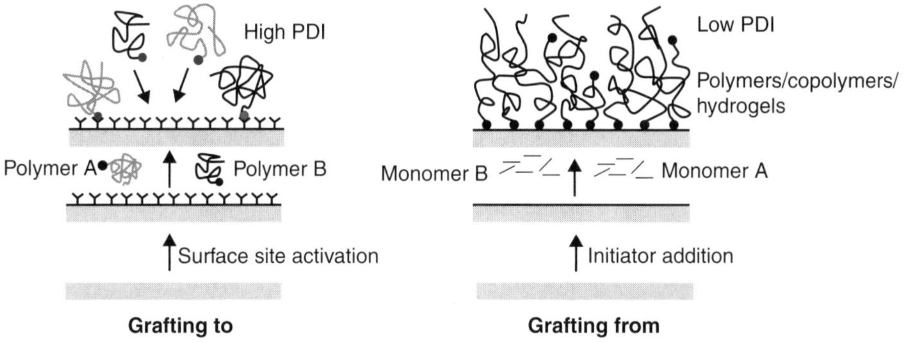

FIGURE 1 Scheme explaining the difference between "grafting to" and "grafting from" surface functionalization methods. *Abbreviation*: PDI, polydispersity.

the reaction of end-functionalized polymers with appropriate surface sites. Initially, grafted polymer layers over these active sites, however, hinder the further attachment of polymer chains because of limited availability of more active sites, thus limiting film thickness and brush density. In the "grafting from" approach, a reactive unit on the surface initiates the polymerization, and consequently the polymer chains grow from the surface. Most "grafting from" polymerization reactions utilize controlled radical polymerization mechanisms. Since monomers diffuse more easily to reactive sites than macromolecules, this approach generally leads to higher grafting densities. A variety of functionalization techniques over gold surface are described in the following text.

"Grafting To" Surface Modification Over Gold Surface

Surface modification of gold particles with stabilizing agents can be achieved by many methods. The thiol gold chemistry is used as the key mechanism for grafting small biomolecules and short-chain, end-functionalized polymeric stabilizers to gold. The "grafting to" fabrication of GNP based sensors and/or delivery systems has been made easy by introducing functional moieties to GNPs that are synthesized by using the one-pot protocol as developed by Brust et al. in 1994 (12). These monolayer-protected clusters of 1.5 to 6 nm are prepared by the reduction of $HAuCl_4$ by sodium borohydride ($NaBH_4$) in the presence of alkanethiol-capping agents. Murray and his colleagues extended Schiffrin's method to diversify the functionality of monolayer-protected clusters to mixed monolayer-protected clusters by using a place-exchange reaction between the thiols (13). Table 1 provides a list of "grafting to" surface-modified particles, as synthesized by various researchers for biorelated diagnostic and therapeutic applications. For further information, the reader is directed to the respective references given in the table for the attachment/reaction chemistry.

Although the synthesis of biomolecule and other short-molecule–protected GNPs is easy, polymer-coated GNPs are currently receiving considerable attention because the polymers are capable of effectively stabilizing nanoparticles by steric effects, controlled growth of the polymers providing a control over the particle size, and monodispersity, thereby providing good stability and uniform properties for biomedical applications. "Grafting to" approaches for the synthesis of such

TABLE 1 Postsynthetic Functionalization Methods Used for the Preparation of Modified Biodiagnostic and Therapeutic Gold Nanoparticles

Type of gold nanoparticles	Functional group attached	References
Biomolecule protected	Peptide	108,109
	Phospholipids	110,111
	Synthetic lipids	112–115
	Microorganism	114,116
	Viruses	117,118
"Green" chemistry	Ionic liquids	119–122
	Polysaccharides: chitosan	123–126
	Polysaccharides: sucrose	127
Dendrimer protected	Poly(amidoamine) based	128–131
	Other dendrimers	132–136
Polymer protected	Linear polymer	137,138
	Hyperbranched polymer	139–141
	Amphiphilic polymer	142,143
	Environmentally responsive	144,145
	PEG functionalized	146–149
	Bioconjugated PEG	150–152

monodisperse particles are not viable. This is due to the effect of steric hindrance on the nonuniform attachment of polymer chains to the gold surface. To overcome this problem, the "grafting from" approach is often preferred.

"Grafting From" or Surface-Initiated Polymerization Over Gold Surfaces

In the "grafting from" approach, a reactive unit on the surface initiates the polymerization, and, consequently, the polymer chains grow from the surface (Fig. 1). This one monomer at a time attachment to the surface of interest leads to a much denser and more uniform polymer-coated surface when compared to that obtained by "grafting to" techniques. While a wide combination of polymeric networks can be obtained via "grafting from" techniques, the same is not viable by using "grafting to" techniques. Surface modification with polymer brushes had been widely used to tailor various surface properties of gold (14,15). Most "grafting from" or surface-initiated polymerization (SIP) reactions work on the principle of controlled radical polymerizations (16). A general mechanism of how controlled radical polymerization renders a uniform, polymer-coated surface is shown in Figure 2. Briefly, an initiator and/or a ligand transforms into a surface-attached radical ($R^{\bullet}$) that thereby undergoes polymer propagation in a controlled manner, whereas the ligand complex radical (in this case, $Z^{\bullet}$ or $ZY^{\bullet}$) participates in controlling the polymerization reaction and finally undergoes termination with the remaining free radicals. A few key elements of the reaction molecules are shown in Figure 2.

There are different types of SIP mechanisms utilized on both planar and particulate gold surfaces. The many types of polymerization mechanism include free radical, cationic, anionic, and ring-opening metathesis polymerization, atom transfer radical polymerization (ATRP), and polymerization using 2,2,6,6-tetramethyl-1-piperidyloxy (11). Although bioconjugation with modified thiols is the most common method for addressing bioapplied gold surfaces, polymerization via "grafting from" techniques affords controlled polymer grafting density and composition. This control is vital in the development of various biosensors and drug

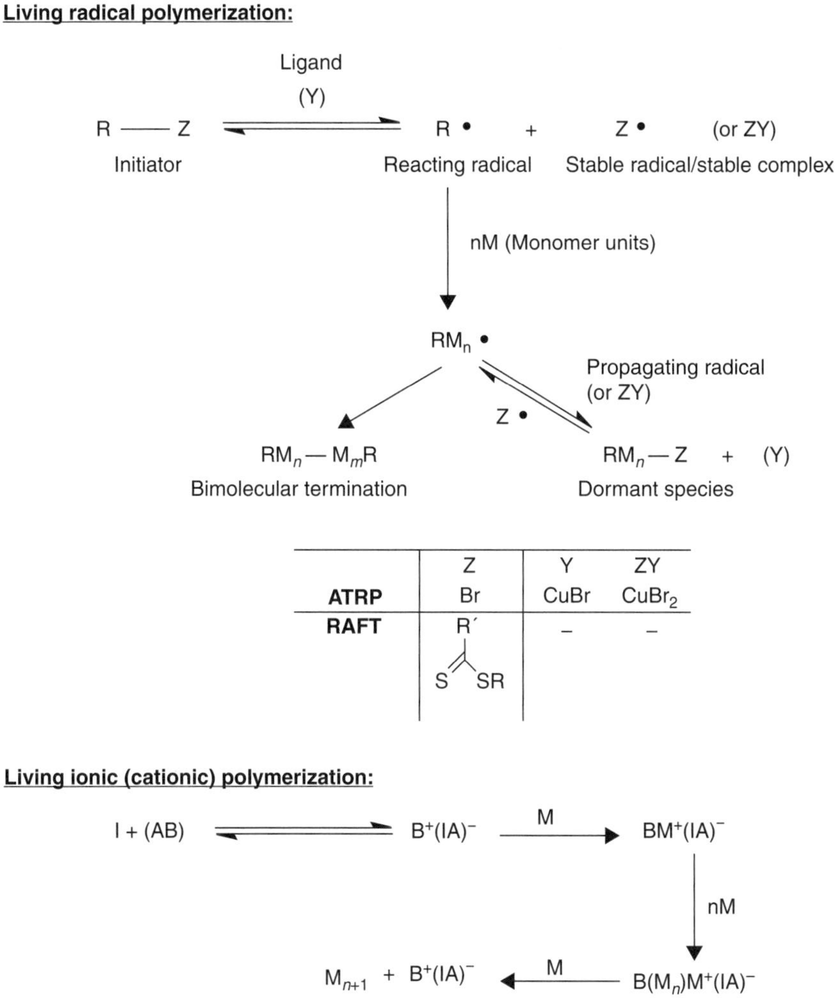

FIGURE 2 Mechanisms of polymerization in living radical and ionic polymerizations.

delivery therapeutic systems for precise biodiagnostic and therapeutic applications. A few of these "grafting from" methods are addressed in the following text.

Living Radical Polymerization

Surface-initiated free radical polymerization (FRP) is used for the polymerization of a large variety of monomers (17,18). For example, Dunér et al. used thermal FRP to synthesize acrylamide (AAm)/acrylate polymer brushes from the Au surfaces that were initially coated with polystyrene-type thin films, derivatized from photolabile groups (19). These AAm and acrylate brushes, in general, are sensitive to environmental stimuli such as temperature and pH. By encompassing biomolecules, which lead to a change in the environmental conditions, these materials on gold surface can be effectively used as biosensors. Although FRP can be used in a variety of polymers, it is often not preferred due to uncontrolled polymer growth leading to coating of varying thickness. The SIP technique in conjunction with a living radical polymerization (LRP) technique is among the most useful synthetic routes to precisely design and functionalize the surface of various materials. Surface-initiated LRP is particularly promising due to its simplicity and versatility.

Living Ionic Polymerization

The use of multiple functionalities on material surfaces enables multiplex usage for various biomedical applications. A control over the polymer thickness, the type of polymer formed (e.g., block copolymer systems), can be controlled via ionic polymerization. Briefly, in this technique, the coinitiator I combines with a Lewis acid or Lewis base (considered also as the initiator) to generate the ionic radical complex $B^+(IA)^-$, which adds up monomer units (M) in propagation steps to produce the polymer (Fig. 2). The Advincula group reported the surface-initiated living anionic polymerization of styrene and diene homopolymers as well as diblock copolymer brushes on gold surfaces (20,21). It used 1,1-diphenylethylene self-assembled monolayer (SAM) as the immobilized precursor initiator from which anionic polymerization of monomer was carried out. Other groups have also reported work related to anionic polymerization on gold surfaces (22). Recently, Ohno et al. synthesized GNPs coated with well-defined, high-density polymer brushes via anionic polymerization (23). Jordan and colleagues have carried out surface-initiated living cationic polymerization over planar gold substrates and functionalized GNPs (24,25). Using these polymerization techniques, dense polymer brushes were prepared in a "one-pot multistep" reaction.

Reversible Addition–Fragmentation Chain Transfer Polymerization

Reversible addition–fragmentation chain transfer (RAFT) polymerization is a versatile, controlled free radical polymerization technique that operates via a degenerative transfer mechanism in which a thiocarbonylthio compound acts as a chain transfer agent (Fig. 2) (26). An advantage of the RAFT polymerization is the synthesis of wide range of polymers with narrow polydispersity and controlled end groups. Poly(*N*-isopropylacrylamide) (PNIPAAm) monolayer–protected clusters on GNPs were synthesized using RAFT polymerization, in which a dithiobenzoate was used as the chain transfer reagent (27). Li et al. synthesized poly(styrene-b-*N*-isopropylacrylamide) (PSt-b-PNIPAAm) with dithiobenzoate terminal group, which was later converted to thiol terminal group by using $NaBH_4$, resulting in a thiol-terminated polymer (PSt-b-PNIPAAm-SH). After PSt-b-PNIPAm-SH reassembled into core–shell micelles in an aqueous solution, GNPs were surface linked onto the micelles in situ through the reduction of gold precursor anions with $NaBH_4$. Thus, temperature-responsive core–shell micelles of PSt-b-PNIPAm with surface-linked GNPs (PSt-b-PNIPAm-GNP) were obtained (28).

Atom Transfer Radical Polymerization

Among the various controlled radical polymerization methods, ATRP has generated high interest because of its versatility in producing controlled polymers with low polydispersity and its compatibility with a variety of functionalized monomers (29). ATRP is relatively tolerant to the presence of water and oxygen, and with the correct choice of catalyst, it can be performed at relatively low temperatures (14). The halogen atom undergoes reversible switching between the oxidation states of the transition metal complexes, thereby reducing the initial radical concentration and also suppressing the bimolecular termination step. This, in addition to the suitable catalyst, increases the control over the polymer growth in ATRP. It has been successfully employed on various surfaces by researchers. It has been used to amplify patterned monolayers of assembled initiators formed using various lithography techniques into polymeric brushes (30–33). For example, Huck and colleagues prepared PNIPAAm-micropatterned domains from mixed SAMs on gold substrates (33,34). Recently, ATRP has been successfully employed for the synthesis of pH-sensitive brushes over gold surfaces (35,36). The combined synthetic route of microcontact printing (μCP) and ATRP has been effectively used in multicomponent polymeric brush patterning by Huck and colleagues (37). In this case, various brushes of different monomer systems have been patterned over gold surface by using repeated μCP followed by ATRP.

Although work has been done using μCP and ATRP as a major method for preparing patterned brushes or copolymer chains for biomedical applications over gold surfaces, minimal research has been carried out for the synthesis of hydrogels over surfaces by using these methods. Hydrogels are hydrophilic, insoluble polymeric networks that have the property of swelling to a high degree when placed in an aqueous or biological medium. By tailoring the various functional groups along the polymer backbone, hydrogels can be designed to be sensitive when subjected to changes in the ambient conditions such as temperature, pH, electric field, or ionic strength. To utilize the environmentally responsive intelligent hydrogel systems over material surfaces for a wide range of applications in the field of biomedical engineering, Chirra et al. (unpublished results, 2008) combined μCP with ATRP to synthesize controlled, temperature-sensitive hydrogel platform over gold surface (Fig. 3). The end result, showing the responsive behavior of a temperature-sensitive hydrogel made up of cross-linker poly(ethylene glycol) (PEG)-400 dimethacrylate (PNIPAAm-co-PEG400DMA) on planar gold surface, is shown in Figure 4. The same strategy of responsive behavior can be used in combination with novel micro- and nanodevices such as MEMS and NEMS for biomedical diagnostic and therapeutic applications.

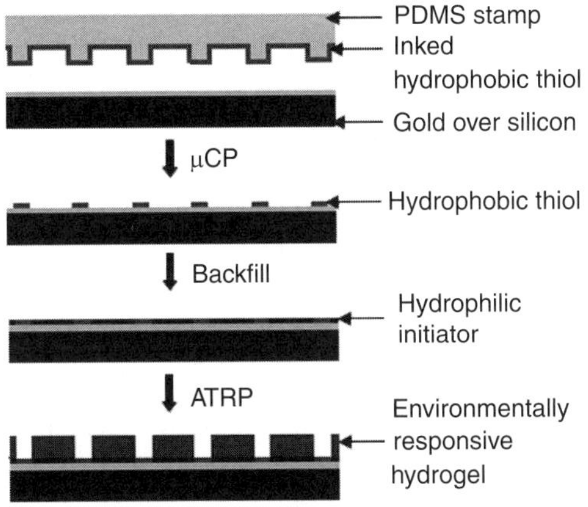

FIGURE 3 Schematic showing the various steps involved in the synthesis of thin hydrogel micropatterns via microcontact printing (μCP) and atom transfer radical polymerization (ATRP). *Abbreviation*: PDMS, polydimethylsiloxane.

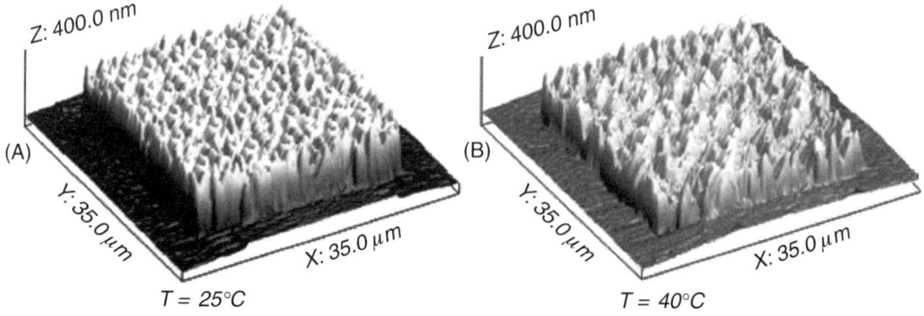

FIGURE 4 Normalized three-dimensional atomic force microscopic images showing the temperature-sensitive behavior of the 90:10 mol% NIPAAm-co-PEG400DMA 25-μm hydrogel square at (**A**) $T = 25°C$, and (**B**) $T = 40°C$. *Source*: From Ref. Chirra et al.

In addition to these SIP techniques, a wide variety of "grafting from" techniques are available. The reader is directed to books compiled by Odian (11) and Matyjaszewski et al. (38).

DIAGNOSTIC APPLICATIONS OF GNPs

Over the past decade, nanotechnology for biomolecular detection has made enormous advancements. Nanoparticles, in particular, have been developed for accurate, sensitive, and selective biosensing devices due to their unique size-related, ease-of-functionalization, and unique physical properties (electrical, optical, electrochemical, and magnetic) (39). GNPs are mostly used in biomedical field as labels for biomolecular detection in the place of conventional molecular fluorophores, where their unique size-tuned optical properties are exploited (40). Some of the biosensing applications of GNPs are detailed in the following text.

Optical GNP Biosensors

In 1996, Mirkin et al. reported the combined optical and melting properties of GNP–oligonucleotide aggregates, which paved way for the development of a plethora of biomolecular optical detection strategies (41). The immediate follow-up work done by researchers by using the colorimetric system worked on the principle of color change observed when a polymeric network of nanoparticles was formed specific to the length of the oligonucleotide that aggregated to the biomolecule of interest (5,42). In addition to the color change, the modified GNP–DNA detection probes also exhibited a sharp melting transition at the detection limit, which was used to distinguish complementary target sequences for one-base mismatch sequences (43,44). While selectivity was achieved with the melting property, the sensitivity was improved 100 times compared with conventional fluorophores by carrying out catalytic silver (Ag) deposition on GNP labels (Fig. 5) (45).

Biobarcode amplification, which is an extension of scanometric nucleotide detection, employs magnetic microparticles with capture probes along with GNPs with both receptor probes and numerous barcode oligonucleotides (46). The presence of a target molecule that matches both the capture and reporter probes forms a magnetic microparticle–GNP sandwich. The scanometric detection of the barcode

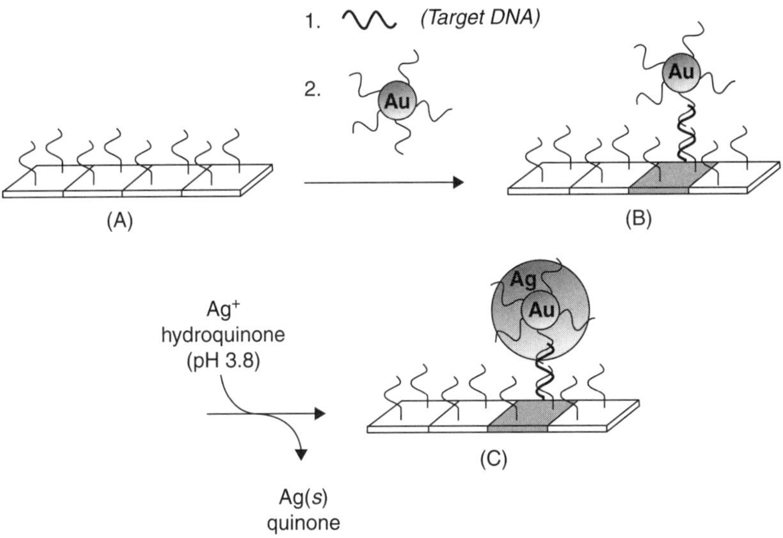

FIGURE 5 Scanometric DNA assay. (**A**) Immobilization of capture probes on the electrode; (**B**) hybridization with target DNA and labeled detection probe; and (**C**) amplification by reductive deposition of Ag followed by scanometric detection. *Source*: From Ref. 45.

nucleotides via Ag amplification gives a detection limit of both DNA and proteins as low as 500 zM.

The surface-enhanced Raman scattering (SERS) of Raman dye–conjugated nanoparticles, along with the narrow spectral characteristics of Raman dyes, was used with GNPs for multiplexed detection of proteins (47) and nucleic acids (48). In both instances, Raman dyes and detection probes are first attached to the GNPs and then were hybridized with the captured targets. A silver coating on the GNP promotes SERS of the dye, and the amplified signal is captured by spectroscopy for fM-level detection. On a similar principle, GNP-coupled surface plasmon resonance effect has been reported for nucleic acid and protein detection (49–51).

A real-time bioaffinity monitoring system based on an angular-ratiometric approach to plasmon resonance light scattering was recently developed by Aslan and colleagues (52). The process depends on the basic principle that as the bioaffinity reaction proceeds with the size increase of GNPs, they deviate from Rayleigh theory and thereby scatter more light in a forward direction relative to the incident geometry. An innovative technique involving a GNP detector–fluorophore was employed by Maxwell and colleagues for the optical detection of nucleic acids (53). In this strategy, the target molecule induces a conformational change of the detector–fluorophore chain, from arch to stretched form and vice versa, thereby either restoring or quenching the fluorophore for optical detection (Fig. 6).

Biomolecule-Conjugated Electrochemical Biosensors

Although direct electrical detection is the simplest method for biosensing (54), a large fraction of the GNP-based sensing research involves electrochemical detection for bioaffinity, specificity, and improved sensitivity reasons. A summary of the recent approaches in the construction of electrochemical biosensors utilizing

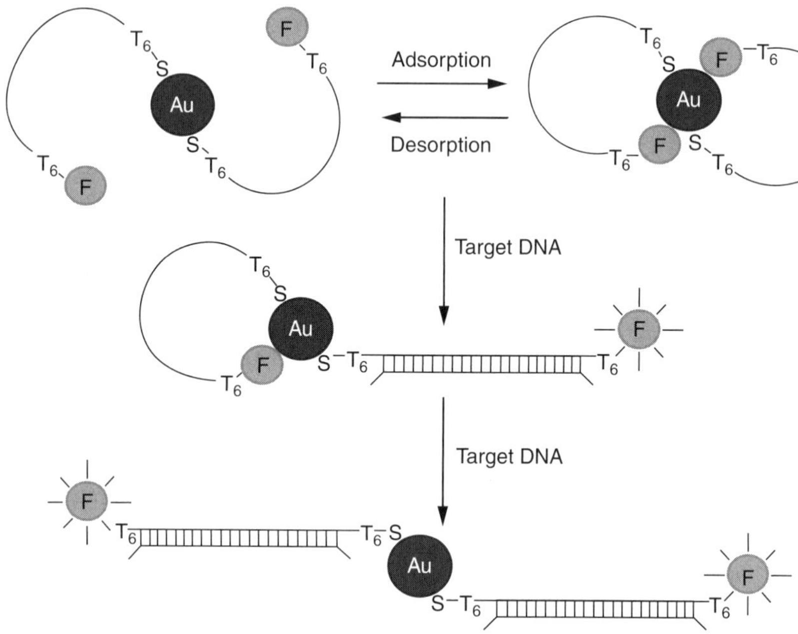

FIGURE 6 Scheme representing the conformational change–induced fluorophore activity for target DNA detection by using a fluorophore-attached gold nanoparticle. *Source*: From Ref. 53.

the various binding and electroenhancing properties of GNPs is best reviewed by Pingarrón et al. (55). Research involving GNPs for electrochemical biosensing is largely focused on enzyme-conjugated electrodes. One of the first electrochemical GNP biosensor was a simple enzymatic glucose biosensor that had the glucose oxidase (GOx), covalently attached to a GNP-modified Au electrode. Mena et al. then modified the GOx sensor for increased sensitivity and lifetime (56). They used a configuration involving colloidal gold, bound to cystamine SAM on a gold electrode for blood glucose sensing. GNP modified with 4-aminothiophenol, conjugated with hepatitis B antibody, was used to detect hepatitis B virus surface antigen via electrochemical impedance spectroscopy (57).

Electrochemical DNA biosensors provide useful analytical tools for the diagnosis of sequence-specific DNA strands. The redox property of GNPs forms the key element for their widespread use as electrochemical labels for nucleic acid detection, thereby assisting in gene analysis, detection of genetic disorders, tissue matching, and forensics. Ozsoz and colleagues detected nucleic acid sequences by measuring the oxide wave current generated from a complementary probe–GNP hybridized to an immobilized target DNA on an electrode(58). In another modified configuration, GNP labels are attached to a single-stranded DNA–binding protein (59). When these capture probes are hybridized to matching targets, the binding of the labeled proteins is hindered and is indicated by the decrease in the Au redox signal. The Au oxide wave technique was modified to a greater level by Kerman and coworkers to not only detect the presence of a single-nucleotide polymorphism but also identify the bases involved in the nucleotide by using GNP-attached monobase

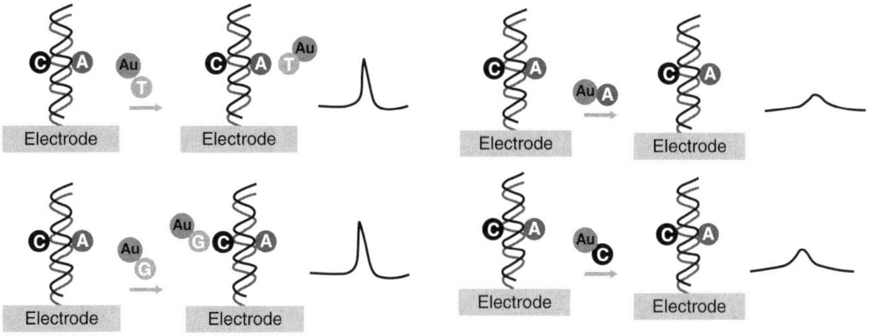

FIGURE 7 Scheme displaying the voltametric identification of the bases involved in single-nucleotide polymorphism by using a monobase-modified gold nanoparticle. *Source*: From Ref. 40.

nucleotide labels (60). Particular base particles attach only in the presence of DNA polymerase to their counterpart in the nucleotide strand. This, in turn, establishes a unique Au oxidation wave that can be detected using the electrode (Fig. 7).

GNP-Composite Biosensors

Electrodes made of composite matrices, encompassing nanoparticles, provide improved enzymatic biosensing and better analytical performances. These devices have an added advantage of low background currents, depending on the type of electrode used (55). A reagentless glucose biosensor, based on the direct electron transfer of GOx, has been constructed (61). While immobilization of the enzyme onto the electrode was done by mixing colloidal gold with the carbon paste components, a layer-by-layer (LBL) technique was used to prepare a GOx amperometric biosensor with a relatively high sensitivity (62). A film, made up of alternating layers of dendrimer-modified GNP and poly(vinylsulfonic acid), was used as the substrate for immobilizing GOx in the presence of bovine serum albumin and glutaraldehyde cross-linker. This was successfully applied as a biosensor for the amperometric detection of glucose at 0.0 V.

α-Fetoprotein (AFP) is an oncofetal glycoprotein that is widely employed as a tumor marker for the diagnosis of germ cell carcinoma and hepatocellular cancer in patients. An immunosensor for AFP detection was prepared by entrapping thionine into Nafion matrix into a membrane, which at the amine interface, assembles GNP layer for the immobilization of AFP antibody. The detection is identified by the linear drop in current across the membrane with increasing concentrations of AFP protein (63). A similar matrix was used by Tang and colleagues for the detection of carcinoma antigen 125 as represented in various tumors (64).

Other composite materials, such as GNP-conjugated carbon nanotube (CNT) composite electrodes, have generated much interest due to their synergistic properties. While hybridized GNPs possess properties related to bioaffinity and inertness, CNTs improve the electrocatalytic ability of the electrode. The electrocatalytic ability toward the electrooxidation of NADH or H_2O_2 has been utilized for the fabrication of a colloidal gold–CNT composite electrode. Significant H_2O_2 response was observed by Manso et al., who then incorporated GOx into the composite matrix for the preparation of a mediatorless glucose biosensor with remarkable sensitivity (65).

Polymer-Hybridized GNP Biosensors

The widely used method for the synthesis of immobilized matrices of biomolecules is via electropolymerization. The advantage of this technique is that by controlling the electrochemical parameters of polymer grown on the electrode surface, it is possible to control the various polymer characteristics, such as film thickness, permeation, and charge transport. GNPs incorporated in the conductive polymer matrix provide enhanced electrochemical activity and conductivity. Furthermore, rapid diffusion of biomolecular substrate and product, using small amount of enzyme, is achieved in the presence of GNPs (55). The main advantage of incorporating GNPs is that the enzyme-degrading electrodeposition process is substituted with a chemical polymerization technique. A simple substitution reaction was used for the synthesis of GNP–polymer hybrids for glucose biosening (66) and H_2O_2 detection (67). Tang and colleagues developed GNP–polymer-based biosensing electrodes for the detection of hepatitis B virus and diphtheria infections. They immobilized the respective antigens on colloidal GNPs, associated with poly(vinyl butyral) polymer on a Nafion–gelatin-coated platinum electrode by using a modified self-assembling and opposite charge adsorption technique (68,69).

Chitosan, a natural polymer exhibiting excellent film-forming and adhesion ability, together with susceptibility to chemical modification, has led to the immobilization of various enzymes over conductive electrodes. Self-assembled, GNP-adsorbed chitosan hydrogels, coated on Au electrodes, were used as GOx and horse radish peroxidase biosensors (70,71). Wu et al. carried out LBL assembly of multilayer films composed of chitosan, GNPs, and GOx over Pt electrodes for amperometric glucose detection (72). Sol–gel technology, a conventional technique, is used for the preparation of three-dimensional networks of hybrid biosensors made up of encapsulated biomolecules and GNPs. GNPs immobilized by silica gel act as highly sensitive, nanostructured, electrochemical biosensors and had been used for the construction of various polymer-based hybrid biosensors (73).

THERAPEUTIC APPLICATIONS OF GNPs

The inert and nontoxic property of gold along with the ready addition of biological molecules and antibodies capitalizing on the thiol–gold chemistry has rendered gold applicable in a variety of therapeutic systems, ranging from drug and biomolecular delivery to hyperthermia to active and passive targeting (Fig. 8). GNPs have recently emerged as an attractive platform for the delivery of small drug molecules and large biomolecules to specific targets (74). The release of these therapeutic agents can be triggered by cellular chemical [e.g., glutathione (GSH)] (75), pH (76), or external (e.g., light) (77) stimuli. Furthermore, the external stimuli via light can be controlled by exploiting the plasmon resonance effect of GNPs (10–100 nm) for precise drug delivery applications and hyperthermia (78,79). The various biomedical therapeutic applications of GNPs are detailed in the following text.

Therapy via Active and Passive In Vivo Targeting

The disadvantage of conventional drug administration is that the drug typically does not localize to the target site but is systemically distributed. Nanoparticles, on the other hand, can aid in the targeted accumulation of drugs. The success of any in vivo medical application depends on the ability of a nanocarrier to arrive at the targeted tissues after administration into the circulatory system. While passive targeting can be used for certain applications (e.g., extravasation through leaky

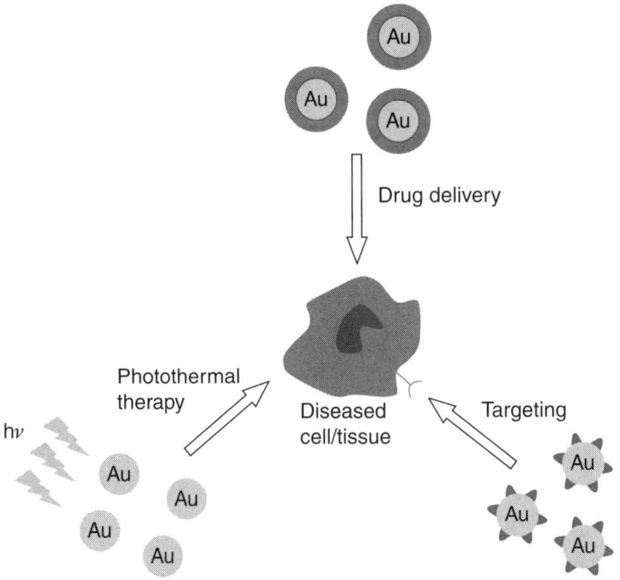

FIGURE 8 Various therapeutic applications of gold nanoparticles.

vasculature in tumors), active targeting employs the use of ligands on the carrier surface for specific recognition by the cell surface receptors (80). GNPs offer excellent drug delivery systems for in vivo applications due to their tunable size and versatile surface functionalization properties.

GNPs can be designed to be big enough to be retained in the liver and the spleen but small enough to pass through other organs (80). Passive targeting with these particles, along with radiotherapy, has been used for the treatment of liver cancer, in which the sinus endothelium of liver has openings of 150 nm in diameter. O'Neal et al. used the fact that tumor vasculature is more permeable than healthy tissues to concentrate gold nanoshells with diameters of 130 nm for photoablative therapy in mice (81).

Active targeting, which employs a ligand or antibody functionalized particle, has been successfully used by many researchers to treat certain diseases. Although these ligands are specific to target receptors, the presence of the particle in the circulatory system provides possibility of normal host immune response. To overcome this issue, PEG-coated stealth particles are widely employed (3). Pun and colleagues injected PEGylated nanoparticles, unmodified nanoparticles, and galactose-attached nanoparticles of varying sizes to explain the effect of size and functionalization over gold for active targeting (82). Two-hour postinjection data of the respective particles are shown in Figure 9. It is observed from the blood circulation data that PEGylation increased blood circulation lifetime, and since galactose-coated particles actively target liver cells, these particles had higher filtration inside the liver. Hepatocyte cell–specific targeting rendered by the galactose ligand can be potentially used for drug delivery.

More specific active targeting for treatment was shown by Dixit et al. They recently employed folic acid–conjugated GNPs (10 nm) with PEG spacer for the

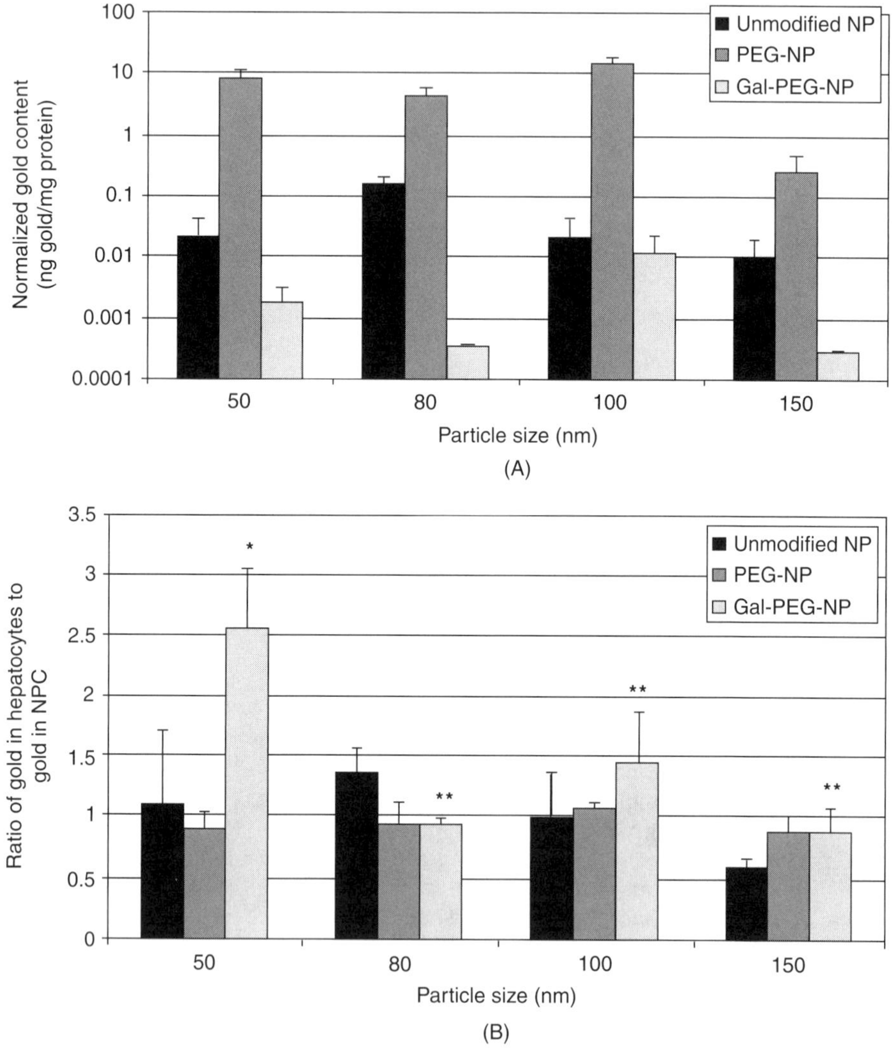

FIGURE 9 The functionalization of gold nanoparticles with poly(ethylene glycol) (PEG) and galactose ligand for specific active targeting. Two-hour postinjection data for (**A**) blood content and (**B**) liver distribution of the injected PEGylated nanoparticles, unmodified nanoparticles, and galactose-attached nanoparticles of varying sizes. *Abbreviations*: Gal, galactose; NP, nanoparticle. *Source*: From Ref. 82.

cellular uptake of tumor cells that exhibit folate receptors (83). In addition to the folic acid conjugates, Wu and colleagues loaded the GNPs with methotrexate (MTX), which is another tumor cell targeting folate-receptor-binding ligand and an anticancer drug, for the inhibition of lung tumor growth in mice (84). Once the MTX-conjugated particles got bound to the tumor cells, MTX was activated via cellular uptake and caused programmed cell death. Active targeting using PEGylated gold colloids was studied by Paciotti et al., who conjugated tumor necrosis factor (TNF) to colloidal Au-PEG (74). An extended application of this work was also done with the help of grafted paclitaxel, an anticancer drug, onto colloidal Au-PEG-TNF for the treatment of colon carcinoma tumors (74).

A plethora of therapeutic delivery systems were developed using the active and passive targeting strategies. The following discussion gives a better classification of the various therapeutic systems based on the type of molecular package delivered via GNPs.

GNP-BASED DRUG DELIEVRY

One of the main advantages of using gold particles at the nanoscale is its surface-to-volume ratio. GNPs can be loaded on their surface to form drug delivery systems. These systems can be used for chemotherapeutic delivery to tumor cells. The enhanced permeation and retention effect as provided by passive targeting has been used for treating carcinogenic tumors. The surface-to-volume ratio was utilized to prepare passive targeting GNPs conjugated with chemotherapeutic paclitaxel (85) and diatomic cytotoxic singlet oxygen ($^1O_2$), as well as nitric oxide (NO) GNP reservoirs (86,87). Triggered release of NO was shown by Schoenfisch and colleagues. Nitric oxide was released from water-soluble nanocontainers when a pH stimulus (pH = 3) was given to these drug delivery systems (76). Since tumor tissues have mild acidic environment, NO-carrying GNPs can potentially be used effectively to treat cancer. A separate section on temperature-based triggered release is described later in the "Hyperthermia using GNPs" section.

While external stimuli such as pH and temperature prove effective in triggered release, internal cellular signals/chemicals can provide an enhanced control on drug release. GSH, an antioxidant that protects the cell from toxins such as free radicals, in its reduced form has been used to mediate activated release of prodrugs from GNPs. The release is caused by a simple difference in the molar concentration of GSH inside the cell and that of thiols of GSH in the blood plasma/outside the cell. This differential causes the cleavage of the disulfide bond, which is the most commonly employed linkage between drugs and GNPs, thereby releasing the drug into the system (88). Researchers employed mixed-monolayer particles of size 2 nm, composed of tetra(ethylene glycol)ylated cationic ligands (TTMA) and fluorogenic ligands (HSBDP), and loaded them with a model hydrophobic drug BODIPY (a dye) for cellular drug delivery (75,89). TTMA facilitated transport across the cell membrane, and the fluorophore assisted in probing the drug release. Figure 10 shows the controlled release of the dye in mouse embryonic fibroblast cells, with varying concentrations of glutathione monoester (GSH-OEt) (0, 5, and 20 mM). GSH-OEt crosses across the fibroblast cell membrane and is hydrolyzed to GSH, which cleaves the BODIPY as well as the fluorophore of the conjugated nanoparticle via place-exchange reaction and the fluorophore is triggered, thereby following the drug release kinetics.

GNP-BASED BIOMOLECULAR DELIVERY

GNPs are easy to tune in terms of size and functionality, thereby contributing as a useful platform for efficient recognition and delivery of biomolecules. Their success in delivering peptides, proteins, or nucleic acids (DNA and RNA) is explained in the following text.

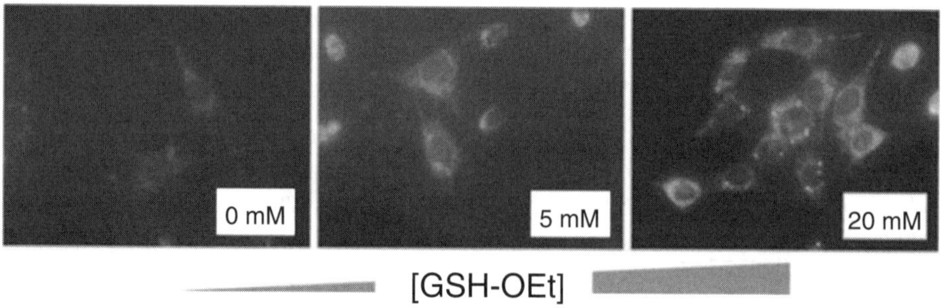

FIGURE 10 Fluorescence images showing dose-dependent release of payloads in mouse embryonic fibroblast cells. Glutathione monoester (GSH-OEt) was used as an external stimulus to release HSBDP from gold nanoparticles. *Source*: From Ref. 75.

Gene therapy is the treatment of genetic as well as acquired diseases by the insertion of genes into the cell or tissues. Although an adenovirus vector vehicle for gene delivery has been successfully used for gene therapy, safety issues concerned with unpredictable viral cytotoxicity and immune responses have minimized viability of gene therapy (90,91). Synthetic DNA delivery vehicles have to be effective in protecting the nucleic acid from degradation, efficiently enter the cell, and release the functional nucleic acid to the cell nucleus. Small GNPs with high surface-to-volume ratio prove to be successful candidates for gene therapy. Along with maximized payload/carrier ratio, GNPs can be functionalized to behave in a nontoxic and hydrophobic manner for efficient transfection. Rotello and colleagues used functionalized GNPs with cationic quaternary ammonium groups to show the effective binding of plasmid DNA to the cell, protection of DNA from enzymatic digestion, release of bound DNA by GSH treatment in cuvettes, and in gene delivery in mammalian 293T cells (92–95). The increase in hydrophobicity of the GNPs enhances its transfection efficiency, thereby contributing to better cellular internalization. Thomas and Klibanov prepared hybrid GNP–polymer transfection vectors by using branched polyethyleimine (molecular weight = 2 kDa) and showed that the increase in hydrophobicity increased the potency of the conjugate in monkey kidney (Cos-7) cells by approximately 12-folds as compared with the polymer itself (96). Recently, Liu and colleagues fabricated β-cyclodextrin end-capped, oligo(ethylenediamino)-modified GNPs to deliver plasmid DNA into breast cancer cells (MCF-7) (97).

Polycationic materials are famous for condensing and transporting DNA inside the cell. A recent breakthrough by Mirkin and colleagues, however, showed that DNA-loaded GNPs carrying a large negative surface potential were successfully internalized for cellular uptake and were stable against enzymatic digestion (98). RNA interference mechanism, which inhibits gene expressions on the translation or transcription of specific genes, has revolutionized gene therapy strategy with the help of thiolated nucleic acids. RNA-modified GNPs were coated with PEG-block-poly(2-(*N*,*N*-dimethylamino)ethylmethacrylate) copolymer for cellular delivery in HuH-7 cells (99). In addition to nucleic acids, GNPs are also used directly as nanocarriers for peptides and proteins. Electrostatic interaction between an anionic protein and a cationic tetraalkyl ammonium–functionalized GNPs has been utilized effectively for protein delivery (100,101). Figure 11 depicts the effectiveness of using chitosan-coated GNPs for transmucosal insulin delivery. As observed, the

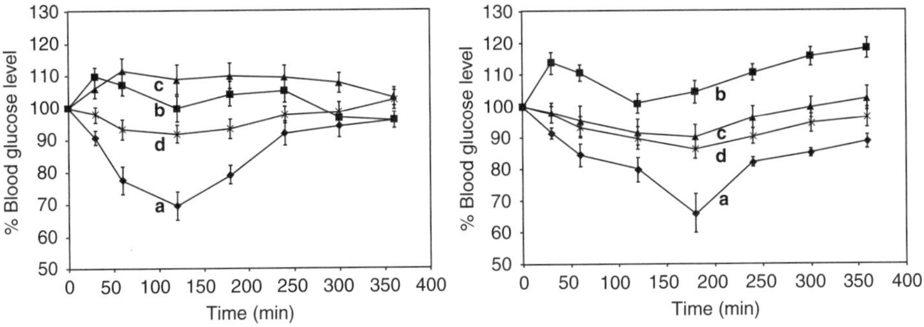

FIGURE 11 (*Left*): Percentage reduction in blood glucose level after oral administration (dose 50 IU/kg, $n = 6$). (*Right*): Percentage reduction in blood glucose level after nasal administration (dose 10 IU/kg, $n = 6$); a, insulin-loaded chitosan-reduced gold nanoparticles; b, blank chitosan-reduced nanoparticles; c, insulin solution; and d, insulin in chitosan solution. *Source*: From Ref. 101.

insulin-conjugated GNPs had a better drug influence than even direct conventional insulin solution, due to the enhanced permeation and retention effect.

HYPERTHERMIA USING GNPs

Hyperthermia is the process of using heat in a controlled manner for therapeutic treatment at elevated temperatures. The plasmon resonance light effect of GNPs, in which the electrons of gold resonate with that of the incoming radiation causing them to both absorb and scatter light, has been used to generate heat. By varying the relative thickness of the core and shell layers, the plasmon resonance effect of the nanoshell can be shifted dramatically toward the infrared (IR) region (Fig. 12) for introducing plasmon resonance–assisted heating. This physical

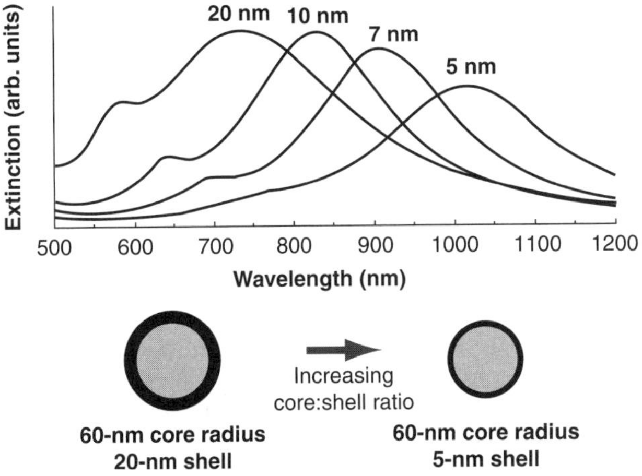

FIGURE 12 Optical resonances of gold shell–silica core nanoshells as a function of core–shell ratio (core diameter = 120 nm). The various profiles are characteristic of the shell thickness. *Source*: From Ref. 153.

property of GNPs to locally heat, when irradiated with light in the therapeutic window (800–1200 nm), can be potentially harnessed to either destroy local tissue or release payload therapeutics. The plasmon resonance for GNPs occurs at around 520 nm, whereas for complex shapes such as nanorods, this can be shifted from visible spectrum to the near infrared (NIR) region. This is advantageous because body tissue is semitransparent to NIR light, and thereby can be used for implantable therapeutics (81).

Halas and coworkers have utilized the photothermal effect of GNPs for hyperthermia treatment (102). They also performed simultaneous imaging of cells with the help of actively targeting fluorophore-conjugated GNPs for photoablative therapy (103). They have done follow-up work on this system by using PEG-sheathed gold-on-silica nanoshells for passive targeting. They were able to destroy breast cancer cells. In this case, NIR irradiation led to a rise in the temperature of the cancer cells to 45°C. El-Sayed and colleagues used citrate-stabilized GNPs of size 30 nm for photoablative therapy (104). The epidermal growth factor receptor–coated particles targeted HSC3 cancer cells (human oral squamous cell carcinoma), thereby increasing the viability of the particles available for heating the cancer cells.

Photoactivated drug release by plasmonically active particles was performed by Caruso and colleagues (105,106). Microcapsules (polymer gel matrix)-encapsulating fluorescein-labeled dextrans and lysozyme were doped with NIR light-responsive gold nanospheres (gold-on-gold-sulfide core-shell particles). When a laser of wavelength 1064 nm was applied to these microcapsules, the GNPs heated up and released the fluorescein-labeled dextran and lysozyme via rupture of the shells (Fig. 13) to destroy the bacterium *Micrococcus lysodeikticus*. Skirtach et al. extended this strategy for cancer treatment via a laser-induced release of encapsulated drug inside living cells (78).

A remote-controlled drug delivery strategy based on nanocomposite hydrogels was developed by West and Halas (107). Copolymers of NIPAAm and AAm exhibit a lower critical solution temperature (LCST) that is slightly above body temperature. As the temperature exceeds LCST, the polymer collapses, which can be used to release components soluble in the imbibed water. Gold nanoshells were incorporated in such a poly(NIPAAm-co-AAm) matrix and were then loaded with proteins of varying molecular weight. The triggered release of the proteins by using a laser of wavelength 1064 nm for multiple burst is shown in Figure 14.

CONCLUSIONS

It is clear that gold has been an important material in medical applications and that it continues to find new and unique applications, especially at the nanoscale. For most applications, it is critical that the gold surfaces are functionalized in a controlled manner. Here, methods for the functionalization of gold surfaces were highlighted, with a focus on "grafting from" methodologies built around SIPs. In addition, recent research activities in applying gold structures (specifically nanoparticles) in diagnostic and therapeutic applications were presented. Gold has a long history of applications at the nanoscale, but only recently applications in medical fields have grown exponentially in part due to the development of novel methods for functionalization. The history is long, but only the surface of potential applications in nanomedicine has been scratched.

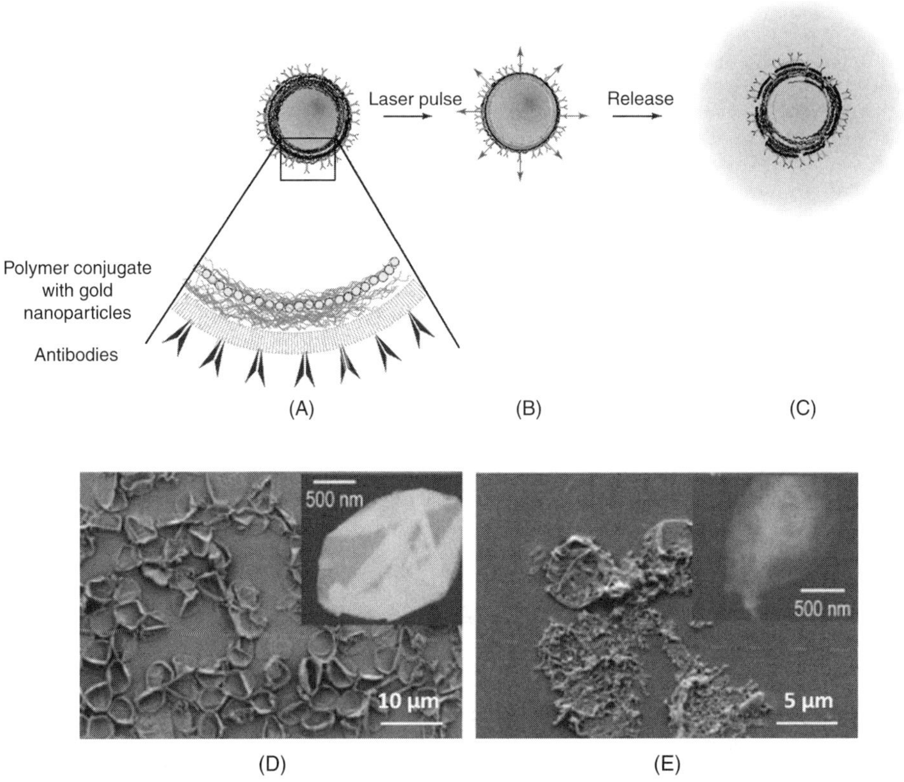

FIGURE 13 Scheme explaining the use of light for releasing encapsulated materials. (**A**) Preparation of microcapsule by layer-by-layer process; (**B**) application of laser to heat up the particles; (**C**) rupture of the shell and release of molecules upon light irradiation. Scanning electron microscopic images of the capsules (**D**) before laser irradiation and (**E**) after irradiation. *Source*: From Ref. 105.

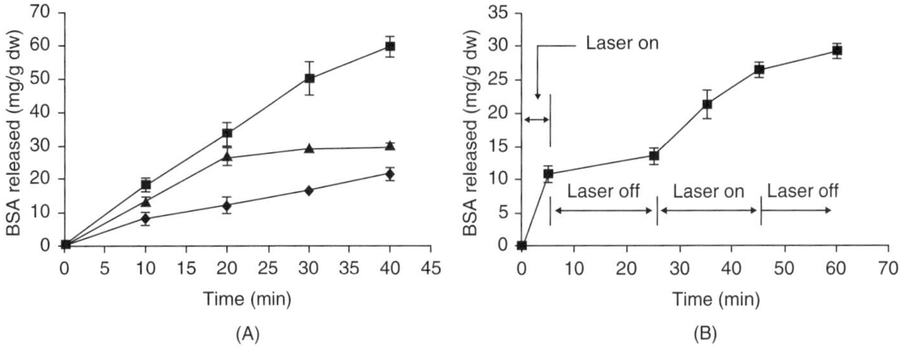

FIGURE 14 (**A**) Release of bovine serum albumin (BSA) from nonirradiated (diamond), irradiated NIPAAm-co-AAm hydrogel (triangle), and irradiated nanoshell-composite hydrogels (square). (**B**) Release of BSA from nanoshell-composite hydrogels in response to sequential multiple irradiations at 1064 nm. *Abbreviation*: dw, dry weight. *Source*: From Ref. 107.

REFERENCES

1. Ware M. Photographic printing in colloidal gold. 1994; 42:157–161.
2. Daniel MC, Astruc D. Gold nanoparticles: Assembly, supramolecular chemistry, quantum-size-related properties, and applications toward biology, catalysis, and nanotechnology. Chem Rev 2004; 104:293–346.
3. Love JC, Estroff LA, Whitesides GM, et al. Self-assembled monolayers of thiolates on metals as a form of nanotechnology. Chem Rev 2005; 105:1103–1170.
4. Cortie MB. The weird world of nanoscale gold. Gold Bull 2004; 37:12–19.
5. Elghanian R, Storhoff JJ, Mirkin CA, et al. Selective colorimetric detection of polynucleotides based on the distance-dependent optical properties of gold nanoparticles. Science 1997; 277:1078–1081.
6. Bendayan M. Worth its weight in gold. Science 2001; 291:1363–1365.
7. Pekka P. Theoretical chemistry of gold. Angew Chem Int Ed Engl 2004; 43:4412–4456.
8. Frens G. Controlled nucleation for the regulation of the particle size in monodisperse gold suspensions. Nature (Lond) Phys Sci 1973; 241:20.
9. Turkevich J, Stevenson PC, Hillier J. A study of the nucleation and growth processes in the synthesis of colloidal gold. Discuss Faraday Soc 1951; 11:55–75.
10. Schmid G. Large clusters and colloids. Metals in embryotic state. Chem Rev 1992; 92:1709–1727.
11. Odian G. Principles of Polymerization, 4th ed. Hoboken, NJ: John Wiley & Sons, 2004;198–349.
12. Brust M, Walker M, Schriffin DJ, et al. Synthesis of thiol-derivatised gold nanoparticles in a two-phase liquid-liquid system. J Chem Soc Commun 1994:801–802.
13. Templeton AC, Wuelfing WP, Murray RW. Monolayer-protected cluster molecules. Acc Chem Res 2000; 33:27–36.
14. Advincula RC. Surface initiated polymerization from nanoparticle surfaces. J Dispers Sci Technol 2003; 24:343–361.
15. Edmondson S, Osborne VL, Huck WTS. Polymer brushes via surface-initiated polymerizations. Chem Soc Rev 2004; 33:14–22.
16. Prucker O, Ruhe J. Synthesis of poly(styrene) monolayers attached to high surface area silica gels through self-assembled monolayers of azo initiators. Macromolecules 1998; 31:592–601.
17. Huang W, Baker GL, Bruening ML, et al. Surface-initiated thermal radical polymerization on gold. Langmuir 2001; 17:1731–1736.
18. Schmidt R, Zhao T, Green JB, et al. Photoinitiated polymerization of styrene from self-assembled monolayers on gold. Langmuir 2002; 18:1281–1287.
19. Dunér X, Henrik A, Teodor A, et al. Surface-confined photopolymerization of pH-responsive acrylamide/acrylate brushes on polymer thin films. Langmuir 2008; 24:7559–7564.
20. Advincula R, Zhou Q, Park M, et al. Polymer brushes by living anionic surface initiated polymerization on flat silicon (SiO_x) and gold surfaces: Homopolymers and block copolymers. Langmuir 2002; 18:8672–8684.
21. Sakellariou G, Advincula R, Mays J, et al. Homopolymer and block copolymer brushes on gold by living anionic surface-initiated polymerization in a polar solvent. J Polym Sci Part A: Polym Chem 2006; 44:769–782.
22. Jordan R, Ulman A, Kang JF, et al. Surface-initiated anionic polymerization of styrene by means of self-assembled monolayers. J Am Chem Soc 1999; 121:1016–1022.
23. Ohno K, Koh Km, Tsujii Y, et al. Synthesis of gold nanoparticles coated with well-defined, high-density polymer brushes by surface-initiated living radical polymerization. Macromolecules 2002; 35:8989–8993.
24. Jordan R, Ulman A. Surface initiated living cationic polymerization of 2-oxazolines. J Am Chem Soc 1998; 120:243–247.
25. Jordan R, West N, Ulman A, et al. Nanocomposites by surface-initiated living cationic polymerization of 2-oxazolines on functionalized gold nanoparticles. Macromolecules 2001; 34:1606–1611.

26. Mayadunne RTA, Rizzardo E, Chiefari J, et al. Living radical polymerization with reversible addition-fragmentation chain transfer (RAFT polymerization) using dithiocarbamates as chain transfer agents. Macromolecules 1999; 32: 6977–6980.
27. Shan J, Nuopponen M, Jiang H, et al. Preparation of poly(*N*-isopropylacrylamide)-monolayer-protected gold clusters: Synthesis methods, core size, and thickness of monolayer. Macromolecules 2003; 36:4526–4533.
28. Li J, He W, Sun X. Preparation of poly(styrene-co-*N*-isopropylacrylamide) micelles surface-linked with gold nanoparticles and thermo-responsive ultraviolet-visible absorbance. J Polym Sci Part A: Polym Chem 2007; 45:5156–5163.
29. Matyjaszewski K, Miller PJ, Shukla N, et al. Polymers at interfaces: Using atom transfer radical polymerization in the controlled growth of homopolymers and block copolymers from silicon surfaces in the absence of untethered sacrificial initiator. Macromolecules 1999; 32:8716–8724.
30. Matyjaszewski K, Dong H, Jakubowski W, et al. Grafting from surfaces for "everyone": ARGET ATRP in the presence of air. Langmuir 2007; 23:4528–4531.
31. Kim JB, Bruening ML, Baker GL. Surface-initiated atom transfer radical polymerization on gold at ambient temperature. J Am Chem Soc 2000; 122:7616–7617.
32. Takei YG, Aoki T, Sanui K, et al. Dynamic contact angle measurement of temperature-responsive surface properties for poly(*N*-isopropylacrylamide) grafted surfaces. Macromolecules 1994; 27:6163–6166.
33. He Q, Kuller A, Grunze M, et al. Fabrication of thermosensitive polymer nanopatterns through chemical lithography and atom transfer radical polymerization. Langmuir 2007; 23:3981–3987.
34. Jones DM, Huck WTS. Controlled surface-initiated polymerizations in aqueous media. Adv Mater 2001; 13:1256–1259.
35. Treat ND, Ayres N, Boyes SG, et al. A facile route to poly(acrylic acid) brushes using atom transfer radical polymerization. Macromolecules 2006; 39:26–29.
36. Tugulu S, Barbey R, Harms M, et al. Synthesis of poly(methacrylic acid) brushes via surface-initiated atom transfer radical polymerization of sodium methacrylate and their use as substrates for the mineralization of calcium carbonate. Macromolecules 2007; 40:168–177.
37. Zhou F, Zheng Z, Huck WTS, et al. Multicomponent polymer brushes. J Am Chem Soc 2006; 128:16253–16258.
38. Matyjaszewski K, Davis TP. Handbook of Radical Polymerization. Hoboken, NJ: John Wiley & Sons, 2002:28–143.
39. Fortina P, Kricka LJ, Surrey S, et al. Nanobiotechnology: The promise and reality of new approaches to molecular recognition. Trends Biotechnol 2005; 23:168–173.
40. Tansil NC, Gao Z. Nanoparticles in biomolecular detection. Nano Today 2006; 1:28–37.
41. Mirkin CA, Letsinger RL, Mucic RC, et al. A DNA-based method for rationally assembling nanoparticles into macroscopic materials. Nature 1996; 382:607–609.
42. Storhoff JJ, Elghanian R, Mirkin CA, et al. One-pot colorimetric differentiation of polynucleotides with single base imperfections using gold nanoparticle probes. J Am Chem Soc 1998; 120:1959–1964.
43. Taton TA, Mucic RC, Mirkin CA, et al. The DNA-mediated formation of supramolecular mono- and multilayered nanoparticle structures. J Am Chem Soc 2000; 122:6305–636.
44. Jin R, Wu G, Mirkin CA, et al. What controls the melting properties of DNA-linked gold nanoparticle assemblies? J Am Chem Soc 2003; 125:1643–1654.
45. Taton TA, Mirkin CA, Letsinger RL. Scanometric DNA array detection with nanoparticle probes. Science 2000; 289:1757–1760.
46. Nam JM, Stoeva SI, Mirkin CA. Bio-bar-code-based DNA detection with PCR-like sensitivity. J Am Chem Soc 2004; 126:5932–5933.
47. Mulvaney SP, Musick MD, Keating CD, et al. Glass-coated, analyte-tagged nanoparticles: A new tagging system based on detection with surface-enhanced Raman scattering. Langmuir 2003; 19:4784–4790.
48. Cao YC, Jin R, Mirkin CA. Nanoparticles with Raman spectroscopic fingerprints for DNA and RNA detection. Science 2002; 297:1536–1540.

49. He L, Musick MD, Nicewarner SR, et al. Colloidal Au-enhanced surface plasmon resonance for ultrasensitive detection of DNA hybridization. J Am Chem Soc 2000; 122:9071–9077.
50. Lyon LA, Musick MD, Natan MJ. Colloidal Au-enhanced surface plasmon resonance immunosensing. Anal Chem 1998; 70:5177–5183.
51. Hsu H-Y, Huang Y-Y. RCA combined nanoparticle-based optical detection technique for protein microarray: A novel approach. Biosens Bioelectron 2004; 20:123–126.
52. Aslan K, Holley P, Davies L, et al. Angular-ratiometric plasmon-resonance based light scattering for bioaffinity sensing. J Am Chem Soc 2005; 127:12115–12121.
53. Maxwell DJ, Taylor JR, Nie S. Self-assembled nanoparticle probes for recognition and detection of biomolecules. J Am Chem Soc 2002; 124:9606–9612.
54. Park S-J, Taton TA, Mirkin CA. Array-based electrical detection of DNA with nanoparticle probes. Science 2002; 295:1503–1506.
55. Pingarrón JM, Yáñez-Sedeño P, González-Cortés A. Gold nanoparticle-based electrochemical biosensors. Electrochim Acta 2008; 53:5848–5466.
56. Mena ML, Yáñez-Sedeño P, Pingarrón JM. A comparison of different strategies for the construction of amperometric enzyme biosensors using gold nanoparticle-modified electrodes. Anal Biochem 2005; 336:20–27.
57. Wang M, Wang L, Wang G, et al. Application of impedance spectroscopy for monitoring colloid Au-enhanced antibody immobilization and antibody-antigen reactions. Biosens Bioelectron 2004; 19:575–582.
58. Ozsoz M, Erdem A, Kerman K, et al. Electrochemical genosensor based on colloidal gold nanoparticles for the detection of Factor V Leiden mutation using disposable pencil graphite electrodes. Anal Chem 2003; 75:2181–2187.
59. Kerman K, Morita Y, Takamura Y, et al. Modification of *Escherichia coli* single-stranded DNA binding protein with gold nanoparticles for electrochemical detection of DNA hybridization. Anal Chim Acta 2004; 510:169–174.
60. Kerman K, Saito M, Morita Y, et al. Electrochemical coding of single-nucleotide polymorphisms by monobase-modified gold nanoparticles. Anal Chem 2004; 76:1877–1884.
61. Liu S, Ju H. Reagentless glucose biosensor based on direct electron transfer of glucose oxidase immobilized on colloidal gold modified carbon paste electrode. Biosens Bioelectron 2003; 19:177–183.
62. Crespilho FN, Emilia Ghica M, Florescu M, et al. A strategy for enzyme immobilization on layer-by-layer dendrimer-gold nanoparticle electrocatalytic membrane incorporating redox mediator. Electrochem Commun 2006; 8:1665–1670.
63. Zhuo Y, Yuan R, Chai Y, et al. A reagentless amperometric immunosensor based on gold nanoparticles/thionine/Nafion-membrane-modified gold electrode for determination of [alpha]-1-fetoprotein. Electrochem Commun 2005; 7:355–360.
64. Tang D, Yuan R, Chai Y. Electrochemical immuno-bioanalysis for carcinoma antigen 125 based on thionine and gold nanoparticles-modified carbon paste interface. Anal Chim Acta 2006; 564:158–165.
65. Manso J, Mena ML, Pingarrón J, et al. Electrochemical biosensors based on colloidal gold-carbon nanotubes composite electrodes. J Electroanal Chem 2007; 603:1–7.
66. Xian Y, Hu Y, Liu F, et al. Glucose biosensor based on Au nanoparticles-conductive polyaniline nanocomposite. Biosens Bioelectron 2006; 21:1996–2000.
67. Gao F, Yuan R, Chai Y, et al. Amperometric hydrogen peroxide biosensor based on the immobilization of HRP on nano-Au/Thi/poly(*p*-aminobenzene sulfonic acid)-modified glassy carbon electrode. J Biochem Biophys Methods 2007; 70:407–413.
68. Tang DP, Yuan R, Chai YQ, et al. Novel potentiometric immunosensor for hepatitis B surface antigen using a gold nanoparticle-based biomolecular immobilization method. Anal Biochem 2004; 333:345–350.
69. Tang D, Yuan R, Chai Y, et al. Preparation and application on a kind of immobilization method of anti-diphtheria for potentiometric immunosensor modified colloidal Au and polyvinylbutyral as matrixes. Sens Actuators B Chem 2005; 104:199–206.
70. Luo X-L, Xu J-J, et al. A glucose biosensor based on chitosan-glucose oxidase-gold

nanoparticles biocomposite formed by one-step electrodeposition. Anal Biochem 2004; 334:284–289.
71. Luo X-L, Xu J-J, Zhang Q, et al. Electrochemically deposited chitosan hydrogel for horseradish peroxidase immobilization through gold nanoparticles self-assembly. Biosens Bioelectron 2005; 21:190–196.
72. Wu B-Y, Hou S-H, Yin F, et al. Amperometric glucose biosensor based on layer-by-layer assembly of multilayer films composed of chitosan, gold nanoparticles and glucose oxidase modified Pt electrode. Biosens Bioelectron 2007; 22:838–844.
73. Jena BK, Raj CR. Electrochemical biosensor based on integrated assembly of dehydrogenase enzymes and gold nanoparticles. Anal Chem 2006; 78:6332–6339.
74. Paciotti GF, David GI, Tamarkin KL, et al. Colloidal gold nanoparticles: A novel nanoparticle platform for developing multifunctional tumor-targeted drug delivery vectors. Drug Dev Res 2006; 67:47–54.
75. Hong R, Forbes NS, Rotello VM, et al. Glutathione-mediated delivery and release using monolayer protected nanoparticle carriers. J Am Chem Soc 2006; 128:1078–1079.
76. Polizzi MA, Stasko NA, Schoenfisch MH. Water-soluble nitric oxide-releasing gold nanoparticles. Langmuir 2007; 23:4938–4943.
77. Gang Han, You C-C, Rotello VM, et al. Light-regulated release of DNA and its delivery to nuclei by means of photolabile gold nanoparticles. Angew Chem Int Ed Engl 2006; 45:3165–3169.
78. Skirtach AG, Oliver J, Karen K, et al. Laser-induced release of encapsulated materials inside living cells. Angew Chem Int Ed Engl 2006; 45:4612–4617.
79. Loo C, Lin A, Hirsch L, et al. Nanoshell-enabled photonics-based imaging and therapy of cancer. Technol Cancer Res Treat 2004; 3:33–40.
80. Moghimi SM, Hunter AC, Murray JC. Nanomedicine: Current status and future prospects. FASEB J 2005; 19:311–330.
81. O'Neal DP, Hirsch LR, West JL, et al. Photo-thermal tumor ablation in mice using near infrared-absorbing nanoparticles. Cancer Lett 2004; 209:171–176.
82. Bergen JM, Archana P, Pun H, et al. Gold nanoparticles as a versatile platform for optimizing physicochemical parameters for targeted drug delivery. Macromol Biosci 2006; 6:506–516.
83. Dixit V, Bossche J, Sherman DM, et al. Synthesis and grafting of thioctic acid-PEG-folate conjugates onto Au nanoparticles for selective targeting of folate receptor-positive tumor cells. Bioconjug Chem 2006; 17:603–609.
84. Chen Y-H, Tsai C-Y, Huang P-Y, et al. Methotrexate conjugated to gold nanoparticles inhibits tumor growth in a syngeneic lung tumor model. Mol Pharm 2007; 4:713–722.
85. Gibson JD, Khanal BP, Zubarev ER. Paclitaxel-functionalized gold nanoparticles. J Am Chem Soc 2007; 129:11653–11661.
86. Hone DC, Walker PI, Evans-Gowing R, et al. Generation of cytotoxic singlet oxygen via phthalocyanine-stabilized gold nanoparticles: A potential delivery vehicle for photodynamic therapy. Langmuir 2002; 18:2985–2987.
87. Mocellin S, Bronte V, Nitti D, et al. Nitric oxide, a double edged sword in cancer biology: Searching for therapeutic opportunities. Med Res Rev 2007; 27:317–352.
88. Saito G, Swanson JA, Lee K-D. Drug delivery strategy utilizing conjugation via reversible disulfide linkages: Role and site of cellular reducing activities. Adv Drug Deliv Rev 2003; 55:199–215.
89. Sapsford KE, Berti L, Mendintz IL. Materials for fluorescence resonance energy transfer analysis: Beyond traditional donor-acceptor combinations. Angew Chem Int Ed Engl 2006; 45:4562–4589.
90. Yeh P, Perricaudet M. Advances in adenoviral vectors: From genetic engineering to their biology. FASEB J 1997; 11:615–623.
91. Check E. Gene therapy: A tragic setback. Nature 2002; 420:116–118.
92. McIntosh CM, Martin CT, Rotello VM, et al. Inhibition of DNA transcription using cationic mixed monolayer protected gold clusters. J Am Chem Soc 2001; 123:7626–7629.
93. Han G, Martin CT, Rotello VM, et al. Stability of gold nanoparticle-bound DNA toward biological, physical, and chemical agents. Chem Biol Drug Des 2006; 67:78–82.

94. Han G, Chari NS, Rotello VM, et al. Controlled recovery of the transcription of nanoparticle-bound DNA by intracellular concentrations of glutathione. Bioconjug Chem 2005; 16:1356–1359.
95. Sandhu KK, McIntosh CM, Rotello VM, et al. Gold nanoparticle-mediated transfection of mammalian cells. Bioconjug Chem 2002; 13:3–6.
96. Thomas M, Klibanov AM. Conjugation to gold nanoparticles enhances polyethylenimine's transfer of plasmid DNA into mammalian cells. Proc Natl Acad Sci USA 2003; 100:9138–9143.
97. Wang H, Chen Y, Li XY, et al. Synthesis of oligo(ethylenediamino)-beta-cyclodextrin modified gold nanoparticle as a DNA concentrator. Mol Pharm 2007; 4:189–198.
98. Rosi NL, Giljohann DA, Thaxton CS, et al. Oligonucleotide-modified gold nanoparticles for intracellular gene regulation. Science 2006; 312:1027–1030.
99. Oishi M, Nakaogami J, Ishii T, et al. Smart PEGylated gold nanoparticles for the cytoplasmic delivery of siRNA to induce enhanced gene silencing. Chem Lett 2006; 35:1046–1047.
100. Verma A, Simard JM, Rotello VM, et al. Tunable reactivation of nanoparticle-inhibited beta-galactosidase by glutathione at intracellular concentrations. J Am Chem Soc 2004; 126:13987–13991.
101. Bhumkar D, Joshi H, Sastry M, et al. Chitosan reduced gold nanoparticles as novel carriers for transmucosal delivery of insulin. Pharm Res 2007; 24:1415–1426.
102. Hirsch LR, Stafford RJ, Bankson JA, et al. Nanoshell-mediated near-infrared thermal therapy of tumors under magnetic resonance guidance. Proc Natl Acad Sci USA 2003; 100:13549–13554.
103. Hirsch LR, Halas NJ, West JL, et al. A whole blood immunoassay using gold nanoshells. Anal Chem 2003; 75:2377–2381.
104. Huang X, Qian W, El-Sayed IH, et al. The potential use of the enhanced nonlinear properties of gold nanospheres in photothermal cancer therapy. Lasers Surg Med 2007; 39:747–753.
105. Radt B, Smith A, Caruso F. Optically addressable nanostructured capsules. Adv Mater 2004; 16:2184–2189.
106. Angelatos AS, Radt B, Caruso F. Light-responsive polyelectrolyte/gold nanoparticle microcapsules. J Phys Chem B 2005; 109:3071–3076.
107. West JL, Halas NJ. Applications of nanotechnology to biotechnology: Commentary. Curr Opin Biotechnol 2000; 11:215–217.
108. Aili D, Enander K, Rydberg J, et al. Aggregation-induced folding of a de novo designed polypeptide immobilized on gold nanoparticles. J Am Chem Soc 2006; 128:2194–2195.
109. Slocik JM, Morley OS, Rajesh RN. Synthesis of gold nanoparticles using multifunctional peptides. Small 2005; 1:1048–1052.
110. He P, Urban MW. Phospholipid-stabilized Au-nanoparticles. Biomacromolecules 2005; 6:1224–1225.
111. He P, Zhu X. Phospholipid-assisted synthesis of size-controlled gold nanoparticles. Mater Res Bull 2007; 42:1310–1315.
112. Zhang L, Sun X, Song Y, et al. Didodecyldimethylammonium bromide lipid bilayer-protected gold nanoparticles: Synthesis, characterization, and self-assembly. Langmuir 2006; 22:2838–2843.
113. Cheng W, Dong S, Wang E. Synthesis and self-assembly of cetyltrimethylammonium bromide-capped gold nanoparticles. Langmuir 2003; 19:9434–9439.
114. Isaacs SR, Cutler EC, Park JS, et al. Synthesis of tetraoctylammonium-protected gold nanoparticles with improved stability. Langmuir 2005; 21:5689–5692.
115. Cheng W, Wang E. Size-dependent phase transfer of gold nanoparticles from water into toluene by tetraoctylammonium cations: A wholly electrostatic interaction. J Phys Chem B 2004; 108:24–26.
116. Du L, Jiang H, Liu X, et al. Biosynthesis of gold nanoparticles assisted by *Escherichia coli* DH5[α] and its application on direct electrochemistry of hemoglobin. Electrochem Commun 2007; 9:1165–1170.

117. Loo L, Guenther RH, Basnayake VR, et al. Controlled encapsidation of gold nanoparticles by a viral protein shell. J Am Chem Soc 2006; 128:4502–4503.
118. Slocik JM, Naik RR, Stone MO, et al. Viral templates for gold nanoparticle synthesis. J Mater Chem 2005; 15:749–753.
119. Scheeren CW, Machado G, Dupont J, et al. Nanoscale Pt(0) particles prepared in imidazolium room temperature ionic liquids: Synthesis from an organometallic precursor, characterization, and catalytic properties in hydrogenation reactions. Inorg Chem 2003; 42:4738–4742.
120. Dupont J, Fonseca GS, Umpierre AP, et al. Transition-metal nanoparticles in imidazolium ionic liquids: Recyclable catalysts for biphasic hydrogenation reactions. J Am Chem Soc 2002; 124:4228–4229.
121. Itoh H, Naka K, Chujo Y. Synthesis of gold nanoparticles modified with ionic liquid based on the imidazolium cation. J Am Chem Soc 2004; 126:3026–3027.
122. Kim KS, Demberelnyamba D, Lee H. Size-selective synthesis of gold and platinum nanoparticles using novel thiol-functionalized ionic liquids. Langmuir 2004; 20:556–560.
123. Huang H, Yang X. Synthesis of chitosan-stabilized gold nanoparticles in the absence/presence of tripolyphosphate. Biomacromolecules 2004; 5:2340–2346.
124. Wang B, Chen K, Jiang S, et al. Chitosan-mediated synthesis of gold nanoparticles on patterned poly(dimethylsiloxane) surfaces. Biomacromolecules 2006; 7:1203–1209.
125. dosSantos DS, Goulet PJG, Pieczonka NPW, et al. Gold nanoparticle embedded, self-sustained chitosan films as substrates for surface-enhanced Raman scattering. Langmuir 2004; 20:10273–10277.
126. Miyama T, Yonezawa Y. Aggregation of photolytic gold nanoparticles at the surface of chitosan films. Langmuir 2004; 20:5918–5923.
127. Qi Z, Zhou H, Matsuda N, et al. Characterization of gold nanoparticles synthesized using sucrose by seeding formation in the solid phase and seeding growth in aqueous solution. J Phys Chem B 2004; 108:7006–7011.
128. Garcia ME, Baker LA, Crooks RM. Preparation and characterization of dendrimer-gold colloid nanocomposites. Anal Chem 1999; 71:256–258.
129. Esumi K, Hosoya T, Suzuki A, et al. Spontaneous formation of gold nanoparticles in aqueous solution of sugar-persubstituted poly(amidoamine) dendrimers. Langmuir 2000; 16:2978–2980.
130. Kim YG, Oh SK, Crooks RM. Preparation and characterization of 1–2 nm dendrimer-encapsulated gold nanoparticles having very narrow size distributions. Chem Mater 2004; 16:167–172.
131. Haba Y, Kojima C, Harada A, et al. Preparation of poly(ethylene glycol)-modified poly(amido amine) dendrimers encapsulating gold nanoparticles and their heat-generating ability. Langmuir 2007; 23:5243–5246.
132. Xuping Sun. One-step synthesis and size control of dendrimer-protected gold nanoparticles: A heat-treatment-based strategy. Macromol Rapid Commun 2003; 24:1024–1028.
133. Brusatin G, Abbotto A, Innocenzi SP, et al. Poled sol-gel materials with heterocycle push-pull chromophores that confer enhanced second-order optical nonlinearity. Adv Funct Mater 2004; 14:1160–1166.
134. Wang R, Yang J, Seraphin S, et al. Dendron-controlled nucleation and growth of gold nanoparticles. Angew Chem Int Ed Engl 2001; 40:549–552.
135. Nakao S, Torigoe K, Kon-No K, et al. Self-assembled one-dimensional arrays of gold-dendron nanocomposites. J Phys Chem B 2002; 106:12097–12100.
136. Kim M-K, Jeon Y-M, Jeon WS, et al. Novel dendron-stabilized gold nanoparticles with high stability and narrow size distribution. Chem Commun 2001:667–668.
137. Hussain I, Graham S, Wang Z, et al. Size-controlled synthesis of near-monodisperse gold nanoparticles in the 1–4 nm range using polymeric stabilizers. J Am Chem Soc 2005; 127:16398–16399.
138. Teranishi T, Kiyokawa I, Miyake M. Synthesis of monodisperse gold nanoparticles using linear polymers as protective agents. Adv Mater 1998; 10:596–599.

139. Wan D, Fu Q, Huan J. Synthesis of amphiphilic hyperbranched polyglycerol polymers and their application as template for size control of gold nanoparticles. J Appl Polym Sci 2006; 101:509–514.
140. Perignon N, Marty JD, Mingotaud AF, et al. Hyperbranched polymers analogous to PAMAM dendrimers for the formation and stabilization of gold nanoparticles. Macromolecules 2007; 40:3034–3041.
141. Duan H, Nie S. Etching colloidal gold nanocrystals with hyperbranched and multivalent polymers: A new route to fluorescent and water-soluble atomic clusters. J Am Chem Soc 2007; 129:2412–2413.
142. Kang Y, Taton TA. Core/shell gold nanoparticles by self-assembly and crosslinking of micellar, block-copolymer shells. Angew Chem Int Ed Engl 2005; 44:409–412.
143. Zubarev ER, Xu J, Sayyad A, et al. Amphiphilic gold nanoparticles with V-shaped arms. J Am Chem Soc 2006; 128:4958–4959.
144. Kim JH, Lee TR. Thermo- and pH-responsive hydrogel-coated gold nanoparticles. Chem Mater 2004; 16:3647–3651.
145. Zheng P, Jiang X, Zhang X, et al. Formation of gold at polymer core-shell particles and gold particle clusters on a template of thermoresponsive and pH-responsive coordination triblock copolymer. Langmuir 2006; 22:9393–9396.
146. Wuelfing WP, Gross SM, Murray RW, et al. Nanometer gold clusters protected by surface-bound monolayers of thiolated poly(ethylene glycol) polymer electrolyte. J Am Chem Soc 1998; 120:12696–12697.
147. Otsuka H, Akiyama Y, Nagasaki Y, et al. Quantitative and reversible lectin-induced association of gold nanoparticles modified with alpha-lactosyl-omega-mercapto-poly(ethylene glycol). J Am Chem Soc 2001; 123:8226–8230.
148. Otsuka H, Nagasaki Y, Kataoka K. PEGylated nanoparticles for biological and pharmaceutical applications. Adv Drug Deliv Rev 2003; 55:403–419.
149. Shimmin RG, Schoch AB, Braun PV. Polymer size and concentration effects on the size of gold nanoparticles capped by polymeric thiols. Langmuir 2004; 20:5613–5620.
150. Takae S, Akiyama Y, Otsuka H, et al. Ligand density effect on biorecognition by PEGylated gold nanoparticles: Regulated interaction of RCA120 lectin with lactose installed to the distal end of tethered PEG strands on gold surface. Biomacromolecules 2005; 6:818–824.
151. Ishii T, Otsuka H, Kataoka K, et al. Preparation of functionally PEGylated gold nanoparticles with narrow distribution through autoreduction of auric cation by alpha-biotinyl-PEG-block-[poly(2-(*N*,*N*-dimethylamino)ethyl methacrylate)]. Langmuir 2004; 20:561–564.
152. Olivier J-C, Huertas R, Lee HJ, et al. Synthesis of pegylated immunonanoparticles. Pharm Res 2002; 19:1137–1143.
153. Oldenburg SJ, Averitt RD, Westcott SL, et al. Nanoengineering of optical resonances. Chem Phys Lett 1998; 288:243–247.

8 NPDDS for the Treatment of Diabetes

Karthikeyan Subramani

Institute for Nanoscale Science and Technology (INSAT), University of Newcastle upon Tyne, Newcastle upon Tyne, U.K.

INTRODUCTION

Diabetes mellitus, often referred as diabetes, is caused by decrease in insulin secretion by pancreatic islet cells, leading to increase in blood glucose level (hyperglycemia). Diabetes insipidus is a condition characterized by excretion of large amounts of severely diluted urine which cannot be reduced when fluid intake is reduced. This is caused due to the deficiency of antidiuretic hormone, also known as vasopressin, secreted by the posterior pituitary gland. Diabetes mellitus is characterized by excessive weight loss, increased urge for urination (polyuria), increased thirst (polydipsia), and an excessive desire to eat (polyphagia) (1). Diabetes mellitus has been classified as (*i*) type 1, or insulin-dependent diabetes, (*ii*) type 2, or non–insulin-dependent diabetes, and (*iii*) gestational diabetes. Type 1 diabetes mellitus is characterized by the loss of insulin-producing β cells of islets of Langerhans in the pancreas, thereby leading to deficiency of insulin. The main cause of this β-cell loss is T-cell–mediated autoimmune attack. Type 1 diabetes in children is termed as juvenile diabetes. Type 2 diabetes mellitus is caused by insulin resistance or reduced insulin sensitivity combined with reduced insulin secretion. The defective responsiveness of body tissues to insulin almost certainly involves the insulin receptor in cell membranes. Gestational diabetes occurs in women without previously diagnosed diabetes who exhibit high blood glucose levels during pregnancy. No specific cause has been identified, but it is believed that the hormones produced during pregnancy reduce a woman's sensitivity to insulin, resulting in high blood sugar levels. Controlling the blood sugar level through modified dietary sugar intake, physical exercise, insulin therapy, and oral medications has been advised for control of type 1 diabetes mellitus. Nanomedicine research over the past few decades has been aimed at the applications of nanoparticles for the treatment of type 1 diabetes mellitus through effective insulin delivery and is discussed in the following sections.

NANOPARTICLES IN THE TREATMENT OF DIABETES

Polymeric Nanoparticles

Polymeric nanoparticles have been used as carriers of insulin (2). These are biodegradable polymers, with the polymer–insulin matrix surrounded by the nanoporous membrane containing grafted glucose oxidase. A rise in blood glucose level triggers a change in the surrounding nanoporous membrane, resulting in biodegradation and subsequent insulin delivery. The glucose/glucose oxidase reaction causes a lowering of the pH in the delivery system's microenvironment. This

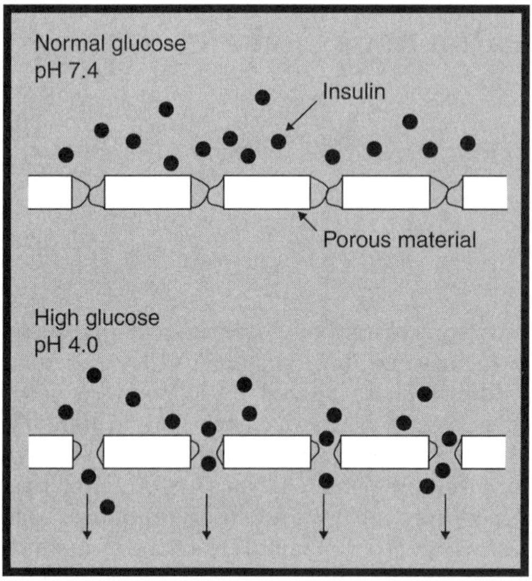

FIGURE 1 Schematic of polymeric nanoparticles with pH-sensitive molecular gates for controlled insulin release triggered by the presence of glucose in blood.

can cause an increase in the swelling of the polymer system, leading to an increased release of insulin. The polymer systems investigated for such applications include copolymers such as *N,N*-dimethylaminoethyl methacrylate (3) and polyacrylamide (4). This "molecular gate" system is composed of an insulin reservoir and a delivery rate–controlling membrane made of poly[methacrylic acid-g-poly(ethylene glycol)] copolymer. The polymer swells in size at normal body pH (pH = 7.4) and closes the gates. It shrinks at low pH (pH = 4) when the blood glucose level increases, thus opening the gates and releasing the insulin from the nanoparticles (Fig. 1) (5). These systems release insulin by swelling caused due to changes in blood pH. The control of the insulin delivery depends on the size of the gates, the concentration of insulin, and the rate of gates' opening or closing (response rate). These self-contained polymeric delivery systems are still under research, whereas the delivery of oral insulin with polymeric nanoparticles has progressed to a greater extent in the recent years.

Oral Insulin Administration by Using Polysaccharides and Polymeric Nanoparticles

The development of improved oral insulin administration is very essential for the treatment of diabetes mellitus to overcome the problem of daily subcutaneous injections. Insulin, when administered orally, undergoes degradation in the stomach due to gastric enzymes (6). Therefore, insulin should be enveloped in a matrix-like system to protect it from gastric enzymes. This can be achieved by encapsulating the insulin molecules in polymeric nanoparticles. In one such study, calcium phosphate–poly(ethylene glycol)–insulin combination was combined with casein (a milk protein) (7). The casein coating protects the insulin from the gastric enzymes (Fig. 2). Due to casein's mucoadhesive property, the formulation remained

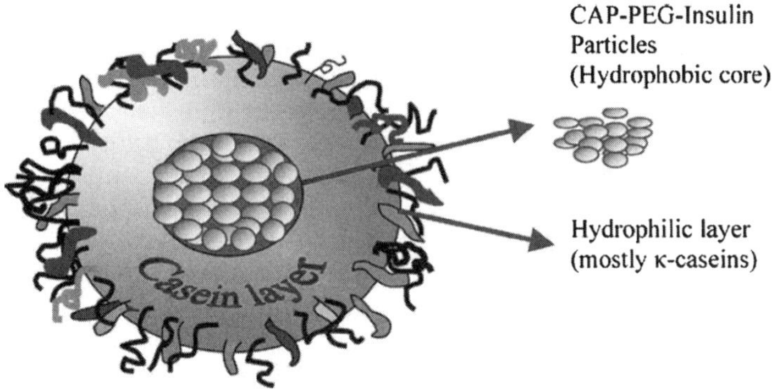

FIGURE 2 Schematic of a calcium phosphate–PEG–insulin–casein oral insulin delivery system. *Abbreviations*: CAP, calcium phosphate; PEG, poly(ethylene glycol). *Source*: From Ref. 7.

concentrated in the small intestine for a longer period, resulting in slower absorption and longer availability in the bloodstream. In another study, insulin-loaded polymeric nanoparticles were used in the form of pellets for oral delivery of insulin in diabetic rats (8). The results showed a significant decrease in blood sugar level following the administration of insulin through the buccal route. Temperature-sensitive nanospheres made from poly(*N*-isopropylacrylamide) and poly(ethylene glycol) dimethacrylate were shown to protect the loaded insulin from high temperature and high shear stress; such a polymeric system can be an effective carrier for insulin (9). Polysaccharides, such as chitosan, dextran sulfate, and cyclodextrin, have been used to deliver the insulin molecules with polymeric nanoparticles as carrier systems. Although chitosan was used for nasal delivery of insulin, it has also been tested for oral delivery (10). The in vivo results in a diabetic rat model with insulin-loaded chitosan/poly(γ-glutamic acid) were shown to effectively reduce the blood glucose level (11). A combination of dextran sulfate–chitosan nanoparticles was shown to be an effective pH-sensitive delivery system, and the release of insulin was governed by the dissociation mechanism between the polysaccharides (12). Dextran sulfate combined with polyethylenimine nanoparticles was shown to exhibit a high level of insulin entrapment and an ability to preserve insulin structure and biological activity in vitro (13). Poly(methacrylic acid)–based nanoparticles that are encapsulated in cyclodextrin–insulin complex have also been reported as an effective oral delivery system (14). Over recent years, different polymeric nanoparticles made of poly(isobutyl cyanoacrylate) (15), poly(lactide-co-glycolide) (16), poly(-ε-caprolactone) (17), pluronic/poly(lactic acid) block copolymers (18), and poly(lactide-co-glycolide) nanoparticles embedded within poly(vinyl alcohol) hydrogel (19) have been reported. The encapsulation of insulin into mucoadhesive alginate/chitosan nanoparticles was shown to be a key factor in the improvement of oral absorption and oral bioactivity in diabetic rats (20). These approaches substantiate the potential use of polymeric nanoparticles in oral administration of insulin, thereby bypassing the enzymatic degradation in the stomach.

Insulin Delivery Through Inhalable Nanoparticles

Inhalable, polymeric nanoparticle-based drug delivery systems have been tried earlier for the treatment of tuberculosis (21). Such approaches can be directed toward insulin delivery through inhalable nanoparticles. Insulin molecules can be encapsulated within the nanoparticles and can be administered into the lungs by inhaling the dry powder formulation of insulin. The nanoparticles should be small enough to avoid clogging up the lungs but large enough to avoid being exhaled. Such a method of administration allows the direct delivery of insulin molecules to the bloodstream without undergoing degradation. A few studies have been done to test the potential use of ceramic nanoparticles (calcium phosphate) as drug delivery agents (22,23). Porous hydroxyapatite nanoparticles have also been tested for the intestinal delivery of insulin (24). Preclinical studies in guinea pig lungs with insulin-loaded poly(lactide-co-glycolide) nanospheres demonstrated a significant reduction in blood glucose level with a prolonged effect over 48 hours when compared with insulin solution (25). Insulin-loaded poly(butyl cyanoacrylate) nanoparticles when delivered to the lungs of rats were shown to extend the duration of hypoglycemic effect over 20 hours when compared with pulmonary administration of insulin solution (26). The major factors limiting the bioavailability of nasally administered insulin include poor permeability across the mucosal membrane and rapid mucociliary clearance mechanism that removes the nonmucoadhesive formulations from the absorption site (27). To overcome these limitations, mucoadhesive nanoparticles made of chitosan/tripolyphosphate (28) and starch (29) have been evaluated. These nanoparticles showed good insulin-loading capacity, providing the release of 75% to 80% insulin within 15 minutes after administration.

BioMEMS for Insulin Delivery

Implantable biological micro electro mechanical systems (BioMEMS) can be used as insulin pumps for controlled release of insulin when there is an increase in blood glucose level (30). Another proposed BioMEMS device has a drug reservoir compartment filled with insulin molecules (31). Biosensors and nonporous membranes with pores of 6-nm diameter are located in the exterior to detect the changes in blood glucose level and for insulin release. A review about the fabrication of a glucose-sensitive microvalve MEMS device for insulin delivery discusses extensively about the research attempts done in the past few decades (32). Another implantable, polymer-based micropump system with integrated biosensors for optimal insulin delivery without user intervention has been described in a recent study (33). Microfabrication techniques have taken the miniaturization science to the nanoscale level. Microneedles have also been reported as effective transdermal systems for insulin delivery (34). The concept of an assembled biocapsule consisting of two micromachined membranes bonded together to form a cell-containing cavity bound by membranes with nanopores was reported earlier (Fig. 3). Microfabricated membranes with 18-nm pore size (Fig. 4) were shown to be sufficiently permeable to small biomolecules, such as oxygen, glucose, and insulin (35). Such biocapsules can be incorporated into BioMEMS devices for insulin delivery and diabetes treatment. While the nanopores were designed to be permeable to glucose, insulin, and other metabolically active products, the pores were small enough to prevent the passage of larger cytotoxic cells, macrophages, antibodies, and complement (36) (Fig. 5).

Other Nanoparticulate Systems for Insulin Delivery and Diabetes-Associated Symptoms Treatment

Other than the ceramic and polymeric nanoparticles, gold nanoparticles have also been tested as insulin carriers. Gold nanoparticles synthesized with chitosan as a

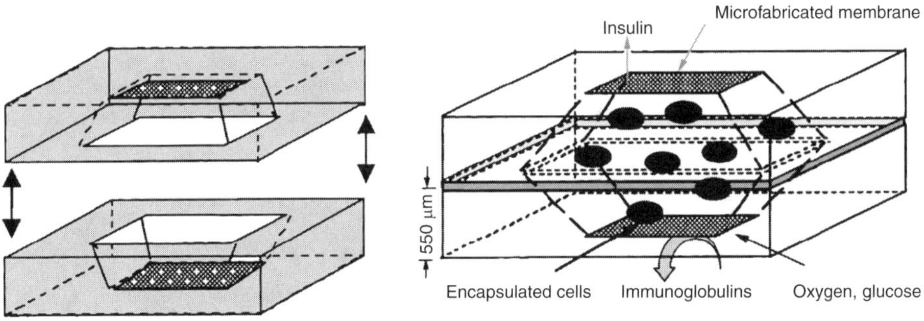

FIGURE 3 Schematic of an assembled biocapsule consisting of two micromachined membranes bonded together to form a cell-containing cavity bounded by membranes. *Source*: From Ref. 35.

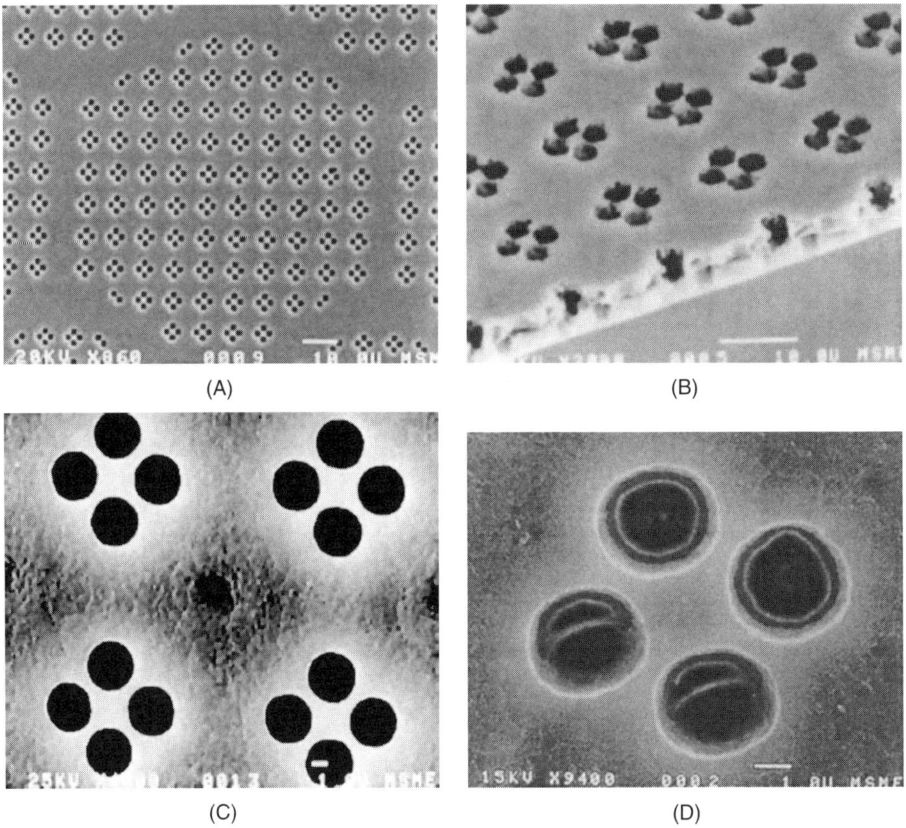

FIGURE 4 Scanning electron micrographs of the microfabricated membrane of 18-nm pore size: (**A**) top view of pore array; (**B**) cross-sectional view of membrane; (**C**) cluster of 2 × 2-mμ entry ports; and (**D**) magnified view of 2 × 2-mμ entry ports with 18-nm channels underneath. *Source*: From Ref. 35.

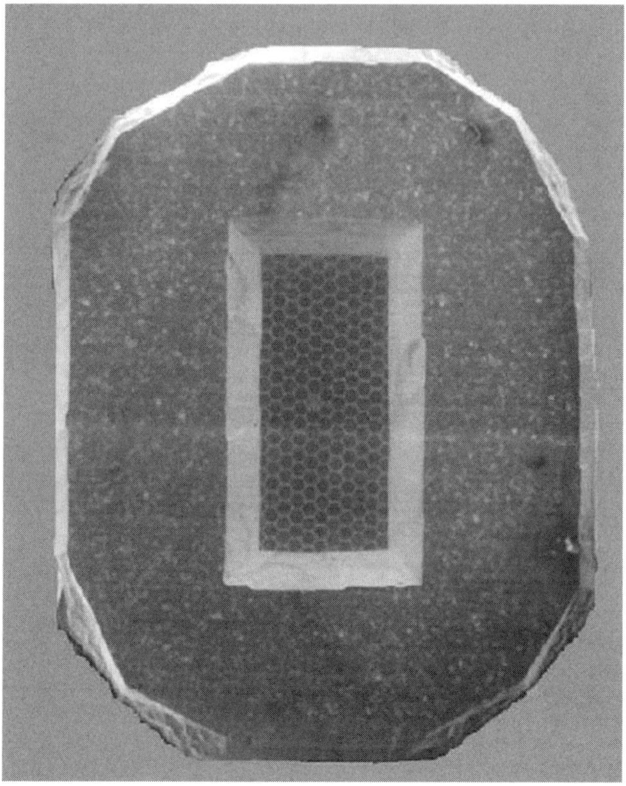

FIGURE 5 A micrograph of a biocapsule membrane with 24.5-nm pores. *Source*: From Ref. 36.

reducing agent were tested as a carrier for insulin (37). The nanoparticles showed long-term stability in terms of aggregation and good insulin loading of 53%. The use of chitosan served dual purpose by acting as a reducing agent in the synthesis of gold nanoparticles and also promoting the penetration and uptake of insulin across the oral and nasal mucosa in diabetic rats. The study concluded that oral and nasal administration of insulin-loaded, chitosan-reduced gold nanoparticles improved pharmacodynamic activity of insulin. Dextran nanoparticle–vitamin B_{12} combination has been tested to overcome the gastrointestinal degradation of vitamin B_{12}–peptide–protein drug conjugates (38). These nanoparticles were found to protect the entrapped insulin against gut proteases. Dextran nanoparticle–vitamin B_{12} combination showed a release profile that was suitable for oral delivery systems of insulin.

Diabetes causes a lot of systemic complications. The associated conditions are inflammatory diseases of skin and gums, diabetic retinopathy (eyes), diabetic neuropathy (nervous system), heart diseases, kidney diseases, delayed wound healing, and many more. Nanoparticulate systems have also been tested for the treatment of these associated conditions. Nanoparticle-based ocular drug delivery systems have been already described in the past decade (39,40). The recent years have seen the advancement in applications of nanoparticles made of polyacrylic acid (41), polylactide (42), and chitosan (43) for ophthalmic drug delivery. The scientific community is working toward utilizing nanoparticle-based drug delivery systems for the treatment of diabetes-associated complications.

LIMITATIONS OF NANOPARTICLES AND CONCLUSION

The science and knowledge that the scientific community has today about nanotechnology and its potential versatile applications are based only on the research work done in the laboratories. These research studies are being conducted to understand how matter behaves at the nanoscale level. Factors and conditions governing the behavior of macrosystems do not really apply to the nanosystems. The major limitations and technological hurdles faced by nanotechnology and its applications in the field of drug delivery should be addressed (44,45). Scientific community has not yet understood completely how the human body would react to these nanoparticles and nanosystems, which are acting as drug carriers. Nanoparticles have larger surface area when compared to their volume. Friction and clumping of the nanoparticles into a larger structure is inevitable, which may affect their function as a drug delivery system. Due to their minute size, these drug carriers can be cleared away from the body by the body's excretory pathways. When these are not excreted, larger nanoparticles can accumulate in vital organs, causing toxicity leading to organ failure. Recent study in mice revealed that tissue distribution of gold nanoparticles is size dependent, with the smallest nanoparticles (15–50 nm) showing the most widespread organ distribution including blood, liver, lung, spleen, kidney, brain, heart, and stomach (46). Liposomes have certain drawbacks, such as being captured by the human body's defense system. The drug-loading capacity of liposomes is being tested by researchers and still remains inconclusive. All previous studies resulted in posttreatment accumulation of the nanoparticles in skin and eyes. Gold nanoparticles tend to accumulate in bone joints and organs. Once the nanoparticles are administered into the human body, they should be controlled by an external control, preventing them from causing adverse effects. These drug delivery technologies are in various stages of research and development. It is expected that these limitations can be overcome and the discoveries to come into practical use within the next 5 to 10 years.

REFERENCES

1. Kuzuya T, Nakagawa S, Satoh J, et al. Report of the Committee on the classification and diagnostic criteria of diabetes mellitus. Diabetes Res Clin Pract 2002; 55(1):65–85.
2. Attivi D, Wehrle P, Ubrich N, et al. Formulation of insulin-loaded polymeric nanoparticles using response surface methodology. Drug Dev Ind Pharm 2005; 31(2):179–189.
3. Kost J, Horbett TA, Ratner BD, et al. Glucose-sensitive membranes containing glucose oxidase: Activity, swelling, and permeability studies. J Biomed Mater Res 1985; 19:1117–1133.
4. Ishihara K, Kobayashi M, Shinohara I. Control of insulin permeation through a polymer membrane with responsive function for glucose. Makromol Chem Rapid Commun 1983; 4:327.
5. Dorski CM, Doyle FJ, Peppas NA. Preparation and characterization of glucose-sensitive P(MAA-g-EG) hydrogels. Polym Mater Sci Eng Proc 1997; 76:281–282.
6. Morishita M, Morishita I, Takayama K, et al. Novel oral microspheres of insulin with protease inhibitor protecting from enzymatic degradation. Int J Pharm 1992; 78(1–3):1–7.

7. Morcol T, Nagappan P, Nerenbaum L, et al. Calcium phosphate-PEG-insulin-casein (CAPIC) particles as oral delivery systems for insulin. Int J Pharm 2004; 277(1–2):91–97.
8. Venugopalan P, Sapre A, Venkatesan N, et al. Pelleted bioadhesive polymeric nanoparticles for buccal delivery of insulin: Preparation and characterization. Pharmazie 2001; 56(3):217–219.
9. Leobandung W, Ichikawa H, Fukumori Y, et al. Preparation of stable insulin-loaded nanospheres of poly(ethylene glycol) macromers and *N*-isopropyl acrylamide. J Control Release 2002; 80(1–3):357–363.
10. Pan Y, Li YJ, Zhao HY, et al. Bioadhesive polysaccharide in protein delivery system: Chitosan nanoparticles improve the intestinal absorption of insulin in vivo. Int J Pharm 2002; 249(1–2):139–147.
11. Lin YH, Mi FL, Chen CT, et al. Preparation and characterization of nanoparticles shelled with chitosan for oral insulin delivery. Biomacromolecules 2007; 8(1):146–152.
12. Sarmento B, Ribeiro A, Veiga F, et al. Development and characterization of new insulin containing polysaccharide nanoparticles. Colloids Surf B Biointerfaces 2006; 53(2):193–202.
13. Tiyaboonchai W, Woiszwillo J, Sims RC, et al. Insulin containing polyethylenimine-dextran sulfate nanoparticles. Int J Pharm 2003; 255(1–2):139–151.
14. Sajeesh S, Sharma CP. Cyclodextrin-insulin complex encapsulated polymethacrylic acid based nanoparticles for oral insulin delivery. Int J Pharm 2006; 325(1–2):147–154.
15. Mesiha MS, Sidhom MB, Fasipe B. Oral and subcutaneous absorption of insulin poly(isobutylcyanoacrylate) nanoparticles. Int J Pharm 2005; 288(2):289–293.
16. Cui FD, Tao AJ, Cun DM, et al. Preparation of insulin loaded PLGA-Hp55 nanoparticles for oral delivery. J Pharm Sci 2007; 96(2):421–427.
17. Damgé C, Maincent P, Ubrich N. Oral delivery of insulin associated to polymeric nanoparticles in diabetic rats. J Control Release 2007; 117(2):163–170.
18. Xiong XY, Li YP, Li ZL, et al. Vesicles from pluronic/poly(lactic acid) block copolymers as new carriers for oral insulin delivery. J Control Release 2007; 120(1–2):11–17.
19. Liu J, Zhang SM, Chen PP, et al. Controlled release of insulin from PLGA nanoparticles embedded within PVA hydrogels. J Mater Sci Mater Med 2007; 18(11), 2205–2210.
20. Sarmento B, Ribeiro A, Veiga F, et al. Alginate/chitosan nanoparticles are effective for oral insulin delivery. Pharm Res 2007; 24(12):2198–2206.
21. Pandey R, Sharma A, Zahoor A, et al. Poly(DL-lactide-co-glycolide) nanoparticle-based inhalable sustained drug delivery system for experimental tuberculosis. J Antimicrob Chemother 2003; 52(6):981–986.
22. Cherian AK, Rana AC, Jain SK. Self-assembled carbohydrate-stabilized ceramic nanoparticles for the parenteral delivery of insulin. Drug Dev Ind Pharm 2000; 26(4):459–463.
23. Corkery K. Inhalable drugs for systemic therapy. Respir Care 2000; 45(7):831–835.
24. Paul W, Sharma C. Porous hydroxyapatite nanoparticles for intestinal delivery insulin. Trends Biomed Artif Organs 2001; 14(2):37–38.
25. Kawashima Y, Yamamoto H, Takeuchi H, et al. Pulmonary delivery of insulin with nebulized DL-lactide/glycolide copolymer (PLGA) nanospheres to prolong hypoglycemic effect. J Control Release 1999; 62(1–2):279–287.
26. Zhang Q, Shen Z, Nagai T. Prolonged hypoglycemic effect of insulin-loaded polybutylcyanoacrylate nanoparticles after pulmonary administration to normal rats. Int J Pharm 2001; 218(1–2):75–80.
27. O'Hagan DT, Illum L. Absorption of peptides and proteins from the respiratory tract and the potential for development of locally administered vaccine. Crit Rev Ther Drug Carrier Syst 1990; 7(1):35–97.
28. Grenha A, Seijo B, Remuñán-López C. Microencapsulated chitosan nanoparticles for lung protein delivery. Eur J Pharm Sci 2005; 25(4–5):427–437.
29. Jain AK, Khar RK, Ahmed FJ, et al. Effective insulin delivery using starch nanoparticles as a potential trans-nasal mucoadhesive carrier. Eur J Pharm Biopharm 2008; 69(2):426–435.
30. Maillefer D, Gamper S, Frehner B, et al. A high-performance silicon micropump for disposable drug delivery systems. 14th IEEE Int Conf MEMS Tech Digest 2001:413–417.

31. Leadership Medica. Biomedical technology. Available at: http://www.leadershipmedica.com/scientifico/sciesett02/scientificaing/7ferrarie/7ferraring.htm. Accessed September 2008.
32. Ziaie B, Baldi A, Lei M, et al. Hard and soft micromachining for BioMEMS: Review of techniques and examples of applications in microfluidics and drug delivery. Adv Drug Deliv Rev 2004; 56(2):145–172.
33. Wong E, Bigdeli A, Biglari-Abhari M. A conducting polymer-based self-regulating insulin delivery system. Int J Sci Res 2006; 16:235–239.
34. Martanto W, Davis SP, Holiday NR, et al. Transdermal delivery of insulin using microneedles in vivo. Pharm Res 2004; 21(6):947–952.
35. Desai TA, Hansford D, Ferrari M. Characterization of micromachined silicon membranes for immunoisolation and bioseparation applications. J Membr Sci 1999; 159(1–2):221–231.
36. Tao SL, Desai TA. Microfabricated drug delivery systems: From particles to pores. Adv Drug Deliv Rev 2003; 55:315–328.
37. Bhumkar DR, Joshi HM, Sastry M, et al. Chitosan reduced gold nanoparticles as novel carriers for transmucosal delivery of insulin. Pharm Res 2007; 24(8):1415–1426.
38. Chalasani KB, Russell-Jones GJ, Yandrapu SK, et al. A novel vitamin B_{12}-nanosphere conjugate carrier system for peroral delivery of insulin. J Control Release 2007; 117(3):421–429.
39. Zimmer A, Kreuter J. Microspheres and nanoparticles used in ocular delivery systems. Adv Drug Deliv Rev 1995; 16(1):61–73.
40. Gurny R, Boye T, Ibrahim H. Ocular therapy with nanoparticulate systems for controlled drug delivery. J Control Release 1985; 2:353–361.
41. De TK, Rodman DJ, Holm BA, et al. Brimonidine formulation in polyacrylic acid nanoparticles for ophthalmic delivery. J Microencapsul 2003; 20(3):361–374.
42. Bourges JL, Gautier SE, Delie F, et al. Ocular drug delivery targeting the retina and retinal pigment epithelium using polylactide nanoparticles. Invest Ophthalmol Vis Sci 2003; 44(8):3562–3569.
43. Enríquez de Salamanca A, Diebold Y, Calonge M, et al. Chitosan nanoparticles as a potential drug delivery system for the ocular surface: Toxicity, uptake mechanism and in vivo tolerance. Invest Ophthalmol Vis Sci 2006; 47(4):1416–1425.
44. Neuberger T, Schöpf B, Hofmann H, et al. Superparamagnetic nanoparticles for biomedical applications: Possibilities and limitations of a new drug delivery system. J Magn Magn Mater 2005; 293(1):483–496.
45. Kayser O, Lemker A, Trejo NH. The impact of nanobiotechnology on the development of new drug delivery systems. Curr Pharm Biotechnol 2005; 6:3–5.
46. Sonavane G, Tomoda K, Makino K. Biodistribution of colloidal gold nanoparticles after intravenous administration: Effect of particle size. Colloids Surf B Biointerfaces 2008; 66(2):274–280.

9 Nanosystems for Dermal and Transdermal Drug Delivery

Venkata Vamsi Venuganti and Omathanu P. Perumal

Department of Pharmaceutical Sciences, South Dakota State University, Brookings, South Dakota, U.S.A.

INTRODUCTION

Skin is the largest organ in the human body and functions as a protective barrier. The large surface area (1.8 m^2) and easy accessibility of skin make it one of the attractive routes for drug delivery. However, the unique bioarchitecture of skin limits the transport of molecules through it (1). The skin is primarily divided into epidermis, dermis, and subcutaneous tissue. Epidermis is again subdivided into stratum corneum (SC), stratum lucidum, stratum granulosum, stratum germinativum, and stratum basale [Fig. 1(A)]. Stratum basale is the germinal layer which continuously divides to produce new keratinocytes that move to outer layers finally to form a horny layer of dead cells in the SC. It takes 2 to 3 weeks for the migration of keratinocytes from the basal layer to finally shed off from the SC. The skin also has appendages such as hair follicles and sweat pores, which constitute 0.1% of the total skin surface area. The hair follicles originate from the dermis and terminate at the surface of the skin. Although the SC is only 10 to 20 μm thick, it forms a formidable barrier for the exchange of solutes and moisture. It is made up of stacks of keratin-filled corneocytes interdispersed by tightly arranged lipid bilayers (2). The intercellular lipids mainly consist of free fatty acids, ceramides, and cholesterol (2). Molecules can penetrate the skin by three main routes: (*i*) intracellular (across the corneocytes), (*ii*) intercellular lipids, and (*iii*) appendageal [Fig. 1(B)]. The intercellular lipids are the major transport pathways for most drugs, in which the molecule has to pass through successive hydrophilic and hydrophobic domains in the lipid bilayers. On the other hand, the skin appendages serve as a shunt pathway for drug molecules. Since the appendages occupy only a fraction of the skin surface, they contribute very little to the drug transport. However, the appendages constitute a significant pathway for the iontophoretic transport of charged molecules and the penetration of particulate systems (3,4). Once the molecule crosses the SC and subsequently the viable epidermis, it is absorbed through the blood vessels in the dermis [Fig. 1(A)].

Drugs are delivered to and through the skin for the treatment of skin diseases and systemic diseases, respectively. These include various types of formulations/delivery systems such as powders, solutions, sprays, suspensions, emulsions, ointments, creams, pastes, gels, and patches. For dermatological applications, formulations are targeted to different layers of the skin to protect (e.g., sunscreens), enhance the appearance (e.g., cosmetics), or deliver medicaments (e.g., for acne, psoriasis) to the skin (Table 1). The goal of dermatological preparations is to

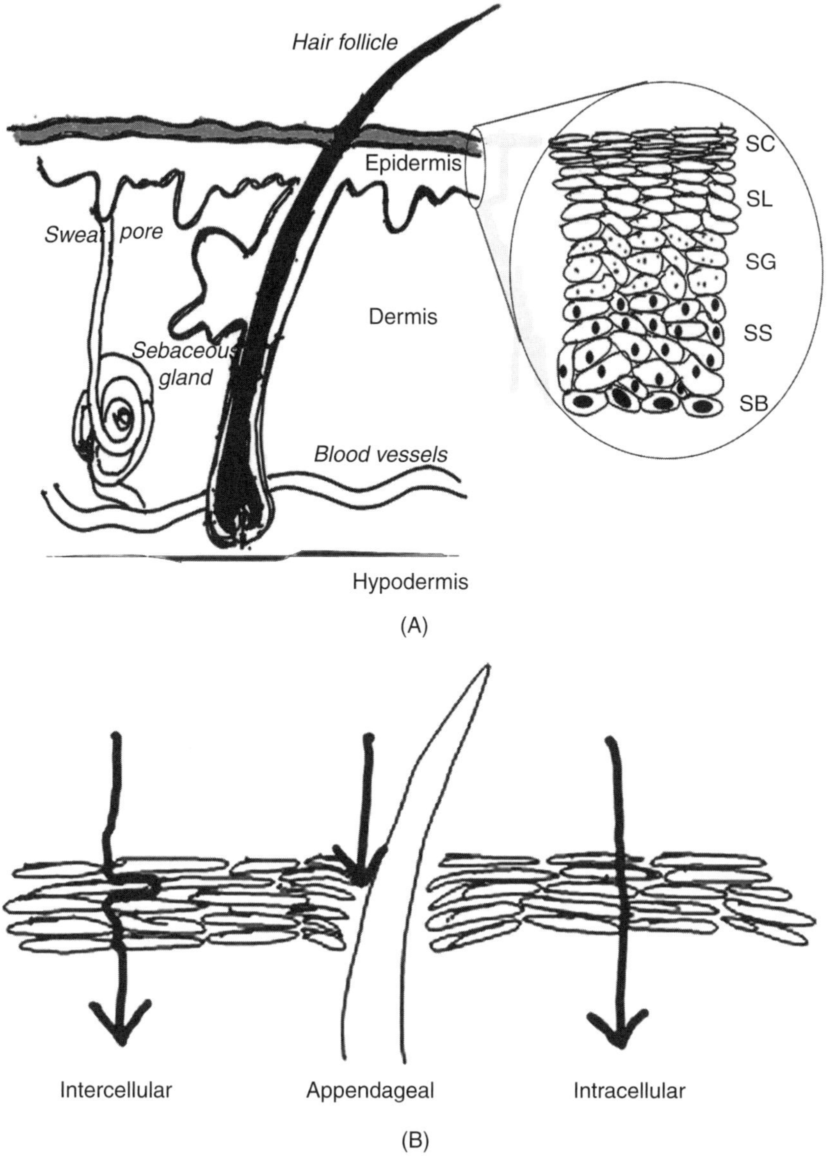

FIGURE 1 (**A**) Structure of the skin. *Abbreviations*: SC, stratum corneum; SL, stratum lucidum; SG, stratum granulosum; SS, stratum spinosum; SB, stratum basale. (**B**) Major routes of skin penetration.

maximize drug retention in the skin, while minimizing drug absorption into the systemic circulation. In contrast, the goal in transdermal systems is to maximize drug absorption in the systemic circulation. The rate and extent of drug penetration into different layers of skin and into systemic circulation are governed by the drug properties and formulation characteristics. Although several models have been

TABLE 1 Target Site for Topical and Transdermal Drug Delivery

Skin layer	Cosmetics/drugs
Stratum corneum	Cosmetics Sunscreens Antimicrobials Skin protectants
Viable epidermis	Anti-inflammatory agents Antiproliferative agents Antihistamines Vaccines Gene therapy
Hair follicles	Antiacne agents Antimicrobials Depilatories Vaccines, gene therapy
Dermis	Local anesthetics, analgesics Drugs for systemic administration

developed to explain drug permeation through the skin (5), most, if not all, are based on the simple Fick's second law of diffusion:

$$\frac{\mathrm{d}M}{\mathrm{d}t} = \frac{DKC_\mathrm{s}}{h}$$

where $\mathrm{d}M/\mathrm{d}t$ is the flux or rate of drug permeation with respect to time, D the diffusion coefficient, K the drug partition coefficient, C_s the drug concentration in the formulation, and h the membrane (skin) thickness. The concentration gradient drives the passive permeation of drug molecules through the skin, whereas the rate and extent of drug permeation are influenced by the physicochemical properties of the drug such as drug solubility in the vehicle, relative solubility of the drug in both the vehicle and the skin (partition coefficient), and molecular size, among others. Both K and C_s mainly depend on the drug property, whereas D and h mainly depend on the membrane (skin) characteristics.

Different theories have been proposed to predict the transport of hydrophilic and lipophilic permeants (6). For the transport of hydrophilic molecules, the pore transport theory has been proposed by Peck et al. (7), in which the SC is described as a porous membrane having an effective pore radius ranging from 15 to 25 Å (i.e., 1.5–2.5 nm) and these pores are positive, neutral, or negatively charged. Other authors have reported pore sizes of 0.4 to 36 nm, as well as larger pores (~100 nm) in the skin (8). The pore estimates vary depending on the size of the permeant used to characterize the pores and the geometry of the measured pore (8). Chemical and physical enhancement methods are believed to increase drug permeation by increasing the effective pore radius and/or the number of pores (7,9). On the other hand, for the transport of lipophilic permeants, both porous and lipoidal pathways have been proposed (10,11). The general physicochemical properties for passive skin permeation have been widely accepted, and all the transdermal products in the market fulfill these criteria (Table 2). At the same time, the stringent requirements imposed by the skin also explains why only a handful of transdermal drugs have reached the market, in spite of intensive research over the last two decades. There is enormous interest in expanding the number of drugs delivered through the skin

TABLE 2 Physicochemical Properties for Passive Skin Permeation

Parameters	Values
Molecular weight	<500 Da
Log *P*	1–3
Aqueous solubility	≥ 1 mg/mL
Hydrodynamic radius	≤ 2 nm
Melting point	$<200^{\circ}$C
Dose	10–40 mg/day

Source: Adapted from Ref. 124.

due to the perceived advantages of high patient compliance, controlled drug delivery, ease of termination of therapy, flexibility in the choice of drug administration site, and ease of altering the dose by controlling the application area and avoidance of first-pass effect (12). This has led to a number of passive and active skin permeation enhancement strategies (13). Passive enhancement strategies include the use of chemical enhancers and prodrug approaches to improve drug partitioning and/or increase drug diffusion through the skin (13). On the other hand, active enhancement strategies use physical methods such as electric current, ultrasound, laser, or mechanical methods (12,13). All these enhancement strategies alter the drug characteristics and/or the skin barrier properties to improve skin permeation.

The recent emergence of nanotechnology has opened up new opportunities to develop nanosystems for topical and transdermal applications. If not all, some of these nanosystems have been specifically developed for skin applications. The nanosystems can be classified based on the properties of the carriers as shown in Figure 2. For drug delivery applications, the nanosystems generally range in size from 1 to 1000 nm. The goal of this chapter is to provide a comprehensive overview of the formulation characteristics, mechanism of skin penetration, applications, and future prospects of nanosystems for dermal and transdermal drug delivery systems.

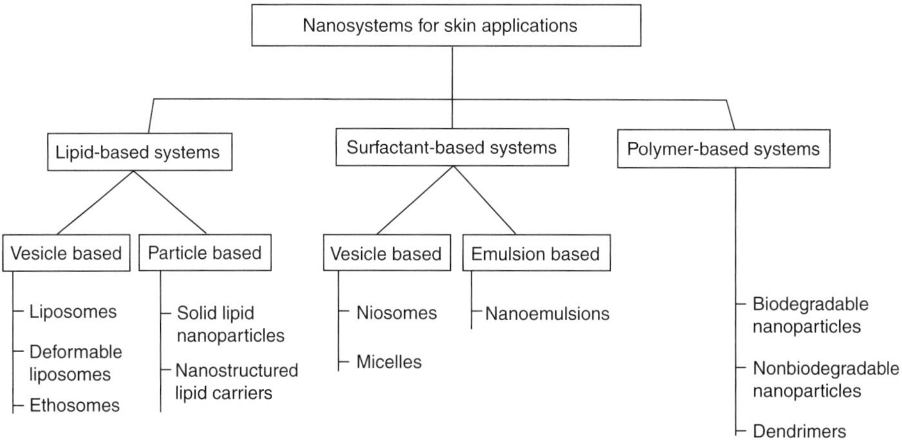

FIGURE 2 Classification of nanosystems used for skin applications.

LIPID-BASED NANOSYSTEMS

Liposomes

Liposomes are thermodynamically stable, spontaneously formed, submicroscopic vesicular structures of amphipathic lipids arranged in one or more concentric bilayers with an entrapped aqueous core (Fig. 3). The lipids are natural or synthetic phosopholipids, which include phosphatidyl–choline (lecithin), phosphatidyl–ethanolamine, phosphatidyl–glycerol, phosphatidyl–serine, and phosphotidyl–

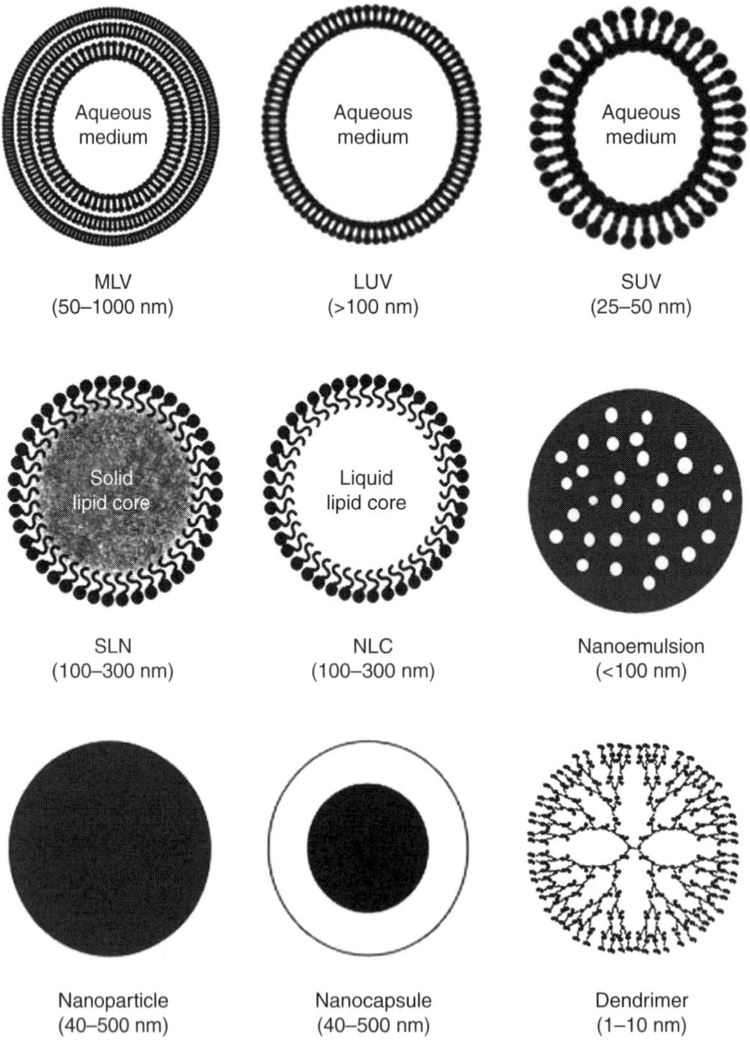

FIGURE 3 Schematic of various nanosystems (not to scale). *Abbreviations*: MLV, multilamellar vesicle; LUV, large unilamellar vesicle; SUV, small unilamellar vesicle; SLN, solid lipid nanoparticle; NLC, nanostructured lipid carrier.

ionositol (14). On mixing with an aqueous medium, the phosphate groups of the phospholipids orient themselves to the hydrophilic environment spontaneously forming unilamellar or multilamellar bilayer vesicles (Fig. 3). The head groups of the phosopholipids have neutral, positive or negative charge. Cholesterol is usually included to improve the stability of the vesicles, impart fluidity to the bilayer membrane, and prevent the leakage of vesicle contents (14). In addition, stearylamine or diacetyl phosphate can be included to impart a net positive or negative charge respectively (15). Hydrophilic drugs are incorporated in the aqueous core, whereas lipophilic drugs are entrapped within the bilayer. In general, hydrophilic drugs have higher encapsulation efficiency than lipophilic drugs (14).

Liposomes range in size from 50 nm to several hundred nanometers. They are classified based on the size of vesicles or number of lipid bilayers into (*i*) small unilamellar vesicles (SUVs) and (*ii*) large unilamellar vesicles (LUVs) (Fig. 3). Based on the number of bilayers, they are classified into (*i*) multilamellar vesicles (MLVs) and (*ii*) unilamellar vesicles (ULVs). The different methods used to prepare liposomes are described in Table 3. In the case of film hydration method, a mixture of lipids is dissolved in organic solvents, which is then dried and redispersed in an aqueous medium to form MLVs (0.05–10 μm). SUVs (50–200 nm) can be prepared by sonication of MLVs by using a probe or bath sonicator (16). Small vesicle sizes can also be achieved by passing the MLVs through a polycarbonate filter under high pressure (extrusion). Multiple passes through the filter produce uniform SUVs with a low polydispersity index (14). Alternatively, SUVs can be prepared by ether injection method (Table 3). LUVs (~100 nm) are prepared using reverse-phase evaporation (Table 3). A larger amount of drug can be incorporated (60–65%) in the LUVs than in the SUVs (14).

The use of liposomes as a topical delivery carrier was first demonstrated by Mezei and Gulasekharam (17). In comparison to topical application of a simple lotion, the liposomes delivered four to five fold higher steroid concentrations in the skin layers. Since then, there has been an exponential growth in the use of liposomes as topical delivery vehicles (15). On the other hand, the use of liposomes as transdermal drug delivery vehicles is debatable (18). Now it is clear that the conventional liposomes have little value for transdermal drug delivery and in fact may decrease the drug penetration (18). However, liposomes can improve the deposition of drugs

TABLE 3 Preparation Methods for Liposomes

Preparation method	Description
Thin-film hydration method	Lipid mixture is dissolved in a suitable organic solvent such as methanol or chloroform. The organic solvent is removed under reduced pressure and lyophilized to remove any traces of solvent. The thin lipid film is redispersed in aqueous medium and sonicated to give a uniform dispersion. Lamellarity and size of the vesicles can be reduced by sonication or extrusion.
Reverse-phase evaporation	A water-in-oil emulsion is prepared by dissolving phospholipids in the organic phase. The organic phase is then removed slowly under reduced pressure. Large unilamellar vesicles with entrapped aqueous core are formed.
Ether injection method	Lipid is dissolved in ether and injected into the aqueous medium at a very slow rate. The large vesicles formed are removed by gel filtration.

intended for topical applications. Skin penetration of liposomes is influenced by their size, composition, lamellarity, and charge (15,18). The mechanism of skin penetration of liposomes has been attributed to the penetration-enhancing properties of phosopholipids, increased skin partitioning of drug, adsorption of liposomes onto the skin, and penetration through transappendageal pathways (15,18). The phase-transition temperature of lipids in the liposomes influences their penetration and interaction with the skin lipids (19). Lipids in the liquid state penetrate deeper than do lipids in the gel state at the skin temperature (19). In the latter case, the liposomes are limited to the SC. The skin penetration of liposomes has a weak dependence on their size (18). This is partly because liposomes may break up during skin penetration, particularly MLVs may lose their bilayers during skin penetration to form ULVs (18). Studies have shown that liposomes adsorb and fuse with the skin surface and form a favorable environment for the partitioning of lipophilic drugs (18–20). On the other hand, intact liposomes can penetrate through the appendageal pathways, fuse with the sebum, and slowly release the drug (20). As a result, liposomes have been used to target therapeutics to the follicles (21). The role of surface charge on the skin penetration of liposomes is not clear, as some studies have shown that cationic liposomes penetrate better than anionic liposomes (22), whereas others have shown that anionic liposomes penetrate better than cationic liposomes (23).

Deformable Liposomes

To address the limited skin penetration of conventional liposomes, Cevc (8) developed a new class of highly deformable (elastic or ultraflexible) liposomes which are termed as Transfersomes®. A distinct feature of these liposomes is the presence of surfactants as an edge activator that destabilizes the vesicle, giving it more flexibility and deformability. Span 60, Span 80, Tween 20, Tween 80, sodium cholate, sodium deoxycholate, dipotassium glycyrrhizinate, and oleic acid have been used as edge activators (18). Transfersomes range in size from 200 to 300 nm. Due to their high deformability, they are believed to squeeze through skin pores (20–30 nm), which are one-tenth of their size and reach deeper layers in the skin (24). The osmotic gradient arising from the hydration difference between the skin surface and the viable epidermis drives the skin penetration of transfersomes (Fig. 4). This is supported by the fact that deformable liposomes penetrate better under nonoccluding conditions (25). Transfersomes have been shown to increase the delivery of drugs into the skin as well as through the skin into systemic circulation for both small and large molecules (25). Interestingly, the use of transfersomes as topical, regional, or systemic delivery depends on the amount of transfersomes applied per surface area of the skin (25). When a small amount is applied, the transfersomes penetrate only to a limited depth, before they are degraded in the skin. On the other hand, when an

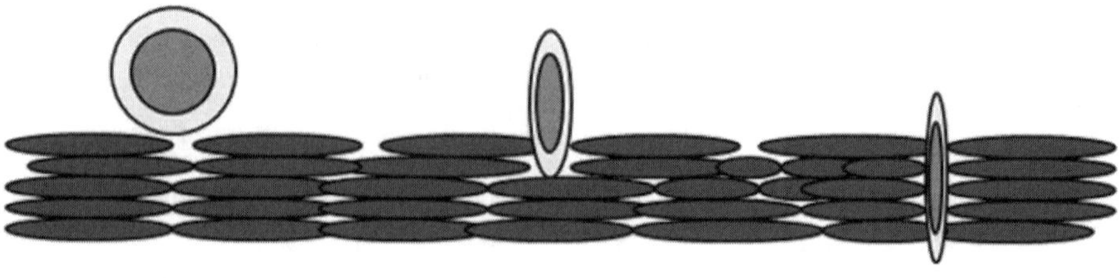

FIGURE 4 The mechanism of skin penetration of deformable vesicles.

intermediate amount is applied, the transfersomes penetrate deeper into the underlying tissues such as the muscle. An increase in the applied amount saturates the skin layers, the subcutaneous tissue, and then the transfersomes enter into the systemic circulation through lymphatic capillaries. Unlike the conventional liposomes, the transfersomes can penetrate the skin intact (24). Furthermore, the transfersomes can interact with lipid bilayers and increase the skin penetration of drugs (18). In contrast to conventional liposomes, the transfersomes do not penetrate through the transappendageal pathways (26). For hydrophilic drugs, the penetration-enhancing effect may play a greater role than for lipophilic drugs (18). On the other hand, for lipophilic drugs, the association of the drug with the intact transfersomes may be important for enhanced skin penetration (18). Verma et al. (27) found an inverse relationship between the size of deformable liposomes and the in vitro skin penetration of entrapped fluorescent dyes. Deformable liposomes larger than 600 nm remained in the SC, whereas flexible liposomes smaller than 300 nm were found to penetrate deeper. An optimal size was found to be 120 nm for maximal skin penetration of these flexible liposomes. Similar to conventional liposomes, the vesicle composition and the membrane fluidity affect the skin penetration characteristics of deformable liposomes (28).

Ethosomes

Ethosomes are ethanol-containing phospholipid vesicles, and they were first reported by Touitou et al. (29). The ethanol concentration in ethosomes usually ranges from 20% to 45%, and it imparts high flexibility and malleability to the vesicles. The lipids and the drug are dissolved in ethanol, followed by mixing them with a constant stream of aqueous solution in a sealed container. Mixing is continued further for a few more minutes to obtain small (90–150 nm) homogenous MLVs. One of the distinct advantages of ethosomes is their small size relative to conventional liposomes, which obviates the need for size reduction (30). An increase in ethanol concentration (from 20% to 45% ethanol) generally decreases the vesicle size (29). However, a high concentration of ethanol leads to the interdigitation of the lipid bilayers and destabilization of the vesicles. Very high encapsulation efficiencies of lipophilic drugs can be achieved due to the enhanced solubility from the presence of ethanol (29). Unlike transfersomes, ethosomes can enhance drug delivery through skin under both nonoccluding and occluding conditions (31). Ethosomes act by multiple mechanisms to enhance drug permeation through the skin (Fig. 5). This includes the release of ethanol from the ethosomes, which, in turn, increases the

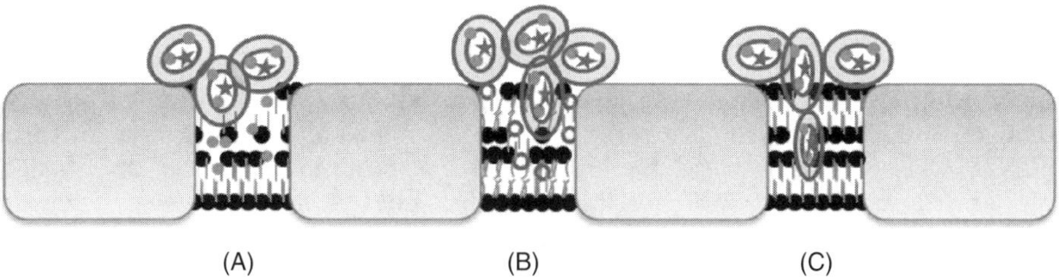

FIGURE 5 Mechanisms of ethosomal skin penetration: (**A**) release of ethanol from the ethosomes (solid circles) and fluidizes the skin lipids to increases skin permeation; (**B**) release of lipids (open circles) from ethosomes and/or interaction of ethosomes with skin lipids; (**C**) ethosomes squeeze through the skin lipid bilayer. Drug molecules are shown as asterisks inside the vesicles.

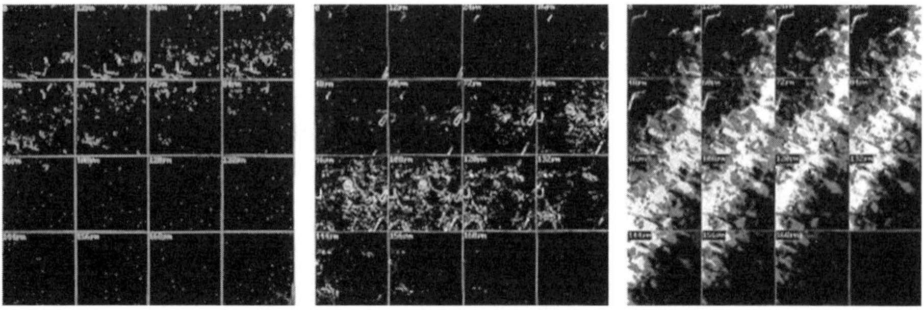

(A) (B) (C)

FIGURE 6 Confocal laser scanning micrographs of mouse skin after the application of the fluorescent probe D-289 in (**A**) conventional liposomes (**B**) 30% hydroethanolic solution, and (**C**) ethosomes. The images are optical sections of *XYZ* scan in which the image was scanned in increments of 10 to 20 μm. *Source*: From Ref. 30.

fluidity of skin lipids. However, the penetration enhancement seen with ethosomes is more than from a simple hydroalcoholic solution (30). Therefore, an additional mechanism is the direct interaction of ethosomes with the skin lipids to cause penetration enhancement (Fig. 5). Furthermore, ethanol provides soft flexible characteristics to aid in the skin penetration of intact ethosomes (18,29). The drug may be released by the fusion of the ethosomes with skin lipids in deeper layers, leading to systemic drug absorption (29). Caclein, a hydrophilic fluorescent probe, from the ethosomes penetrated to a depth of 160 μm in excised mouse skin compared with 80 and 60 μm from hydroalcoholic solution (30% ethanol) and conventional liposomes, respectively (32). Lipophilic fluorophore penetrated to a depth of 140 μm in mouse skin from both ethosomes and hydroalcoholic solution (Fig. 6). However, the fluorescence intensity was significantly higher with the ethosomes (29).

Lipid Nanoparticles

Unlike the flexible vesicular carriers, the lipid nanoparticles are made of solid lipids or a combination of solid and liquid lipids (33). These are known as solid lipid nanoparticles (SLNs) and nanostructured lipid carriers (NLCs). They are prepared by using "generally recognized as safe" lipids, which do not melt at room or body temperature (34). The main difference between SLNs and NLCs is the composition and physical state of the lipids (34). In the case of SLNs, the core consists of a solid lipid, while NLCs have a liquid lipid core (Fig. 3). NLCs consist of a mixture of solid and liquid lipids, but SLNs are composed mainly of solid lipids. The inclusion of liquid lipids in NLCs provides flexibility in modulating drug encapsulation and drug release (35). The drug is incorporated between the voids in the crystal lattice of SLNs (36). On the other hand, the NLC matrix has more imperfections and, hence, higher drug encapsulation is possible (37). High encapsulation has been achieved for lipophilic drugs (90–98%), whereas the encapsulation efficiency for hydrophilic molecules is relatively low (20–30%) (35). The location of the drug in the core or inside the shell of lipid nanoparticles depends on the nature of the lipid, drug properties, the solubility of the drug in the lipid, and the preparation method (35). Several highly purified lipids, such as tristearin, have been used for preparing SLNs

(36), whereas mixtures of mono-, di-, and triacylglycerols including monoacids and poly(acid acyl) glycerols are used for the preparation of NLCs (37).

Lipid nanoparticles are prepared by hot or cold high-pressure homogenization (35). In both techniques, the solid lipid is melted and the drug is dissolved or dispersed in the lipid matrix. In hot homogenization method, a preemulsion is formed by mixing the melted lipid with the drug (dissolved or dispersed) and a hot surfactant solution under high shear. This mixture is then homogenized by applying a pressure of 200 to 500 bars and two to three homogenization cycles. Subsequently, the hot o/w nanoemulsion is cooled at or below room temperature to form the lipid nanoparticles (35). In cold homogenization, the drug is dissolved or dispersed in the molten lipid, followed by rapid cooling using liquid nitrogen or dry ice. The drug is homogeneously distributed in the lipid matrix due to rapid cooling. This is followed by grinding the particles in a ball or mortar mill, which results in micron-sized lipid particles (50–100 μm). These lipid microparticles are then dispersed in an aqueous emulsifier solution. Subsequently, this dispersion is subjected to high-pressure homogenization at or below room temperature to produce lipid nanoparticles. Other techniques that have been used to produce SLNs include solvent emulsification–evaporation, emulsification–diffusion technique, and phase inversion (35). The SLNs are usually in the size range of 100 to 300 nm.

Lipid nanoparticles have superior physical stability compared with liposomes and other disperse systems (34). In liposomes and emulsions, the drug can diffuse and partition between the oil and the aqueous phase. In contrast, the solid lipid shell minimizes the partition of the drug into the aqueous medium and thus prevents drug leakage and degradation (35,36). The SLNs offer many advantages over other lipid carriers for skin. The increased surface area from the lipids in a nanoparticulate form increases their adhesiveness to the skin (38). Also, the lipid nanoparticles form an occluding film when applied to the skin (38). This leads to increased skin hydration, which, in turn, reduces the corneocyte packing and increases skin penetration. Furthermore, the lipids can interact with the skin lipids and act as penetration enhancers (35). Lipid nanoparticles can also be used to control the drug release in dermatological formulations (36). The drug is released by diffusion through the lipid matrix, and the release can be modulated by altering the composition of the lipids in SLNs or NLCs (35). Lipid nanoparticles can be incorporated in conventional topical dosage forms such as lotions, creams, and gels (39). The consistency of the formulation can be easily varied from free-flowing solids to semisolids or liquids depending on the proportion of the lipid particles in the formulation (35,39). Since lipid nanoparticles are composed of skin-compatible lipids, they are devoid of skin irritation or sensitization issues.

SURFACTANT-BASED NANOSYSTEMS

Niosomes

Niosomes are self-assembled, submicron vesicles of nonionic surfactants with closed bilayer structures similar to liposomes (40). However, they are much more stable and less expensive than liposomes. The hydrophobic part of the surfactant face toward the core, whereas the hydrophilic groups interface with the surrounding aqueous medium. Niosomes can be constructed by using a variety of amphiphiles, which possess a hydrophilic head group and a hydrophobic tail. Usually, the alkyl group chain length of the hydrophobic tail is between C_{12} and

C_{18} and, in certain cases, consists of a single steroidal chain (40). On the other hand, the hydrophilic head groups include glycerol, ethylene oxide, polyhydroxy groups, crown ethers, sugars, and amino acids (40). The most commonly used nonionic surfactants include Span and Brij surfactants. Cholesterol and their derivatives are included in niosomes (usually in a 1:1 molar ratio) as steric stabilizers to prevent aggregation (41). Cholesterol also prevents the phase transition of niosomes from the gel to liquid state and thereby reduces drug leakage from the niosomes. The stability of the niosomes can be further improved by the addition of charged molecules such as dicetyl phosphate, which prevents aggregation by charge repulsion (40). Generally, an increase in surfactant/lipid level increases the drug encapsulation efficiency in niosomes (41).

Niosome preparation requires some energy in the form of elevated temperature and/or shear (41). The majority of the methods involve hydration of a mixture of surfactant/lipid at elevated temperature, followed by size reduction using sonication, extrusion, or high-pressure homogenization. Finally, the unentrapped drug is removed by dialysis, centrifugation, or gel filtration. Niosomes prepared by hydration methods usually are in micron size range (40). Size reduction by sonication and/or extrusion results in niosomes of 100 to 200 nm, whereas microfluidizer or high-pressure homogenizer can achieve niosomes of 50 to 100 nm (40). The smaller size of niosomes is achieved at the cost of reduced drug loading. Furthermore, the smaller niosomes are relatively more unstable than larger ones and, therefore, require stabilizers to prevent aggregation (41). Both hydrophilic and hydrophobic drug molecules have been encapsulated in niosomes by using either dehydration–rehydration technique or the pH gradient within and outside the niosomes (40,41). The rate of drug release from the niosome is dependent on the surfactant type and its phase-transition temperature. For example, the release of carboxy fluorescein, a water-soluble fluorescent dye from Span niosomes, was in the following decreasing order: Span 20 > Span 40 > Span 60 (i.e., the release decreased with an increase in alkyl chain length of the surfactant) (40).

Niosomes exhibit different morphologies and size depending on the type of nonionic surfactants and lipids. Discoid and ellipsoid vesicles ($\sim$60 μm in diameter) with entrapped aqueous solutes are formed when hexadecyl diglycerol ether is solubilized by Solulan C24 [cholesteryl-poly(24-oxyethylene ether)] (42). Polyhedral niosomes are formed when cholesterol content is low in the same system (43). Polyhedral niosomes are thermoresponsive and release the encapsulated drug when heated above 35°C (40). This can be useful for sunscreen formulations in which the sunscreen can be released on exposure to sun (40). Niosomes have been shown to penetrate the skin and enhance the permeation of drugs (44). Span niosomes showed significantly higher skin permeation and partitioning of enoxacin than those shown by liposomes and the free drug (44). Niosomes (100 nm) are mainly localized in the SC, but some can penetrate deeper layers of the skin. The niosomes dissociate and form loosely bound aggregates, which then penetrate to the deeper strata (40). Furthermore, the skin penetration has been attributed to the flexibility of niosomes, and this is supported by the fact that a decrease in cholesterol content increases the drug penetration through the skin (45). Furthermore, the nonionic surfactant can also modify the intercellular lipid structure in the SC to enhance skin permeability (46). In addition, adsorption and fusion of niosomes with the skin surface increase the drug's thermodynamic activity, leading to enhanced skin penetration (46). In vitro studies have found that the chain length of alkyl polyoxyethylene in niosomes did not affect the cell proliferation of human keratinocytes, but ester bond was found to be more toxic than ether bond in the surfactants (47).

Nanoemulsion

Nanoemulsion (Fig. 3) is a thermodynamically stable and visually clear disperse system of oil and water with a high proportion of surfactants. Generally, the droplet size of these systems is less than 100 nm and they flow easily (48). Nanoemulsion is transparent, stable, and spontaneously formed, whereas a macroemulsion is milky and nonstable that requires some energy to form (49). The formation of nanoemulsion is dependent on a narrow range of oil, water, surfactant, and cosurfactant concentration ratios (48). A cosurfactant is commonly used to lower the interfacial tension and fluidize the interfacial surfactant (48–50). Nonionic and zwitterionic surfactants are the first line of choice for emulsion-based systems (51). Structurally, nanoemulsions biphasic with oil or water as the continuous phase, depending on the phase ratios (48). As nanoemulsion is in a dynamic state and the phases are interchangeable, it is difficult to characterize these systems, unlike other disperse systems. As these systems have water and oil phases, both hydrophilic and lipophilic drugs can be delivered using nanoemulsions (48,49). The surfactants in the system can act on the intercellular lipid structure and increase skin permeation (48). On the other hand, the oil phase may act as an occluding agent and can increase skin hydration (51). Drug release from the nanoemulsions depends on whether the drug is in the internal or external phase (52). Nanoemulsions have been found to produce higher skin penetration than macroemulsions (53). In contrast, a comparative study of macroemulsions and nanoemulsions found no significant difference in the skin penetration of tetracaine (54). The emulsion droplets may collapse or fuse with the skin components, and thus the size of the emulsion may have a minimal effect on skin penetration. On the other hand, nanoemulsions have also been shown to penetrate through the hair follicles (55).

POLYMER-BASED NANOSYSTEMS

Polymeric Nanocapsules/Nanoparticles

Polymers can be used to form nanocapsules or nanoparticles in which the drug is entrapped in the core or dispersed in the polymeric matrix, respectively (Fig. 3). Furthermore, the drug can be adsorbed, complexed, or conjugated to the surface of nanoparticles. Unlike the other systems discussed so far, these are relatively rigid nanosystems. Various types of biodegradable and nondegradable polymers can be used for the preparation of these nanosystems. Some of the polymers that have been used for topical or transdermal drug delivery include poly(lactide-co-glyocolide), polymethacrylate, poly(butyl cyanoacrylate), poly(ε-caprolactone), and chitosan (56–60). Recently, poly(vinyl alcohol)–fatty acid copolymers and tyrosine-derived copolymers have also been used for preparing nanocapsules or nanoparticles for skin applications (61,62).

Nanoparticles or nanocapsules can be prepared by either solvent evaporation or solvent displacement procedures (63). In solvent evaporation technique, the polymer is dissolved in an organic phase, such as dichloromethane or ethyl acetate. This organic phase is then dispersed in an aqueous phase containing the surfactant and emulsified by sonication or high-pressure homogenization. Subsequently, the organic phase is removed by evaporation under reduced pressure or continuous stirring to form polymeric nanoparticles (63). In this method, a lipophilic drug is loaded in the polymeric matrix by dissolving the drug in the organic phase. In solvent displacement method, the polymer is dissolved in a water-miscible organic solvent and injected into an aqueous medium with stirring in the presence of the surfactant as a stabilizer (63). Water-miscible organic solvents such as ethanol, acetonitrile, and acetone are used. The rapid diffusion of the organic solvent through the aqueous phase with the dissolved polymer at the interface leads to the formation of nanoparticles.

The skin penetration of polymeric nanoparticles is restricted to the SC, whereas the follicular penetration appears to be the major transport pathway for particles that penetrate deeper in the skin. Only a few studies have investigated the size-dependent penetration of polymeric nanoparticles into the skin. Alvarez-Roman et al. (64) studied the skin penetration and distribution of fluorescent-labeled 20- and 200-nm polystyrene nanoparticles through porcine skin. No particles were found in the corneocytes or intercorneocyte spaces. On the other hand, there was a size- and time-dependent accumulation of particles in the follicular regions, where 20-nm particles accumulated more than the 200-nm particles. Similar skin penetration behavior was also observed in excised human skin (65). The 40-nm particles were found to penetrate deeper in the follicles and also further penetrate into the epidermal Langerhans cells present at the infundibulum of hair follicles. On the other hand, the larger particles (750 and 1500 nm) did not penetrate into the follicles. In this regard, hair follicles can be used as a reservoir for drug delivery to localize the drug to the hair follicles or deliver the drug to the surrounding epidermal cells (4). This was found tape-stripping studies in human volunteers by using fluorescent-labeled poly(lactide-co-glycolide) nanoparticles (300–400 nm). The nanoparticles were detected in hair follicles even 10 days after application, while the particles on the SC were cleared after 24 hours. The nanoparticles are slowly cleared from the hair follicles by sebum secretions and the migration of particles to nearby cells and through the lymphatic system (4). The surface charge on the polymeric nanoparticles also influences their permeation through the skin. In a comparative study using positive, negative, and neutral latex nanoparticles (50–500 nm), it was found that only the negatively charged nanoparticles were able to penetrate the SC and reach the viable epidermis (66). The authors attributed the higher penetration to the charge repulsion between the negatively charged skin lipids and the carboxylate groups in the negatively charged nanoparticles (66). Only 50- and 500-nm particles were found to penetrate the skin. The larger surface of the smaller 50-nm particles and the high charge density in 500-nm particles were attributed to their higher skin penetration (65).

Dendrimers

Dendrimers are a relatively new class of nanocarriers. They are perfectly branched, spherical polymers (Fig. 3) with core–shell nanoarchitecture, and they are often termed as unimolecular micelles (67). The particle size of the dendrimers is from 1 to 10 nm. One of the distinct features of dendrimers is their large number of surface functional groups that can carry a high drug payload and also undergo multivalent interactions with the biological membranes (67). Due to their unique architecture, drugs can be encapsulated inside the core (nanocontainers), complexed, or conjugated to the surface functional groups (nanoshells). Unlike the

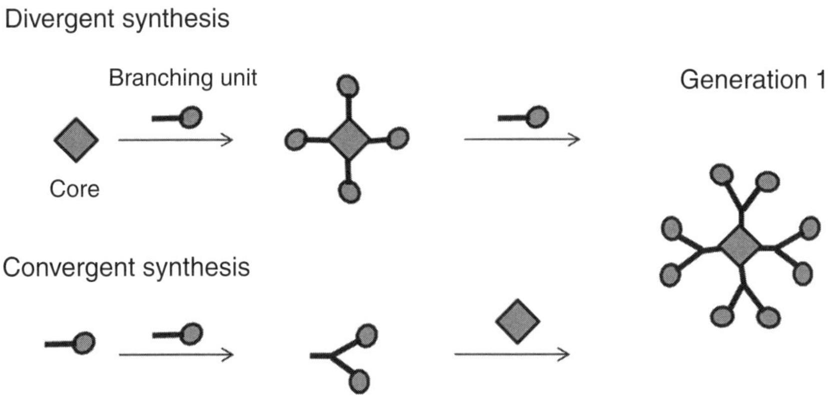

FIGURE 7 Schematic of dendrimer synthesis by divergent and convergent methods.

linear polymers, dendrimers are monodisperse systems and can be synthesized by divergent or convergent approaches (Fig. 7). The surface functional groups in the dendrimers can be tailored for various drug delivery applications (67,68). The number of branches and surface functional groups increases with each dendrimer generation. Poly(amidoamine) (PAMAM) and poly(propyleneimine) dendrimers are commercially available. Many studies have been reported with dendrimers in cell cultures and other routes of administration (68), but very few studies have explored dendrimers for skin mediated drug delivery. Studies in our laboratory with fluorescent-labeled PAMAM dendrimers have shown that the penetration of dendrimers is dependent on their surface charge and generation. Cationic dendrimers were found to penetrate deeper (40–60 μm) in the skin compared to other dendrimers (Fig. 8). Although the skin penetration pathway is not fully characterized, it is mainly localized to the SC, viable epidermis, and hair follicles (69). It has been found to increase the skin penetration of both hydrophilic and lipophilic drug molecules (70–73). The possible mechanisms include increased drug solubility, increased skin partitioning, and the penetration-enhancing effect through their interaction with the skin lipids (70–73). Since the skin penetration of dendrimers is

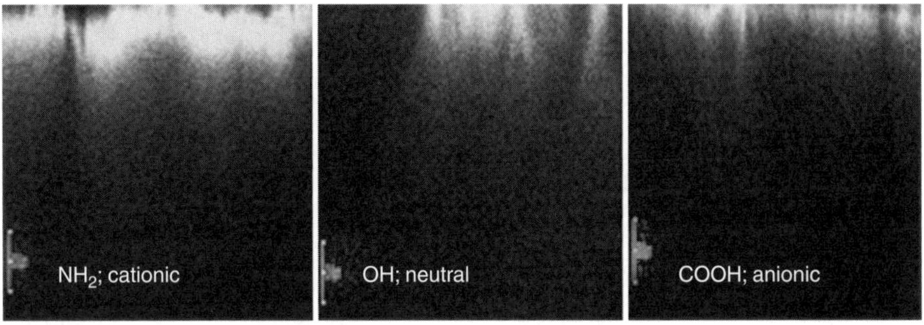

FIGURE 8 Confocal laser scanning micrographs of excised porcine skin after the application of polyamidoamine dendrimer of varying surface groups. Optical sections were taken in *XZ* plane from the surface to 100 μm in 1 μm increments; vertical bar represents 20 μm.

limited to the epidermis (69), it would be expected to show minimal skin irritation or toxicity. However, further studies are required to clarify their mechanism of skin penetration.

APPLICATIONS

Cosmetics/Sunscreens/Skin Protectants

Most of the nanosystems that are used for skin drug delivery applications have their origins from the cosmetic industry. In particular, liposomes and lipid nanoparticles are widely used in cosmetic products for their moisturizing and smoothening effect on the skin (37,74). Furthermore, they can be used to deliver skin protectants, antioxidants, and skin-whitening agents. Vesicular systems can be used to deliver hydrophilic and hydrophobic cosmetic agents and improve their skin retention and sustain the release of these agents. Table 4 provides a representative list of cosmetic agents delivered using various nanosystems. Inorganic sunscreens, such as titanium dioxide and zinc oxide, derive their sunscreen functionality from their particulate nature. Titanium dioxide reflects and scatters UV light most efficiently at a size of 60 to 120 nm (75), whereas zinc oxide has an optimal size of 20 to 30 nm for scattering UV light (76). The particle size of inorganic sunscreens is optimized to maximize their functionality, and often they are coated with silicon

TABLE 4 Representative List of Cosmetics/Skin Protectants Delivered by Using Nanocarriers

Nanoparticulate system	Cosmetic/sunscreen ingredient	Use	Reference
Liposomes	Tretinoin	Inflammatory diseases	125
	Glycolic acid	Exfoliant	126
	Sodium ascorbyl phosphate	UV protectant	127
	Ethylhexyl methoxycinnamate	Sunscreen	128
SLN®	α-Lipoic acid	Antiaging	129
	Coenzyme Q10	Antioxidant	130,131
	N,*N*-Diethyl-*m*-toluamide	Insect repellent	132
	Ferulic acid	Sunscreen	133
	Isotretinoin	Inflammatory diseases	134
	γ-Cyhalothrin	Insecticide	135
	Juniper oil	Antiacne	136
	3,4,5-Trimethoxybenzoyl chitin	Sunscreen	80
	Retinoic acid	Antiacne	137–139
	Inorganic sunscreens	Sunscreen	140
Nanostructured lipid carriers	Ascorbyl palmitate	Antioxidant	141
	Coenzyme Q10	Antioxidant	142
Poly(lactide-co-glycolide) nanospheres	2-Ethylhexyl-*p*-methoxycinnamate	Sunscreen	81
	Ascorbic acid	Cosmetic applications	143
Poly(lactic acid) nanoparticles	Octyl methoxycinnamate	Sunscreen	144
Poly(vinyl alcohol) nanoparticles	Benzophenone	Sunscreen	61
Poly(ε-caprolactone) nanocapsules	Octyl methoxycinnamate	Sunscreen	82

oil to improve the dispersion in sunscreen products (77). On the other hand, organic sunscreens such as benzene, salicylate, and cinnamate derivatives function as molecular sunscreens by absorbing UV light and dispersing it as lower energy radiation (78). The functionality of organic sunscreens can be improved by encapsulating them in various nanosystems (79) in which the nano-encapsulated sunscreen can function as both particulate and organic sunscreens. Furthermore, the encapsulation improves skin retention and reduces systemic absorption of sunscreens. In addition, the nanosystem can protect the organic sunscreen from photoxidation and enhance sun protection factor by sustaining the release from the nanosystem (79). Lipid nanoparticles have been widely used to deliver sunscreen agents. SLNs improved the skin protection from UV radiation, in which the SLNs themselves act as a sunscreen in addition to acting as a carrier for sunscreens (80). Polymeric nanoparticles, made of poly(vinyl alcohol) substituted with various saturated fatty acids, including myristic, palmitic, stearic, and behenic acids, were used to limit the skin penetration of benzophenone-3 (61). Photodegradation of *trans*-2-ethylhexyl-*p*-ethoxycinnamate is reduced from 52.3% to 35.3% when encapsulated in poly(lactide-co-glycolide) nanoparticles (81). In a similar manner, nanocapsules made of poly(ε-caprolactone) were used to protect octyl methoxycinnamate (82).

Drug Molecules

The target site for skin diseases such as dermatitis, acne, psoriasis, and skin carcinomas lay past the SC. In such cases, the active agent has to penetrate to a depth of 20 to 200 μm in the skin (83). The rigid colloidal carriers can spread only on the skin surface or may penetrate few layers in the SC bilayers, whereas deformable vesicles have been shown to penetrate the skin both into dermal layers and into systemic circulation (8). Therefore, deformable liposomes, ethosomes, and niosomes have been widely explored for topical and transdermal applications. Drugs that have been delivered using various nanosystems are listed in Table 5. Dexamethasone in transfersomes decreased rat paw edema by 82.3%, whereas conventional liposomes and ointment formulation decreased it by 38.3% and 25.3%, respectively (84). The concentration of estradiol was significantly increased in the epidermis by using transfersomes compared with an aqueous solution (85). Transfersomes produced a threefold increase in methotrexate penetration across excised pig skin compared with an aqueous solution and conventional liposomes (86). Touitou et al. (29) showed that the skin permeation of minoxidil from ethosomes was 45- and 35-fold higher than that obtained by 30% hydroalcoholic solution and 100% alcohol, respectively. Similarly, ethosomes resulted in 30-fold higher testosterone levels in 24 hours compared with commercial testosterone patch (29). Acyclovir delivered from ethosomes was significantly higher than commercial cream formulation (87). Ethosomes also enhanced the skin permeation of polar molecules such as trihexyphenidyl HCl and propanolol in vivo in mice (30,88). Fang et al. (44) studied the effects of different nonionic surfactants in niosomes for the transdermal delivery of estradiol. The estradiol flux was in the following order: Tween 20 > Span 60 > Span 85 > Span 40. SLNs increased the SC penetration of coenzyme Q10 in comparison with isopropanol or liquid paraffin (89). Of the various systems, transfersomes are promising for topical/transdermal delivery of small molecules, with several of them in early or advanced clinical trials. Ketoprofen transfersomes are in phase III clinical trials for treating peripheral chronic pain such as osteoarthritis (90). Transfersomes retained most of the drug in the tissues below the skin, unlike the commercial

TABLE 5 Representative List of Small Drugs Delivered by Using Nanocarriers

Nanosystem	Drug	Reference
Liposomes	Triamcenalone acetonide	17
	Cyclosporine	145
Deformable vesicles	Estradiol	146
	Cyclosporine	147
	5-Fluorouracil	148
	Diclofenac sodium	149
	Methotrexate	86
	Ketotifen	150
Ethosomes	Acyclovir	87
	Propranolol	88
	Testosterone	29
	Minoxidil	29
	Zidovudine	151
	Ketotifen	150
Niosomes	Enoxacin	44
	Estradiol	152
	Dithranol	153
Polymeric nanoparticles	Flufenamic acid	56
	Minoxidil	59
Dendrimers	Indomethacin	70
	Diflunisal, ketoprofen	71
	5-Fluorouracil	72,73

formulation. As a result, ketoprofen transfersomes was found to have much lower systemic exposure (90).

Protein Delivery

Among the various nanosystems, transfersomes and ethosomes appear to be promising for systemic delivery of proteins (Table 6). Several preclinical and clinical studies have shown the feasibility of transdermal delivery of insulin by using transfersomes (Transferulin®) (91). Insulin in transfersomes produced a comparative pharmacokinetic profile to subcutaneously injected insulin (91). The normoglycemia lasted for 16 hours, with a single application of Transferulin. However, transfersomes may not be suitable for producing peak insulin concentrations (due to their relatively long lag time of 6 hours) but can be used as a sustained insulin delivery system. This is because the insulin associated with the carrier has to

TABLE 6 Representative List of Macromolecules Delivered by Using Nanocarriers

Nanosystem	Macromolecule	Reference
Cationic liposomes	pDNA-encoding luciferase	154
Transfersomes	Insulin	91
	Gap junction protein	96
	Hepatitis B surface antigen	155
Cationic transfersomes	pDNA-encoding hepatitis B surface antigen	107,108
Ethosomes	Hepatitis B surface antigen	97
Biphasic lipid vesicles	Insulin	92
	pDNA-encoding herpes virus type 1 glycoprotein D	109

dissociate and accumulate in both the skin and the subcutaneous tissue before entering blood through lymph (91). Alternatively, biphasic vesicles have been developed for the systemic delivery of proteins through the skin (92). The protein is entrapped in a w/o microemulsion, which is, in turn, encapsulated in a lipid vesicle. Biphasic vesicles of insulin have been shown to reach steady state glucose levels within 6 to 8 hours and the effect of insulin lasted for 75 hours in diabetic rats (92). The biphasic vesicles dissociate in the skin and release the insulin, which is then taken up into the systemic circulation through lymph. One of the advantages of the biphasic vesicles is that drugs can be loaded in both the microemulsion and the vesicle (92).

Transcutaneous immunization (TCI) is a novel strategy that induces robust mucosal and secretary responses upon application of antigen and adjuvant on the skin (93). TCI utilizes the potent bone marrow–derived dendritic cells (DCs, Langerhans cells) present in the epidermal layers of skin. Although the Langerhans cells form only 1% of keratinocytes, they cover 25% of the total skin area (93). When these cells are activated, they migrate to the draining lymph nodes and induce strong antigen-specific responses by B and T lymphocytes (94). One of the significant advantages of TCI is the ability to produce both systemic and mucosal immunity unlike the parenteral vaccines (94). However, the large molecular size of the antigens limits their skin penetration. Transfersomes and ethosomes have been shown to penetrate the skin intact to present the antigen to DCs and elicit an antigen-specific immune response (95). Gap junction protein loaded in transfersomes elicited antigen-specific antibody titers that were equivalent to subcutaneous injection (96). In a similar manner, ethosomes were used to immunize the mice with hepatitis B surface antigen (97). Results showed that the ethosomal system produced more robust immune response compared to intramuscular injection of hepatitis B surface antigen suspension or topically applied hydroalcohilic solution.

Gene Delivery

An attractive alternative to protein is to deliver the gene of interest to the epidermal cells, which can then express the protein. This approach has advantages over protein delivery (98): (*i*) DNA can provide a more continuous supply of protein within the skin and possibly serve as a bioreactor for systemic delivery of proteins and (*ii*) DNA is relatively more stable than the proteins. Cutaneous gene therapy is particularly attractive owing to the multitude of potential disease states in the skin, such as infectious (herpes), proliferative (psoriasis), and invasive (carcinoma) diseases (99). Topical gene therapy can be easily confined to the affected area, thus reducing the likelihood of systemic toxicity. Moreover, the assessment of efficacy by visual inspection or biopsy is immeasurably more practical for the skin than any other organ. However, there are several key physical and enzymatic barriers that gene-based medicines have to overcome before producing a therapeutic effect (99). The delivery of high-molecular-weight, negatively charged, naked DNA through the negatively charged skin is difficult, if not impossible. Furthermore, DNA has to be protected from enzymatic breakdown during its transport through skin. Once inside the skin, DNA has to be taken up by the cells and withstand the lysosomal breakdown for successful transfection (98,99). Cationic lipids have been used to condense the negatively charged DNA (98). The physiochemical properties of the lipoplexes such as particle size, charge density, and stability of the complex influence the skin transport and subsequent cell uptake. A mixture of cationic lipid 1,2-dioleoyl-3-trimethylammoniumpropane (DOTAP) and neutral fusogenic lipid dioleylphosphatidyl ethanolamine (DOPE) has been used to form lipoplexes. DOTAP is a cationic lipid that condenses the DNA, while DOPE serves as a helper lipid to fuse with the lysosomal membrane and release the DNA into cytosol (98). The size of the lipoplexes is usually in the range of 100 to 500 nm and is dependent on lipid/DNA ratio. Generally, the size decreases with an increase in lipid/DNA ratio due to the increased condensation of DNA at higher lipid/DNA ratios (98). Lipoplexes have been found to mainly localize to the follicular regions in the skin and hence can be used to treat perifollicular diseases such as alopecia (100). The transfection was found to depend on the hair growth cycle (100). On the other hand, the penetration of lipoplexes through other pathways is rather limited, and except for a very few studies (101), gene expression in epidermal cells was found only after chemical or physical disruption of the SC (98,102).

A new class of cationic Gemini surfactants has been explored for topical gene delivery (103). These surfactants are composed of two ionic head groups which are attached to their hydrocarbon tails [e.g., *N*,*N'*-bis(dimethylhexadecyl)-1,3-propanediammonium dibromide]. The Gemini surfactants are combined with DOPE and dipalmitoyl-*sn*-glycerol-phosphatidylcholine to form Gemini nanoparticles (140–500 nm). The inclusion of Gemini surfactants resulted in a highly condensed form of DNA and enhanced the transfection through the intact skin (in the viable epidermis and the dermis). No skin irritation was observed, unlike the conventional cationic liposomes (104). Similarly, niosomes of pDNA-encoding IL-1 showed higher expression than cationic phosopholipids in vivo in rat pups (105). Wu et al. (55) used nanoemulsions of plasmid-encoding interferon α_2-cDNA for gene transfection in hair follicles. In contrast, no gene expression was found for naked DNA or lipoplexes. The higher transfection of nanoemulsions was attributed to their small size (~30 nm) and the penetration-enhancing effect of nonionic surfactants in the emulsions. Dendrimer–DNA complexes (dendriplexes) coated onto solid matrices, such as poly(lactide-co-glycolide) and collagen films, resulted in GFP expression in vivo in denuded mice, whereas lipoplexes did not result in gene transfection from these films (106). Thus, dendriplexes offer the additional advantage of gene transfection in the solid state unlike liposomes, which can be used to transfect only from a liquid matrix.

Mahor et al. (107) and Wang et al. (108) have shown that pDNA-encoding hepatitis B surface antigen complexed to cationic transferosomes produced significantly higher antibody titers and cytokines than do naked DNA after topical application (Table 6). Babiuk et al. (109) used biphasic lipid vesicles as a carrier for the plasmid-encoding bovine herpes virus type 1 glycoprotein D. Topical administration of this formulation produced specific immune response in the lymph nodes. Cui et al. (1110) used plasmid DNA–encoding, β-galactosidase–coated, cationic, wax-based nanoparticles to induce antigen-specific immunity. Nanoparticles were prepared using microemulsion method, with emulsifying wax as the oil phase and hexadecyltrimethyl ammonium bromide as a cationic surfactant. Topical application of this formulation showed 16-fold higher antigen-specific titers than did naked pDNA in mice (111). Recently, a novel nanoparticle known as DermaVir was reported, which consists of a mixture of plasmid DNA, polyethyleneimine, mannose, and glucose (112). This was designed based on the fact that pathogens enter the cell membranes through the mannose receptor. DermaVir was used to topically deliver the gene encoding simian–human immunodeficiency virus (SIV) in mice. It was taken up by lipid carriers and drained to the lymph nodes to produce an immune response. When DermaVir was tested in SIV-infected monkeys, it improved the survival in comparison to untreated controls (112).

FUTURE DIRECTIONS AND SUMMARY

The skin represents a unique biological membrane, not only in its remarkable barrier characteristics but also for the transport of nanosystems. In spite of an increasing number of reports on the use of various nanosystems for the skin, there is no clear consensus on the optimal size for skin penetration. The difference in the skin type and experimental conditions makes it difficult to compare the results. However, based on the studies so far, it is fairly accurate to state that the influence of particle size on skin penetration is a complex interplay of physicochemical properties of the drug, physicochemical properties of the carrier including the shape and the formulation matrix in which the carrier is incorporated. A close look at the molecular size of the marketed passive transdermal drugs reveals that they all have a very small hydrodynamic radius of 0.75 to 1.6 nm (77). This suggests that the passive permeation can be expected for particles in the similar size range. However, as discussed earlier, the estimate of the skin pore size varies from 2 to 30 nm (8). Furthermore, the presence of appendages provides an additional transport pathway that is not available in other biological membranes. Although this pathway accounts for only 0.1% of the skin surface, it appears to be an important pathway for particulate systems (4). The diameter of the pilosebaceous opening varies from 10 to 75 μm, whereas the diameter of the hair follicle varies from 0.3 to 0.75 μm (4). Rigid vesicles, or particles, deposit in the hair follicles. Studies using excised skin obtained from different anatomical sites showed that the highest penetration was found with particles that are comparable to the thickness of the hair shaft (0.75 μm) (4). In general, smaller particles were found to penetrate deeper into the follicles from the pumping action caused by the hair movement that continuously occurs in living tissues. In addition to serving as a site for localized drug delivery for the treatment of perifollicular diseases, the follicles can also serve as a long-term reservoir for delivery to surrounding cells in the skin (4).

On the other hand, the intercorneocyte penetration of particles is dictated by the flexibility of the carriers and their interaction with the intercellular lipids. In case of soft colloidal particles such as liposomes, it is often questioned whether the intact particles penetrate through the SC (18). Most of these soft colloidal particles seem to collapse at the surface of the skin, and the components then interact and enhance the skin permeation of drugs (18). On the other hand, ultradeformable liposomes appear to be a unique carrier, in which the remarkable deformability of the vesicles results in the penetration of intact vesicles deep into the skin under osmotic gradient (24). Arguably, newer methods are required to understand the skin penetration of ultradeformable vesicles. The surface charge on the nanosystems also plays a significant role in skin penetration. It is known that the skin carries a negative charge at the physiological pH due to the carboxyl residues from skin proteins and lipids (113). Surprisingly, even negatively charged liposomes and polymeric nanoparticles have been found to penetrate the skin very well. This is attributed to the charge repulsion with the carboxyl groups of the lipids in the pores (66). On the other hand, in case of dendrimers, the positively charged dendrimers penetrated better than the negatively charged or neutral dendrimers (Fig. 8). The difference may be due to the other skin–carrier interactions, in addition to the charge interactions. The vehicle used for the nanosystems can also have a significant influence on the skin penetration. Titanium dioxide nanoparticles were found to penetrate deeper into the SC from an oily emulsion, as opposed to an aqueous emulsion (114).

In general, lipid carriers have been found to produce higher penetration enhancement for hydrophilic drugs than for lipophilic drugs (18). The optimal particle size differs for the skin penetration of lipophilic and hydrophilic solutes. In a comparative study (27) using lipophilic and hydrophilic dyes in liposomes of varying sizes (73–810 nm), the highest skin penetration was seen with 71 nm particles for the lipophilic dye and 120 nm particles for the hydrophilic dye. This difference is also partly attributed to the difference in the skin penetration pathways for hydrophilic and lipophilic molecules. Essa et al. (115) found that in the case of ultradeformable liposomes, the shunt pathway plays a predominant role in the skin penetration of mannitol whereas the intercellular lipid pathway was predominant for estradiol. The authors used a novel, in vitro human skin sandwich model to study the role of shunt pathway (115). In this technique, the SC excised from the same donor skin is placed over the full epidermis based on the hypothesis that the top SC layer would block all the shunt pathways in the lower epidermal membrane. Therefore, the drug has to penetrate only through nonshunt pathways. The skin sandwich model is a useful technique to characterize the transport pathways of other nanosystems. Furthermore, comparative studies between different nanosystems in a single skin model can clarify the role of size, charge, shape, and other properties on the skin penetration of nanocarriers.

The data generated from animal skin should be carefully extrapolated to human skin since the animal skin differs in their composition and follicular density (77). In general, the rank-order correlation for skin permeation is rabbit skin > rat skin > pig skin > monkey skin > human skin. The commonly used rodent skin is at least nine times more permeable than human skin, whereas pig skin is four times more permeable than human skin (116). It is also important to note that the skin diseases can alter the barrier integrity vis-à-vis the skin penetration of nanosystems. For example, in case of eczema, the skin barrier integrity is lowered, whereas in case of inflammation and psoriasis, the skin thickness is increased (77). The nanosystems are removed by routine desquamation of the SC layer, sebum clearance, and clearance through the lymphatic and blood vessels in the dermis. Techniques such as skin stripping, microdialysis, and/or spectroscopic [Fourier transform (FT) infrared, FT-Raman) are useful to understand the skin disposition of nanosystems. The skin has received a lot of attention from the toxicological perspective as a potential route for the systemic exposure of nanomaterials, particularly with respect to sunscreen agents (77). Although debatable, studies have repeatedly shown that rather than the size, the intrinsic toxicity of the material used in the nanosystems is important (77). However, some of the components of nanosystems, such as surfactants, can produce skin irritation. On the other hand, it is important to understand the immunogenicity potential of nanoparticulate systems considering the abundance of Langerhans cells in the skin.

Stability of the nanosystem is another important criterion for drug delivery. Generally, the lipid vesicles are unstable and suffer from drug leakage and fusion of the vesicles on storage (14). On the other hand, SLNs, NLCs, and polymeric nanoparticles have excellent stability (Table 7). Furthermore, the polymeric nanoparticles and lipid nanoparticles are better in terms of sustaining the drug release over other systems (Table 7). Of the various nanosystems, SLNs, liposomes, and nanoemulsions appear promising for topical drug delivery, whereas ethosomes

TABLE 7 Comparative Account of Different Nanosystems[a]

			Loading efficiency					
Nanosystem	Size	Stability	Hydrophilic drug	Lipophilic drug	Sustained drug release	Irritation potential	Topical delivery	Systemic delivery
Conventional liposomes	+++	+	+++	++	+	+++	+++	+
Deformable liposomes	++	++	++	++	++	+++	+++	+++
Ethosomes	++	++	++	++	+	++	+++	+++
Niosomes	+++	++	++	++	++	+	++	++
Nanoemulsions	+++	++	++	++	++	+	+++	++
Solid lipid particles	++	+++	+	+++	+++	+++	+++	+
Polymeric nanoparticles	++	+++	++	++	+++	++	++	+

Abbreviations: +, poor; ++, moderate; +++, good.
[a]The smaller size is denoted by +++, whereas the larger size is denoted by +.

and ultradeformable liposomes offer good potential for both topical and systemic drug delivery (Table 7). In addition to passive delivery, these nanosystems can be combined with active skin-enhancement strategies to further enhance drug delivery through the skin.

To this end, charged liposomes and polymers can be used as carriers for electrical enhancement methods such as iontophoresis. Iontophoresis increased the flux of estradiol from ultradeformable liposomes by 15 times over a simple drug solution (115). Fang et al. (117) reported that the incorporation of positively charged stearylamine into liposomes decreased the enoxacin permeation whereas the incorporation of negatively charged dicetyl phosphate increased the skin permeation of enoxacin in the presence of iontophoresis. The authors also investigated the stability of liposomes after current application and showed that the liposomes were stable after 6 hours of current application and there was no leakage of drug from the vesicles (117). Furthermore, iontophoresis also prevented the fusion of vesicles. In another study, iontophoresis enhanced the follicular delivery of adriamycin from cationic liposomes (118). Brus et al. (119) reported increased penetration of oligonculeotide–polyethyleneimine complex across excised human skin by using iontophoresis. Electroporation has also been used to enhance the skin permeation of drugs encapsulated in liposomes. Sen et al. (120) showed that anionic phospholipids enhanced the transdermal transport of molecules in the presence of electroporation. Furthermore, the phospholipids were shown to accelerate the barrier recovery after electroporation (121). Physical methods can create additional pathways as well as widen the existing pores in the skin for the penetration of nanosystems. Low-frequency ultrasound increased the depth of skin penetration of quantum dots (20 nm) to up to 60 μm in excised porcine skin (122). Furthermore, ultrasound also significantly increased the penetration of quantum dots within the intercellular lipid regions of the SC (122). Microneedles that create micropores in the SC have been shown to deliver polystyrene nanospheres of 25 and 50 nm across excised human skin (123). Thus, the application of nanosystems can be further expanded in combination with physical enhancement methods, leading to new opportunities for drug delivery through the skin. To conclude, some of the nanosystems are already in the market and many more products can be expected in the near future.

REFERENCES

1. Scheuplein RJ, Blank IH. Permeability of the skin. Physiol Rev 1971; 51:702–747.
2. Downing DT. Lipid and protein structure in the permeability barrier of mammalian epidermis. J Lipid Res 1992; 33:301–313.
3. Cullander C, Guy RH. Sites of iontophoretic current flow into the skin: Identification and characterization with the vibrating probe electrode. J Invest Dermatol 1991; 97: 55–64.
4. Lademann J, Richter H, Schaefer UF, et al. Hair follicles – A long term reservoir for drug delivery. Skin Pharmacol Physiol 2006; 19:232–236.
5. Roberts MS, Anissimov YG. Mathematical models in percutaneous absorption. In: Bronough RL, Maibach HI, eds. Percutaneous Absorption: Cosmetic Mechanisms, Methodology, 4th ed. Boca Raton, FL: Taylor & Francis, 2005:1–44.
6. Mitragotri S. Modeling skin permeability to hydrophilic and hydrophobic solutes based on four permeation pathways. J Control Release 2003; 86:69–92.

7. Peck KD, Ghanem AH, Higuchi WI. Hindered diffusion of polar molecules through and effective pore radii estimates of intact and ethanol treated human epidermal membrane. Pharm Res 1994; 11:1306–1314.
8. Cevc G. Transfersomes, liposomes and other lipid suspensions on the skin: Permeation enhancement, vesicle penetration and transdermal drug delivery. Crit Rev Ther Drug Carrier Syst 1996; 13:257–388.
9. Li SK, Ghanem AH, Peck KD, et al. Pore induction in human epidermal membrane during low to moderate voltage iontophoresis. A study using AC iontophoresis. J Pharm Sci 1998; 88:419–442.
10. Hatanaka T, Shimoyama M, Sugibayashi K, et al. Effect of vehicle on the skin permeability of drugs: Polyethylene glycol 400-water and ethanol-water binary solvents. J Control Release 1993; 23:247–260.
11. Potts RO, Guy RH. Predicting skin permeability. Pharm Res 1992; 9:663–669.
12. Prausnitz MR, Mitragotri S, Langer R. Current status and future potential of transdermal drug delivery. Nat Rev Drug Discov 2004; 3:115–124.
13. Barry BW. Model mechanisms and devices to enable successful transdermal drug delivery. Eur J Pharm Sci 2001; 14:101–114.
14. Vemuri S, Rhodes CT. Preparation and characterization of liposomes as therapeutic delivery systems: A review. Pharm Acta Helv 1995; 70:95–111.
15. Fang JY, Hwang TL, Huang YL. Liposomes as vehicles for enhancing drug delivery via skin routes. Curr Nanosci 2006; 2:55–70.
16. Mozafari MR. Liposomes: An overview of manufacturing techniques. Cell Mol Biol Lett 2005; 10:711–719.
17. Mezei M, Gulasekharam V. Liposomes: A selective drug delivery system for the topical route of administration: Gel dosage form. J Pharm Pharmacol 1982; 34:473–474.
18. Elsayed MMA, Abdallah OY, Naggar VF, et al. Lipid vesicles for skin delivery of drugs. Reviewing three decades of research. Int J Pharm 2007; 332:1–6.
19. Bouwstra JA, Honeywell-Nguyen PL. Skin structure and mode of action of vesicles. Adv Drug Deliv Rev 2002; 54:S41–S55.
20. El Maghraby GM, Williams AC, Barry BW. Can drug-bearing liposomes penetrate intact skin? J Pharm Pharmacol 2006; 58:415–429.
21. Lieb LM, Ramachandran C, Egbaria K, et al. Topical delivery enhancement with multilamellar liposomes into pilosebaceous units; part I: In vitro evaluation using fluorescent techniques with the hamster ear model. J Invest Dermatol 1992; 99:108–113.
22. Jung S, Otberg N, Thiede G, et al. Innovative liposomes as a transfollicular drug delivery system: Penetration into porcine hair follicles. J Invest Dermatol 2006; 126:1728–1732.
23. Ogiso T, Yamaguchi T, Iwaki M, et al. Effect of positively and negatively charged liposomes on skin penetration of drugs. J Invest Dermatol 2001; 9:49–59.
24. Cevc G, Blume G. Lipid vesicles penetrate into intact skin owing to the transdermal osmotic gradients and hydration force. Biochim Biophys Acta 1992; 1104:226–232.
25. Cevc G. Lipid vesicle and other colloids as drug carriers on the skin. Adv Drug Deliv Rev 2004; 56:675–711.
26. El Maghraby GM, Williams AC, Barry BW. Skin hydration and possible shunt route penetration in controlled estradiol delivery from ultradeformable and standard liposomes. J Pharm Pharmacol 2001; 53:1311–1322.
27. Verma DD, Verma S, Blume G, et al. Particle size of liposomes influences dermal delivery of substances into skin. Int J Pharm 2003; 258:141–151.
28. El Maghraby GM, Williams AC, Barry BW. Skin delivery of oestradiol from lipid vesicles. Importance of liposome structure. Int J Pharm 2000; 204:159–169.
29. Touitou E, Dayan N, Bergelson L, et al. Ethosomes – Novel vesicular carriers: Characterization and delivery properties. J Control Release 2000; 65:403–418.
30. Dayan N, Touitou E. Carriers for skin delivery of trihexyphenidyl HCl: Ethosomes vs liposomes. Biomaterials 2000; 21:1879–1885.
31. Elsayed MM, Abdallah OY, Naggar VF, et al. Deformable liposomes and ethosomes. Mechanism of enhanced skin delivery. Int J Pharm 2006; 322:60–66.

32. Touitou E, Godin B, Dayan N, et al. Intracellular delivery mediated by an ethosomal carrier. Biomaterials 2001; 22:3053–3059.
33. Muller RH, Radtke M, Vissing SA. Solid lipid nanoparticles (SLN) and nanostructured lipid carriers (NLC) in cosmetic and dermatological preparations. Adv Drug Deliv Rev 2002; 54:S131–S55.
34. Saupe A, Wissing SA, Lenke A, et al. Solid lipid nanoparticles (SLN) and nanostructured lipid carriers (NLC) – Structural investigations on two different carrier systems. Biomed Mater Eng 2005; 15:393–402.
35. Sauto EB, Muller RH. Lipid nanoparticles (solid lipid nanoparticles and nanostructured lipid carriers) for cosmetic, dermal and transdermal applications. In: Thassu D, Deleers M, Pathak Y, eds. Nanoparticulate Drug Delivery Systems. New York: Informa Health Care, 2007:213–233.
36. Muller RH, Mader K, Gohla S. Solid lipid nanoparticles (SLN) for controlled drug delivery – A review of the state of the art. Eur J Pharm Biopharm 2000; 50:161–177.
37. Muller RH, Peterson RD, Hommoss A, et al. Nanostructured lipid carriers (NLC) in cosmetic dermal products. Adv Drug Deliv Rev 2007; 59:522–530.
38. Wissing S, Lippacher A, Muller RH. Investigations on the occlusive properties of solid lipid nanoparticles (SLN). J Cosmet Sci 2001; 52:313–324.
39. Lippacher A, Muller RH, Mader K. Preparation of semisolid drug carriers for topical application based on solid lipid nanoparticles. Int J Pharm 2001; 214:9–12.
40. Uchegbu IF, Vyas SP. Nonionic surfactant based vesicles (niosomes) in drug delivery. Int J Pharm 1998; 172:33–70.
41. Florence AT. Nonionic surfactant vesicles. Preparation and characterization. In: Gregoriadis G, ed. Liposome Technology, Vol 2. Boca Raton, FL: CRC Press, 1993:157–176.
42. Uchegbu IF, Bouwstra JA, Florence AT. Large disk-shaped structures (discomes) in non-ionic surfactant vesicle to micelle transitions. J Phys Chem 1992; 96:10548–10553.
43. Uchegbu IF, Schatzlein A, Vanlerberghe G, et al. Polyhedral non-ionic surfactant vesicles. J Pharm Pharmacol 1997; 49:606–610.
44. Fang JY, Hong CT, Chiu WT, et al. Effect of liposomes and niosomes on skin permeation of enoxacin. Int J Pharm 2001; 219:61–72.
45. Vanhal D, Vanrensen A, Devinger T, et al. Diffusion of estradiol from non-ionic surfactant vesicles through human stratum corneum in vitro. STP Pharm Sci 1996; 6:72–78.
46. Schreier H, Bouwstra J. Liposomes and niosomes as topical drug carriers: Dermal and transdermal drug delivery. J Control Release 1994; 30:1–15.
47. Hofland HEJ, Bouwstra JA, Ponec M, et al. Interactions of non-ionic surfactant vesicles with cultured keratinocytes and human skin in vitro. A survey of toxicological aspects and ultrastructural changes in stratum corneum. J Control Release 1991; 16:155–167.
48. Heuschkel S, Goebel A, Neubert RHH. Microemulsions – Modern colloidal carrier for dermal and transdermal drug delivery. J Pharm Sci 2008; 97:603–631.
49. Kreilgaard M. Influence of microemulsions on cutaneous drug delivery. Adv Drug Deliv Rev 2002; 54:S77–S98.
50. Kogan A, Garti N. Microemulsions as transdermal drug delivery vehicles. Adv Colloid Interface Sci 2006; 123–126:369–385.
51. Amselem S, Friedman D. Submicron emulsions as drug carriers for topical administration. In: Benita S, ed. Submicron Emulsion in Drug Targeting and Delivery, Vol 9. Amsterdam: Harwood Academic Publishing, 1998:153–173.
52. Labor CB, Flynn GL, Weiner N. Formulation factor affecting release of drug from formulations 1. Effect of emulsion type upon in vitro delivery of ethyl *p*-amino benzoate. J Pharm Sci 2000; 83:1525–1528.
53. Friedman DI, Schwarz JS, Weisspapir M. Submicron emulsion vehicle for enhanced transdermal delivery of steroidal and non-steroidal anti-inflammatory drugs. J Pharm Sci 1995; 84:324–329.
54. Izquierdo P, Wiechers JW, Escribano E, et al. A study on the influence of emulsion droplet size on the skin penetration of tetracaine. Skin Pharmacol Physiol 2007; 20:263–270.

55. Wu H, Ramachandran C, Bielinske AU, et al. Topical transfection using plasmid DNA in a water-in-oil nanoemulsion. Int J Pharm 2001; 221:23–34.
56. Luengo J, Weiss B, Schneider M, et al. Influence of nanoencapsulation on human skin transport of flufenamic acid. Skin Pharmacol Physiol 2006; 19:190–197.
57. Alverez-Roman R, Berre G, Guy RH, et al. Biodegradable polymer nanocapsules containing a sunscreen agent: Preparation and photoprotection. Eur J Pharm Biopharm 2001; 52:191–195.
58. Capier MJ, Kreufer J. Effect of nanoparticles on transdermal drug delivery. J Microencapsul 1991; 8:369–374.
59. Shim J, Seok KH, Park WS, et al. Transdermal delivery of minoxidil with block copolymer nanoparticles. J Control Release. 2004; 97:477–484.
60. Cui Z, Mumper RJ. Chitosan based nanoparticles for topical genetic immunization. J Control Release 2001; 75:409–419.
61. Luppi B, Cerchiara T, Bigucci F, et al. Polymeric nanoparticles composed of fatty acids and polyvinyl alcohol for topical application of sunscreens. J Pharm Pharmacol 2004; 56:407–411.
62. Sheihet L, Chandra P, Batheja P, et al. Tyrosine-derived nanospheres for enhanced topical penetration. Int J Pharm 2008; 350:312–319.
63. Soppimath KS, Aminabhavi TM, Kulkarni AR, et al. Biodegradable polymeric nanoparticles as drug delivery devices. J Control Release 2001; 70:1–20.
64. Alvarez-Roman R, Naik A, Kalia YN, et al. Skin penetration and distribution of polymeric nanoparticles. J Control Release 2004; 99:53–62.
65. Vogt A, Combadiere B, Hadam S, et al. 40 nm, but not 740 or 1500 nm, nanoparticles enter epidermal $CD1^+$ cells after transcutaneous application on human skin. J Invest Dermatol 2006; 126:1316–1322.
66. Kohli AK, Alpar HO. Potential use of nanoparticles for transcutaneous vaccine delivery: Effect of particle size and charge. Int J Pharm 2004; 275:13–17.
67. Esfand R, Tomalia DA. Poly(amidoamine) (PAMAM) dendrimers: From biomimicry to drug delivery and biomedical applications. Drug Discov Today 2001; 6:427–436.
68. Kitchens KM, El-Sayed M, Ghandehari H. Transepithelial and endothelial transport of poly(amidoamine) dendrimers. Adv Drug Deliv Rev 2005; 57:2163–2176.
69. Kraeling ME, Ogunsola OA, Bronaugh RL. In Vitro Penetration of Dendrimer Nanoparticles into Human Skin. Itinerary Planner. Seattle, WA: Society of Toxicology, 2008:abstr no. 1032.
70. Chauhan AS, Sridevi S, Chalasani KB, et al. Dendrimer-mediated transdermal delivery: Enhanced bioavailability of indomethacin. J Control Release 2003; 90:335–343.
71. Yiyun C, Na M, Tongwen X, et al. Transdermal delivery of non-steroidal anti-inflammatory drug mediated by polyamidoamine (PAMAM) dendrimers. J Pharm Sci 2007; 96:595–562.
72. Venuganti VK, Perumal OP. Effect of poly(amidoamine) PAMAM dendrimer on the skin permeation of 5-fluorouracil. Int J Pharm 2008; 361:230–238.
73. Venuganti VK, Perumal OP. Poly(amidoamine) dendrimers as skin penetration enhancers: Influence of charge, generation and concentration. J Pharm Sci. Published online October 20, 2008.
74. Weiner N, Lieb L, Niemiec S, et al. Liposomes: A novel topical delivery system for pharmaceutical and cosmetic applications. J Drug Target 1994; 2:405–410.
75. Popov AP, Lademann J, Priezzhev AV, et al. Effect of size of TiO_2 nanoparticles embedded into stratum corneum on ultraviolet-A and ultraviolet-B sun-blocking properties of the skin. J Biomed Opt 2005; 10:1–9.
76. Cross SE, Innes B, Roberts MS, et al. Human skin penetration of sunscreen nanoparticles in vitro assessment of a novel micronized zinc oxide formulation. Skin Pharmacol Physiol 2007; 20:148–154.
77. Nohynek GJ, Lademann J, Ribaud C, et al. Grey goo on the skin? Nanotechnology, cosmetic and sunscreen safety. Crit Rev Toxicol 2007; 37:251–277.
78. Wolf R, Wolf D, Morganti P, et al. Sunscreens. Clin Dermatol 2001; 19:452–459.

79. Fairhurst D, Mitchnick M. Submicron encapsulation of organic sunscreens. Cosmet Toil 1995; 110:47–50.
80. Song C, Liu S. A new healthy sunscreen system for human: Solid lipid nanoparticle as carrier of 3,4,5-trimethoxybenzoylchitin and the improvement by adding vitamin E. Int J Biol Macromol 2005; 36:116–119.
81. Perugini P, Simeoni S, Scalia S, et al. Effect of nanoparticle encapsulation on the photostability of the sunscreen agent, 2-ethylhexyl-*p*-methoxycinnamate. Int J Pharm 2002; 246:37–45.
82. Jimenez MM, Pelletier J, Bobin MF, et al. Poly-epsilon-caprolactone nanocapsules containing octyl methoxycinnamate: Preparation and characterization. Pharm Dev Technol 2004; 9:329–339.
83. Lebwohl M, Herrmann LG. Impaired skin barrier function in dermatologic disease and repair with moisturization. Cutis 2005; 76(6 suppl):7–12.
84. Jain S, Jain P, Umamaheswari RB, et al. Transfersomes – A novel vesicular carrier for enhanced transdermal delivery: Development, characterization, and performance evaluation. Drug Dev Ind Pharm 2003; 29:1013–1026.
85. El Maghraby GMM, Williams AC, Barry BW. Oestradiol skin delivery from deformable liposomes: Refinement of surfactant concentration. Int J Pharm 2000; 196:63–74.
86. Trotta M, Peira E, Carlotti ME, et al. Deformable liposomes for dermal administration of methotrexate. Int J Pharm 2004; 270:119–125.
87. Horwitz E, Pisanty S, Czerninski R, et al. A clinical evaluation of a novel liposomal carrier for acyclovir in the topical treatment of recurrent herpes labialis. Oral Surg Oral Med Oral Pathol Oral Radiol Endod 1999; 87:700–705.
88. Kirjavainen M, Urtti A, Valjakka-Koskela R, et al. Liposome-skin interactions and their effects on the skin permeation of drugs. Eur J Pharm Sci 1999; 7:279–286.
89. Pardeike J, Müller RH. Penetration enhancement and occlusion properties of coenzyme Q10-loaded NLC. AAPS J 2006; 8(suppl 2), abstract no. 1662.
90. Idea AG. Available at: http://www.idea-ag.de. Accessed October 2008.
91. Cevc G. Transdermal drug delivery of insulin with ultradeformable carriers. Clin Pharmacokinet 2003; 42:461–474.
92. King MJ, Michel D, Foldvari M. Evidence for lymphatic transport of insulin by topically applied biphasic vesicles. J Pharm Pharmacol 2003; 55:1339–1344.
93. Glenn GM, Kenney RT. Mass vaccination: Solutions in the skin. Curr Top Microbiol Immunol 2006; 304:247–268.
94. Giudice EL, Campbell JD. Needle-free vaccine delivery. Adv Drug Deliv Rev 2006; 58:68–89.
95. Vyas SP, Khatri K, Mishra V. Vesicular carriers constructs for topical immunization. Expert Opin Drug Deliv 2007; 4:341–348.
96. Paul A, Cevc G, Bachhawat BK. Transdermal immunization with an integral membrane component, gap junctional protein by means of ultradeformable drug carriers, transfersomes. Vaccine 1998; 16:188–195.
97. Mishra D, Mishra PK, Dubey V, et al. Systemic and mucosal immune response induced by transcutaneous immunization using hepatitis B surface antigen-loaded modified liposomes. Eur J Pharm Sci 2008; 33:424–433.
98. Foldvari M, Babiuk S, Badea I. DNA delivery for vaccination and therapeutics through skin. Curr Drug Deliv 2006; 3:17–28.
99. Jensen TG. Cutaneous gene therapy. Ann Med 2007; 39:108–115.
100. Domashenko A, Gupta S, Costarelis G. Efficient delivery of transgenes to human hair follicle progenitor cells using topical lipoplex. Nat Biotechnol 2000; 18:420–423.
101. Birchall J, Marichal C, Campbell L, et al. Gene expression in an intact ex-vivo skin tissue model following percutaneous delivery of cationic liposome-plasmid DNA complexes. Int J Pharm 2000; 197:233–238.
102. Meykadeh N, Mirmohammadsadegh A, Wang Z, et al. Topical application of plasmid DNA to mouse and human skin. J Mol Med 2005; 83:897–903.
103. Wettig SD, Badea I, Donkuru M, et al. Structural and transfection properties of amine-substituted Gemini surfactant-based nanoparticles. J Gene Med 2007; 9:649–658.

104. Badea I, Wettig S, Verrall R, et al. Topical non-invasive gene delivery using Gemini nanoparticles in interferon-γ-deficient mice. Eur J Pharm Biopharm 2007; 65:414–422.
105. Raghavachari N, Fahl WE. Targeted gene delivery to skin cells in vivo: A comparative study of liposomes and polymers as delivery vehicles. J Pharm Sci 2002; 91:615–622.
106. Bielinska AU, Yen A, Wu HL, et al. Application of membrane-based dendrimer/DNA complexes for solid phase transfection in vitro and in vivo. Biomaterials 2000; 21:877–887.
107. Mahor S, Rawat A, Dubey PK, et al. Cationic transferosomes based topical genetic vaccine against hepatitis B. Int J Pharm 2007; 340:13–19.
108. Wang J, Hu J, Li F, et al. Strong cellular and humoral immune responses induced by transcutaneous immunization with HBsAg DNA-cationic deformable liposome complex. Exp Dermatol 2007; 16:724–729.
109. Babiuk S, Baca-Estrada ME, Pontarollo R, et al. Topical delivery of plasmid DNA using biphasic lipid vesicles (Biphasix). J Pharm Pharmacol 2002; 54:1609–1614.
110. Cui Z, Baizer L, Mumper RJ. Intradermal immunization with novel plasmid DNA-coated nanoparticles via a needle-free injection device. J Biotech 2003; 102:105–115.
111. Cui Z, Mumper RJ. Topical immunization using nanoengineered genetic vaccines. J Control Release 2002; 81:173–184.
112. Lisziewicz J, Kelly L, Lori F. Topical DermaVir vaccine targeting dendritic cells. Curr Drug Deliv 2006; 3:83–88.
113. Burnette, RR. Ongpipattanakul, B. Characterization of permeselective properties of skin. J Pharm Sci. 1987; 76:765–773.
114. Bennat C, Muller Goymann CC. Skin penetration and stabilization of formulation containing microfine titanium dioxide and a physical UV filter. Int J Cosmet Sci 2000; 22:271–283.
115. Essa EA, Bonner MC, Barry BW. Human skin sandwich for assessing shunt route penetration during passive and iontophoretic drug and liposome delivery. J Pharm Pharmacol 2002; 54:1481–1490.
116. Magnusson BW, Walters KA, Roberts MS. Veterinary drug delivery: Potential for skin penetration enhancement. Adv Drug Deliv Rev 2001; 50:205–227.
117. Fang JY, Sung KC, Linc HH, et al. Transdermal iontophoretic delivery of enoxacin from various liposome-encapsulated formulations. J Control Release 1999; 60:1–10.
118. Han I, Kim M, Kim J. Enhanced transfollicular delivery of adriamycin with liposome and iontophoresis. Exp Dermatol 2004; 13:86–92.
119. Brus C, Santi P, Colombo P, et al. Distribution and quantification of polyethyleneimine oligonucleotide complexes in human skin after iontophoretic delivery using confocal scanning laser microscopy. J Control Release 2002; 84:171–181.
120. Sen A, Zhao YL, Hui SW. Saturated Anionic Phospholipids enhance transdermal transport by electroporation. Biophys J 2002; 83:2064–2073.
121. Essa EA, Bonner MC, Barry BW. Electroporation and ultradeformable liposomes; human skin barrier repair by phospholipid. J Control Release 2003; 92:163–172.
122. Paliwal S, Menon GK, Mitragotri S. Low frequency sonophoresis: Ultrastructural basis for stratum corneum permeability assessed using quantum dots. J Invest Dermatol 2006; 126:1095–1101.
123. McAllister DV, Wang PM, Davis SP, et al. Microfabricated needles for transdermal delivery of macromolecules and nanoparticles: Fabrication methods and transport studies. Proc Natl Acad Sci U S A 2003; 100:13755–13760.
124. Naik A, Kalia YN, Guy RH. Transdermal drug delivery: Overcoming skin's barrier function. Pharm Sci Technol Today 2000; 3:318–326.
125. Sinico C, Manconi M, Peppi M, et al. Liposomes as carriers for dermal delivery of tretinoin: In vitro evaluation of drug permeation and vesicle-skin interaction. J Control Release 2005; 103:123–136.
126. Perugini P, Genta I, Pavanetto F, et al. Study on glycolic acid delivery by liposomes and microspheres. Int J Pharm 2000; 196:51–61.
127. Foco A, Gasperlin M, Kristl J. Investigation of liposomes as carriers of sodium ascorbyl phosphate for cutaneous photoprotection. Int J Pharm 2005; 291:21–29.

128. Ramon E, Alonso C, Coderch L, et al. Liposomes as alternative vehicles for sun filter formulation. Drug Deliv 2005; 12:83.88.
129. Souto EB, Muller RH. Gohla. A novel approach based on lipid nanoparticles (SLN®) for topical delivery of α-lipoic acid. J Microencapsul 2005; 22:581–592.
130. Bunjes H, Drechsler M, Koch MH, et al. Incorporation of the model drug ubidecarenone into solid lipid nanoparticles. Pharm Res 2001; 18:287–293.
131. Wissing SA, Muller RH, Manthei L, et al. Structural characterization of Q10-loaded solid lipid nanoparticles by NMR spectroscopy. Pharm Res 2004; 21:400–405.
132. Iscani Y, Hekimoglu S, Sargon MF, et al. DEET-loaded solid lipid particles for skin delivery: In vitro release and skin permeation characteristics in different vehicles. J Microencapsul 2006; 23:315–327.
133. Souto EB, Anselmi C, Centini M, et al. Preparation and characterization of *n*-dodecyl-ferulate-loaded solid lipid nanoparticles (SLN®). Int J Pharm 2005; 295:261–268.
134. Liu J, Hub W, Chena H, et al. Isotretinoin loaded solid lipid nanoparticles with skin targeting for topical delivery. Int J Pharm 2007; 328:191–195.
135. Frederiksen HK, Kristensen HG, Pedersen M. Solid lipid microparticle formulations of the pyrethroid gamma-cyhalothrin-incompatibility of the lipid and the pyrethroid and biological properties of the formulations. J Control Release 2003; 86:243–252.
136. Gavini E, Sanna V, Sharma R, et al. Solid lipid microparticles (SLM) containing juniper oil as anti-acne topical carriers: Preliminary studies. Pharm Dev Technol 2005; 10:479–487.
137. Jee JP, Lim SJ, Park JS, et al. Stabilization of all-trans retinol by loading lipophilic antioxidants in solid lipid nanoparticles. Eur J Pharm Biopharm 2006; 63:134–139.
138. Lim SJ, Lee MK, Kim CK. Altered chemical and biological activities of all-trans retinoic acid incorporated in solid lipid nanoparticle powders. J Control Release 2004; 100:53–61.
139. Castro GA, Orefice RL, Vilela JM, et al. Development of a new solid lipid nanoparticle formulation containing retinoic acid for topical treatment of acne. J Microencapsul 2007; 24:395–407.
140. Villalobos-Hernandez JR, Muller-Goymann CC. Novel nanoparticulate carrier system based on carnauba wax and decyl oleate for the dispersion of inorganic sunscreens in aqueous media. Eur J Pharm Biopharm 2005; 60:113–122.
141. Teeranachaideekul V, Muller RH, Junyaprasert VB. Encapsulation of ascorbyl palmitate in nanostructured lipid carriers (NLC) – Effects of formulation parameters on physicochemical stability. Int J Pharm 2007; 340:198–206.
142. Teeranachaideekul V, Souto EB, Junyaprasert VB, et al. Cetyl palmitate-based NLC for topical delivery of coenzyme Q10 – Development, physicochemical characterization and in vitro release studies. Eur J Pharm Biopharm 2007; 67:141–148.
143. Stevanovi M, Savi J, Jordovi B, et al. Fabrication, in vitro degradation and the release behaviours of poly(DL-lactide-co-glycolide) nanospheres containing ascorbic acid. Colloids Surf B Biointerfaces 2007; 59:215–223.
144. Vettor M, Perugini P, Scalia S, et al. Poly(D,L-lactide) nanoencapsulation to reduce photoinactivation of a sunscreen agent. Int J Cosmet Sci 2008; 30:219–27.
145. Egbaria K, Ramachandran C, Weiner N. Topical application of liposomally entrapped cyclosporine evaluated by in vitro diffusion studies with human skin. Skin Pharmacol Physiol 1991; 4:21–28.
146. El Maghraby GM, Williams AC, Barry BW. Skin delivery of oestradiol from deformable and traditional liposomes: Mechanistic studies. J Pharm Pharmacol 1999; 51:1123–1134.
147. Guo J, Ping Q, Sun G, et al. Lecithin vesicular carriers for transdermal delivery of cyclosporin A. Int J Pharm 2000; 194:201–207.
148. El Maghraby GM, Williams AC, Barry BW. Skin delivery of 5-fluorouracil from ultradeformable and standard liposomes in-vitro. J Pharm Pharmacol 2001; 53:1069–1077.
149. Boinpally RR, Zhou SL, Poondru S, et al. Lecithin vesicles for topical delivery of diclofenac. Eur J Pharm Biopharm 2003; 56:389–392.
150. Elsayed MM, Abdallah OY, Naggar VF, et al. Deformable liposomes and ethosomes as carriers for skin delivery of ketotifen. Pharmazie 2007; 62:133–137.

151. Jain S, Umamaheswari R, Bhadra D, et al. Ethosomes: A novel vesicular carrier for enhanced transdermal delivery of an anti HIV agent. Ind J Pharm Sci 2004; 66:72–81.
152. Fang JY, Yu SY, Wu PC, et al. In vitro skin permeation of estradiol from various proniosome formulations. Int J Pharm 2001; 215:91–99.
153. Agarwal R, Katare OP, Vyas SP. Preparation and in vitro evaluation of liposomal/niosomal delivery systems for antipsoriatic drug dithranol. Int J Pharm 2001; 228: 43–52.
154. Shi Z, Curiel DT, Tang DC. DNA-based non-invasive vaccination onto the skin. Vaccine 1999; 17:2136–2141.
155. Mishra D, Dubey V, Asthana A, et al. Elastic liposomes mediated transcutaneous immunization against hepatitis B. Vaccine 2006; 24:4847–4855.

10 In Vitro Evaluation of NPDDS

R. S. R. Murthy

Pharmacy Department, The M. S. University of Baroda, Vadodara, India

INTRODUCTION

The development of regulatory framework for the regulation of nanomaterials is critical to the future of virtually every potential application of what is commonly referred to as nanotechnology. It is vital that the regulatory process be coherent and avoid mistakes made in developing regulatory frameworks for recent innovations, such as nanotechnology and agricultural biotechnology, to ensure the development of new uses, as well as public confidence (1). Although standards of care have been established, accurate prediction of the effects, both therapeutic and toxic, of a given therapeutic system on a given patient is frustrated by a host of cellular resistance mechanisms that yield disappointing differentials between in vitro predictions and in vivo results (2). Computational models may bridge the gap between the two, producing highly realistic and predictable therapeutic results. The power of such models over in vitro monolayer and even spheroid assays lies in their ability to integrate the complex in vivo interplay of phenomena such as diffusion through lesion, heterogeneous lesion growth, apoptosis, necrosis, and cellular uptake, efflux, and target binding. This chapter covers in vitro drug release process from particulate (micro/nano) drug carriers. The discussion is about nanoparticle cell interactions; various techniques used for immunoassays are discussed in later parts of this book.

DRUG RELEASE FROM PARTICULATE DRUG CARRIERS

To develop a successful nanoparticulate system, both drug release and polymer biodegradation are important consideration factors. In general, drug release rate depends on (*i*) solubility of drug; (*ii*) desorption of the surface-bound/adsorbed drug; (*iii*) drug diffusion through the nanoparticle matrix; (*iv*) nanoparticle matrix erosion/degradation; and (*v*) combination of erosion/diffusion process. Thus, solubility, diffusion, and biodegradation of the matrix materials govern the release process. In the case of nanospheres, where the drug is uniformly distributed, the release occurs by diffusion or erosion of the matrix under sink conditions. If the diffusion of the drug is faster than matrix erosion, the mechanism of release is largely controlled by a diffusion process. The rapid initial release or "burst" is mainly due to drug particles over the surface, which diffuse out of the drug polymer matrices (3).

Kinetics of Drug Release from Micro/Nanoparticles

Kinetics of drug release is an important evaluation parameter. The knowledge of the mechanism and kinetics of drug release from these microparticlulate systems indicates their performance and gives proof of adequateness of their design. Drug release from microcapsules and micro/nanoparticles involve mass transfer phenomenon involving diffusion of the drug from higher to low concentration regions in the surrounding liquid. Drug release data is applied basically for (*i*) quality control; (*ii*) understanding of physicochemical aspects of drug delivery systems; (*iii*) understanding release mechanisms; and (*iv*) predicting behavior of systems in vivo.

However, there are difficulties in modeling drug release data, as there is a great diversity in the physical form of micro/nanocapsules/particles with respect to size, shape, arrangement of the core and the coat, properties of core-like solubility, diffusivity, partition coefficient, properties of coat-like porosity, tortuosity, thickness, crystallinity, inertness, etc. In addition, there are problems in translating kinetics of drug release from "micro" products of perfect geometry to various irregular micro/nanosystems (4).

Factors Influencing Drug Release

There are various factors that influence drug release, discussed as follows:

1. *Permeation*: It is the process whereby the drug is transported through one or more polymeric membranes corresponding to the coating material which acts as the barrier to drug release. Permeation depends on crystallinity, nature of polymer, degree of polymerization, presence of fillers and plasticizers, matrix properties such as thickness, porosity, tortuosity, diffusion layer, etc. (5). Permeation may be reduced by the incorporation of dispersed solids, fillers, waxy sealants, and others.
2. *Diffusion*: It is the movement of drug across concentration gradient until equalization takes place. Governed by Fick's first law, where flux is given as follows:

$$J = \frac{\mathrm{d}M}{\mathrm{d}t} = -DA\frac{\mathrm{d}C}{\mathrm{d}x} \tag{1}$$

where $\mathrm{d}M$ = mass of the drug diffused in time $\mathrm{d}t$; D = diffusion coefficient; A = diffusion area; $\mathrm{d}x$ = the diffusion layer thickness; and $\mathrm{d}C/\mathrm{d}x$ = the concentration gradient. Negative slope indicates movement from higher to lower concentration. Here D and C are assumed to be constant. But when C varies with distance and time, the equation is given as Fick's second law:

$$J = \left(\frac{\partial c}{\partial t}\right)_x = D\left(\frac{\mathrm{d}^2 c}{\mathrm{d}x^2}\right)_t \tag{2}$$

Upon integration, we get:

$$\left(\frac{\mathrm{d}c}{\mathrm{d}t}\right) = D\left[\frac{\mathrm{d}^2C}{\mathrm{d}x^2} + \frac{\mathrm{d}^2C}{\mathrm{d}y^2} + \frac{\mathrm{d}^2C}{\mathrm{d}z^2}\right] = DV^{-2}C \tag{3}$$

This implies that the rate of change in concentration of the volume element is proportional to the rate of change of concentration gradient at that region of the field. Diffusion coefficient (D) is a measure of the rate of drug movement

Diffusion coefficient (6) depends on various factors such as (*i*) temperature (Arrhenius equation); (*ii*) molecular weight of the molecule; (*iii*) radius (for small, electrically neutral, spherical molecules); (*iv*) plasticizer concentration; (*v*) size of the penetrant, (*vi*) position of the drug in the microsphere; and (*vii*) interaction between the polymer and the drug.

With regard to drug diffusion through microcapsules/microparticles, drug transport involves dissolution of the permeating drug in the polymer and diffusion across the membrane; thus,

$$J = \frac{DKA\Delta C}{l_m} \tag{4}$$

where ΔC = concentration difference on either side of the membrane; l_m = membrane thickness; and K = partition coefficient of the drug toward the polymer. Often, DK = permeability coefficient and DK/l_m = permeability when l_m is not known; D/l_m = permeability constant (7).

3. *Partition coefficient*: Partition coefficient between polymer solvents is referred to as $K_{o/w}$. It varies if solvent solubility varies. As the value of K becomes very large, flux tends to become diffusion controlled. The value of K may vary if drug concentration continuously varies (e.g., weak acid or base) in changing pH conditions or when drug binds with some component, and in that case, Fick's law would not be followed.
4. *Drug solubility*: As diffusion depends on concentration gradient, drug solubility in the penetrant becomes important and then drug release becomes dissolution dependent for sparingly soluble drugs. This can be expressed in various ways:

   A. The Noyes-Whitney equation (8)

   $$\frac{dC}{dt} = k(C_s - C) \tag{5}$$

   where dC/dt = amount of drug released per unit time; k = dissolution rate constant; C_s = saturation solubility in solvent; C = concentration in solvent at time t; and

   $$k = \frac{D_s A}{V l_b} \tag{6}$$

   where D_s = diffusion coefficient of the solvent; V = volume of the solution; and l_b = boundary layer thickness.

   By substituting the value of k in equation (5), we get

   $$\left[\frac{dC}{dt}\right] = \frac{D_s A}{V l_b}(C_s - C) \tag{7}$$

   thus, water-soluble drugs will be released faster than the hydrophobic ones.

   B. Si-Nang and Carlier (9) modified this equation for drug release from microcapsules

   $$\left(\frac{dC}{dt}\right) = \frac{D_s A' K'}{V l_m} \tag{8}$$

   where A' = internal surface area of coating. K' includes porosity and tortuosity terms.

   C. Khanna et al. (10) modified the Noyes-Whitney equation and applied it for characterizing the dissolution of chloramphenicol from epoxy resins.

   $$W_0^{\frac{1}{3}} - W_t^{\frac{1}{3}} = kat \tag{9}$$

where W_0 = initial weight of particles; W_t = weight at time t; and a = surface weight fraction at time t. In this case, the plot of $\sqrt[3]{W_t}$ versus t gives a straight line and the value of k can be obtained from the slope.

D. For weakly acidic and basic drugs, the influence of pH on solubility is given by the Handersson-Hasselbach equation:

$$\text{For weak acids, pH} = \text{p}k_\text{a} + \log\left(\frac{S - S_0}{S_0}\right) \tag{10}$$

$$\text{For weak base, pH} = \text{p}k_\text{a} + \log\left(\frac{S_0}{S - S_0}\right) \tag{11}$$

where S = saturation solubility of the solute; S_0 = intrinsic solubility of the solute.

5. *Coating area and thickness*: Flux is proportional to the area. Hence, as the size decreases, drug release increases.

   In addition, flux $\propto 1/l$; so, as the thickness decreases, flux also increases due to reduced diffusional path length. Other factors include type and amount of matrix material, size and density of the microparticle, presence of additives or adjuvants, extent of polymerization, denaturation, cross-linking or hardening, diffusion temperature, diffusion medium, its polarity, presence of enzymes, etc.

Empiric Models of Drug Release

Kinetics of drug release from microparticulates can be understood from various models based on their nature. However, simple empiric models are often used in place of complex models, which are discussed in the following text.

Exponential Equation

Diffusional exponent approach has been given by Peppas and colleagues (11,12). It is applicable for hydrating or eroding systems in which D is not constant, thereby giving anomalous diffusion.

$$\frac{M_t}{M_0} = kt^n \tag{12}$$

where M_t/M_0 = fractional mass of drug released at time t; and n = diffusional exponent. Values greater than 0.5 indicate non-Fickian or anomalous diffusion, which is usually found in swellable systems. For Fickian release, $n = 0.5$ for the planar surface and $n = 0.432$ for spheres. For non-Fickian or anomalous diffusion, $n > 0.5$

Biexponential Equation

$$\frac{M_t}{M_0} = 1 - \left[A\exp(-K_1 t) + B\exp(-K_2 t)\right] \tag{13}$$

where M_t/M_0 = fractional mass of drug released at time t; A and B = constants; and k_1 and k_2 = rate constants for two lifetime exponents into which decay function is being decomposed. The two exponents consist of rapid or burst phase and slow or sustained release phase, respectively.

Diffusion Through Interfacial Barrier

For example, in emulsified systems, where, the Guy equation (13) is followed:

$$\frac{M_t}{M_0} = 3KT \tag{14}$$

where K = reduced interfacial rate constant; $K = K_1 / D$; and

$$T = \frac{Dt}{r^2}$$

where D = diffusion coefficient; t = time; and r = radius of the particle. This implies that at short times, zero-order release is obtained. At long times where $T \gg 1$, release is given by a single exponential decay.

$$\frac{M_t}{M_0} = 1 - \exp\left(-\frac{3k_1 t}{r^2}\right) \tag{15}$$

where k_1 = true interfacial rate constant. On converting equation (15), we get In

$$\left(1 - \frac{M_t}{M_0}\right) = -\frac{3k_1 t}{r^2}$$

which is the equation for a straight line. Hence, plot of $(1 - M_t/M_0)$ versus t will give the value of k_1 from the slope.

Nowadays, drug release kinetics are determined and better understood from their nature, depending on whether they are reservoir-, matrix-, or sandwich-type systems.

Reservoir-Type Devices (Microcapsules) (14–18)

Various equations have been given depending on different situations.

Case 1

Assuming that thermodynamic activity of the core material is constant within the microcapsule, which is spherical and has inert homogeneous coating, steady-state release rate is derived from Fick's first law of diffusion.

$$\frac{\mathrm{d}M_t}{\mathrm{d}t} = 4\pi DK\Delta C \frac{r_o r_i}{r_o - r_i} \tag{16}$$

where $\mathrm{d}M_t/\mathrm{d}t$ = fractional mass of drug released at time t; and r_o and r_i are the outer and inner radii of the coat.

- If right-hand side parameters are constant, drug release rate will be zero order.
- However, drug release rate decreases as coating thickness $(r_o - r_i)$ increases.
- If $r_o >>> r_i$, then the equation becomes

$$\frac{\mathrm{d}M_t}{\mathrm{d}t} = 4\pi DK\Delta C r_i \tag{17}$$

- Hence, drug release rate is dependent on the coating thickness. However, when the ratio of r_o to r_i is $\gg 4$, further increase in size will not significantly affect drug release (14).

Case 2

If thermodynamic activity of the core is not constant, then release rate is exponential or first order

$$\frac{dM_t}{dt} = -\frac{2\pi DK\left(r_i^2 + r_o^2\right)(C_{it} - C_{ot})}{(r_o - r_i)} \tag{18}$$

or

$$\frac{dM_t}{dt} = -\frac{2\pi DK\left(r_i^2 + r_o^2\right)(M_{it}V_o - M_{ot}V_i)}{V_i V_o (r_o - r_i)} \tag{19}$$

where C_{ot}, C_{it} = drug concentrations in the outer and inner compartments after time t, respectively; M_{it}, M_{ot} = mass of the drug remaining in the reservoir and sink at any time t, respectively; and V_i, V_o = volume of the internal reservoir and outer sink, respectively. By integration, we get:

$$\frac{dM_{it}}{dt} = \frac{-M_{tot}2\pi DK\left(r_i^2 + r_o^2\right)}{V_i(r_o - r_i)} \exp\left[-\frac{2\pi DKt\left(r_i^2 + r_o^2\right)(V_o + V_i)}{V_i V_o (r_o - r_i)}\right] \tag{20}$$

This indicates exponential release rate, which decreases as both coating thickness and time increases.

From this, we can calculate the time required for 50% drug release, $t_{1/2}$, where:

$$t_{1/2} = -\frac{V_i V_o (r_o - r_i)}{2\pi DK\left(r_i^2 + r_o^2\right)(V_o + V_i)} \ln\left(\frac{V_o - V_i}{2V_o}\right) \tag{21}$$

Lag Time and Burst Effect

A. If product is tested immediately after preparation, as fluid takes time to penetrate and attain concentration gradient, there will a delay or lag time, t_1 is given by Crank's equation as,

$$t_1 = (r_o - r_i)^2\, 6D \tag{22}$$

This can be used to find D at a particular time, and vice versa, if film thickness is known.

B. If the product is stored for a long period of time before testing or has surface-associated drug, it shows burst effect, leading to the initial overdosage.

Thus, the time necessary to reach steady state depends on coating thickness and D. The burst time, t_b, is

$$t_b = \frac{(r_o - r_i)^2}{3D} \tag{23}$$

Monolithic Devices (Microparticles) (19–22)

Monolithic or matrix systems are those in which the core is uniformly dispersed throughout the matrix polymer. The drug release kinetics will depend on whether the drug is dissolved or dispersed in the polymer.

Case 1

When the drug is dissolved in the matrix polymer, the release equations give series equations, which are simplified into early stage and later stage approximations.

Early stage approximation is given by the Baker-Lonsdale equation (14):

$$\frac{M_t}{M_\infty} = 6\left(\frac{Dt}{r^2\pi}\right)^{1/2} - \frac{3Dt}{r^2} \tag{24}$$

where M_∞ = drug dissolved in the polymer; M_t = drug released at time t; r = radius of particle; and D = diffusion coefficient. This equation is valid for $(M_t/M_0) < 0.4$, that is, for first 40% drug release.

Now, $Dt/r^2 = \Gamma$. Then, the equation will convert to

$$\frac{M_t}{M_0} = \sigma\sqrt{\frac{\Gamma}{\Pi}} - 3\Gamma \tag{25}$$

Total drug release can be obtained by the integration of the equation:

$$\frac{d(M_t/M_\infty)}{dt} = 3\left(\frac{D}{r^2\pi^t}\right)^{1/2} - \frac{3D}{r^2} \tag{26}$$

This is the equation for a straight line and shows that release is inversely proportional to the radius, r, of the microparticles.

Later time approximation is given as follows:

$$\frac{M_t}{M_\infty} = \left(1 - \frac{6}{\pi^2}\right)\exp\left[\frac{\pi^2 Dt}{r^2}\right] \tag{27}$$

which is valid for $(M_t/M_\infty) > 0.6$, that is, for remaining 60% of drug release. Integration of the equation gives

$$\frac{d(M_t/M_\infty)}{dt} = \frac{6Dt}{r^2}\exp\left[-\frac{\pi^2 Dt}{r^2}\right] \tag{28}$$

Thus, it shows that later drug release is exponential.

Case 2

When the drug is dispersed in the coat, that is, the drug is insoluble in and is uniformly dispersed throughout the matrix

A. The Baker-Lonsdale equation derived from the Higuchi equation for homogeneous, spherical matrix

$$\frac{d(M_t/M_\infty)}{dt} = \frac{3C_m D}{r_o^2 C_0}\left[\frac{(1 - M_t/M_\infty)^{\frac{1}{3}}}{1 - (1 - M_t/M_\infty)^{\frac{1}{3}}}\right] \tag{29}$$

where C_m = drug dissolved in the membrane; C_0 = initial drug concentration.

This is valid when $C_m <<< C_{tot}$, that is, not more than 1% of the drug is present in the membrane. This equation assumes that the solid drug is dissolved from surface layer first and next layer dissolves only after the first layer is exhausted.

B. Simple, uniform equations hold only for spherical, homodisperse systems. However, the majority of microparticulate systems are heterogeneous and drug release is complex. For irregular particles,

$$\frac{\mathrm{d}m}{\mathrm{d}t} = \frac{A}{2}\left[\frac{DC_{\mathrm{m}}}{t}(2C_{\mathrm{tot}} - C_{\mathrm{s}})\right]^{\frac{1}{2}} \tag{30}$$

where A = cross-sectional area of matrix; C_{tot} = total drug loading in the polymer matrix; C_{s} = saturation solubility of drug; and C_{m} = concentration of the drug. Integrating the equation, we get

$$Q = \frac{M}{A} = [DC_{\mathrm{m}}(2C_{\mathrm{tot}} - C_{\mathrm{s}})t]^{\frac{1}{2}} \tag{31}$$

where Q = mass of the drug released per unit area A at time t; and C_{tot} should be at least equal to $10C_{\mathrm{s}}$. The equation implies that drug release is proportional to the square root of time and is simplified to

$$Q = k_1 t^{1/2} \tag{32}$$

This is the modified Higuchi equation.

C. For planar surface of granular-type matrix having irregular, clustered, or aggregated particles, the following equations holds:

$$Q = \left[\frac{D\in}{\Gamma}(2C_{\mathrm{tot}} - C_{\mathrm{s}})C_{\mathrm{s}}t\right]^{\frac{1}{2}} \tag{33}$$

where $\in$ = porosity; Γ = tortuosity; C_{tot} = total concentration of drug in matrix; and C_{s} = saturation concentration.

Pore Effects

A. Flux of the drug passing through pores of certain microcapsules can be a major source of drug release, independent of coating and controlled by the rate of dissolution of the core in case of large pores. But in the case of very fine pores, resistance is offered by coating to mass transport.

Drug release in this case is given by the Flynn equation (21):

$$\frac{D_{\mathrm{p}}}{D_{\mathrm{s}}} = \left(1 - \frac{r_{\mathrm{s}}}{r_{\mathrm{p}}}\right)^2\left[1 - 2.104\left(\frac{r_{\mathrm{s}}}{r_{\mathrm{p}}}\right) + 2.09\left(\frac{r_{\mathrm{s}}}{r_{\mathrm{p}}}\right)^3 - 0.95\left(\frac{r_{\mathrm{s}}}{r_{\mathrm{p}}}\right)^5\right] \tag{34}$$

where D_{p} and D_{s} = diffusivity in the fine pores and free solution; and r_{p}, r_{s} = radius of pores and spheres. This equation shows that the release of large molecules is slowed to a much greater extent by fine pores than of smaller ones. In microcapsules produced by interfacial polycondensation, porous membranes act as semipermeable coatings, allowing the transport of low molecular weight solutes but retaining high molecular weight compounds. This concept forms the principle of artificial cells (22).

B. For the release of a water-soluble drug from porous hydrophobic matrix impermeable to the drug, transport occurs exclusively within water-filled pores. The equation is given as follows:

$$Q_{ss} = \frac{A \in D_{eff} C_{is}}{l} \tag{35}$$

where C_{is} = concentration of saturated drug solution inside reservoir; Q_{ss} = steady-state drug release flux out of the membrane; with area = A, porosity = $\in$, and thickness = l. If $D_{eff} = FD_{iw}$, then,

$$Q_{ss} = \frac{A \in FD_{iw} C_{is}}{l} \tag{36}$$

where F = formation factor, which includes pore geometry and topology; D_{iw} = aqueous diffusion coefficient of the drug.

Boundary Layer Effects

Stagnant boundary layer of the drug concentrated in contact with surface can hinder drug release by diffusion. The effect is more marked for drugs with low aqueous solubility, in which their concentration in unstirred layer can tend toward the drug solubility with resultant loss in driving force for diffusion. This is more important in microcapsules with thin coating.

Erodible and Biodegradable Systems

In many biodegradable polymeric micro/nanoparticulate systems, such as poly(lactic acid), poly(glycolic acid), poly(lactide-co-glycolide), polyanhydrides, proteins, and polysaccharides, drug release is controlled by erosion of matrix rather than diffusion of the drug within the matrix equation. Erosion of the polymer can be caused by a variety of mechanisms. Typically, these are cross-linked cleavage, ionization of the pendant groups, and backbone cleavage (23). Erosion rates are controlled by crystallinity of the polymers, glass transition temperature (T_g), polymer chain length and copolymer composition, etc. (24,25). Controlled release from protein or polysaccharide particles can be achieved by varying the size or degree of cross-linking (26,27).

Theoretically, it is essential that the rate-determining step occurs at the boundary between the unaffected region of the microspheres and the degraded material. Various equations applied can be (28):

Drug release from the surface-eroding homogeneous particle is given by fractional drug release at time t, that is,

$$\frac{M_t}{M_0} = 1 - \left(1 - \frac{k_e t}{C_0 r}\right)^3 \tag{37}$$

where k_e = erosion rate constant; r = initial radius of sphere; and C_0 = drug concentration in the sphere.

Generally, bulk eroding particles lose no mass till critical chain length (molecular weight = 15,000) after which breakdown is rapid (29). However, complications arise in case of (*i*) nonsphericity of particles, (*ii*) nonhomogeneity of matrix, (*iii*) porosity, (*iv*) tortuosity of pores, and (*v*) phase separation of drug within the polymer.

For nonspherical particles, the equation is given by

$$\frac{M_t}{M_0} = 1 - \left(1 - \frac{k_e t}{C_0 a}\right)^n \tag{38}$$

where $a = \frac{1}{2} \times$ thickness of slabs or radius of the sphere; n denotes the shape term. For slabs, $n = 1$, and for cylinders and spheres, $n = 3$.

Swelling Controlled System

For hydrophobic matrix, the drug is released as the matrix swells and polymer chains relax. Drug release generally precedes polymer biodegradation in such systems.

Factors affecting drug release kinetics include tortuosity, porosity, diffusion coefficient, solubility, etc. Diffusion is Fickian under equilibrium conditions but non-Fickian during the swelling process.

For thin, glassy slab, drug release profile obtained under countercurrent diffusion of a swelling agent is given by the following empiric equation:

$$\frac{M_t}{m_\infty} = Kt^n \tag{39}$$

where K = matrix constant depending on D; n = constant depending on polymer swelling characteristics and relaxation rate of swelling front. For hydrophilic matrices, $n = 1$, indicating zero order release, which is highly desirable for most ideal controlled release forms.

Methods of Measurement of Drug Release

United States Pharmacopoeia (USP) methods are generally used to evaluate drug release profiles of conventional and novel drug delivery systems of macro size by using any of the USP-recommended dissolution test apparatus. In case of micro- and nanoparticulate systems, these apparatus are not usable due to the following reasons:

A. Difficulty to achieve sink conditions with nanoparticles having a very high surface area in the existing USP methods.
B. Difficulty to separate dissolved drug from undissolved particulates while sampling.
C. Need for specific enzymes to release the drug from biodegradable polymeric particulates (colon-specific microparticulates).
D. Need for unconventional conditions of pH or temperature for specialized nanoparticulates (pH/temperature-sensitive nanoparticles).

In addition, the purpose of carrying out the drug release studies from the micro/nanoparticulates is generally to understand the rate and mechanism of drug release rather than as a routine quality control method as used in the case of conventional dosage forms. Hence, the methods of drug release study from micro/nanoparticulates are highly individualistic.

Various methods that can be used to study the in vitro release of the drug are (*i*) side-by-side diffusion cells with artificial or biological membranes; (*ii*) dialysis bag diffusion technique; (*iii*) reverse dialysis bag technique; (*iv*) agitation followed by ultracentrifugation/centrifugation; and (*v*) ultrafiltration or centrifugal ultrafiltration techniques.

Usually, the release study is carried out by controlled agitation followed by centrifugation. Due to the time-consuming nature and technical difficulties encountered in the separation of nanoparticles from release media, the dialysis technique is generally preferred. Various researchers have proposed different methods with one common strategy of using synthetic membrane bag with specified porosity to hold the sample. The bag containing the sample is immersed in the recipient fluid, which is stirred at a specified rpm. The samples are withdrawn at regular intervals and are analyzed for the drug content. Some reports by various workers on the methods adopted to determine the release profile are summarized in the following text.

The in vitro drug release profiles of didanosine (DDI)-loaded gelatin nanoparticles (GNPs) and mannose-conjugated nanoparticles (MN-GNPs) were studied by Jain et al. (30) by using a dialysis membrane. The release behavior of the drug from the gelatin matrix showed a biphasic pattern that is characterized by an initial burst, followed by a slower sustained release. The GNPs formulation showed 72.6% ± 4.5% drug release in 120 hours, whereas the MN-GNPs formulation showed 64.2% ± 4.8% drug release at the end of 120 hours. An initial burst drug release (22.0% ± 0.6% and 16.0% ± 1.9%) for plain and MN-GNPs was observed until 8 hours after administration, followed by a constant slow drug release until day 5.

It is evident that the method of drug incorporation has an effect on its release profile. If the drug is loaded by incorporation method, the system has a relatively small burst effect and better sustained release characteristics (31). If the nanoparticle is coated by the polymer, the release is then controlled by diffusion of the drug from the core across the polymeric membrane. The membrane coating acts as a barrier to release; therefore, the solubility and diffusivity of the drug in the polymer membrane becomes determining factor in drug release. Furthermore, release rate can also be affected by ionic interaction between the drug and the addition of auxiliary ingredients. When the drug is involved in interaction with auxiliary ingredients to form a less water soluble complex, then the drug release can be very slow with almost no burst release effect (32); whereas if the addition of auxiliary ingredients [e.g., addition of poly(ethylene oxide)–poly(propylene oxide) (PEO-PPO) block copolymer to chitosan] reduces the interaction of the model drug bovine serum albumin with the matrix material (chitosan) due to competitive electrostatic interaction of PEO-PPO with chitosan, then an increase in drug release could be observed (33).

CONCLUSION

These in vitro release characteristics are very important in the formulation of micro/nanoparticulate drug delivery systems. Depending on the drug–polymer interaction, several mathematical models are discussed based on the type and mechanism of drug release from the micro/nanoparticulate drug delivery systems. Techniques used to study the release profiles are also discussed.

REFERENCES

1. Mansour M, Foley JD, Lardner LLP. The emerging regulatory framework for nanomaterials. Nanomedicine: Nanotechnology, Biology and Medicine; 2(4):281.

2. Cristini V. Predicting drug pharmacokinetics and effect in vascularized tumors using computer simulation. Nanomedicine: Nanotechnology, Biology and Medicine; 2(4): 276.
3. Dhawan S, Singla AK, Sinha VR. Evaluation of mucoadhesive properties of chitosan microspheres prepared by different methods. AAPS PharmSciTech 2004; 5(4):article 67.
4. Clive Washington. Drug release from microparticulate systems. In: Benita S, ed. Microencapsulation: Methods and Industrial Applications. New York: Marcel Dekker, 1996;156–175.
5. Si-Nang L, Carlier PF, Delort P, et al. Determination of coating thickness of microcapsules and influence upon diffusion. J Pharm Sci 1973; 62:452–455.
6. Lippoid BC. Solid dosage forms: Mechanism of drug release. In: Polderman J, ed. Formulation and Preparation of Dosage Forms. Amsterdam, The Netherlands: Elsevier, 1977;215–236.
7. Crank J. The Mathematics of Diffusion, 2nd ed. Gloucestershire, U.K.: Clarendon Press, 1975:414.
8. Noyes AA, Whitney WR. The rate of solution of solid substances in their own solutions. J Am Chem Soc 1897; 19:930–934.
9. Si-Nang L, Carlier PF. Some physical chemical aspects of diffusion from microcapsules. In: Nixon JR, ed. Microencapsulation. New York: Marcel Dekker, 1976:185–192.
10. Khanna SC, Soliva M, Speiser P. Epoxy resin beads as a pharmaceutical dosage form; part II: Dissolution studies and release of drug. J Pharm Sci 1969; 58:1385–1388.
11. Sinclair GW, Peppas NA. Analysis of non-Fickian transport in polymers using simplified exponential equation. J Membr Sci 1984; 17:329–331.
12. Ritger PL, Peppas NA. A simple equation for description of solute release; part II: Fickian and anomalous release from swellable devices. J Control Release 1987; 2:37–42.
13. Guy RH, Hadgraft J, Kellaway IW, et al. Calculation of drug release rates from spherical particles. Int J Pharm 1982; 11:199–207.
14. Baker RW, Lonsdale HK. Controlled release: Mechanism and rates. In: Tanquary AC, Lacey RE, eds. Controlled Release of Biologically Active Agents. New York: Plenum Press, 1974:15–71.
15. Yeo Y, Kinam P. Characterization of reservoir-type microcapsules made by the solvent, exchange method. AAPS PharmSciTech 2004; 5(4):article 52.
16. Rhine WD, Hseih DST, Langer R. Polymers for sustained macromolecular release: Procedures to fabricate reproducible delivery systems and control release kinetics. J Pharm Sci 1980; 69:265–270.
17. Lui H, Magron P, Bouzon J, et al. Spherical dosage form with a core and a shell, experiments and modeling. Int J Pharm 1988; 45:217–227.
18. Lu SM, Chen SR. Mathematical analysis of drug release from a coated particle. J Control Release 1993; 23:105–121.
19. Higuchi T. Mechanism of sustained action medication: Theoretical analysis of rate of release of solid drugs dispersed in solid matrices. J Pharm Sci 1963; 52:1145–1149.
20. Fessi H, Marty JP, Puisieux F, et al. Square root of time dependence of matrix formulations with low drug content. J Pharm Sci 1982; 71:749–752.
21. Flynn GL, Yalkowsky SH, Roseman TJ. Mass transport phenomena and models: Theoretical concepts. J Pharm Sci 1974; 63:479–510.
22. Chang TMS. Artificial Cells. Springfield, IL: CC Thomas, 1972.
23. Langer RS. Polymeric delivery systems for controlled drug release. Chem Eng Commun 1980; 6:1–48.
24. Izumikawa S, Yoshioka S, Aso Y, et al. Preparation of PLA microspheres of different morphology and its effect on drug release rate. J Control Release 1991; 15:133–140.
25. Le Corre P, Le Guevello P, Gajan V. Preparation and characterization of bupivacaine-loaded PLGA microspheres. Int J Pharm 1994; 107:41–49.
26. Rubino OP, Kowalsky R, Swarbrick J. Albumin microspheres as a drug delivery system: Relation among turbidity ratio, degree of crosslinking and drug release. Pharm Res 1993; 10:1059–1065.

27. Jayakrishnan A, Knepp WA, Goldberg EP. Casein microspheres: Preparation and evaluation as a carrier for controlled drug delivery. Int J Pharm 1994; 106:221–228.
28. Hopfenberg HB. Controlled release from erodible slabs, cylinders and spheres. In: Paul DR, Harris FW, eds. Controlled Release Polymeric Formulations. Washington, DC: American Chemical Society, 1976:182–194.
29. Pitt CG, Gratzl MM, Kimmel GL, et al. Aliphatic polyesters; part II: The degradation of PLA, polycaprolactone and their copolymers in vivo. Biomaterials 1981; 2:215.
30. Jain SK, Gupta Y, Jain A, et al. Mannosylated gelatin nanoparticles bearing an anti-HIV drug didanosine for site-specific delivery. Nanomedicine 2008; 4:41–48.
31. Tomofumi Y, Hiraku O, Yoshiharu M. Sustained release ketoprofen microparticles with ethylcellulose and carboxymethylethylcellulose. J Control Release 2001; 75(3):271–282.
32. Kang J, Schwendeman SP. Comparison of the effects of $Mg(OH)_2$ and sucrose on the stability of bovine serum albumin encapsulated in injectable poly(D,L-lactide-co-glycolide) implants. Biomaterials 2002; 23(1):239–245.
33. Aiedeh, K, Taha, MO. Synthesis of chitosan succinate and chitosan phthalate and their evaluation as suggested matrices in orally administered, colon-specific drug delivery systems. Arch Pharm (Weinheim), 1999; 332(3):103–107.

11 In Vitro Characterization of Nanoparticle Cellular Interaction

R. S. R. Murthy
Pharmacy Department, The M. S. University of Baroda, Vadodara, India

Yashwant Pathak
Department of Pharmaceutical Sciences, Sullivan University College of Pharmacy, Louisville, Kentucky, U.S.A.

INTRODUCTION

The study of nanoparticle cell integration involves various techniques based on the nature of the nanoparticle and the cells involved. Some general methods and instrumentation used for cytomic study are discussed in this chapter. Flow cytometry uses the principles of light scattering, light excitation, and emission of fluorochrome molecules to generate specific multiparameter data from particles and cells in the size range of 0.5 to 40 μm. Cells are hydrodynamically focused in a sheath of phosphate-buffered saline (PBS) before intercepting an optimally focused light source (1). Lasers are most often used as a light source in flow cytometry.

As cells or particles of interest intercept the light source, they scatter light, and fluorochromes are excited to a higher energy state. This energy is released as a photon of light with specific spectral properties unique to different fluorochromes. Commonly used fluorescent dyes and their excitation and emission spectra are given in Figure 1 (2). These images also include the most common laser light sources with their multiple lines of emission. One unique feature of flow cytometry is that it measures fluorescence per cell or particle.

Both scattered light and emitted light from cells and particles are converted to electrical pulses by optical detectors. Collimated (parallel light waveforms) light is picked up by confocal lenses focused at the intersection point of cells and the light source. Light is sent to different detectors with optical filters. For example, a 525-nm band-pass filter placed in the light path prior to the detector will allow only "green" light into the detector. The most common type of detector used in flow cytometry is the photomultiplier tube (PMT).

The electrical pulses originating from light detected by the PMT are then processed by a series of linear and log amplifiers. Logarithmic amplification is most often used to measure fluorescence in cells. This type of amplification expands the scale for weak signals and compresses the scale for "strong" or specific fluorescence signals.

After the different signals or pulses are amplified, they are processed by an analog-to-digital converter (ADC), which, in turn, allows for events to be plotted on a graphical scale (one- or two-parameter histograms). Flow cytometry data outputs are stored in the computer as listmode and/or histogram files.

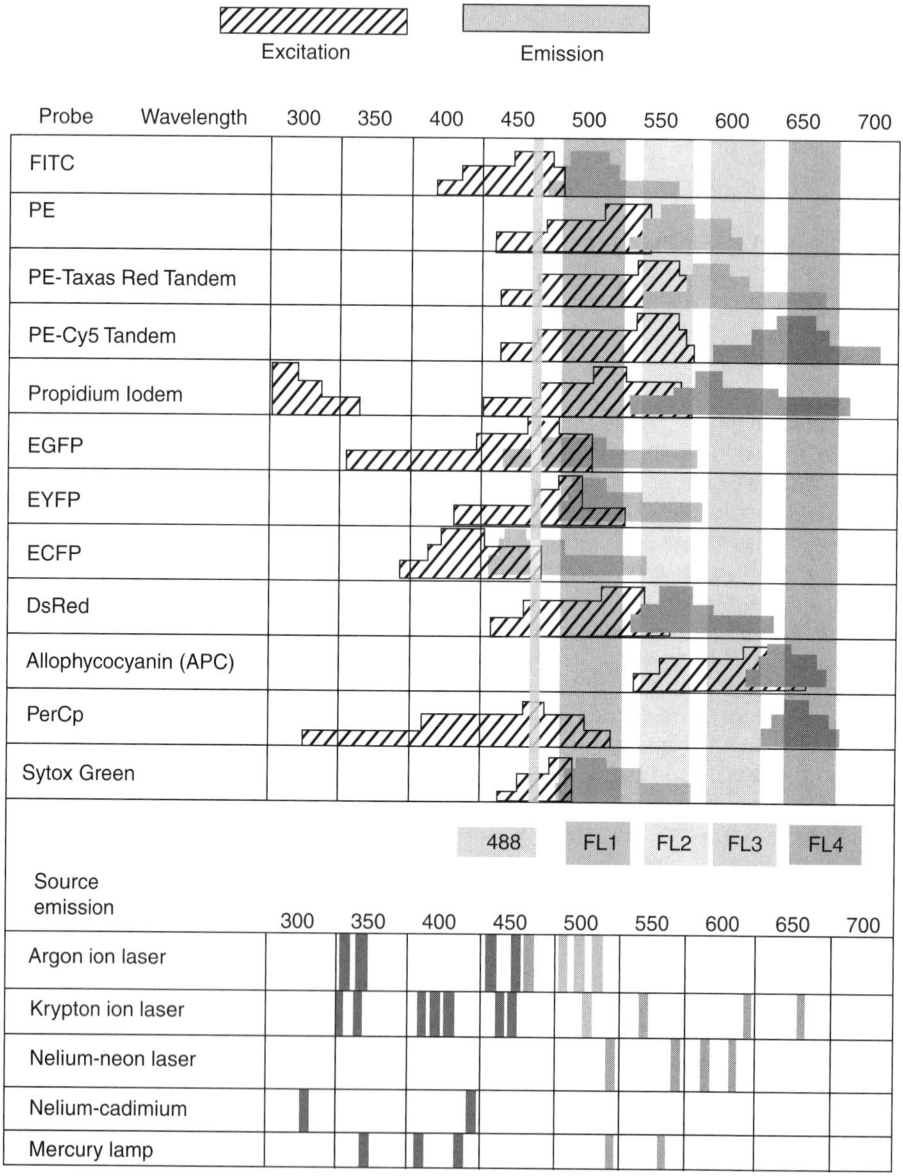

FIGURE 1 Fluorescence spectra of commonly used fluorochromes. Excitation spectra are represented by the gray lines, while emission spectra are in black. The bottom part of the table summarizes the emission wavelengths of various light sources used in flow cytometry. The 488-nm line of the argon ion laser is extended over the spectra. *Abbreviations*: PE, phycoerythrin; EGFP, enhanced green fluorescent protein. *Source*: From Ref. 2.

Histogram Files

Histogram files can be in the form of one-parameter or two-parameter files. Histogram files consist of a list of the events corresponding to the graphical display specified in your acquisition protocol.

One-Parameter Histograms

A one-parameter histogram is a graph of cell counts on the y-axis and the measurement parameter on the x-axis. All one-parameter histograms have 1024 channels. These channels correspond to the original voltage generated by a specific "light" event detected by the PMT. In other words, the ADC assigns a channel number based on the pulse height for individual events. Therefore, brighter and specific fluorescence events will yield a higher pulse height, and thus a higher channel number, when displayed as a histogram.

Two-Parameter Histograms

A graph representing two measurement parameters, on the x-axis and the y-axis, and cell count height on a density gradient is similar to a topographical map. You can select 64 or 256 channels on each axis of two-parameter histograms. Particle counts are shown either by dot density or by contour plot.

Listmode Data Files

Listmode files consist of a complete listing of all events corresponding to all the parameters collected, as specified by one's acquisition protocol. It allows you to collect specified parameters (i.e., FLS, FL1, FL2, etc.), and how these parameters are displayed. Once the data are collected and written into a listmode file, one can replay the file using either the specific protocol used for collection or any other program specifically designed for the analysis of flow cytometry data.

Laser Scanning Cytometry

Although laser scanning cytometry (LSC) technology has a flow cytometry heritage, it is not limited to analyzing cells in fluids. Instead, the technology allows automated analysis of solid-phase samples, including adherent cultured cells, tissue sections, cancer tissue imprints, and cytology smears, preserving the sample along with the exact position of each measured sample. This important feature allows the researcher to automatically return to visually inspect and interrogate specific cells having defined genetic, biochemical, or morphological properties, or to remeasure specimens after re-treating them with reagents or drugs.

LSC allows users to

- generate stoichiometric data and analyze heterogeneous cell populations;
- visualize specimens conventionally either with powerful laser scanning imaging techniques or with a microscope;
- view any number of individual events in the population data;
- reanalyze the same cells under varying conditions;
- measure and localize cellular constituents;
- detect molecular constituents in the surrounding environment and correlate their presence to cell processes;
- simultaneously study cultured cells at the individual and colony levels; and
- perform automated analysis on tissue sections and tissue microarrays.

High Content Screening Assays

One of the many formats of cell-based assays include high content screening (HCS) assays, in which typical multiplexed targets of interest in compound-treated cells are imaged and analyzed in an automated fashion. A typical HCS assay can provide data on such multiple parameters as cell density, cellular morphology and size, localization, and fluorescent intensity of the targets (usually more than one) in a cell. An HCS assay is usually performed in a microplate with 96 or 384 wells, allowing for the screening of a large number of compounds against one or multiple targets of interest. This not only allows for a more efficient use of reagents and other resources but also provides for direct and easy cross correlation of compound effects on multiple cellular targets from the same experiment. HCS assays are especially useful in studying cytotoxicity of compounds, because it allows for multiplexing targets of relevance for cytotoxicity such as nuclear morphology, permeability, membrane potential, pH, mitochondrial transmembrane potential, apoptosis, and changes in cellular concentration of Ca^{2+} and other ions or oxidative stress. By using an appropriate combination of fluorescent reagents, a single HCS experiment can provide valuable data for many of these targets. These cell-based HCS assays being rapid also act as a valuable first step before studying the toxicological effects of compounds in animal testing. A typical HCS experiment in a 96-well plate can provide useful data in about 3 hours postexposure of cells to nanoparticles, depending on the nature of cellular targets selected for studying. Also, all of the steps involved in sample preparation for HCS can be automated, providing for a very efficient way to screen for a variety of targets or nanoparticles in a short period of time.

Using an HCS approach, Vasudevan et al. (3) studied the toxicity of nanotubes in two human tissue–derived cell lines: (*i*) A549, a lung cancer–derived cell line, and (*ii*) HepG2, a liver carcinoma–derived cell line. The targets/biological indicators of toxicity we chose include cell membrane permeability, nuclear morphology, mitochondrial transmembrane potential, and induction of apoptosis. Our results show that single-walled nanotubes are more potent than multiwalled nanotubes or C60 fullerene in affecting the mitochondrial transmembrane potential in the two cell lines studied.

Confocal Laser Scanning Microscopy

Confocal laser scanning microscopy (CLSM or LSCM) (4) is a technique for obtaining high-resolution optical images. The key feature of confocal microscopy is its ability to produce in-focus images of thick specimens, a process known as *optical sectioning*. Images are acquired point-by-point and reconstructed with a computer, allowing three-dimensional reconstructions of topologically complex objects. The principle of confocal microscopy was originally patented by Marvin Minsky (5), but it took another 30 years and the development of lasers for CLSM to become a standard technique toward the end of the 1980s.

In a confocal laser scanning microscope, a laser beam passes through a light source aperture and then is focused by an objective lens into a small (ideally diffraction limited) focal volume within a fluorescent specimen. A mixture of emitted fluorescent light and reflected laser light from the illuminated spot is then recollected by the objective lens. A beam splitter separates the light mixture by allowing only the laser light to pass through and reflecting the fluorescent light into the detection apparatus. After passing a *pinhole*, the fluorescent light is detected by a photodetection device (a PMT or an avalanche photodiode), transforming the light signal into an electrical one that is recorded by a computer (6).

The detector aperture obstructs the light that is not coming from the focal point, as shown by the dotted gray line in the image. The out-of-focus light is suppressed; most of the returning light is blocked by the pinhole, resulting in sharper images than those obtained from conventional fluorescence microscopy techniques, and permitting one to obtain images of various *z*-axis planes (also known as *z stacks*) of the sample. The detected light originating from an illuminated volume element within the specimen represents 1 pixel in the resulting image. As the laser scans over the plane of interest, a whole image is obtained pixel-by-pixel and line-by-line, whereas the brightness of a resulting image pixel corresponds to the relative intensity of detected fluorescent light. The beam is scanned across the sample in the horizontal plane with one or more (servo-controlled) oscillating mirrors. This scanning method usually has low reaction latency and the scan speed can be varied. Slower scans provide a better signal-to-noise ratio, resulting in better contrast and higher resolution. Information can be collected from different focal planes by raising or lowering the microscope stage. The computer can generate a three-dimensional picture of a specimen by assembling a stack of these two-dimensional images from successive focal planes (7).

Fluorescence Confocal Microscopy

Fluorescence confocal microscopy (FCM) is a version of confocal microscopy in which the inspected specimen is doped with a high-quantum yield fluorescent dye that strongly absorbs at the wavelength of the exciting laser beam. Excited dye molecules fluoresce at somewhat longer wavelength. The difference between the fluorescence and absorption wavelengths is called the Stokes shift. If the Stokes shift is sufficiently large, the exciting and fluorescence signals can be efficiently separated by filters, so that only the fluorescence light would reach the detector. If the specimen is heterogeneous, the concentration of the fluorescent probe is coordinate dependent, which results in a high-contrast image. The FCM technique makes it possible to visualize features in living cells and tissues; it is successfully applied in flow cytometry and even for single-molecule detection.

The illuminated voxel is a diffraction-limited spot within the specimen produced by a focused laser beam. Fluorescence light passes through the pinhole aperture located in the focal plane that is conjugate to the illuminated point of the specimen. The signal that reaches the detector from the regions above and below the voxel is of much weaker intensity since the corresponding beams diverge and cover an area much larger than the area of the pinhole. Two basic features, namely, (*i*) illumination of a single voxel at a time and (*ii*) blocking out-of-the-voxel fluorescence signal, improve the resolution of FCM as compared with an ordinary microscope.

Laser Capture Microdissection

Laser capture microdissection (LCM) is a method for isolating pure cells of interest from specific microscopic regions of tissue sections (8,9). A transparent transfer film is applied to the surface of a tissue section. Under a microscope, the thin tissue section is viewed through the glass slide on which it is mounted and microscopic clusters of cells are selected for isolation. When the cells of choice are in the center of the field of view, the operator pushes a button that activates a near infrared laser diode integral with the microscope optics. The pulsed laser beam activates a precise spot on the transfer film, fusing the film with the underlying cells of choice. The transfer film with the bonded cells is then lifted off the thin tissue section, leaving all unwanted cells behind.

The LCM process does not alter or damage the morphology and chemistry of the sample collected, nor the surrounding cells. For this reason, LCM is a useful method of collecting selected cells for DNA, RNA, and/or protein analyses. It can be performed on a variety of tissue samples, including blood smears, cytologic preparations (10), cell cultures, and aliquots of solid tissue. Frozen and paraffin-embedded archival tissue may also be used (11). On formalin- or alcohol-fixed, paraffin-embedded tissues, DNA and RNA retrieval has been successful but protein analysis is not possible.

ASSESSMENT OF STERILITY OF NANOPARTICLES

Nanoparticles are assessed for sterility and related parameters such as endotoxin and *Mycoplasma*. General methods of their assessment are given in this section.

Detection of Endotoxin

A quantitative detection of gram-negative bacterial endotoxin in nanoparticle preparations by an end-point Limulus amebocyte lysate (LAL) assay is described by NCL (NCL Method STE-1, Version 1.0) (12). The protocol is based on QCL-1000 kit insert manufactured by BioWhittaker/Cambrex Corporation (Walkersville, MD) and the US FDA Guideline, stating "validation of the LAL test as an end-product endotoxin test for human and animal parenteral drugs, biological products, and medical devices." Gram-negative bacterial endotoxin catalyzes the activation of proenzyme in the LAL, and the activated enzyme then catalyzes the splitting of *p*-nitroaniline from the colorless substrate. The released *p*-nitroaniline is measured photometrically at 405 nm after reaction is terminated with a stop reagent (25% v/v glacial acetic acid or 10% SDS in water). The concentration of endotoxin in a sample is in direct proportion with absorbance and is calculated from a standard curve.

Sample Preparation

Study samples are reconstituted in either LAL reagent water or sterile, pyrogen-free PBS to a final concentration of 1.0 mg/mL and the pH is adjusted between 6.0 and 8.0 with sterile NaOH or HCl.

Experimental Procedure

Dispense 50 μL of LAL reagent water blanks (four wells), calibration standards (two wells/each), controls (two wells/each), and unknown samples (two wells/each) into appropriate wells prewarmed to 37 ± 1°C microplate. Add 50 μL of LAL reagent to all wells containing blanks, calibration standards, controls, and unknown samples. Incubate for 10 minutes at 37 ± 1°C. Add 100 μL of prewarmed substrate solution and incubate at a nominal temperature of 37 ± 1°C for another 6 minutes. Add 100 μL of the stop solution to the samples in the microplate and read absorbance at 405 nm.

Assay Acceptance Criteria

Linear regression algorithm is used to build standard curves. Precision (%CV) and accuracy (PDFT) of each calibration standard and quality control should be within 25%. At least three calibration standards should be available in order for assay to be considered acceptable. The correlation coefficient of the standard curve must be at least 0.980. If quality control measures fail to meet the acceptance criterion, runs should be repeated. Precision of the study sample should be within 25%.

If sample interference is detected, then analysis of diluted sample should be performed. The dilution of the study sample should not exceed minimum valid dilution (MVD), which is calculated according to the following formula: MVD = [0.5 (0.06) EU/mL × 1.0 mg/mL)/(0.1 EU/mL)].

Sample Acceptance Criteria

The following limits are approved by the US FDA Devices: 0.5 EU/mL, except for the device in contact with CSF, for which limit is 0.06 EU/mL. For parenteral drugs, limit is equal to K/M, that is, 5.0 EU/kg/maximum human (rabbit) dose/kg administered in a 1-hour period. For parenteral drugs administered intrathecally, limit is equal to K/M, that is, 0.2 EU/kg/maximum human dose/kg administered in a 1-hour period. The nanoparticle formulations will be treated as devices for acceptance/rejection, unless data for K/M formula are available.

Microbial Contamination Assay

Quantitative determination of microbial contamination in a nanoparticle preparation is described by NCL (NCL Method STE-2, Version 1.0) (13). The protocol includes tests for yeast, mold, and bacteria using Millipore sampler devices.

Experimental Procedure

Test nanoparticles are reconstituted in sterile PBS to a final concentration of 1 mg/mL. The pH of the study sample is adjusted with sterile NaOH or HCl as necessary to be within pH range from 6 to 8. Remove the sampler from its plastic bag under sterile conditions, and then remove a paddle from case and apply 1 mL of nanoparticle preparation or dilution onto the surface of a filter. Allow liquid to absorb, and then recap the paddle. Incubate for 72 hours at a nominal temperature of 35°C. Remove paddle from case and examine for the appearance of colonies. Perform colonies count. Report results according to the following formula: Number of Colonies × Dilution = CFU/mL

Acceptance Criteria

The test result is acceptable if positive and negative controls are acceptable. Test nanomaterial is acceptable if it demonstrates negative results.

Detection of *Mycoplasma*

A two-stage nested polymerase chain reaction (PCR) assay system was described by Gopalkrishna et al. (14), who amplified the 16S–23S rRNA spacer region sequences of *Mycoplasma* and *Acholeplasma* infections in cell cultures and virus stocks. The samples were screened for the presence of *Mycoplasma* by nested PCR, using two sets of outer and inner primers, that amplifies 16S–23S rRNA. PCR and restriction fragment length polymorphism (RFLP) assay was used to detect and identify most of the species-specific mycoplasmas involved in contaminants. Infected cultures detected by PCR-RFLP were further treated with BM-cyclin (5 mug/mL) and passaged for three times and tested for *Mycoplasma* infections by PCR-RFLP. The results showed that *Mycoplasma pirum* and *Mycoplasma orale* infections could be detected by nested PCR. However, species specificity was identified with RFLP of *Vsp*I, *Cla*I, and *Hin*dIII restriction enzymes.

Kramer (15) developed an optical biosensor based on fluorescence resonance energy transfer (FRET) for the detection of these pathogens, specifically for *M. capricolum*. Three different antipeptide antibodies (E, F_1, and F_2) were conjugated with AF-546 dye (donor). The antibodies were then complexed to protein G labeled with AF-594 dye (acceptor). When these detection complexes were bound to bacteria or a specific peptide, a measurable change in acceptor emission was observed owing to change in the three-dimensional conformation of the antibody. α-pep F_1 had the most distinct change at 12%, so it was used in two different systems (organic fluorophores and a gold nanoparticle quenching system) and immobilized onto optical fibers. Fiber optic biosensors measure conformational changes occurring when antibodies bind antigens, and thus provide a viable detection method for mycoplasmas.

Fluorescence microscopy method of detection of *Mycoplasma* was described in a protocol developed by Darlington (16). This procedure describes a method for *Mycoplasma* detection with Hoechst 33258, a fluorescent dye that stains DNA. Staining of particulate DNA matter around the cell nucleus (on the cell surface or in the cytoplasm) is indicative of *Mycoplasma* contamination.

NANOPARTICLE TARGETING STUDIES

Since knowledge of the intracellular distribution and trafficking of drugs and other agents is of paramount importance in pharmaceutical research, various microscopy methods can be used for studying intracellular trafficking. This technique provides a valuable resource for researchers wishing to go beyond a basic description of cellular uptake to characterize intracellular trafficking and targeting in greater depth. In addition to identifying cellular compartments accessed by agents and delivery vehicles, it is also important to understand the kinetics of their transport within complex biological environments, both extracellular and intracellular. In recent years, there have been significant advances in the understanding of these transport processes facilitated by time-lapse imaging and associated computational analyses.

Cell Binding and Transfection Studies

Confocal microscopy is a powerful tool to study cell binding and intracellular trafficking in understanding drug targeting and biochemical screening in drug development. Fluorescence methods are commonly used to assay the binding of drug-like compounds to signaling proteins and other bioparticles. Highly sensitive measurements within nanometer-sized, open-probe volumes can be achieved by confocal epi-illumination including fluorescence correlation spectroscopy and its related techniques. A typical biochemical application of fluorescence correlation spectroscopy is a ligand-binding assay. CLSM can also be used to image fluorescent-labeled neurons since this has major advantages over wide-field microscopy, as these neurons are relatively thick.

Ligand-Binding Studies

One of the most powerful tools for receptor research and drug discovery is the use of receptor–ligand affinity screening. The methods adopted earlier involved the use of radioactive ligands to identify a binding event; however, there are numerous limitations involved in the use of radioactivity. These limitations have led to the development of highly sensitive, nonradioactive alternatives such as AcroWell™, a patented low background fluorescence membrane and sealing process together with a filter plate design that is compatible with robotic systems to investigate receptor–ligand interactions (17). Using europium-labeled galanin, Valenzano et al. (18) demonstrated that saturation binding experiments can be performed with low background fluorescence and signal-to-noise ratios that rival traditional radioisotopic techniques while maintaining biological integrity of the receptor–ligand interaction.

Europium-labeled galanin was prepared with the DELFIA europium labeling kit from PerkinElmer Life Sciences Wallac (Turku, Finland) (#1244-302). Membranes from recombinant D98/Raji or human embryonic kidney (HEK)293e cells expressing galanin receptor subtype 1 (GalRl) were prepared in hypotonic buffer (2.5 mM of $MgCl_2$, 50 mM of HEPES, pH 7.4) and quantified with the Bio-Rad (Hercules, CA) protein assay reagent (#500-0006) with bovine serum albumin (BSA) as the standard. Europium-labeled galanin-binding reactions were conducted with 10 μg of membrane protein in a final volume of 60 μL of binding buffer [25 μM of EDTA, 0.2% BSA, 3% poly(ethylene glycol) (PEG-3350) in hypotonic buffer].

Saturation binding experiments are conducted with 0.012 to 6 nM of europium-labeled galanin in the absence and presence of the indicated concentrations of DMSO. Nonspecific binding was determined in the presence of 1.0 μM of unlabeled galanin (Sigma Chemical Co., St. Louis, MO). Mock screening and dose-displacement experiments were conducted with 0.2 nM of europium-labeled galanin in the presence of 5% DMSO. All reactions were performed in 96-well, V-bottom polypropylene plates for 2 hours at room temperature. Reactions were terminated by rapid filtration of the binding reaction (50 μL) through AcroWell filter plates, followed by three filtration washes with 300 μL of hypotonic buffer. Enhancement solution (150 E μL/well) was added to each well and time-resolved fluorescence signal was quantified.

Saturation binding experiments can also be performed with ^{125}I-labeled galanin in 60 μL of hypotonic buffer supplemented with 0.2% BSA and 0.1% bacitracin. After incubation for 2 hours at room temperature, binding reactions were transferred (50 μL) to a 96-well harvest plate, harvested, and washed with hypotonic buffer (3 × 300 μL). Wells were resuspended with 50 μL of Optiphase® Supermix and counted on a Trilux 1450 MicroBeta© (PerkinElmer Wallac).

In Vitro Cell Transfection Studies

Transfection describes the introduction of foreign material into eukaryotic cells with a virus vector or other means of transfer. The term "transfection" for nonviral methods is most often used in reference to mammalian cells. Transfection of animal cells typically involves opening transient pores or "holes" in the cell plasma membrane, to allow the uptake of material. Genetic material (such as supercoiled plasmid DNA or siRNA constructs), or even proteins such as antibodies, may be transfected. In addition to electroporation, transfection can be carried out by mixing a cationic lipid with the material to produce liposomes, which fuse with the cell plasma membrane and deposit their cargo inside.

There are various methods of introducing foreign DNA into a eukaryotic cell. Many materials have been used as carriers for transfection, which can be divided into three kinds: (cationic) polymers, liposomes, and nanoparticles. One of the economical methods is transfection by calcium phosphate, originally discovered by Graham and colleagues in 1973 (19,20). A HEPES-buffered saline (HeBS) solution containing phosphate ions is combined with a calcium chloride solution containing the DNA to be transfected. When the two are combined, a fine precipitate of the positively charged calcium and the negatively charged phosphate is formed, binding the DNA to be transfected on its surface. The suspension of the precipitate is then added to the cells to be transfected (usually a cell culture grown in a monolayer). By a process not entirely understood, the cells take up some of the precipitate, and, with it, the DNA. Another method is the use of cationic polymers such as diethylaminoethyl–dextran or polyethylenimine. The negatively charged DNA binds to the polycation and the complex is taken up by the cell via endocytosis.

Other methods use highly branched organic compounds, the so-called dendrimers, to bind the DNA to be transfected and get it into the cell. A very efficient method is the inclusion of the DNA to be transfected in liposomes, such as small, membrane-bounded bodies that are in some ways similar to the structure of a cell and can actually fuse with the cell membrane, releasing the DNA into the cell. For eukaryotic cells, lipid cation–based transfection is more typically used because the cells are more sensitive.

A direct approach to transfection is the gene gun, in which DNA is coupled to a nanoparticle of an inert solid (commonly gold), which is then "shot" directly into the target cell's nucleus. DNA can also be introduced into cells with viruses as carriers.

Other methods of transfection include nucleofection, electroporation, heat shock, magnetofection, and proprietary transfection reagents such as Lipofectamine, Dojindo Hilymax, Fugene, jetPEI, Effectene, or DreamFect.

In vitro transfection study is generally performed with suitable cell lines. The cells are normally grown in minimal essential medium supplemented with 10% fetal bovine serum, 100 mM of L-glutamine, and 100 mM of an antibiotic (penicillin, streptomycin, amphotericin B) at 37°C, 5% CO_2, and then trypsinized and seeded in 24-well plates (5×10^4 cells/well) and grown for 24 hours. One day before the transfection, the hepatocellular carcinoma (HCC) cell line HCCLM6 is seeded in 6-well plates with a concentration of 2×10^5 cells/well. When the cells reached 80% confluence, the medium is discarded. Sterile PBS (0.01 mol/L, pH 7.4) is added to each well for washing and then discarded. The nanoparticles are added and then nonserum RPMI-1640 is added. The plates are then incubated at 37°C and 95% air/5% CO_2 for 16 hours. Following this, the medium is removed and the wells are washed with PBS. Fresh medium (RPMI-1640 containing 10% fetal bovine serum) is then added to each well. The cell culture plate is incubated at 37°C and 5% CO_2 for another 24 hours, 48 hours, 72 hours, and 1 week (21). The cells are then digested to make single-cell suspension, fixed in 95% ethanol solution at 4°C, and run on flow cytometry by 490-nm excitation to detect the green light of GFP of transfected cells. The transfection efficiency is determined as the percentage of the transfected cells against all cells counted.

In vitro transfection results reported by Dastan and Turan (22) showed that the cell type is an important limitation for DNA delivery with chitosan polymers. HEK293 cells were more efficiently transfected than HeLa and fibroblastic 3T3 cells. Chitosan–DNA microparticles prepared with 0.75% chitosan solution more effectively transfected HEK293 cells.

Determination of Transfection Efficiency by Flow Cytometry

Flow cytometry is also used to determine the transfection efficiency quantitatively. After transfection for 24, 48, and 72 hours, the cells are rinsed three times with PBS, each for 1 minute. The cells are then digested to make single-cell suspension, fixed in 95% ethanol solution at 4°C, and run on flow cytometry by 490-nm excitation to detect the green light of GFP of transfected cells. The transfection efficiency is determined as the percentage of the transfected cells against all cells counted.

Cellular Uptake Studies

Conventional/Traditional Methods

The necessity of in vivo evaluation of any particulate drug delivery system is based on the differentiation of the action of the free drug and the drug encapsulated in it. Generally, for in vivo evaluation, the drug-loaded carrier is administrated to the animal model of disease by the desired route and the drug concentration in blood levels is measured at predetermined intervals by sensitive assay methods. Computer programs are available to analyze pharmacokinetic data on single as well as multicompartment models. The in vivo distribution of the drug in different organs/tissues can also be studied by sacrificing the test animals, followed by analyzing the desired tissues for drug concentration and other pharmacokinetic parameters.

Pharmacoscintigraphy

Noninvasive techniques, such as nuclear medicine techniques, and magnetic resonance imaging and spectroscopy have been utilized for monitoring drug pharmacology. γ-Scintigraphy and positron emission tomography are basic nuclear medicine tools to investigate life: its function and structure in health and disease. Pharmacoscintigraphy is the application of nuclear medicine techniques for tracing the radiolabeled ingredient of the active component of a drug/device or the formulation excipient. It can provide vital information regarding the extent, rate, site, and mode of drug release in animals. In the context of evaluation of microparticulate drug delivery systems, pharmacoscintigraphy appears very promising, as it is noninvasive, permits repeated measurements, and allows the use of same organism as its own pretreatment control. Radiopharmaceuticals are radioactive drugs that, when used for the purpose of diagnosis or therapy, typically elicit no physiological response from the patient. Unlike radiographic procedures, which depend almost entirely on tissue density differences, external imaging of radiopharmaceutical is essentially independent of the density of the target organ.

Basic steps in the assessment of biopharmaceutical and pharmacokinetic parameters by nuclear medicine methods are the design and synthesis of labeled drug, followed by imaging, determination of parameters of interest, and finally data interpretation. Medical diagnostic modalities currently in use include the application of gamma radiation emitting radioactive materials, such as technetium Tc 99m (^{99m}Tc), indium-111 (^{111}I), iodine-125 (^{125}I), iodine-131 (^{131}I), and gallium-67 (^{67}Ga). Nearly 80% of all radiopharmaceuticals used in nuclear medicine are ^{99m}Tc-labeled compounds. In terms of physical properties, ^{99m}Tc is the radionuclide of choice for diagnosis in nuclear medicine.

Pharmacoscintigraphic evaluation of microparticulate drug delivery system has been reported by many researchers and scientists (23–31). Reddy and colleagues (32) studied the influence of administration route on tumor uptake and biodistribution of etoposide-loaded tripalmitin nanoparticles (ETPL) in Dalton's lymphoma tumor–bearing mice. The negatively charged and ^{99m}Tc-radiolabeled ETPL nanoparticles were administered by intravenous, intraperitoneal, and subcutaneous routes and their biodistribution and tumor uptake were evaluated. The tumor concentration of ETPL nanoparticles after subcutaneous injection was 59-fold higher than that obtained after intravenous injection and 8-fold higher than that obtained after intraperitoneal route at 24-hour postinjection. The study signifies the advantage of incorporating etoposide into tripalmitin nanoparticles in controlling its biodistribution and enhancing the tumor uptake by several folds. The study also reveals that, of the three routes investigated, subcutaneous injection is the route of preference for facilitating high tumor uptake and retention.

In Vitro Evaluation with Cell Lines

Uptake by Endothelial Cells

Endothelium is involved in a number of normal and pathophysiological conditions such as angiogenesis, atherosclerosis, tumor growth, myocardial infarction, limb and cardiac ischemia, restenosis, etc. (33–36). Hence, it is considered as an important target for drug or gene therapy and various therapeutic approaches have been investigated to counteract the disease conditions by the modification of the endothelium (37). Vascular endothelial cells in particular are extremely important targets for functional genes because of their large population and contiguity with the bloodstream (38,39). Different delivery systems, including drug conjugates and immunoliposomes, have been studied to actively target therapeutic agents to the endothelium (40,41).

Uptakes of nanoparticles by endothelial cells are studied generally by cell culture methods. Davda and Labhasetwar (34) have proposed strategies for endothelial delivery of nanoparticles by in vitro characterization with human umbilical vein endothelial cell (HUVEC) lines grown in Ham's F-12K medium. The cells were cultured by the method described in the following text. The cell monolayer was washed with sterile PBS (154 mM, pH 7.4) and aspirated. After aspiration, 1.5 mL of trypsin–EDTA was added to the flask. The flask was placed in the incubator at 37°C for 10 minutes to allow cell detachment. Fresh growth medium (20 mL) was added to the flask. The cells were flushed with a 10-mL pipette several times to ensure that all the cells were in suspension and the cell suspension was transferred to a 50-mL Eppendorf tube. To study the cellular uptake of nanoparticles, HUVEC lines were seeded in a 24-well plate at a density of 30,000 cells/well in complete growth medium and allowed to grow for 2 days, with medium changed once.

Cellular Uptake of Nanomedicines

Nanomedicines with a lower size range are preferable to those in the upper submicron and micron sizes to achieve longer circulation half-lives [reduced mononuclear phagocyte system (MPS) uptake] and more efficient cellular uptake (increased internalization). Intracellular uptake of particles can occur by various mechanisms, as described in the following text (42).

Uptake by Phagocytic Cells

The MPS is made up of largely phagocytic cells such as macrophages. Generally, particles >1 micron generate a phagocytic response (43,44). The uptake and transport of IgG-opsonized polystyrene beads of defined size ranging from 0.2 to 3 micron into murine macrophages were investigated by Koval et al. (43). They observed that phagocytosis uptake was size dependent; about 30% of 0.2- to 0.75-μm particles compared with about 80% of 2- and 3-μm particles were taken up. Also, to avoid substantial entrapment by hepatic and splenic endothelial fenestrations and subsequent clearance, carriers should not exceed 200 nm (45). Besides size, which is the focus of this review, other properties of the nanocarriers, such as surface charge and chemistry, can also influence their uptake and subsequent clearance by the cells of the MPS.

Uptake by Nonphagocytic Cells

The internalization of particles by nonphagocytic cells, such as tumor cells, can also happen if particles are about 500 nm (46). The internalization of nanomedicines into the target cells can occur via a diverse range of endocytic pathways, including phagocytosis, macropinocytosis, clathrin-mediated endocytosis, and non–clathrin-mediated (such as caveolae-mediated) endocytosis. Rejman et al. (46) showed that as particle size increased, internalization decreased. There was no cellular uptake of particles N500 nm.

Uptake by Alveolar Macrophages

Recent studies indicate that pulmonary epithelial cells can take up inhaled ultrafine particles, which enter into the circulation. To study this translocation in an in vitro model, three types of pulmonary epithelial cells were examined (47). The integrity of the cell monolayer was verified by measuring the transepithelial electrical resistance (TEER) and passage of sodium fluorescein. Preliminary experiments showed very low TEER value for A549 cells while the TEER values of 1007 ± 300 and 348 ± 62 omega cm^2 were reached for the Calu-3 cell line, using permeable membranes of 0.4- and 3-mμ pore size, respectively. Growing primary rat type II pneumocytes on 0.4-mμ pores, a TEER value of 241 ± 90 omega cm^2 was reached on day 5, but no acceptable high TEER value was obtained for the cell lines grown on 3-mμ pores. Translocation studies conducted with 46-nm fluorescent polystyrene particles through Calu-3 cell line on 0.4-mμ pores showed no translocation, while 6% translocation was observed both for carboxyl- and amine-modified particles on 3-mμ pores.

The uptake of microparticles by the alveolar macrophages harvested from male albino rats (Sprague-Dawley strain, 150 ± 20 g) by bronchoalveolar lavage was studied by Liang et al. (48) and Jain et al. (49). Trachea was exposed along with the lungs under deep anesthesia and the exsanguinations were done by cardiac puncture. The trachea was cut at a point and a polyethylene tube was inserted within it and the lungs were lavaged 10 times with 2-mL aliquots of Ca^{2+}- and Mg^{2+}-free Hank's balanced salt solution. The lavaged cell suspension was centrifuged at 4°C, and the pellet was resuspended in HEPES buffer medium (136 mM of NaCl, 2.2 mM of Na_2HPO_4, 5.3 mM of KCl, 10 mM of HEPES, 5.6 mM of glucose, 1.0 mM of $CaCl_2$, pH 7.4). Cell viability was checked via MTT assay. About 100 mL of the cell suspension, corresponding to a seeding density of 1×10^6 cells/mL, was transferred to 48-well culture plates (Corning Incorporated Life Sciences, MA, USA) having four columns. Ten microliters per well of the formulation at 10 mg of DDI/mL was then added to six wells of columns I, II, and III, respectively. PBS (pH 7.4), unloaded plain gelatin nanoparticles (GNPs), and unloaded mannose-conjugated nanoparticles (MN-GNPs) served as a reference to the loaded formulations. The plates were incubated in a controlled environment at a temperature of 37° ± 1°C for a period of 48 hours. During the incubation of the plates, at appropriate time points of 0, 1, 4, 8, 20, and 48 hours, the cell suspension from each well was transferred to polycarbonate filters (pore size = 0.45 μm). The wells of the cell culture plates were rinsed with 1 mL of PBS (pH 7.4) and the washings were subsequently transferred to the polycarbonate filters. The cells were separated from the medium in the form of a pellet by centrifuging the filters at 4000 rpm for 15 minutes. About 0.5 mL of Triton X-100 was added to the pellet to rupture the cells and the mixture incubated was at 25°C for 5 to 6 hours. Drug uptake was determined by high-performance liquid chromatographic.

Uptake by Cancer Cell Lines

An area of research that is gaining interest is the selective delivery of anticancer drugs and overcoming drug resistance in cancer chemotherapy with nanoscale delivery systems. Major mechanisms that have been proposed include enhanced intracellular concentration of the drug by endocytosis (50), inhibition of multidrug resistance proteins by carrier component materials such as Pluronic® block copolymers (51), adhesion of nanoparticles to the cell surface (50), promotion of other uptake mechanisms such as receptor-mediated cellular internalization (52,53), and increased drug concentrations at the vicinity of target cancer cells (50). Furthermore, both drug and inhibitors of multidrug resistance proteins can be incorporated into the same carrier for simultaneous delivery to the cancer cells. For example, doxorubicin and cyclosporine A encapsulated in poly(alkyl cyanoacrylate) nanoparticles have been demonstrated to reverse resistance synergistically (54).

Bisht et al. (55) used a panel of human pancreatic cancer cell lines to demonstrate comparable in vitro therapeutic efficacy of nanocurcumin to free curcumin by cell viability and clonogenicity assays in soft agar. Curcumin is naturally fluorescent in the visible green spectrum. To study the uptake of curcumin encapsulated in nanoparticles, cells were plated in 100-mm dishes and allowed to grow to subconfluent levels. Thereafter, the cells were incubated with nanocurcumin for 2 to 4 hours and visualized in the FITC channel.

Resistance Development Study In Vitro

Development of in vitro resistance to mustargen or mechlorethamine hydrochloride, *N,N′*-bis(2-chloroethyl)-*N*-nitrosourea (BCNU), and cisplatin [*cis*-diammine dichloro platinum(II)] was studied (56) in two human cell lines, the Raji/Burkitt lymphoma and a squamous cell carcinoma of the tongue. A 10- to 20-fold increase in resistance, relative to the parental line, was achieved in 3 to 4 months of continuous drug exposure. Further increase in selection pressure resulted in cell death, while the removal of the drug led to rapid loss of resistance. The half-life for resistance loss upon the removal of the drug was 2 to 3 months. In the presence of intermittent low concentrations of the alkylating agent, resistance has been maintained in excess of 9 months. Cross-resistance studies performed against HN2, BCNU, cisplatin, phenylalanine mustard, and hydroperoxycyclophosphamide showed a lack of cross-resistance in general.

In Vitro Uptake Study by Spleen Cells

After intravenous injection, the main part of nanoparticles trapped by the spleen is concentrated in the marginal zone. The first step of this capture is the adhesion of the particles to the marginal zone macrophages. As classical techniques of cell suspension preparation did not allow to isolate without damage these actively capturing cells, which are tightly bound to a well-developed reticular meshwork, Demoy et al. (57) designed a tissue slice incubation method, in conditions close to in vivo, to study in vitro the interaction of nanoparticles with these particular macrophages. In

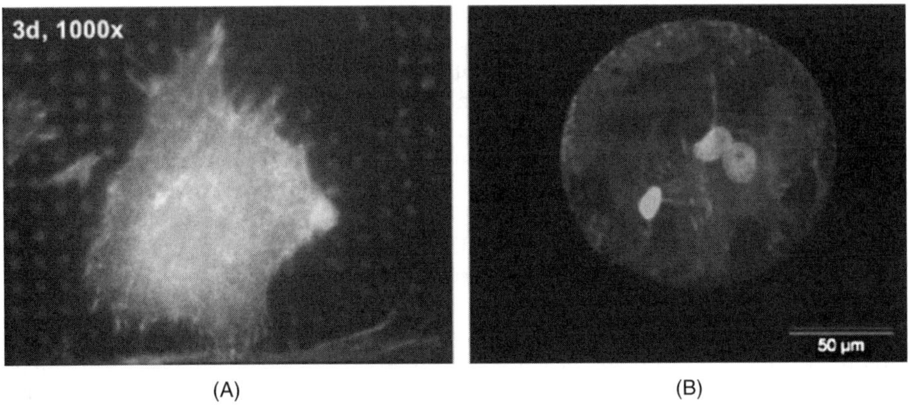

FIGURE 2 (**A**) Fluorescence microscopy pictures showing cells growing micropatterned fibronectin layers (pattern size = 2.5 μm). Fibronectin is stained red, focal contacts are in green. (**B**) MSC cells growing on selective areas of poly(methyl methacrylate) substrates fictionalized with fibronectin. Nuclei are stained in blue, while fibronectin is stained in red. *Source*: From Ref. 58.

a serum-supplemented medium, this in vitro model was able to give similar uptake profile than that obtained after intravenous injection of nanoparticles, thus proving its validity.

Method for spleen preparation has been described by Demoy et al. (57) Spleens of CD mice (25-308) were removed, trimmed of fat and connective tissue, and mounted end-down on a chuck. The spleen block was then immersed in Iscove Modified Dulbecco's Medium (IMDM, Gibco, France). Slices of tissue (250-um thick) were cut with a microtome, transferred to culture trays with 2 mL of IMDM containing 10% fetal calf serum (Gibco, France), and incubated at 37°C, pH 7.4. Fluorescent polystyrene nanoparticles (200-nm diameter, FluoresbriteQ, Polysciences, Warrington, U.K.) were added, trays were gassed with 5% CO_2, sealed in plastic boxes, and gently agitated on a rotating table for 3 hours. The tissue sections were then observed under the fluorescence microscope. Samples were scored as highly positive (+++) when there were a lot of capturing cells with nanoparticles and aggregates covering part of the cytoplasm. When there were few capturing cells with only some isolated fluorescent dots per cell, the samples were scored as moderately positive (+), and when no capturing cells were observed, the samples were scored 0. Representative photographs reported from such an assay are given in Figure 2 (58). The proposed in vitro method may be considered a good tool to investigate the spleen capture of nanoparticulate carriers. Moreover, these results clearly show that the in vivo spleen capture of nanoparticles was not only due to the anatomical location of the macrophages along a vascular sinus, but essentially to the own activity of the marginal zone macrophages. In mice, these cells were identified as large angular cells with high efficiency to phagocyte nanoparticles. It is noteworthy that other spleen macrophage populations took up very few nanoparticles (59).

Cell Viability (MTS) Assays

Growth inhibition is measured by cell proliferation assay, which relies on the conversion of a tetrazolium compound (MTS) to a colored formazan product by the activity of living cells. Generally, 2000 cells/well are plated in 96-well plates and treated with 0, 5, 10, 15, and 20 μM concentrations of free drug and equivalent nanoparticle for 72 hours, at which point the assay is terminated. Relative growth inhibition is compared with vehicle-treated cells measured using the reagent, as described in the manufacturer's protocol. All experiments are set up in triplicate to determine means and standard deviations. Bisht et al. (55) performed cell viability (MTT) assays by similar method with equivalent dosages of free curcumin and nanocurcumin for 72 hours, and cell viability was determined colorimetrically. Four of six cell lines (BxPC3, ASPC-1, PL-11, and XPA-1) demonstrate response to nanocurcumin, while two lines (PL-18 and PK-9) are curcumin resistant.

Colony Assays in Soft Agar

Colony formation in soft agar is assessed for therapy with free drug and equivalent dosage of nanoparticles. Briefly, a mixture of 2 mL of serum-supplemented media and 1% agar containing 5, 10, or 15 μM of the free drug and equivalent nanoparticle is added in a 35-mm culture dish and allowed to solidify (base agar), respectively. Next, on top of the base layer a mixture of serum-supplemented media and 0.7% agar (total 2 mL) containing 10,000 cells in the presence of void polymer, free drug, or nanoparticle is added and is allowed to solidify (top agar); a fourth set of plates contains cells without any additives. Subsequently, the dishes are kept in a tissue culture incubator maintained at 37°C and 5% CO_2 for 14 days to allow for colony growth. All assays are performed in triplicate. The colony assay is terminated at day 14, when plates are stained and colonies counted on ChemiDoc XRS instrument (Bio-Rad, Hercules, CA). Colony assays in soft agar were performed by Bisht et al. (55), comparing the effects of free and nanocurcumin in inhibiting the clonogenicity of the pancreatic cancer cell line.

Electrophoretic Mobility Shift Assay

Nuclear extracts are prepared by standard method reported (55) and 2.5 to 5 μg of nuclear extracts are incubated with approximately 1 μL of labeled oligonucleotide (20,000 cpm) in 20 μL of incubation buffer [10 mM of Tris-HCl, 40 mM of NaCl, 1 mM of EDTA, 1 mM of β-mercaptoethanol, 2% glycerol, 1–2 μg of poly(dI-dC)] for 20 minutes at 25°C. DNA–protein complexes are resolved by electrophoresis in 5% nondenaturing polyacrylamide gels and analyzed by autoradiography. Bisht et al. (55), using electrophoretic mobility shift ("gel shift") assays, analyzed the mechanisms of action of nanocurcumin on pancreatic cancer cell lines and compared the functional pathways impacted by nanocurcumin to what has been previously reported for free curcumin (60–65). A principal cellular target of curcumin in cancer cells is the activated nuclear factor kappa B (NFκB), as inhibition of this seminal transcription factor has been ascribed to many of the pleiotropic effects of curcumin. In electrophoretic mobility shift ("gel shift") assays, they demonstrated that nanocurcumin robustly inhibits NFκB function in pancreatic cancer cell lines BxPC3 and MiaPaCa (Fig. 3) (55).

Cell Adhesion Study

In vivo, mammalian cells interact with one another, triggering diverse intracellular processes that control cell development. Similarly, the surrounding environment, constituted basically by the extracellular matrix (ECM) and soluble factors, causes cells to adapt to it, reprogramming their intracellular machinery. Hence,

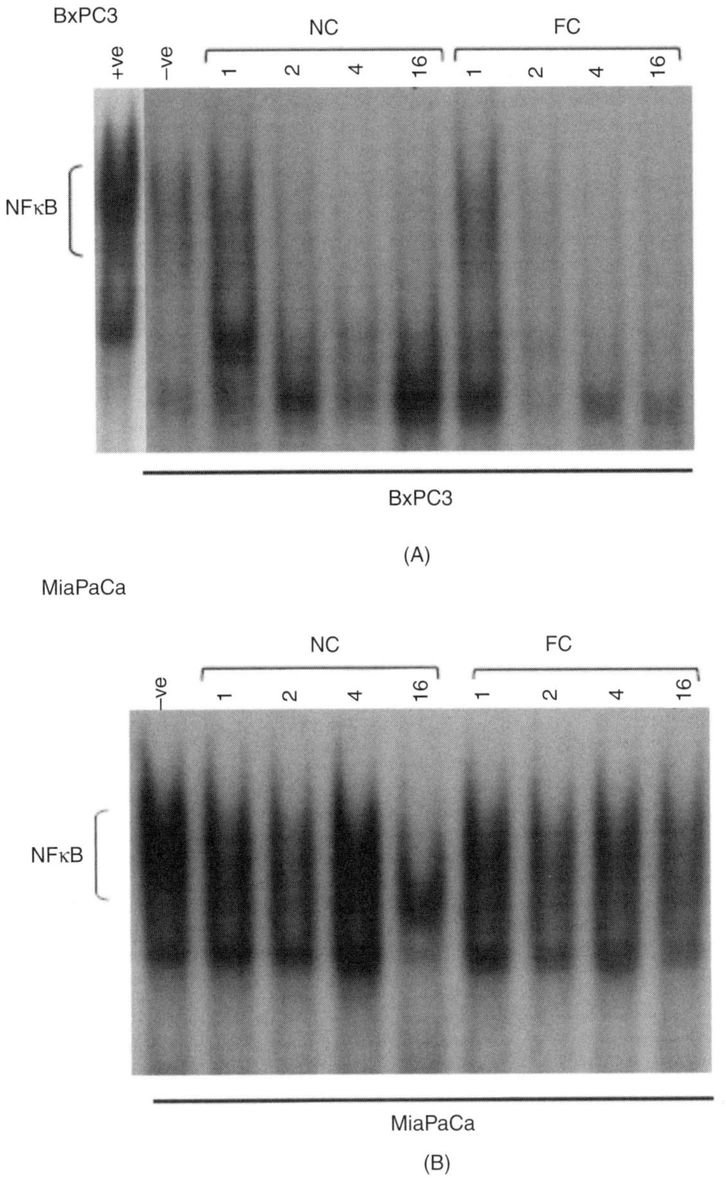

FIGURE 3 Electrophoretic mobility shift assay or "gel shift" assay for the assessment of NFκB inhibition in pancreatic cancer cell lines. Nuclear extracts were prepared from free curcumin (FC) and nanocurcumin (NC)-treated BxPC3 and MiaPaCa cells after 1, 2, 4, and 16 hours (overnight) exposure to the respective formulation. Inhibition of NFκB function is gauged by faster migration (i.e., absence of NFκB binding) of the radiolabeled κ-binding oligonucleotide. In BxPC3 cells, inhibition of NFκB is seen as early as 1 to 2 hours following curcumin exposure in both free and nanoparticulate formulations. In contrast, in MiaPaCa cells, inhibition of binding and consequent gel shift is seen only after overnight incubation in the nanocurcumin-treated cells, while no discernible gel shift is apparent in the free curcumin-treated cells. TPA-activated Jurkat cells were used as positive control and untreated cells as negative control. *Source*: From Ref. 55.

artificial biofunctionalized substrates with nano- and micropatterns might make cells develop according to the substrate design in a noninvasive approach; controlling specifically processes such as cell adhesion, survival, proliferation, migration, and differentiation. The main objective of the work reported by Martínez et al. (64) was to apply novel micro- and nanofabrication techniques and surface modification strategies to generate well-defined topographical and biochemical cues for cell culture. To achieve this goal, micro- and nanostructured polymer substrates have been generated by nanoembossing and their biochemical surface properties modified by microcontact printing, nanoplotting, and dip-pen nanolithography, transferring ECM proteins, which will be attached covalently to the surface. All these will be used to study their influence on cell adhesion, morphology, proliferation and differentiation. Results show that surface fictionalization with adhesion proteins such as fibronectin can be used to selectively attach and confine cells on specific surface locations. When micro- and nanopatterned, fibronectin can also alter cell morphology, cytoskeletal organization, and stress level. On the other hand, surface micro- and nanotopography not only proves special relevance in cell guiding and alignment processes but also greatly affects cell morphology. The combination of both topographical and biochemical features gives very interesting results regarding cell differentiation.

β-Galactosidase Assay

Dastan and Turan (22) used HEK293, Swiss3T3, and HeLa cell lines for β-galactosidase assay. Cells grown at standard culture conditions for 24 hours, thereafter incubated with fresh complete media containing the defined concentration of chitosan–DNA microparticles for 48 hours, are harvested for β-galactosidase assay. Cells are washed with PBS, and then they are detached with trypsin, suspended in PBS, and collected by centrifugation. The cells obtained are lysed in 200 mL of lyses buffer (100 mM of KH_2PO_4/K_2HPO_4, pH 7.5, 0.2% Triton X-100, 1 mM of DTT) by freezing and thawing. The β-galactosidase assay is performed by adding buffer containing 100 mM of KH_2PO_4/K_2HPO_4 (pH 7.5), 10 mM of KCl, 1 mM of $MgSO_4$, and 50 mM of 2-mercaptoethanol to cell lysate and incubating it for 5 minutes at 37°C. *O*-Nitrophenyl-β-D-galactopyranoside substrate solution was then added to the reaction mixture and incubated for 1 to 16 hours at 37°C. After the incubation period, the reaction was terminated by the addition of 1 M of Na_2CO_3 solution and the absorbance of samples was measured with a microtiter dish reader set at 420 nm. The β-galactosidase activity was calculated by the following equations and units of enzyme were expressed as nanomoles of β-galactose formed per minute (modified from Ref. 43). β-Galactosidase activity (U/mg of total protein in lysate) = $[OD_{420}/0.0045 \times \text{assay volume (mL)}]\ \text{min}^{-1}\ \text{mg}^{-1}$.

CONCLUSION

Nanoparticle cellular interactions are of utmost importance. In vitro characterization of such interactions can be done by several techniques. Flow cytometry, LSC, HCS assays, CLSM, FCM, and LCM are some of the techniques used to understand these interactions. Sterility of nanoparticles is challenging due to the nanosize of the particles comparable with the size of the microbial contaminants. Several techniques are discussed for the nanoparticle targeting studies and different assay procedures to characterize them.

REFERENCES

1. Current Protocols in Cytometry, Unit 1, 2, p1.2.2, Supplement 45. New York: John Wiley & Sons, Inc., 2008. Available at: http://biology.berkeley.edu/crl/flow_cytometry_basic.html. Accessed May 2008.
2. Shapiro HM. Practical Flow Cytometry, 3rd ed. Wilmington, DE: Wiley-Liss; 1995: 245.
3. Vasudevan C, Haskins JR, Burmeister WE, et al. Assessing Toxicity of Nanoparticles with In Vitro Cell Based Assay. Pittsburgh, PA: Cellomics, Inc., 2006.
4. Pawley JB, ed. Handbook of Biological Confocal Microscopy, 3rd ed. Berlin, Germany: Springer; 2006.
5. Minsky M. Microscopy apparatus. US patent 3,013,467, 1957.
6. Fellers TJ, Davidson MW. Introduction to Confocal Microscopy. National High Magnetic Field Laboratory, USA: Olympus Fluoview Resource Center, 2007.
7. Patel DV, McGhee CN. Contemporary in vivo confocal microscopy of the living human cornea using white light and laser scanning techniques: A major review. Clin Exp Ophthalmol 2007; 35(1):71–88.
8. Emmert-Buck MR, Bonner RF, Smith PD, et al. Laser capture microdissection. Science 1996; 274(5289):998–1001.
9. Espina V, Heiby M, Pierobon M, et al. Laser capture microdissection technology. Expert Rev Mol Diagn 2007; 7(5):647–657.
10. Orba Y, Tanaka S, Nishihara H, et al. Application of laser capture microdissection to cytologic specimens for the detection of immunoglobulin heavy chain gene rearrangement in patients with malignant lymphoma. Cancer 2003; 99(4):198–204.
11. Kihara AH, Moriscot AS, Ferreira PJ, et al. Protecting RNA in fixed tissue: An alternative method for LCM users. J Neurosci Methods 2005; 148(2):103–107.
12. Detection of Endotoxin Contamination by End Point Chromogenic LAL Assay. NCL Method STE-1, Version 1.0, Nanotechnology Characterization Laboratory, National Cancer Institute. Frederick, MD: SAIC-Frederick, Inc., 2005.
13. Detection of Microbial Contamination Assay. NCL Method STE-2, Version 1.0, Nanotechnology Characterization Laboratory, National Cancer Institute. Frederick, MD: SAIC-Frederick, Inc., 2005.
14. Gopalkrishna V, Verma H, Kumbhar NS, et al. Detection of *Mycoplasma* species in cell culture by PCR and RFLP based method: Effect of BM-cyclin to cure infections. Indian J Med Microbiol 2007; 25(4):364–368.
15. Kramer MAW. Development of a Fret Biosensor to Detect the Pathogen *Mycoplasma capricolum* [doctoral dissertation]. Columbia: Missouri University, 2005. Available at: http://edt.missouri.edu/Fall2005/Thesis/WindsorM-051706-3116/short.pdf.
16. Darlington GJ. Detection of *Mycoplasma* in mammalian cell cultures using fluorescence microscopy, CSH protocols; 2006 (adapted from: Growth and manipulation of cells in culture. In: Spector et al., eds. Cells, chapter 2.). Cold Spring Harbor, NY: Cold Spring Harbor Laboratory Press, 1998.
17. Seeley KA. The AcroWell^{TM} 96 Filter Plate: Low Fluorescence Background Using the DELFIA* System. East Hills, NY: Pall Corporation, 2000.
18. Valenzano KJ, Miller W, Kravitz JN, et al. Development of a fluorescent ligand-binding assay using the AcroWell^{TM} filter Plate. J Biomol Screen 2000; 5(6):455–461.
19. Graham FL, van der Eb AJ. A new technique for the assay of infectivity of human adenovirus 5 DNA. Virology 1973; 52(2):456–467.
20. Bacchetti S, Graham F. Transfer of the gene for thymidine kinase to thymidine kinase-deficient human cells by purified herpes simplex viral DNA. Proc Natl Acad Sci USA 1977; 74(4):1590–1594.
21. Zheng F, Shi X-W, Yang G-F, et al. Chitosan nanoparticle as gene therapy vector via gastrointestinal mucosa administration: Results of an in vitro and in vivo study. Life Sci 2007; 80:388–396.
22. Dastan T, Turan K. In vitro characterization and delivery of chitosan-DNA microparticles into mammalian cells. J Pharm Pharm Sci 2004; 7(2):205–214.

23. Banerjee T, Singh AK, Mitra S, et al. Comparative evaluation of stannous chloride and sodium borohydride as reducing agents for preparation of technetium-99m labeled chitosan nanoparticles. Indian J Nucl Med 2002; 17(4):11.
24. Reddy LH, Sharma RK, Murthy RSR. Tumor retention and biodistribution studies of etoposide loaded tripalmitin nanoparticles in Dalton's lymphoma bearing mice. Alasbimn J 2004; 6:25.
25. Soni S, Babbar AK, Sharma RK, et al. Pharmacoscintigraphic evaluation of Polysorbate 80 coated chitosan nanoparticles for brain targeting. Indian J Nucl Med 2004; 19(4):144.
26. Reddy LH, Sharma RK, Mishra AK, et al. Etoposide incorporated tripalmatin nanoparticles with different surface charge: Formulation, characterization and biodistribution studies. AAPS J 2004; 6(3):e23.
27. Reddy LH, Sharma RK, Murthy RSR. Enhanced tumour uptake of doxorubicin loaded poly(butyl-cyanoacrylate) nanoparticles (DPBC) in mice bearing Dalton's lymphoma tumor. J Drug Target 2004; 12(7):443–451.
28. Sharma RK. The road ahead. In: Health Technology as Fulcrum of Development for the Nation. New Delhi, India: National Institute of Science Communication and Information Resources, 2005:214–218.
29. Banerjee T, Singh AK, Sharma RK, et al. Labeling efficiency and biodistribution of technetium-99m labeled nanoparticles: Interference by colloidal tin oxide particles. Int J Pharm 2005; 289(1–2):189–195.
30. Soni S, Babbar AK, Sharma RK, et al. Delivery of hydrophobised 5-fluorouracil derivative to brain tissue through intravenous route using surface modified nanogel. J Drug Target 2005; 14(2):87–95.
31. Jain S, Sharma RK, Vyas SP. Chitosan nanoparticles encapsulated vesicular systems for oral immunization: Preparation, in vitro and in vivo characterization. J Pharm Pharmacol 2006; 58:303–310.
32. Reddy LH, Sharma RK, Mishra AK, et al. Influence of administration route on the uptake and biodistribution of etoposide loaded tripalmitin nanoparticles in Dalton's lymphoma tumor bearing mice. J Control Release 2005; 105(3):185–198.
33. Dass CR, Su T. Delivery of lipoplexes for genotherapy of solid tumours: Role of vascular endothelial cells. J Pharm Pharmacol 2000; 52:1301–1317.
34. Davda J, Labhasetwar V. Characterization of nanoparticle uptake by endothelial cells. Int J Pharm 2002; 233:51–59.
35. Martin SG, Murray JC. Gene-transfer systems for human endothelial cells. Adv Drug Deliv Rev 2000; 41:223–233.
36. Nabel EG. Biology of the impaired endothelium. Am J Cardiol 1991; 68:6c–8c.
37. Nabel EG. Gene therapy for cardiovascular disease. Circulation 1995; 91:541–548.
38. Yao SN, Wilson JM, Nabel EG, et al. Expression of human factor IX in rat capillary endothelial cells: Toward somatic gene therapy for hemophilia B. Proc Natl Acad Sci USA 1991; 88:8101–8105.
39. Parikh SA, Edelman ER. Endothelial cell delivery for cardiovascular therapy. Adv Drug Deliv Rev 2000; 42:139–161.
40. Scherpereel A, Wiewrodt R, Solomidou CM, et al. Cell-selective intracellular delivery of a foreign enzyme to endothelium in vivo using vascular immunotargeting. FASEB J 2001; 15:416–426.
41. Stahn R, Grittner C, Zeisig R, et al. Sialyl Lewisx-liposomes as vehicles for site-directed, E-selectin-mediated drug transfer into activated endothelial cells. Cell Mol Life Sci 2001; 58:141–147.
42. Otilia M, Koo MS, Rubinstein I, et al. Role of nanotechnology in targeted drug delivery and imaging: A concise review. Nanomedicine 2005; 1:193–212.
43. Koval M, Preiter K, Adles C, et al. Size of IgG opsonized particles determines macrophage response during internalization. Exp Cell Res 1998; 242:265–273.
44. Harashima H, Sakata K, Funato K, et al. Enhanced hepatic uptake of liposomes through complement activation depending on the size of liposomes. Pharm Res 1994; 11:402–406.
45. Moghimi SM, Hunter AC, Murray JC. Long-circulating and target specific nanoparticles: Theory to practice. Pharmacol Rev 2001; 53:283–318.

46. Rejman J, Oberle V, Zuhorn IS, et al. Size-dependent internalization of particles via the pathways of clathrin- and caveolae-mediated endocytosis. Biochem J 2004; 377:159–169.
47. Geys J, Coenegrachts L, Vercammen J, et al. In vitro study of the pulmonary translocation of nanoparticles: A preliminary study. Toxicol Lett 2006; 160(3):218–226.
48. Liang WWX, Deshpande D, Malanga CJ, et al. Oligonucleotide targeting to alveolar macrophages by mannose receptor mediated endocytosis. Biochim Biophys Acta 1996; 1279:227–233.
49. Jain SK, Gupta Y, Jain A, et al. Mannosylated gelatin nanoparticles bearing an anti-HIV drug didanosine for site-specific delivery. Nanomedicine 2008; 4:41–48.
50. Vauthier C, Dubernet C, Chauvierre C, et al. Drug delivery to resistant tumors: The potential of poly(alkyl cyanoacrylate) nanoparticles. J Control Release 2003; 93:151–160.
51. Miller DW, Batrakova EV, Kabanov AV. Inhibition of multidrug resistance-associated protein (MRP) functional activity with Pluronic block copolymers. Pharm Res 1999; 16:396–401.
52. Sapra P, Allen TM. Ligand-targeted liposomal anticancer drugs. Prog Lipid Res 2003; 42:439–462.
53. Mamot C, Drummond DC, Hong K, et al. Liposome-based approaches to overcome anticancer drug resistance. Drug Resist Update, 2003; 6:271–279.
54. Soma CE, Dubernet C, Bentolila D, et al. Reversion of multidrug resistance by co-encapsulation of doxorubicin and cyclosporin A in polyalkylcyanoacrylate nanoparticles. Biomaterials 2000; 21:1–7.
55. Bisht S, Feldmann G, Soni S, et al. Polymeric nanoparticle-encapsulated curcumin ("nanocurcumin"): A novel strategy for human cancer therapy. J Nanobiotechnol 2007; 5:3.
56. Frei E, Cucchi CA, Rosowsky A, et al. Alkylating agent resistance: In vitro studies with human cell lines. Proc Natl Acad Sci USA 1985; 82:2158–2162.
57. Demoy A, Andreux J-P, Weingarten C, et al. In vitro evaluation of nanoparticles spleen capture. Life Sci 1999; 64(15):1329–1337.
58. Martínez E, Ríos-Mondragón I, Pla-Roca M, et al. Cell-surface interactions studies to trigger stem cell differentiation. *Nanomedicine: Nanotechnology, Biology and Medicine*, 3(4):346.
59. Demoy M, Gibaud S, Andreux JP, et al. Splenic trapping of nanoparticles: Complementary approaches for in situ studies. Pharm Res 1997; 14(4):463–468.
60. Li L, Aggarwal BB, Shishodia S, et al. Nuclear factor-kappaB and IkappaB kinase are constitutively active in human pancreatic cells, and their down-regulation by curcumin (diferuloylmethane) is associated with the suppression of proliferation and the induction of apoptosis. Cancer 2004; 101:2351–2362.
61. Aggarwal BB, Shishodia S. Suppression of the nuclear factor-kappaB activation pathway by spice-derived phytochemicals: Reasoning for seasoning. Ann N Y Acad Sci 2004; 1030:434–441.
62. Aggarwal S, Ichikawa H, Takada Y, et al. Curcumin (diferuloylmethane) down-regulates expression of cell proliferation and antiapoptotic and metastatic gene products through suppression of IkappaBalpha kinase and Akt activation. Mol Pharmacol 2006; 69:195–206.
63. Hidaka H, Ishiko T, Furuhashi T, et al. Curcumin inhibits interleukin 8 production and enhances interleukin 8 receptor expression on the cell surface: Impact on human pancreatic carcinoma cell growth by autocrine regulation. Cancer 2002; 95:1206–1214.
64. Shishodia S, Amin HM, Lai R, et al. Curcumin (diferuloylmethane) inhibits constitutive NF-kappaB activation, induces G1–S arrest, suppresses proliferation, and induces apoptosis in mantle cell lymphoma. Biochem Pharmacol 2005; 70:700–713.
65. Bharti AC, Donato N, Aggarwal BB. Curcumin (diferuloylmethane) inhibits constitutive and IL-6-inducible STAT3 phosphorylation in human multiple myeloma cells. J Immunol 2003; 171:3863–3871.

12 In Vitro Blood Interaction and Pharmacological and Toxicological Characterization of Nanosystems

R. S. R. Murthy

Pharmacy Department, The M. S. University of Baroda, Vadodara, India

Yashwant Pathak

Department of Pharmaceutical Sciences, Sullivan University College of Pharmacy, Louisville, Kentucky, U.S.A.

INTRODUCTION

Immunological evaluation includes both immunosuppression and immunostimulation and is applicable to nanoparticles intended to be used as drug candidates and/or as drug delivery platforms. Short-term in vitro assays are developed for quick evaluation of biocompatibility of nanoparticles, which includes analysis of plasma protein binding by polyacrylamide gel electrophoresis (PAGE), hemolysis, platelet aggregation, coagulation, compliment activation, colony-forming unit–granulocyte macrophage (CFU-GM), leukocyte proliferation, phagocytosis, cytokine secretion by macrophages, chemotaxis, oxidative burst, and evaluation of cytotoxic activity of natural killer (NK) cells. In addition to these methods, in vitro test may also include sterility assessment and pyrogen contamination test by Limulus amebocyte lysate assay. These assay cascades are based on several regulatory documents recommended by the U.S. Food and Drug Administration for immunotoxicological evaluation of new investigational drugs, medical devices, and biotechnology derived pharmaceuticals (1–5), as well as ASTM and ISO standards developed for characterization of blood contact properties of medical devises (6–8). Challenges for specific immunological assessment of nanoparticulate materials are summarized in this chapter.

An important aspect of this testing is to ensure the absence of toxicity to blood elements when nanoparticulate delivery systems are injected into the patients. The test protocols are developed in general, and when the same is applied for nanoparticles, there can be specific problems that are not generally anticipated. However, some have reported that interactions would act as guideline in carrying out these tests with nanoparticulates.

PLASMA PROTEIN BINDING

Despite the remarkable pace of development of nanoscience, very little is known about the interaction of nanoscale objects with living systems. In a biological fluid, proteins associate with nanoparticles, and the amount and presentation of the proteins on the surface of the particles lead to an in vivo response. Proteins compete for the nanoparticle surface, leading to a protein corona that largely defines the biological identity of the particles.

Two-dimensional gel electrophoresis (2D-PAGE) method has been described by NCL (NCL Method ITA-4, Version 1.0) (9) for the analysis of nanoparticle interaction with plasma proteins. Nanoparticles are incubated with pooled human plasma derived from healthy donors to allow for protein interaction and binding. Following a separation procedure, bound proteins are eluted from the nanoparticle surface and analyzed by 2D-PAGE. The identity of individual proteins separated by this procedure can be evaluated by mass spectrometry.

In addition to the characterization of proteins involved in binding, knowledge of rates, affinities, and stoichiometries of protein association with, and dissociation from, nanoparticles is also of prime importance for understanding the nature of the particle surface seen by the functional machinery of cells.

Cedervall et al. (10) developed approaches to study these parameters and apply them to plasma and simple model systems, albumin and fibrinogen. A series of copolymer nanoparticles is used with variation of size and composition (hydrophobicity). They also showed that isothermal titration calorimetry (ITC) is suitable for studying the affinity and stoichiometry of protein binding to nanoparticles. They determined the rates of protein association and dissociation by surface plasmon resonance (SPR) technology with nanoparticles that are thiol linked to gold and through size-exclusion chromatography (SEC) of protein–nanoparticle mixtures. This method is less perturbing than centrifugation and is developed into a systematic methodology to isolate nanoparticle-associated proteins. The kinetic and equilibrium binding properties depend on protein identity, as well as particle surface characteristics and size.

A deep understanding of the biological effects of nanoparticles requires knowledge of the equilibrium and kinetic binding properties of proteins (and other molecules) that associate with the particles. However, fundamental prerequisite for nanobiology, nanomedicine, and nanotoxicology, the isolation and identification of particle-associated proteins, is not a simple task. Furthermore, in terms of the biological response, the more abundantly associated proteins may not necessarily have the most profound effect while a less-abundant protein with high affinity and specificity for a particular receptor may play a key role. It is thus essential to develop methods to identify both major and minor particle-associated proteins and to study the competitive binding of proteins with nanoparticles under kinetic or thermodynamic control.

A central methodological problem is to separate free protein from protein bound to nanoparticles, ideally employing nonperturbing methods that do not disrupt the protein–particle complex or induce additional protein binding. The preferred method to isolate and identify the major serum proteins albumin, IgG, and fibrinogen to date has been centrifugation. Due to its high abundance, albumin is almost always observed on particles and may be retrieved even if it has relatively low affinity. Other proteins observed with several particle types in these centrifugation assays are immunoglobulins, apolipoproteins, and α_1-antitrypsin.

The understanding of protein–nanoparticle interactions and their biological consequences could be more advanced if we could find information on the binding affinities, their rate kinetics and stoichiometries for different combinations of proteins and nanoparticles, and ranking of the affinities of proteins that coexist in specific body fluids or cellular compartments. Cedervall et al. (10) implemented a range of methods for studying kinetic and equilibrium parameters of protein–nanoparticle interactions. They used a set of tailored copolymer nanoparticles that allows systematical investigation of how the size and composition (hydrophobicity) of the particles affects their interaction with proteins, relative affinities for different proteins, and rates of association and dissociation. ITC can be used to assess the stoichiometry and affinity of protein binding and SPR studies (in which nanoparticles are linked to gold via a thiol anchor) yield additional data on protein association to and dissociation from nanoparticles.

Isothermal Titration Calorimetry

ITC was investigated for its potential to assess the stoichiometry, affinity, and enthalpy of protein–nanoparticle interaction. Protein is injected into a nanoparticle solution in the sample cell and the difference in heat that needs to be added to the sample and reference cells to keep both cells at the same temperature is monitored. If the reaction is exothermic, less heat needs to be added to the sample cell and a negative signal is obtained. If the concentrations of both the nanoparticles and injected protein are known, data from multiple injections provide information on the number of protein molecules bound per particle, the apparent affinity, and the enthalpy change.

Size-Exclusion Chromatography

SEC was the second method investigated for its potential to reveal quantitative information on protein–nanoparticle interactions. Chromatographic elution profiles of protein mixed with nanoparticles were compared with both protein alone and particles alone. The chromatographic resin, sephacryl S1000 SF, separation range = 5 × 105 to >108 kDa (GE Healthcare Bio-Sciences Ltd, USA) allows protein and nanoparticles to be resolved but not different types of proteins. There is a clear difference in the elution profile of human serum albumin (HSA) mixed with nanoparticles, compared with free albumin, which implies an interaction between the protein and the particles. HSA mixed with 200-nm 85:15 NIPAM/BAM particles elutes earlier than HSA without particles. Results for HSA and 200-nm particles with 85:15, 65:35, and 50:50 NIPAM/BAM reveal that more protein elutes early with the more hydrophilic particles, implying a longer residence time on these nanoparticles. With the most hydrophobic nanoparticles, a large fraction of the protein elutes later than HSA alone. Fibrinogen, with 200-nm 65:35 NIPAM/BAM, elutes with elution times equivalent to the free protein and earlier than HSA on the same particles, suggesting that fibrinogen dissociates at a lower rate.

Cedervall et al. (10) evaluated SEC for its potential to be a nonperturbing method for studying differential protein binding in such complex fluids and revealed that several plasma proteins are preferentially enriched on the nanoparticles. The protein elution profile is distinctly different with and without particles. HSA elutes in different fractions when plasma is mixed with 200-nm 50:50 NIPAM/BAM particles compared with chromatographic runs with plasma alone, indicating that HSA in plasma binds to the particles. In addition, there are at least six plasma proteins that elute earlier with 200-nm 50:50 NIPAM/BAM particles than with plasma without particles. This finding implies that these proteins associate with the particles, and their elution profiles indicate slower exchange than for HSA and many other plasma proteins.

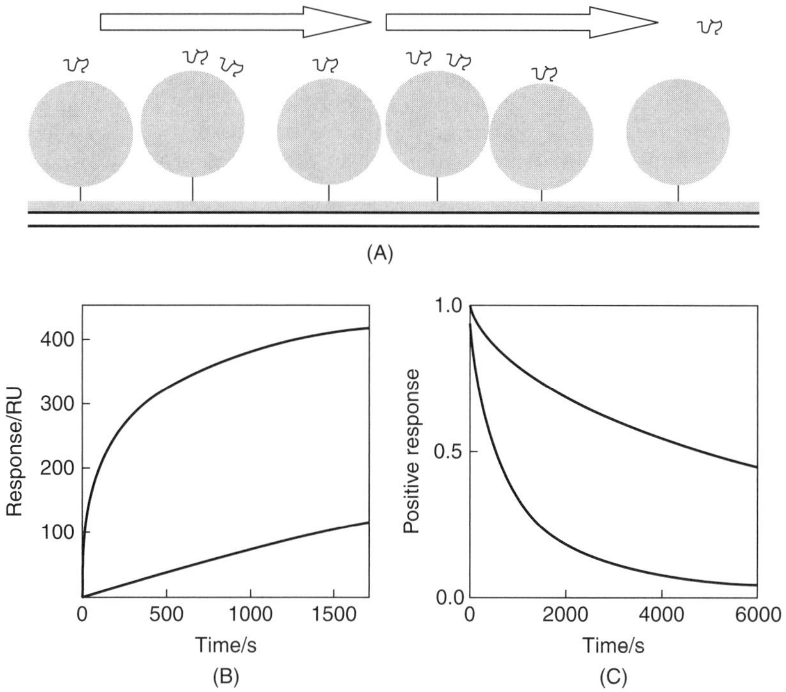

FIGURE 1 Surface plasmon resonance (SPR) studies of plasma–nanoparticle interactions. (**A**) A cartoon of a gold surface with thiol-tethered particles and associated protein over which buffer is flown. (**B** and **C**) SPR data of plasma proteins injected at 60-fold dilution over 70-nm 85:15NIPAM/BAM (blue) or 50:50 NIPAM/BAM (red) for 30 minutes (**B**) followed by buffer flow for 24 hours (**C**, first 6000 seconds shown). *Source*: From Ref. 1.

Surface Plasmon Resonance

SPR is used to measure protein association to and dissociation from nanoparticles. Gold surfaces with thiol-conjugated nanoparticles were used to study the kinetics of association and dissociation of plasma, HAS, and fibrinogen with nanoparticles. Both the association and dissociation rates were found to be clearly dependent on the hydrophobicity of the particles (Fig. 1). The dissociation rate constant for plasma proteins on the 70-nm 85:15 NIPAM/BAM particles was found to be higher than that for plasma proteins on the 70-nm 50:50 NIPAM/BAM particles. A similar difference is seen between the 200-nm particles with 85:15 and 50:50 NIPAM/BAM. SPR studies with pure HSA and fibrinogen show dissociation rate constant consistent with the fast dissociation event, suggesting that these proteins account for the faster of the observed kinetic processes. Again for HSA and fibrinogen, we observe faster dissociation rate from the more hydrophobic particles than from the more hydrophilic particles.

HEMOLYSIS

Hemolysis can lead to life-threatening conditions such as anemia, hypertension, arrhythmia, and renal failure. Protocols have been developed to evaluate hemolytic properties of nanoparticles based on the existing ASTM international standard used to characterize other materials (11,12). Problems identified include interference in the absorption maxima used in standard assay, such as colloid gold nanoparticles of 5- to 50-nm size absorbs at 535 nm, which overlaps the absorption maxima of 540 nm recommended to estimate plasma free hemoglobin (PFH) in the general assay procedure. Hence, the removal of nanoparticles prior to sample evaluation by centrifugation is required. However, nanoparticles near 5-nm size may also sediment hemoglobin (size $\approx$ 5 nm) during centrifugation at very high *g* values, leading to false-negative results. Ultracentrifugation is not feasible for fullerenes and dendrimer particles. Polystyrene nanoparticles in the range from 20 to 80 nm are prepared with surfactants, and the traces remained in the sample, although within the limits, cause damage to red blood cells (RBCs). Problem with the characterization of metal-containing nanoparticles is hemoglobin oxidation leading to wrong optical density (OD) value. Hence, these assay procedures need slight modification depending on the sample to be characterized.

Analysis of Hemolytic Properties of Nanoparticles (NCL Method ITA-1, Version 1.0) (13)

A protocol has been developed by NCL for quantitative colorimetric determination of hemoglobin in whole blood (total blood hemoglobin) and hemoglobin released into plasma (PFH) when blood is exposed to nanoparticles. Hemoglobin and its derivatives, except sulfhemoglobin, are oxidized to methemoglobin by ferricyanide in the presence of alkali. Cyanmethemoglobin is then formed from the methemoglobin by its reaction with cyanide (Drabkin's solution).

The cyanmethemoglobin can then be detected by spectrophotometer set at 540 nm. The hemoglobin standard is used to build a standard curve covering the concentration range from 0.025 to 0.80 mg/mL and to prepare quality control samples at low (0.0625 mg/mL), mid (0.125 mg/mL), and high (0.625 mg/mL) concentrations to monitor assay performance.

The results expressed as percentage of hemolysis are used to evaluate the acute in vitro hemolytic activity of the nanoparticles.

PLATELET AGGREGATION INDUCED BY NANOPARTICLES

Increasing the use of engineered carbon nanoparticles in nanopharmacology for selective imaging, sensor, or drug delivery systems has increased the potential for blood platelet–nanoparticle interactions. Radomski et al. (14) studied the effects of engineered and combustion-derived carbon nanoparticles on human platelet aggregation in vitro and rat vascular thrombosis in vivo. Platelet function was studied by lumiaggregometry, phase-contrast, immunofluorescence, and transmission electron microscopy, flow cytometry, zymography, and pharmacological inhibitors of platelet aggregation.

Method for the analysis of platelet aggregation is described by NCL (NCL Method ITA-2, Version 1.0) (15). Platelet-rich plasma (PRP) is obtained from fresh, pooled human whole blood and incubated with control or test sample for 15 minutes at a nominal temperature of 37°C. After that, PRP is analyzed by Z2 particle count and size analyzer to determine the number of active platelets. Percentage aggregation is calculated by comparing the number of active platelets in a test sample to the one in a control baseline tube.

Platelets were isolated from blood obtained from healthy volunteers and resuspended in Tyrode's solution (2.5×10^8 platelets/mL), as previously described by Radomski and Moncada (16). Platelets were preincubated for 2 minutes at 37°C in a whole-blood ionized calcium lumiaggregometer (Chrono-log, Havertown, PA) prior to the addition of particles (0.2–300 μg/mL). Platelet aggregation was studied for 8 minutes and analyzed with Aggro-Link data reduction system (Chronolog, Havertown, PA) (17–19). The release of ATP was measured by luciferin–luciferase, using lumiaggregometer as previously described by Sawicki et al. (20) and Chung et al. (21). For phase-contrast microscopy, platelet aggregation was terminated at 20% maximal response, as determined with the aggregometer. The samples were fixed by adding an equal volume of 2% glutaraldehyde and 2% paraformaldehyde in 0.1 M of phosphate buffer, pH 7.4, and then incubated for 1 hour at room temperature. Aliquots of each sample were then taken for phase-contrast microscopic examination with an Olympus CKX41 microscope (Olympus America Inc., Melville, NY). The remaining samples were then prepared for transmission electron microscopic examination as per method reported by Alonso-Escolano et al. (22). Ultrathin sections obtained were stained with uranyl acetate and lead citrate in an LKB ultrostainer and examined under a JEM 1010 transmission electron microscope (JEOL Inc., Peabody, MA) at an accelerating voltage of 80 kV.

For immunofluorescence microscopy, cytospins were prepared by centrifuging 120 μL of platelet suspension onto a glass slide in a cytocentrifuge. Slides were allowed to air dry at room temperature and nonspecific binding was blocked by incubation for 30 minutes at room temperature in Dulbecco's PBS containing 10% BSA (DPBS/BSA). Slides were incubated for 60 minutes with anti-MMP-9 (10 μg/mL) antibodies in blocking buffer (DPBS/BSA). IgG (10 μg/mL) was used as an isotype control. After washing with DPBS/BSA, slides were incubated with a 1:300 dilution of anti-mouse IgG conjugated with FITC for 60 minutes. The slides were then washed with PBS and mounted in SlowFade Light Antifade solution (Molecular Probes, Eugene, OR) and examined with a fluorescence imaging microscope.

Flow Cytometry

Flow cytometry was performed on single-stained platelet samples as described previously (16,23). Briefly, platelets (10 μL of suspension) and fluorescent-labeled antibodies (10 μL) containing 0.25 μg of antiactivated GPIIb/IIIa (PAC-1), anti-GPIb, anti-P-selectin (BD Biosciences, San Diego, CA), or 1 μg of anti-integrin β3 (SouthernBiotech, Birmingham, AL) were diluted 10-fold with physiologic saline. Samples and antibodies were incubated in the dark at room temperature for 5 minutes. Platelets were identified by forward and side scatter signals, and 10,000 platelet-specific events were analyzed for cytometric fluorescence.

Zymography (20,21,24)

Zymography was performed by 8% SDS-PAGE with copolymerized gelatin (2 mg/mL). Ten microliters of platelet release was subjected to electrophoresis. Gels were washed in 2.5% Triton X-100 for 1 hour (3×, 20 minutes each) and twice in zymography buffer (20 minutes each wash). Then, the samples were incubated in enzyme assay buffer (25 mM of Tris, pH 7.5, 5 mM of $CaCl_2$, 0.9% NaCl, 0.05% Na_3N) until the matrix metalloproteinase (MMP) activities could be determined. MMP-2 and MMP-9 were identified by their molecular weight and quantified by reference to purified standards.

Vascular Thrombosis

Vascular thrombosis was induced by ferric chloride and the rate of thrombosis was measured, in the presence of carbon particles, with an ultrasonic flow probe. Carbon particles, except C60CS, stimulated platelet aggregation [mixed carbon nanotube $\geq$ single-walled nanotube (SWNT) > multiwalled nanotube > SRM1648] and accelerated the rate of vascular thrombosis in rat carotid arteries with a similar rank order of efficacy. All particles resulted in upregulation of GPIIb/IIIa in platelets. In contrast, particles differentially affected the release of platelet granules, as well as the activity of thromboxane-, ADP-, MMP-, and protein kinase C–dependent pathways of aggregation. Furthermore, particle-induced aggregation was inhibited by prostacyclin and *S*-nitroso-glutathione but not by aspirin. Thus, some carbon nanoparticles and microparticles have the ability to activate platelets and enhance vascular thrombosis. These observations are of importance for the pharmacological use of carbon nanoparticles and pathology of urban particulate matter.

BLOOD COAGULATION (NCL METHOD ITA-12, VERSION 1.0) (25)

The plasma coagulation is assayed in four tests [i.e., prothrombin time (PT), activated partial thromboplastin time (APTT), thrombin time (TT), and reptilase time (RT)]. This assay requires 270 μL of a test nanomaterial.

Experimental Procedure

Place cuvettes into A, B, C, and D test rows on a coagulometer. Add one metal ball into each cuvette and let cuvette with the ball warm for at least 3 minutes before use. Add 100 μL of control or test plasma to a cuvette when testing PT and TT and 50 μL when testing APTT and RT. Prepare three duplicate cuvetts for each plasma sample. For APTT and RT, add 50 μL of PTT-A reagent (APTT test) or Owren-Koller reagent (reptilase test) to plasma samples in cuvettes. Start the timer for each of the test rows by pressing A, B, C, or D timer buttons. Ten seconds before time is up, transfer cuvettes to PIP row and press PIP button to activate the pipettor. When time is up, add coagulation activation reagent to each cuvette and record coagulation time.

Nanoparticles on Coagulatory Changes

Nanomaterials could cause blood coagulation, as modification in surface chemistry has been shown to improve immunological compatibility at the particle–blood interface. Application of poly(vinyl chloride) resin particles resulted in 19% decrease in platelet count, indicating platelet adhesion/aggregation and increased blood coagulation time. The same particle coated with poly(ethylene glycol) did not affect platelet count and also elements of coagulation cascade. Similarly, folate-coated Gd nanoparticles did not aggregate platelets or activate neutrophils (26). Hence, blood coagulation studies with nanoparticles would include studies on the platelet aggregation assay and four coagulation assays measuring PT, APTT, TT, and RT.

For interactions with plasma proteins, high-resolution 2D-PAGE is the method of choice to investigate plasma protein adsorption by the particles. The method has been efficiently used to study plasma protein adsorbed on the surface of stealth poly(cyano acrylate) particles (27), liposome (28,29), solid lipid nanoparticles (30), and iron oxide nanoparticles (31). Proteins commonly identified include antithrombine, C3 component of the compliment, α_2-macroglobulin, heptaglobin, plasminogen, immunoglobulins, albumin, fibrinogen, and apolipoprotein, of which, albumin, immunoglobulins, and fibrinogen are the most abundant.

To investigate the impact of airway exposure to nanoparticles on the coagulatory system, Inoue et al. (32) analyzed coagulatory parameters 24 hours after the intratracheal challenge in mouse. The investigation included intratracheal administration of 14- and 56-nm nanoparticles with and without bacterial endotoxin [lipopolysaccharide (LPS)] in mice. Twenty-four hours postadministration, blood was retrieved from each mouse by cardiac puncture, collected into 3.8% sodium citrate in a ratio of 10:1, and centrifuged at 2500 g for 10 minutes. PT, APTT, fibrinogen, activated protein C (APC), and activity for von Willebrand factor (vWF) ($n = 14$–16 in each group) were measured. Result indicated no significant change in PT among the experimental groups. LPS challenge with or without nanoparticles caused prolongation of APTT compared with vehicle challenge ($p < 0.05$). The fibrinogen level was significantly elevated after LPS challenge ($p < 0.01$ vs. vehicle). LPS significantly decreased APC compared with vehicle ($p < 0.05$). Compared with the vehicle group, LPS showed a significant increase in the level of vWF ($p < 0.05$). General increase in all the parameters was observed in the LPS + nanoparticle group than among the LPS group.

COMPLEMENT ACTIVATION

Complement system represents an innate arm of immune defense and is named so because it complements the antibody-mediated immune response. Three major pathways leading to complement activation have been described (Fig. 2). They are classical pathway, alternative pathway, and lectin pathway. The classical pathway is activated by immune (antigen–antibody) complexes. Activation of the alternative pathway is antibody independent. The lectin pathway is initiated by plasma protein mannose-binding lectin. A complement is a system composed of several components (C1, C2, C9) and factors (B, D, H, I, and P). Activation of either one of the three pathways results in the cleavage of C3 component of the complement.

A protocol for the qualitative determination of total complement activation by Western blot is described by NCL (NCL Method ITA-5, Version 1.0) (33). Here, human plasma is exposed to a test material and subsequently analyzed by PAGE, followed by Western blot with anti-C3-specific antibodies. These antibodies recognize both native C3 component of the complement and its cleaved products. Native C3 and no, or minor, amounts of C3 cleavage products are visualized by Western blot in control human plasma. When a test compound or positive control (cobra venom factor) induces the activation of complement, the majority of C3 component is cleaved and the appearance of C3 cleavage products is documented. This "yes" or "no" protocol is designed for rapid and inexpensive assessment of complement activation. Test nanoparticles found to be positive in this assay will be a subject for more detailed investigation aimed at delineation of specific complement activation pathway.

The ability of injection of nanoparticles for complement activation in experimental animals is the basis of development of nanoparticles as vaccines. Antigen-bearing nanoparticle vaccines were investigated for two novel features: lymph node–targeting and in situ complement activation. Following intradermal injection, interstitial flow transported these ultrasmall nanoparticles (25 nm) highly efficiently into lymphatic capillaries and their draining lymph nodes, targeting half of the

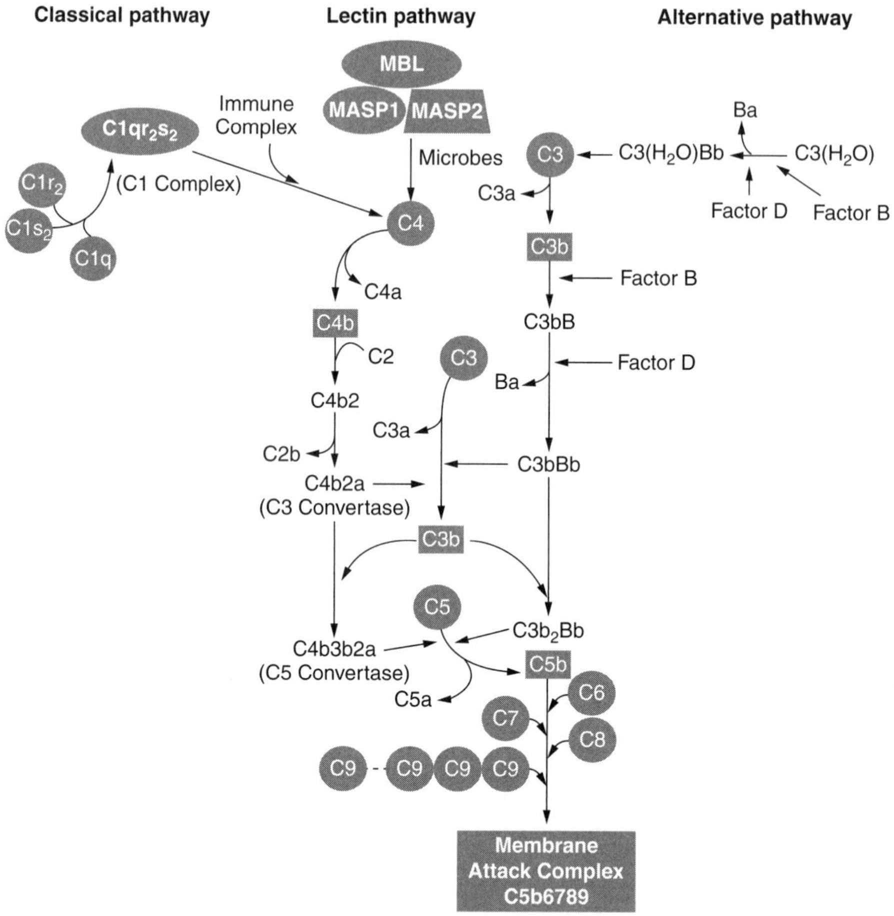

FIGURE 2 Complement activation pathways. *Source*: From Ref. 33.

dendritic cells (DCs) there. Furthermore, surface chemistry of these nanoparticles activated the complement cascade, which spontaneously generated a danger signal in situ and potently activated DCs. With the model antigen ovalbumin (OVA) conjugated to the nanoparticles, Reddy et al. (34) demonstrated humoral and cellular immunity in mice in a highly size- and complement-dependent manner.

Pluronic-stabilized polypropylene sulfide nanoparticles with 25- and 100-nm diameters and fluorescently labeled nanoparticles were synthesized, and a C3a sandwich ELISA was performed to measure complement activation in human serum following the incubation with polyhydroxylated or polymethoxylated nanoparticles. A direct ELISA against OVA was also performed to detect the presence of anti-OVA IgG in mouse serum. Results showed the accumulation of ultrasmall nanoparticles in lymph nodes after subcutaneous injection, while slightly larger ones do not. Polyhydroxylated nanoparticle surfaces activate complement to much higher levels than do polymethoxylated nanoparticles.

Complement activation by core–shell poly(isobutyl cyanoacrylate)–polysaccharide nanoparticles coated with different polysaccharides was investigated by Bertholon et al. (35) by evaluating the conversion of C3 into C3b in serum incubated with nanoparticles. The results showed the cleavage of C3 increased with the size of dextran bound in a "loops" configuration whereas it decreased when dextran was bound in a "brush" configuration. It was explained by an increasing steric repulsive effect of the brush, inducing poor accessibility to OH groups. It was concluded that complement activation was highly sensitive to surface features of the nanoparticles. Type of polysaccharide, configuration on the surface, and accessibility to reactive functions along chains are critical parameters for complement activation.

CELL-BASED ASSAYS: QUANTITATIVE ANALYSIS OF CFU-GM UNITS [NCL METHOD ITA-3, VERSION 1.0 (36)]

The assay employs murine bone marrow (BM). Hematopoietic stem cells of BM proliferate and differentiate to form discrete cell clusters, or colonies. The BM cells are isolated from 8- to 12-week-old mice and cultured in methylcellulose-based medium supplemented with cytokines (mSCF, mIL-3, and hIL-6), either untreated (baseline) or treated with nanoparticles (test). These cytokines promote the formation of granulocyte and macrophage (CFU-GM) colonies (Fig. 3). After 12 days of incubation at 37°C in the presence of 5% CO_2 and 95% humidity, the number of colonies is quantified in baseline and test samples. The percentage of CFU inhibition is then calculated for each test sample.

Experimental Procedure

The experimental protocol described in the technical manual # 28405 (37) developed by StemCell Technologies Inc. is summarized in the following text. Dilute BM cells isolated with Iscove's medium supplemented with 2% FBS to 4×10^5 cells/mL. Add 150 μL of cell suspension and 150 μL of Iscove's medium with 2% FBS (baseline), PBS (negative control), cisplatin (positive control), or nanoparticles (test sample) to 3 mL of MethoCult medium. Vortex tubes to ensure all cells and medium components are mixed thoroughly. Let tubes stand for 5 minutes to allow bubbles to dissipate. Attach a 16-gauge blunt-ended needle to a 3-mL syringe; place the needle below the surface of the solution and draw up approximately 1 mL. Gently

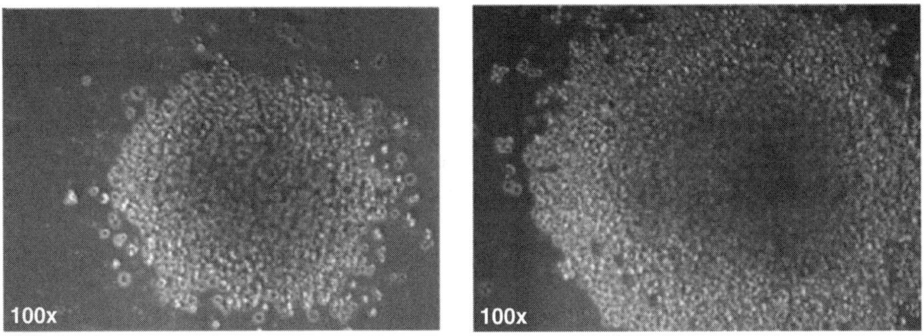

(A) (B)

FIGURE 3 Colony-forming unit–granulocyte macrophage colony. *Source*: From Ref. 36.

depress the plunger and expel medium completely. Repeat until no air space is visible. Draw up MethoCult medium with cells into the syringe and dispense 1.1 mL per 35-mm dish. Distribute the medium evenly by gently tilting and rotating each dish. Place cultures in an incubator maintained at 37°C, 5% CO_2, and 95% humidity. Incubate for 12 days. On day 12, remove dishes from the incubator to identify and count colonies. A representative slide showing the formation of granulocyte and macrophage (CFU-GM) colonies is given in Figure 3. The normal value of CFU-GMs for C57BL6 mice at 8 to 12 weeks of age is 64 ± 16.

Chitosan–poly(aspartic acid)–5-fluorouracil nanoparticles synthesized by ionic gelatification were assayed for CFU-GMs in male BABL/c nude mice induced with human gastric carcinoma. Results showed a significant decrease in the number of CFU-GM formation (38).

Effect of sanazole, and sanazole combining with γ-ray radiation on the colony formation ratio of granuloid/macrophage-committed progenitor cell (CFU-GM), was studied with mice BM in vitro (39). CFU-GMs were separated from mice marrow and incubated at 37°C, 5% CO_2; then, at exponential growth phase, they were exposed to sanazole at a series of dosages and 60Co-γ radiation dose individually and in combination for evaluating their colony formation ratio. After exposure, CFU-GMs were incubated for 7 days and the ratio of colony formation was counted. Results showed that sanazole possesses considerable cytotoxicity to CFU-GM colony formation in vitro and the toxicity enhanced with increasing sanazole concentration. Radiation also showed an inhibitory action on CFU-GMs; the saturation dose and D_0 were 2 and 0.72 Gy, respectively. Their combined inhibitory action on CFU-GMs was stronger than that of each other alone. But the SER of sanazole in each group indicated that there was no cooperation between sanazole and irradiation.

LEUKOCYTE PROLIFERATION ASSAY (NCL METHOD ITA-6, VERSION 1.2) (40)

This assay is adopted to assess the effect of nanoparticle formulation on the basic immunological function of human lymphocytes (i.e., measurement of lymphocytes proliferative responses). Lymphocytes are isolated from pooled human blood anticoagulated with Li-heparin with Ficoll-Paque™ PLUS solution. The isolated cells are incubated with or without phytohemaglutinin (PHAM) in the presence or absence of nanoparticles. The assay, therefore, allows for the measurement of nanoparticles' ability to induce proliferative response of human lymphocytes or to suppress that induced by PHAM.

Procedure

Isolated cells were adjusted for their concentration to 1×10^6 cells/mL with complete RPMI medium. Dispense 100 μL of controls and test samples per well on a round-bottomed, 96-well plate. Prepare duplicate wells for each sample. One hundred microliters of cell suspension per well was then dispensed and shaken gently to allow all components to mix. There is no limit on the number of donors used in this test. It is advised to test each nanoparticle formulation with cells derived from at least three donors. Incubate for 3 days in a humidified 37°C, 5% CO_2 incubator, and then centrifuge for 5 minutes at 700 *g*. Aspirate medium, leaving cells and approximately 50 μL of medium behind and add 150 μL of fresh medium to each well. Add 50 μL of 3-(4,5-dimethyl-)2-thiazolyl-2,5-diphenyl-2*H*-tetrazolium bromide (MTT) to all wells. Cover in an aluminum foil and incubate in a humidified 37°C, 5% CO_2 incubator for 4 hours. Remove the plate from the incubator and spin at 700 *g* for 5 minutes. Aspirate media and MTT. Add 200 μL of DMSO and 25 μL of glycine buffer to all wells and then read at 570 nm on a plate reader

Percentage coefficient of variation (CV) should be calculated for each control or test according to the following formula:

$$\%\text{CV} = (\text{SD}/\text{Mean}) \times 100.$$

Percentage viability is calculated as follows:

$$\%\ \text{Cell proliferation} = (\text{Mean OD}_{\text{sample}} - \text{Mean OD}_{\text{negative control}}) \times 100$$

$$\%\ \text{Proliferation inhibition} = (\text{Mean OD}_{\text{positive control}} - \text{Mean OD}_{\text{positive control}} + \text{Nanoparticles}) \times 100/\text{Mean OD}_{\text{positive control}}$$

MACROPHAGE/NEUTROPHIL FUNCTION

Alveolar macrophages are key cells in both dealing with particles deposited in the lungs and determining the subsequent response to that particle exposure. Nanoparticles are considered a potential threat to the lungs, and the mechanism of pulmonary response to nanoparticles is currently under intense scrutiny. Type II alveolar epithelial cells have previously been shown to release chemoattractants, which can recruit alveolar macrophages to sites of particle deposition.

Chemotaxis Assay (NCL Method ITA-8, Version 1.0) (41)

This method provides a rapid, quantitative measure of the chemoattractant capacity of a nanoparticulate material. Leukocyte recruitment is a central component of the inflammatory process, both in physiological host defense and in a range of prevalent disorders with an inflammatory component. In response to a complex network of proinflammatory signaling molecules (including cytokines, chemokines, and prostaglandins), circulating leukocytes migrate from the bloodstream to the site of inflammation. This assay represents an in vitro model, in which promyelocytic leukemia cells HL-60 are separated from control chemoattractants or test nanoparticles by a 3-μm filter; the cell migration through the filter is then monitored and a number of migrated cells are quantified with fluorescent dye calcein AM. The assay requires 1.5 mL of a test nanomaterial. HL-60 cells are prepared before the experiment by expanding cells in T75 flasks until they are about 80% to 90% confluent (approximately 3–5 days before the experiment). One day before the experiment, count cells with trypan blue. If the cells viability is 95% to 100%, pellet cells for 8 minutes at 120 *g* in a 15-mL tube, resuspend the cells in starving medium, and incubate overnight at 37°C in a humidified incubator (95% air, 5% CO_2). On the day of the experiment, count cells again and adjust concentration to 1×10^6 viable cells/mL in the starving medium. The cell viability should be at least 90%.

Experimental Procedure

Insert a fresh filter plate into a feeding tray and set it aside. Add 150 μL of positive control, negative control, and test nanomaterial in the starving medium into a fresh feeding tray. Add 50 μL of the cells suspension per well of Multi-Screen filter plate (50,000 cells/well). Avoid generating bubbles while adding cells to the wells. Gently assemble the Multi-Screen filter plate and the feeding tray containing controls and test particles. Cover the plate and incubate for 4 hours at 37°C in a humidified incubator (5% CO_2, 95% air). During incubation, prewarm PBS to 37°C and equilibrate calcein AM to room temperature. Prepare working solution of calcein AM by adding 10 μL of stock calcein AM (1 mg/mL) to 2.503 mL of 1× PBS to yield 4 μM/mL solution. After 4 hours of incubation, remove chemotaxis assay plates from the incubator and gently remove the Multi-Screen filter plate and discard it. Add 50 μL of 1× PBS and 50 μL of calcein AM working solution to appropriate wells and 150 μL of 1× PBS plus 50 μL of calcein AM working solution to reagent background control wells on the feeding tray as outlined in the example template as follows. Incubate this calcein plate for 1 hour at 37°C, transfer 180 μL of solutions from the calcein plate to corresponding wells on a Nunc optical bottom plate, and read the Nunc plate on the fluorescent plate reader at 485/535 nm.

Percentage CV should be calculated for each control or test according to the following formula:

$$\% \text{ CV} = (\text{SD}/\text{Mean}) \times 100.$$

$$\text{Background chemotaxis} = \text{Mean FU}_{\text{SM/CAM wells}} - \text{Mean FU}_{\text{SM/PBS wells}} - \text{Mean FU}_{\text{reagent background control wells}}.$$

$$\text{Sample chemotaxis} = \text{Mean FU}_{\text{TS/CAM wells}} - \text{Mean FU}_{\text{TS/PBS wells}} - \text{Mean FU}_{\text{reagent background control wells}}.$$

Barlow et al. (42) assessed the responses of a type II epithelial cell line (L-2) to both fine and nanoparticle exposures in terms of the secretion of chemotactic substances capable of inducing macrophage migration. The adherent murine monocytic macrophage cell line J774.2 was grown in 25-cm^2 tissue culture flasks in RPMI-1640 medium supplemented with 1% L-glutamine, 1% penicillin/streptomycin, and 10% heat-inactivated FBS. Culture flasks were stored in a humidified incubator at 37°C and 5% CO_2. Cell counts and viability were assessed with an improved Neubauer hemocytometer and trypan blue exclusion. All cells that were found to be nonadherent in the culture flasks were discarded by washing prior to use. Fine carbon black, fine titanium dioxide, and nanoparticle carbon black and titanium dioxide were used in the study. Serial dilutions (62.5–2000 μg/mL by mass dose) of each type of particles were prepared in serum-free RPMI and sonicated in a water bath sonicator for 5 minutes before use.

Cells were seeded at 40,000 cells/well in a 96-well plate and incubated for 24 hours in RPMI-1640 supplemented with 10% FBS. After 24 hours, the medium was removed from the cells and replaced with appropriate particle concentrations and incubated for a further 24 hours. Following particle treatments, the medium was removed from the cells and centrifuged for 30 minutes at 15,000 *g* to remove the particles.

Macrophage Chemotaxis Assay

A reusable 96-well Neuroprobe chemotaxis chamber was utilized in these studies. Each sample (30 μL) was loaded, in triplicate, into the bottom wells of the chamber. A Neuroprobe polycarbonate filter (pore size = 5 μM) was inserted between the layers. J774.2 macrophages (2×10^5) in 200 μL of serum-free RPMI-1640 were added to the top of each well. The chamber was incubated at 37°C in 5% CO_2 for 6 hours and the filter was removed and washed 3× with PBS on the upper side to remove nonmigrated macrophages. The filter was stained with a Romanowsky (Diff-Quick) stain. The optical density of each well on the filter was read at 540 nm in a Dynex multiwell plate reader. Increasing absorbance correlates with the increasing number of macrophages moving through the filter. Both fine and nanoparticle carbon black treatment of L-2 cells did induce significant increases in macrophage migration when compared with another negative control, medium incubated with the particles alone.

Phagocytosis Assay (NCL Method ITA-9, Version 1.0) (43)

The method to evaluate nanoparticle internalization by phagocytic cells needs about 600 μL (concentration = 2 mg/mL) of a test nanomaterial. This method, however, may not be applicable for certain types of nanomaterials. For example, nanoparticles with fluorescent capabilities such as quantum dots (generally studies using confocal microscopy or flow cytometry). Modification(s) of this procedure or change in detection dye may be required for particles that demonstrate interference with luminol-dependent chemiluminescence.

Experimental Procedure

Place empty 96-well white test plates inside the reader chamber of the plate reader and warm it at 37°C. Adjust cell concentration to 1×10^7/mL by spinning cell suspension down and reconstituting in complete RPMI-1640 medium. Keep at room temperature. Add 100 μL of controls and test nanoparticles in PBS to appropriate wells. Prepare three duplicate wells for each sample and two duplicate wells for positive and negative controls. Add 100 μL of working luminol solutions in PBS to each well containing the sample. Do not forget to add luminol to two "luminol-only" control wells; keep the plate warm during sample aliquoting. Plate 100 μL of cell suspension per well on a 96-well white plate. Start kinetic reading on a luminescence plate reader immediately. Percentage CV is used to control precision and calculated for each control or test sample according to the following formula:

$$\%CV = (SD/Mean) \times 100.$$

$$\text{Fold phagocytosis induction (FPI)} = \text{Mean RLU}_{\text{sample}}/\text{Mean RLU}_{\text{negative control}}.$$

FPI of the positive control observed during assay qualification is 400 or less.

Cytokine Induction Assay (NCL Method ITA-10, Version 1.2) (44)

The method to evaluate the effect of nanoparticle formulation on cytokine production by peripheral blood mononuclear cells is described here. Lymphocytes are isolated from human blood and anticoagulated with Li-heparin, using Ficoll-Paque PLUS solution. The cells are then incubated with or without LPS in the presence or absence of nanoparticles for 24 hours. After this incubation step, cell culture supernatants are collected and analyzed by cytometry beads array for the presence of interleukin (IL)-1β, tumor necrosis factor α, IL-12, IL-10, IL-8, and IL-6. The assay, therefore, allows for the measurement of nanoparticles' ability to either induce cytokines or suppress cytokines induced by LPS.

This assay requires 1800 μL of nanoparticles dissolved/resuspended in complete culture medium; for example, three 100 μL of replicates per sample were analyzed in duplicate, 600 μL per set with cells derived from one donor. For the original screen, we recommend to use as high concentration of nanoparticles in the sample as possible. The following issues have to be considered when selecting the concentration:

A. Solubility of nanoparticles in a biocompatible buffer
B. pH within physiological range
C. Availability of nonmaterial
D. Stability

Human lymphocytes are isolated by a standard procedure. In brief, place freshly drawn blood into 15- or 50-mL conical centrifuge tube, add equal volume of PBS at room temperature, and mix well. Slowly layer 3 mL of the Ficoll-Paque PLUS solution and then layer on 4 mL of blood/PBS mixture. It is essential to hold the tube at 45° to maintain Ficoll–blood interface. Centrifuge for 30 minutes at 900 *g*, 18°C to 20°C, without brake. Using sterile pipette, remove upper layer containing plasma and platelets and discard it. Using a fresh sterile pipette, transfer the mononuclear cell layer into another centrifuge tube and wash cells by adding excess of HBSS (approximately 3× the volume of mononuclear layer) and centrifuging for 10 minutes at 400 *g*, 18°C to 20°C. Discard supernatant and repeat wash step one more time. Resuspend cells in complete RPMI-1640 medium. Count an aliquot of cells and determine viability by trypan blue exclusion.

Experimental Procedure

Dispense 100 μL of blank medium (baseline), negative control, positive control, and test samples per well on a 96-well plate. Prepare duplicate wells for each sample. Prepare LPS + nanoparticles wells by combining 50 μL of 2 μg/mL of LPS and 50 μL of 2× concentrated nanoparticles. Add 100 μL of cell suspension (2×10^6 cells/mL) per well. Gently shake the plate to allow all components to mix. Repeat these steps for cells obtained from each individual donor. It is advised to test each nanoparticle formulation with cells derived from at least three donors. Incubate 24 hours in a humidified 37°C, 5% CO_2 incubator. Collect supernatants into 0.5-mL centrifuge tubes and spin in a microcentrifuge at a maximum speed for 5 minutes. Transfer supernatants into fresh tubes and either analyze fresh or store at −80°C for future analysis.

On the day of analysis, thaw supernatants at room temperature, and then place them on ice. Dilute culture supernatants (1:5) with assay buffer provided with the human inflammation kit. Follow BD Biosciences kit instructions to prepare an assay standard curve, cytokine detection beads, and cytometer calibration beads. In a Falcon 5-mL tube, combine 50 μL of cytokine detection beads with 50 μL of PEDetection reagent and one of the following: 50 μL of assay buffer (reagent blank), 50 μL of calibration standard, or 50 μL of culture supernatant. Cover tubes and incubate in the dark for 3 hours. During this incubation step, perform calibration of flow cytometer. At the end of incubation, add 1 mL of wash buffer provided with the kit to each tube, and centrifuge for 5 minutes at 200 *g*, then collect and discard supernatants. Add 300 μL of wash buffer provided with the kit to each tube, vortex, and analyze on flow cytometer. Data obtained from flow cytometer are analyzed with CBA software (BD Biosciences). The software calculates mean fluorescent intensity (MFI) for each sample, builds the standard curve with a 4-parameter regression model, and calculates concentration of each cytokine based on sample MFI response extrapolated from the corresponding standard curve.

Oxidative Burst (Nitric Oxide Production) (NCL Method ITA-7, Version 1.0) (45)

This method describes the quantitative determination of nitrite (NO_2^-) concentration, a stable oxidative end product of the antimicrobial effector molecule nitric oxide in the cell culture medium. The protocol is also used to evaluate capability of nanomaterials to induce nitric oxide production by macrophages. Nitric oxide secreted by macrophages has a half-life of a few seconds, as it interacts with a number of different molecular targets, resulting in cytotoxicity. In the presence of oxygen and water, nitric oxide generates other reactive nitrogen oxide intermediates and ultimately decomposes to form NO_2^- and nitrate (NO_3^-). The measurement of NO_2^- in the tissue culture medium with the Griess reagent provides a surrogate marker and quantitative indicator of nitric oxide production. The murine macrophage cell line RAW 264.7 is used as a model in this assay. The upper limit of quantification is 250 μM and the lower limit of quantification is 1.95 μM.

Experimental Procedure

Plate 1000 μL of cell suspension (1×10^5/mL in complete medium) per well on 24-well plates. Prepare triplicate wells for each sample and duplicate wells for each control. Always leave one cell-free well per nanoparticle per plate. These wells will be used to assay potential nanoparticle interference with assay. Incubate this culture plate for 24 hours in a humidified 37°C, 5% CO_2 incubator. Remove the culture medium and add 500 μL of study samples, controls, or medium blank to appropriate wells. Incubate the culture plate again for 48 ± 1 hour in a humidified 37°C, 5% CO_2 incubator. To a fresh 96-well plate, add 50 μL/well of reagent blank (culture medium used to prepare calibration standards and quality controls), calibration standards, quality controls, and medium from each well of the culture plate. Load duplicate wells for each sample and control. This is a NO test plate.

In a separate tube, combine equal volumes of Griess reagent A (1% solution of sulfanilamide in 2.5% H_3PO_4) and Griess reagent B (1% solution of naphthylethylenediamine dihydrochloride in 2.5% H_3PO_4). Add 100 μL of each Griess reagent per well of NO test plate. Place the plate on a shaker for 2 to 3 minutes to allow all ingredients to mix, and measure absorbance at 550 nm. Percentage CV is used to control precision and calculated for each control or test sample according to the following formula: [%CV = (SD/Mean) × 100]. Percentage difference from theoretical (PDFT) is used to control accuracy of the assay calibration standards and quality controls, and it is calculated according to the following formula:

$$\text{PDFT} = [(\text{Calculated NaNO}_2 \text{ concentration} - \text{Theoretical NaNO}_2 \text{ concentration})]/(\text{Theoretical NaNO}_2 \text{ concentration})] \times 100\%.$$

CYTOTOXIC ACTIVITY OF NK CELLS

NK cells are a type of cytotoxic lymphocytes that constitute a major component of the innate immune system. These cells play a major role in the rejection of tumors and cells infected by viruses. The cells kill by releasing small cytoplasmic granules of proteins, called perforin and granzyme, which cause the target cell to die by apoptosis or necrosis. NK cells, morphologically classified as large granular lymphocytes, are important effector lymphocytes of innate immunity. Functionally, they exhibit cytolytic activity against a variety of allogeneic targets in a nonspecific, contact-dependent, nonphagocytotic process that does not require prior sensitization to an antigen. These cells also have a regulatory role in the immune system through the release of cytokines, which, in turn, stimulate other immune functions.

NK cells can be distinguished from T lymphocytes by the expression of distinct phenotypic markers such as $CD16^+$, $CD56^+$ (human NK cells only), and lack of rearranged T-cell receptor gene products. However, in the mouse, expression of DX5/CD49b and NK1.1 (only in $NK1.1^+$ mouse strains) is considered a best phenotypic marker for NK cells. Recent development of specific antibodies to the human and mouse NKG2D suggest that all NK cells also express this marker.

Strong cytolytic activity and the potential for autoreactivity of NK cells are tightly regulated. NK cells must receive an activating signal, which can come in a variety of forms, the most important of which are listed in the following sections.

Cytokines

The cytokines play a crucial role in NK-cell activation. As these are stress molecules, released by cells upon viral infection, they serve to signal the presence of viral pathogens to the NK cells.

Fc Receptor

NK cells, along with macrophages and several other cell types, express the Fc receptor (FcR) molecule, an activating biochemical receptor that binds the Fc portion of antibodies. This allows NK cells both to target cells against which a humoral response has been mobilized and to lyse cells through antibody-dependent cellular cytotoxicity.

Activating and Inhibitory Receptors

Aside from the FcR, NK cells express a variety of receptors that serve to either activate or suppress their cytolytic activity. These receptors bind to various ligands on target cells, both endogenous and exogenous, and have an important role in regulating the NK-cell response.

The measurement of NK cells toxicity against tumor or virus-infected cells, especially in cases with small blood samples, requires highly sensitive methods. Ogbomo et al. (46) reported a coupled luminescent method (CLM) based on glyceraldehyde-3-phosphate dehydrogenase release from injured target cells to evaluate the cytotoxicity of IL-2-activated NK cells against neuroblastoma cell lines. In contrast to most other methods, CLM does not require the pretreatment of target cells with labeling substances that could be toxic or radioactive. The effective killing of tumor cells is achieved by low effector/target ratios ranging from 0.5:1 to 4:1. CLM provides a highly sensitive, safe, and fast procedure for measurement of NK-cell activity with small blood samples such as those obtained from pediatric patients (46).

The effect of change of native immune-adhering function (ENIAF) in self-plasma of patients with hematologic and lymphoid neoplasms, as well as its effect on the killing activity of NK cells, was studied by Zhang et al. (47). The whole blood was anticoagulated with citric acid. Five microliters of precipitated RBCs and 500 μL of plasma of patients or controls were directly mixed with 750 μL of quantitative K562 cells at 37°C for 30 minutes. One K562 cell attached by one or more erythrocytes was counted as one rosette; the ratio of rosettes was calculated. Using K562 cells as target cells, the killing activity of NK cells isolated from normal persons was detected by MTT assay; the change of the killing activity was observed after adding RBCs. The results indicated that the ratio of rosettes formed by RBCs of 21 normal controls and K562 cells was 15.3% ± 6.4% and the ratio of rosettes formed by RBCs of 24 patients and K562 cells was 7.6% ± 7.0%. The ability of ENIAF in patients with hematologic and lymphoid neoplasms was significantly lower than that in healthy individuals ($t = 3.61$, $p < 0.001$). The killing rate of NK cells in peripheral blood of normal individuals ranged from 67% to 71% without adding RBCs, and it increased by 14.7% ± 5.2% after adding RBCs of normal controls but decreased by 4.3% ± 7.6% with RBCs of patients. The study concluded that the ENIAF of RBCs in patients with hematopoietic and lymphoid neoplasms decreases, accompanying with the reduction of the killing activity of NK cells to K562 cells, so to detect change of ENIAF may be helpful for the assessment of the immunological function of patients with hematopoietic and lymphoid neoplasms.

IN VITRO PHARMACOLOGICAL AND TOXICOLOGICAL ASSESSMENTS

Nanotechnology involves the creation and manipulation of materials at nanoscale levels to create products that exhibit novel properties. Recently, nanomaterials such as nanotubes, nanowires, fullerene derivatives (buckyballs), and quantum dots have received enormous attention to create new types of analytical tools for biotechnology and life sciences (48–50). Although nanomaterials are currently being widely used in modern technology, there is a serious lack of information concerning the human health and environmental implications of manufactured nanomaterials (51,52). The major toxicological concern is the fact that some of the manufactured nanomaterials are redox active (53) and some particles transport across cell membranes and especially into mitochondria (54). One of the few relevant studies was with single-walled carbon nanotubes in mice (55), which demonstrated that carbon nanotube products induced dose-dependent epithelioid granulomas in mice and, in some cases, interstitial inflammation in the animals of the 7-day postexposure groups. The recent study by Oberdöster indicated that nanomaterials (fullerenes, C60) induced oxidative stress in a fish model (56).

A limited number of in vitro studies have also been performed to assess the toxicities of the nanoparticles with different cellular systems and test methods (57–59). However, published toxicity data are still considered inadequate to earn a full understanding of the potential toxicity of these nanoparticles. In vitro pharmacological and toxicological studies have been conducted on a variety of cells, including perfused organs, tissue slices, and cell culture based on a single cell line or a combination of cell lines. The cell cultures are prepared with primary cells freshly derived from organ or tissue sources. In vitro models generally allow the examination of biochemical mechanisms under controlled conditions, including specific toxicological pathways that may occur in target organs and tissues. Mechanistic endpoints used for the in vitro assessment provide information on the potential mechanisms of cell death and may also identify compounds that may cause chronic toxicities that often result from sublethal mechanisms that may not cause overt toxicity in cytotoxicity assays.

Nanoparticle toxicity includes common mechanistic paradigms such as oxidative stress, apoptosis, and mitochondrial dysfunction. Using an appropriate model, chemotherapeutic efficacy can be examined in vitro and, in certain cases, targeting of chemotherapeutic agent may be demonstrated, using optimized treatment/washout schemes in cell lines expressing the targeted receptor. Although nanoparticle metabolism or enzyme induction is yet to be demonstrated, certain nanoparticles with appropriate chemistries are believed to be subjected to phase I/II metabolism, as demonstrated by induction studies using cell-based microsomal and/or recombinant enzyme systems.

SPECIAL CONSIDERATIONS IN IN VITRO EXPERIMENTS WITH NANOPARTICLES

Many of the standard methods used to evaluate biocompatibility of new molecular and chemical entities are fully applicable to nanoparticles. However, existing test protocols may require further development and laboratory validation before they become available for routine testing. Such special considerations are as follows:

1. Nanoparticles could interfere with assay, spectral measurements, and inhibition/enhancement of enzyme reactions and absorbance of reagents to nanoparticle surfaces (59–61).
2. For results of in vitro assays with nanoparticles, it is important to recognize that dose–response relationship will not always follow a classical linear pattern. These atypical dose–response relationships have previously been attributed to shift between the different mechanisms underlying the measured response.
3. The results of nanoparticle assessment are also subjected to the impact of dose metric, sample preparation, and experimental conditions. For example, surface area or particle number may be a more appropriate metric than mass when comparing data generated for different sized particles. This has been shown to be the case for 20- and 250-nm titanium dioxide nanoparticles, in which lung inflammation in rats, as assessed by the percentage of neutrophils in lung lavage fluid, correlated with total surface area compared with mass (62).
4. Experimental conditions in study design could alter the extent of response and are required to be controlled. Investigation of functionalized fullerenes in human T lymphocytes in vitro showed enhanced response by photoexcitation (63).

Cytotoxicity

Cell viability of adherent cell lines can be assessed by a variety of methods (64). These methods fall under four major categories:

1. Loss of membrane integrity
2. Loss of metabolic activity
3. Loss of monolayer adherence
4. Cell cycle analysis

These viability assays can be of much importance to identify cell line susceptibility, nanoparticle toxicity, and potentially give clue as to the type (cytostatic/cytotoxic) and location of cellular injury.

Membrane Integrity Assays

Membrane integrity assays (MIAs) are important in estimating the measure of cellular damage. Some cationic particles, such as amine-terminated dendrimers, exhibited toxic effects by disrupting the cell membrane when tested by MIA (65). Membrane integrity measurement includes the trypan blue exclusion assay and the lactate dehydrogenase (LDH) leakage assay (66,67). The LDH leakage assay is generally selected because of its sensitivity and suitability for high throughput screening in 96-well plate formats.

Cell Viability Assay Using MTT

Assays that measure metabolic activity include tetrazolium dye reduction, ATP, and ^{3}H-thymidine incorporation assay. The MTT reduction assay is generally selected to measure the metabolic activity, as it does not use radioactivity and historically has been proven sensitive and reliable. In this assay, MTT – a yellow, water-soluble tetrazolium dye – is metabolized by the live cells to purple, water-soluble formazan crystals. Formazans can be dissolved in DMSO and quantified by measuring the absorbance of the solution at 550 nm. Comparison between spectra of samples of untreated and nanoparticle-treated cells can provide a relative estimate of cytotoxicity (68).

The new generation of tetrazolium dye that form water-soluble formazans (e.g., XTT) is now used to avoid solubilization step that is required in a traditional MTT assay. However, an intermediate electron acceptor is required to stabilize these unstable analogues to overcome assay variability. Furthermore, the net negative charge of these newer analogues limits cellular uptake, resulting in extracellular reduction (69). MTT, with a net positive charge, readily crosses cell membrane and is reduced intracellularly, primarily in the mitochondria.

The traditional MTT assay are reported to be a better choice to assess cell viability in nanoparticle toxicity experiments, as nanoparticles have been shown to interact with cell membrane and could potentially interfere with the reduction of the newer generation analogues via transplasma membrane electron transport. Analytes that are antioxidants or/are substrate inhibitors of drug efflux pumps have also been shown to interfere with the MTT assay (70,71).

The evaluation of cytotoxicity by the MTT assay in RAW cells (mouse macrophase cell lines) was reported by Bhattarai et al. (72). Briefly, RAW cell suspensions containing 1×10^4 cells/well in DMEM-containing 10% FBS were distributed in a 96-well plate and incubated in a humidified atmosphere containing 5% CO_2 at 37°C for 24 hours (73,74). The cytotoxicity of samples was evaluated in comparison with control cells. Cells were incubated for an additional 24 hours after the addition of defined concentration of the analyte. The mixture was replaced with fresh medium containing 10% FBS. Then, 20 μL of MTT solution (5 mg/mL in 1 × PBS) was added to each well. The plate was incubated for an additional 4 hours at 37°C. Next, MTT-containing medium was aspirated off and 150 μL of DMSO was added to dissolve the crystals formed by living cells. Absorbance was measured at 490 nm, using a microplate reader (ELX 800; BIO-TEK Instruments, Inc., USA). The cell viability (%) was calculated according to the following equation:

$$\text{Cell viability (\%)} = [\text{OD}_{490(\text{sample})}/\text{OD}_{490(\text{control})}] \times 100.$$

Cytotoxicity study of paclitaxel-loaded particles or Taxol® was conducted with human colon adenocarcinoma cell lines, HT-29 cells, and Caco-2 cells by the MTT assay (75). A similar study was conducted for coumarin-loaded particles incorporating (*i*) vitamin E TPGS and (*ii*) poly(vinyl alcohol) by using HT-29 cells to demonstrate the protecting ability of vitamin A against cytotoxicity. Confocal images of the cells taken after incubation demonstrate the protective ability of vitamin A against cytotoxicity.

Hussain et al. (76) used BRL 3A immortal rat liver cell line in their study as an in vitro model to assess nanocellular toxicity. The toxicity endpoints [MTT, LDH, reactive oxygen species (ROS), and GSH] that were selected in this study represent vital biological functions of the mammalian system as well as provide a general sense of toxicity in a relatively short time.

Loss of Monolayer Adherence Test

Loss of monolayer adherence to plating surface is often used as a marker of cytotoxicity. Monolayer adherence is commonly measured by staining for total protein, following the fixation of adherent protein. This simple assay is often a very sensitive indicator of loss of cell viability (55). The sulforhodamine B total protein-staining assay was selected for the determination of monolayer adherence. The assay is especially suitable for high throughput screening, as fixed, stained microplates can be stored for extended period prior to measurement (77).

Cell Cycle Analysis

Cell cycle analysis is conducted with propidium iodide staining of DNA and flow cytometry (78). The method can determine the effect of nanoparticle treatment on cell cycle progression as well as cell death. Nanoparticles, such as carbon nanotubes, have been shown to cause G1 cell cycle arrest in human embryonic kidney (HEK) cells, with a corresponding decrease in the expression of G1-associated cdks and cyclins.

In Vitro Target Organ Toxicity

Toxicity screening for environmental exposure of nanoparticles has been reported (79) involving environmentally relevant exposure routes. However, in addition to in vitro examination of the so-called portal of entry tissues, a need for inclusion of target organs is also warranted. The liver and kidneys are generally selected as ideal candidates for these in vitro target organ toxicity studies since these organs are considered to be involved in accumulation, processing, and eventually clearance of nanoparticles.

The liver is basically responsible for reticuloendothelial capture of nanoparticles, often due to phagocytosis of Kupfer cells for hepatic clearance of parenterally administered nanoparticles such as fullerenes, dendrimers, and quantum dots (80,81). In addition to accumulation, nanoparticles are shown to have detrimental effects on the liver function ex vivo and on hepatic morphology (82).

Sprague-Dawley rat hepatic primary cells and human hepatoma HepG2 are generally used, since long time, for in vitro hepatic target organ toxicity assays due to their abundant availability and high metabolic activity (83). They are also chosen for toxicological studies, since hepatic primary cells in culture are more reflective of in vivo hepatocytes with regard to enzyme expression and specialized functions (84).

Pharmacokinetic studies of parenterally administered carbon nanotubes in rodents have shown the urinary excretion as the principal mechanism of clearance (85–87). A variety of engineered nanoparticles, particularly doxorubicin-loaded cyanoacrylate nanoparticles, showed increased renal distribution and thus increased kidney toxicity (88–90). Kidney injury has been demonstrated in many other nanoparticles such nano-zinc particles in which severe histological alterations are observed in murine kidneys (91,92).

The porcine renal proximal tubule cell lines LLC-PK1 were selected as model kidney cell lines, as these cell lines are adherent, which can simplify sample preparation and can be propagated in a 96-well plate format suitable for high throughput screening (93). These cell lines were used in variety of in vitro assays to evaluate cytotoxicity, mechanistic toxicology and pharmacology, etc.

Oxidative Stress

The generation of free radicals by nanomaterials of ambient or industrial origin is well documented (94,95). However, engineered nanomaterials such as fullerenes and polystyrene nanoparticles have also been shown to generate oxidative stress (56,96,97). The unique surface chemistries, large surface area, and redox active or catalytic contaminants of nanoparticles can facilitate ROS generation (98). For example, photoexcitation is observed with fullerenes due to its ability to perform electron transfer (phase I pathway) or energy transfer (phase II pathway) reactions with molecular oxygen (99), resulting in the formation of superoxide anion radical or singlet oxygen, respectively. The superoxide anion radical then generates additional ROS species by reactions, such as dismutation and Fenton chemistry, resulting in cell injury.

Biomarkers of nanoparticle-induced oxidative stress often measured include ROS, lipid peroxidation products, and GSH/GSSG ratio (100). Measurement of ROS, such as hydrogen peroxide, is conducted by fluorescent dichlorodihydrofluoroscein (DCFH) assay (101). DCFH-DA is an ROS probe that undergoes intracellular deacetylation, followed by ROS-mediated oxidation to a fluorescent species, with excitation wavelength of 485 nm and emission wavelength of 530 nm, respectively. DCFH-DA can also be used to measure ROS generation in the cytoplasm and cellular organelles such as mitochondria. The same method was reported by Wang and Joseph (102), with minor modifications. Cells were incubated with 20 μM of DCFH-DA for 30 minutes in a 96-well plate. After DCFH-DA-containing medium was removed, the cells were washed with PBS and treated with Ag (15 and 100 nm) in exposure media for 6 hours. At the end of exposure, dichlorofluorescein fluorescence was determined at excitation wavelength of 485 nm and emission wavelength of 530 nm, respectively. Data are reported as fold increase in fluorescence intensity relative to control. Control cells cultured in Ag-free media (50 and 100 nm) were run in parallel to the treatment groups.

The thiobarbituric acid reactive substances (TBARS) assay is used for the measurement of lipid peroxidation products, such as lipid hydroperoxides and aldehydes. Malondialdehyde (MDA), a lipid peroxidation product, combines with thiobarbituric acid in a 1:2 ratio to form a fluorescent adduct, which is measured at 521 nm (excitation) and 552 nm (emission), and TBARS levels are expressed as MDA equivalents (103).

The evaluation of glutathione homeostasis is done by the dithionitrobenzene (DTNB) assay. In the DTNB assay, reduced GSH interacts with 5,5′-thiobis (2-nitrobenzoic acid) to form the colored product 2-nitro-5-thiobenzoic acid, which is measured at 415 nm. Oxidized glutathione (GSSG) is then reduced by glutathione reductase to form reduced GSH, which is again measured by the preceding method. Pretreatment with thiol-masking reagent, 1-methyl-4-vinyl-pyridinium trifluoromethane sulfonate, prevents GSH measurement, resulting in the measurement of GSSG alone (104).

Apoptosis and Mitochondrial Dysfunction

Nanoparticle-induced cell death can occur by either necrosis or apoptosis, processes that can be distinguished both morphologically and biochemically. Apoptosis, *morphologically*, is characterized by perinuclear partitioning of condensed chromatin and budding of the cell membrane (105). The ability of nanoparticles, such as dendrimers and carbon nanotubes, to induce apoptosis has been demonstrated by in vitro studies (106,107). In vitro exposure of macrophages, such as mouse RAW 264.7 cells, to cationic dendrimers led to apoptosis confirmed by morphological observation and the evidence of DNA cleavage. Pretreatment of cells with a caspase inhibitor (zVAD-fmk) reduced the apoptotic effect of the cationic dendrimer (108). Apoptosis has also been observed in cultured HEK293 cells and T lymphocytes treated with single-walled carbon nanotubes and MCF-7 breast cancer cells treated with quantum dots (106,109).

Apoptosis in mammalian cells can be initiated by four potential pathways: (*i*) mitochondrial pathway, (*ii*) death receptor–mediated pathway, (*iii*) ER-mediated pathway, and (*iv*) granzyme B–mediated pathway (110). Fluorometric protease assay was reported by Gurtu et al. (111) to estimate caspase-3 activation in liver and kidney cells. This assay quantifies caspase-3 activation in vitro by measuring the cleavage of DEVD-7-amino-4-trifluoromethyl coumarin (AFC) to free AFC, which emits a yellow-green fluorescence (λ_{max} = 505 nm). This simple and rapid assay is generally used as an initial apoptosis screen before conducting cellular morphological studies by nuclear staining techniques to detect perinuclear chromatin, or agarose gel electrophoresis to detect DNA laddering (112).

As discussed earlier, nanoparticles have been shown to induce oxidative stress and ROS generation. ROS-mediated pathway induce mitochondrial permeability transition, which is a plausible apoptosis mechanism for nanoparticles. For instance, ambient ultrafine particulates have been shown to translocate to the mitochondria of the murine macrophage cells RAW 264.7, causing structural damage and altered mitochondrial permeability (95). Polar compounds (e.g., quinone contaminant) fractionated from the ultrafine particulates were demonstrated to induce mitochondrial dysfunction and apoptosis in the RAW 264.7 cells (113). However, this link between oxidative stress, mitochondrial dysfunction, and apoptosis has also been observed in man-made nanoparticles such as quantum dots and metals (76,109), water-soluble fullerenes derivatives (54,114), chitosan nanoparticles (115), and in various in vitro models. Mitochondrial dysfunction can also result from several other mechanisms, including uncoupling of oxidative phosphorylation, damage to mitochondrial DNA, disruption of electron transport chain, and inhibition of fatty acid β-oxidation (116).

Methods used to detect mitochondrial dysfunction include measurement of ATPase activity (via luciferin–luciferase reaction), oxygen consumption (via polarographic technique), morphology (via electron microscopy), and membrane potential (via fluorescent probe analysis) (117). The loss of mitochondrial membrane potential in rat hepatic primaries, HepG2, and LLC-PK1 cell lines are measured by the 5,5′,6,6′-tetrachloro-1,1′,3,3′-tetraethyl benzimidazolcarbocyanine iodide (JC1) assay (118). This assay does not require mitochondrial isolation or use of any specialized equipment, as here the fluorescent dye partitions to the mitochondrial matrix as a result of the membrane potential and so the concentration of JC1 in the matrix results in an aggregation that fluoresces at 590 nm (red). The loss of membrane potential causes the dye to dissipate from the matrix and can be measured in its monomeric state at emission wavelength 527 nm (green). Thus, the degree of mitochondrial membrane depolarization is measured as the proportion of green to red fluorescence.

Mitochondrial membrane potential measurement can be an index of toxicity, which can be determined by the uptake study of rhodamine 123 according to the method of Wu et al. (119). Cells were exposed to different concentrations of Ag (15 and 100 nm) for 24 hours. After 24 hours of exposure, cells were incubated with rhodamine 123 for 30 minutes in a 96-well plate, and then the cells were washed with PBS. The fluorescence was determined at excitation wavelength 485 nm and emission wavelength 530 nm. Control cells cultured in Ag-free media (50 and 100 nm) were run in parallel to the treatment groups. The fluorescence intensity value of control cells (nanoparticle-free medium at 0 hour) was taken as 100% and then calculated as the percentage of reduction of fluorescence in nanoparticle-exposed cells.

CONCLUSION

Although standards of care for many nanoparticle delivery systems have been established, accurate prediction of the effects, both therapeutic and toxic, of a given drug on a given patient is frustrated by disappointing differentials between in vitro predictions and in vivo results. Computational models may provide a much needed bridge between the two, producing highly realistic in vitro models upon which alternate therapies may be conducted. The power of such models over in vitro monolayer and even spheroid assays lies in their ability to integrate processes over a multitude of scales, approximating the complex in vivo interplay of phenomena such as heterogeneous vascular delivery of drugs and nutrients, diffusion through lesion, heterogeneous lesion growth, apoptosis, necrosis, and cellular uptake, efflux, and target binding.

REFERENCES

1. FDA/CDER. Guidelines for industries – Immunotoxicological evaluation of investigational new drugs. Available at: http://www.fda.gov/Cder/guidance/. 2002. Accessed May 2008.
2. FDA/CDER. ICH S8 – Immunotoxicological studies for human pharmaceuticals (draft). Available at: http://w.w.w.fda.gov/Cder/guidance/. 2004. Accessed May 2008.
3. FDA/CBER/CDER. Guidelines for industries – Developing medical imaging drug and biological products; part 1: Conducting safety assessments. Available at: http://w.w.w.fda.gov/Cder/guidance/. 2004. Accessed May 2008.
4. FDA/CDER/CBER. Guidelines for industries – ICH S6: Preclinical safety evaluation of biotechnology derived pharmaceuticals. Available at: http://w.w.w.fda.gov/Cder/guidance/. 1997. Accessed May 2008.
5. FDA/CDRH. Guidelines for industries and FDA reviewers. Immunotoxicological testing guidance. Available at: http://w.w.w.fda.gov/Cdrh/guidance.html. May 1999. Accessed May 2008.
6. ANSI/AAMI/ISO. 10993-4: Biological evaluation of medical devices; part 4: Selection of tests for interaction with blood. 2002.
7. ASTM. Standard Practice F1906-98: Evaluation of immune responses in biocompatibility testing using ELISA tests, lymphocyte proliferation, and cell migration. 2003.
8. ASTM. Standard Practice F 748-98: Selecting generic biological test methods for materials and devices. New York: American National Standards Institute.
9. Plasma Protein Binding Assay. NCL Method ITA-4, Version 1.0, Nanotechnology Characterization Laboratory, National Cancer Institute. Frederick, MD: SAIC-Frederick, 2005.

10. Cedervall T, Lynch I, Lindman S, et al. Understanding the nanoparticle-protein corona using methods to quantify exchange rates and affinities of proteins for nanoparticles. PNAS 2007; 104(7):2050–2055.
11. Assay cascade protocols, Nanotechnology Characterization Laboratory, National Cancer Institute. Frederick, MD: SAIC-Frederick, 2005.
12. ASTM. F756-00: Standard practice for assessment of hemolytic properties of materials of materials, 2000.
13. Analysis of Hemolytic Properties of Nanoparticles. NCL Method ITA-1, Version 1.0, Nanotechnology Characterization Laboratory, National Cancer Institute. Frederick, MD: SAIC-Frederick, 2005.
14. Radomski A, Jurasz P, Alonso-Escolano D, et al. Nanoparticle-induced platelet aggregation and vascular thrombosis. Br J Pharmacol 2005; 146(6):882–893.
15. Analysis of Platelet Aggregation. NCL Method ITA-2, Version 1.0, Nanotechnology Characterization Laboratory, National Cancer Institute Frederick, MD: 21702SAIC-Frederick, 2005.
16. Radomski M, Moncad S. An improved method for washing of human platelets with prostacyclin. Thromb Res 1983; 30:383–389.
17. Radomski A, Stewart MW, Jurasz P, et al. Pharmacological characteristics of solid-phase von Willebrand factor in human platelets. Br J Pharmacol 2001; 134:1013–1020.
18. Radomski A, Jurasz P, Sanders EJ, et al. Identification, regulation and role of tissue inhibitor of metalloproteinase-4 (TIMP-4) in human platelets. Br J Pharmacol 2002; 137:1330–1338.
19. Jurasz P, Alonso D, Castro-Blanco S, et al. Generation and role of angiostatin in human platelets. Blood 2003; 102:3217–3223.
20. Sawicki G, Salas E, Murat J, et al. Release of gelatinase A during platelet activation mediates aggregation. Nature 1997; 386:616–619.
21. Chung AW, Jurasz P, Hollenberg MD, et al. Mechanisms of action of proteinase-activated receptor agonists on human platelets. Br J Pharmacol 2002; 135:1123–1132.
22. Alonso-Escolano D, Strongin AY, Chung AW, et al. Membrane type-1 matrix metalloproteinase stimulates tumour cell-induced platelet aggregation: role of receptor glycoproteins. Br J Pharmacol 2004; 141:241–252.
23. Jurasz P, Stewart MW, Radomski A, et al. Role of von Willebrand factor in tumour cell-induced platelet aggregation: Differential regulation by NO and prostacyclin. Br J Pharmacol 2001; 134:1104–1112.
24. Fernandez-Patron C, Martinez-Cuesta MA, Salas E, et al. Differential regulation of platelet aggregation by matrix metalloproteinases-9 and -2. Thromb Haemost 1999; 82:1730–1735.
25. Analysis of Blood Coagulation. NCL Method ITA-12, Version 1.0, Nanotechnology Characterization Laboratory, National Cancer Institute. Frederick, MD: SAIC-Frederick, 2005.
26. Oyewumi MO, Yokel RA, Jay M, et al. Comparison of cell uptake, biodistribution and tumor retention of folate coated and PEG coated gadolinium nanoparticles in tumor bearing mice. J Control Release 2004; 95:613–626.
27. Gref R, Luck M, Quellec P, et al. "Stelth" corona-core nanoparticles surface modified by poly ethylene glycol (PEG): Influence of the corona (PEG chain length and surface density) and of the core composition on phagocytic uptake and plasma protein adsorption. Colloids Surf B 2000; 18:301–313.
28. Diederichs JE. Plasma protein adsorption patterns on liposomes: Establishment of analytical procedure. Electrophoresis 1996; 17:607–611.
29. Harnisch S, Muller RH. Plasma protein adsorption patterns on emulsion for parenteral administration: Establishment of a protocol for two-dimensional polyacrylamide electrophoresis. Electrophoresis 1998; 19:349–354.
30. Goppert TM, Muller RH. Alternative sample preparation prior to two-dimensional electrophoresis protein analysis on solid lipid nanoparticles. Electrophoresis 2004; 25:134–140.
31. Thode K, Luck M, Semmler W, et al. Determination of plasma protein adsorption on magnetic iron oxides: Sample preparation. Pharm Res 1997; 14:905–910.

32. Inoue K-I, Takano H, Yanagisawa R, et al. Effects of airway exposure to nanoparticles on lung inflammation induced by bacterial endotoxin in mice. Environ Health Perspect 2006; 114(9):1325–1331.
33. Compliment Activation Assay. NCL Method ITA-5, Version 1.0, Nanotechnology Characterization Laboratory, National Cancer Institute. Frederick, MD: SAIC-Frederick, 2005.
34. Reddy ST, van der Vlies AJ, Simeoni E, et al. Exploiting lymphatic transport and complement activation in nanoparticle vaccines. Eur Cell Mater 2007; 14(3):103.
35. Bertholon I, Vauthier C, Laberre D. Complement activation by core-shell poly(isobutylcyanoacrylate)-polysaccharide nanoparticles: Influences of surface morphology, length, and type of polysaccharide. Pharm Res 2006; 23(6):1313–1323.
36. Mouse Granulocyte-Macrophage Colony-Forming Unit Assay (CFU-GM Assay). NCL Method ITA-3, Version 1.0, Nanotechnology Characterization Laboratory, National Cancer Institute. Frederick, MD: SAIC-Frederick, 2005.
37. Mouse Colony-Forming Cell Assays Using MethoCult. Technical Manual. Canada: StemCell Technologies Inc. Cat No. 28405.
38. Zhang D-Y, Shen X-Z, Wang J-Y, et al. Preparation of chitosan-polyaspartic acid-5-fluorouracil nanoparticles and its anti-carcinoma effect on tumor growth in nude mice. World J Gastroenterol 2008; 14(22):3554–3562.
39. Effect of sanazole on CFU-GM from mice bone marrow in vitro. Chin J Med Guide. 2004; 6:2.
40. Leukocyte Proliferation Assay. NCL Method ITA-6 Version 1.2, Nanotechnology Characterization Laboratory, National Cancer Institute. Frederick, MD: SAIC-Frederick, 2005.
41. Chemotaxis Assay. NCL Method ITA-8 Version 1.2, Nanotechnology Characterization Laboratory, National Cancer Institute. Frederick, MD: SAIC-Frederick, 2005.
42. Barlow PG, Clouter-Baker A, Donaldson K, et al. Carbon black nanoparticles induce type II epithelial cells to release chemotoxins for alveolar macrophages. Part Fibre Toxicol 2005; 2:11.
43. Phagocytosis Assay. NCL Method ITA-9 Version 1.2, Nanotechnology Characterization Laboratory, National Cancer Institute. Frederick, MD: SAIC-Frederick, 2005.
44. Cytokine Inductive Assay. NCL Method ITA-10 Version 1.2, Nanotechnology Characterization Laboratory, National Cancer Institute. Frederick, MD: SAIC-Frederick, 2005.
45. Oxidative Burst (NO Production). NCL Method ITA-7 Version 1.2, Nanotechnology Characterization Laboratory, National Cancer Institute. Frederick, MD: SAIC-Frederick, 2005.
46. Ogbomo H, Hahn A, Geiler J, et al. NK sensitivity of neuroblastoma cells determined by a highly sensitive coupled luminescent method. Biochem Biophys Res Commun 2005; 339(1):375–379.)
47. Zhang N-H, Feng M, Lu Y-J, et al. Erythrocyte native immune adhering function (ENIAF) in patients with hematopoietic and lymphoid neoplasms and its effect on NK cell killing activity. Zhongguo Shi Yan Xue Ye Xue Za Zhi 2007; 15(5):1037–1041.
48. Bruchez M, Moronne M, Gin P, et al. Semiconductor nanocrystals as fluorescent biological labels. Science 1998; 281:2013–2016.
49. Taton T, Mirkin C, Letsinger R. Scanometric DNA array detection with nanoparticle probes. Science 2000; 289:1757–1760.
50. Cui Y, Wei Q, Park H, et al. Nanowire nanosensors for highly sensitive and selective detection of biological and chemical species. Science 2001; 293:1289–1292.
51. Warheit DB. Nanoparticles: Health impacts? Mater Today 2004; 7:32–35.
52. Adams LK, Lyon DY, Alvarez PJJ. Comparative eco-toxicity of nanoscale TiO_2, SiO_2, and ZnO water suspensions. Water Res 2006; 40:3527–3532.
53. Colvin VL. The potential environmental impact of engineered nanomaterials. Nat Biotechnol 2003; 21:1166–1170.
54. Foley S, Crowley C, Smaihi M, et al. Cellular localisation of a water-soluble fullerene derivative. Biochem Biophys Res Commun 2002; 294:116–119.

55. Lam C-W, James JT, McCluskey R, et al. Pulmonary toxicity of single-wall carbon nanotubes in mice 7 and 90 days after intratracheal instillation. Toxicol Sci 2004; 77:126–134.
56. Oberdöster E. Manufactured nanomaterials (fullerenes, C_{60}) induce oxidative stress in the brain of juvenile largemouth bass. Environ Health Perspect 2004; 112:1058–1062.
57. Cai R, Hashimoto K, Itoh K, et al. Photokilling of malignant cells with ultrafine TiO_2 powder. Bull Chem Soc Jpn 1991; 64:1268–1273.
58. Dunford R, Salinaro A, Cai L, et al. Chemical oxidation and DNA damage catalysed by inorganic sunscreen ingredients. FEBS Lett 1997; 418:87–90.
59. Sayes CM, Wahi R, Kurian PA, et al. Correlating nanoscale titania structure with toxicity: A cytotoxicity and inflammatory response study with human dermal fibroblasts and human lung epithelial cells. Toxicol Sci 2006; 92:174–185.
60. Ueng T-H, Kang J-J, Wang H-W, et al. Suppression of microsomal cytochrome P450-dependent monooxygenases and mitochondrial oxidative phosphorylation by fullerenol, a polyhydroxylated fullerene C_{60}. Toxicol Lett 1997; 93(1):29–37.
61. Shcharbin D, Jokiel M, Klajnert B, et al. Effect of dendrimers on pure acetyl cholinesterase activity and structure. Bioelectrochemistry 2006; 68(1):56–59.
62. Oberdörster G, Finkelstein JN, Johnston C, et al. Acute pulmonary effect of ultrafine particles in rat and mice. Res Rep Health Eff Inst 2000; 5–74.
63. Rancan F, Rosan S, Boehm F, et al. Cytotoxicity and photocytotoxicity of a dendritic C_{60} mono-adduct and a malonic acid C_{60} tris-adduct on Jurkat cells. J Photochem Photobiol B Biol 2002; 67(3):157–162.
64. Mickuviene I, Kirveliene V, Juodka B. Experimental survey of non-clonogenic viability assays for adherent cells in vitro. Toxicology in vitro 2004; 18(5):639–648.
65. Hong S, Bielinska AU, Mecke A, et al. Interaction of poly(amidoamine) dendrimers with supported lipid bilayers and cells: Hole formation and the relation to transport. Bioconjug Chem 2004; 15(4):774–782.
66. Decker T, Lohmann-Matthes M-L. A quick and simple method for the quantitation of lactate dehydrogenase release in measurements of cellular cytotoxicity and tumor necrosis factor (TNF) activity. J Immunol Methods 1988; 115(1):61–69.
67. Korzeniewski C, Callewaert DM. An enzyme-release assay for natural cytotoxicity. J Immunol Methods 1983; 64(3):313–320.
68. Alley MC, Scudiero DA, Monks A, et al. Feasibility of drug screening with panels of human tumor cell lines using a microculture tetrazolium assay. Cancer Res 1988; 48(3):589–601.
69. Berridge MV, Herst PM, Tan AS. Tetrazolium dye as tool in cell biology. Biotechnol Annu Rev 2005; 11:127–152.
70. Natarajan M, Mohan S, Martinez BR, et al. Antioxidant compounds interfere with the MTT assay. Cancer Detect Prev 2000; 24:405–414.
71. Vellonen K-S, Honkakoski P, Urtti A. Substrates and inhibitors of efflux proteins interfere with the MTT assay in cells and may lead to underestimation of drug toxicity. Eur J Pharm Sci 2004; 23(2):181–188.
72. Bhattarai SR, Kc RB, Kim SY, et al. *N*-Hexanoyl chitosan stabilized magnetic nanoparticles: Implication for cellular labeling and magnetic resonance imaging. J Nanobiotechnol 2008; 6:1.
73. Van de Coevering R, Kreiter R, Cardinali F, et al. An octa-cationic core-shell dendrimer as a molecular template for the assembly of anionic fullerene derivatives. Tetrahedron Lett 2005; 46:3353–3356.
74. Tomalia DA, Naylor AM, Goddart WA. Starburst dendrimers-molecular level control of size, shape surface chemistry, topology and flexibility from atoms to macroscopic matter. Angew Chem Int Ed Engl 1990; 29:138–175.
75. Win KY, Mu L, Wang C-H, et al. Nanoparticles of biodegradable polymers for cancer chemotherapy. Presented at the Summer Bioengineering Conference; Sonesta Beach Resort, Key Biscayne, FL; June 25–29, 2003.
76. Hussain SM, Hess KL, Gearhart JM, et al. In vitro toxicity of nanoparticles in BRL 3A rat liver cells. Toxicol In Vitro 2005; 19:975–983.

77. Voigt W. Sulforhodamine B assay and chemosensitivity. Methods Mol Med 2005; 110:39–48.
78. Tuschl H, Schwab CE. Flow cytometric method s used as screening tests for basal toxicity of chemicals. Toxicol In Vitro 2004; 18(4):483–491.
79. Oberdöster G, Maynard A, Donaldson K, et al. Principles for characterizing the potential human health effects from exposure to nanomaterials: Elements of a screening strategy. Part Fibre Toxicol 2005; 2:8.
80. Cagle DW, Kennel SJ, Mirzadeh S, et al. In vivo studies of fullerene-based materials using endohedral metallofullerene radiotracers. PNAS 1999; 96:5182–5187.
81. Ogawara K-I, Furumoto K, Takakura Y, et al. Surface hydrophobicity of particles is not necessarily the most important determinant in their in vivo disposition after intravenous administration in rats. J Control Release 2001; 77(3):191–198.
82. Fernádez-Urrusuno R, Fattal E, Porquet D, et al. Evaluation of liver toxicological effects induced by polyalkylcyanoacrylate nanoparticles. Toxicol Appl Pharmacol 1995; 130(1):272–279.
83. Regoli F, Winston GW, Mastrangelo V, et al. Human-derived cell lines to study xenobiotic metabolism. Chemosphere 1998; 37:2773–2783.
84. Wang K, Shindoh H, Inoue T, et al. Advantages of in vitro cytotoxicity testing by using primary rat hepatocytes in comparison with established cell lines. J Toxicol Sci 2002; 27(3):229–237.
85. Gharbi N, Pressac M, Tomberli V, et al. In: Maggini M, Martin N, Guldi DM, eds. Fullerenes, 2000: Functionalized Fullerenes, Vol. 9. Pennington NJ: The Electrochemical Society Inc., 2000; 240–243.
86. Lee CC, MacKay JA, Fréchet JMJ, et al. Designing dendrimers for biological applications. Nat Biotechnol 2005; 23(12):1517–1526.
87. Wang H, Wang J, Deng X, et al. Biodistribution of carbon single-wall carbon nanotubes in mice. J Nanosci Nanotechnol 2004; 4:1019–1024.
88. Nigavekar SS, Sung LY, Llanes M, et al. 3H dendrimer nanoparticle organ/tumor distribution. Pharm Res 2004; 21:476–483.
89. Manil L, Couvreur P, Mahieu P. Acute renal toxicity of doxorubicin (Adriamycin) loaded cyanoacrylate nanoparticles. Pharm Res 1995; 12:85–87.
90. Manil L, Davin JC, Duchenne C, et al. Uptake of nanoparticles by rat glomerular mesangial cells in vivo and in vitro._ Pharm Res 1994; 11(8):1160–1165.
91. Shiohara A, Hoshino A, Hanaki K, et al. On the cytotoxicity caused by quantum dots. Microbiol Immunol 2004; 48:669–675.
92. Wang B, Feng W-Y, Wang T-C, et al. Acute toxicity of nano- and micro-scale zinc powder in healthy adult mice. Toxicol Lett 2006; 161:115–123.
93. Toutain H, Morin JP. Renal proximal tubule cell culture for studying drug induced nephrotoxicity and modulation of phenotype expression by medium components. Ren Fail 1992; 14:371–383.
94. Tao F, Gonzalez-Flecha B, Kobzik L. Reactive oxygen species in pulmonary inflammation by ambient particulates. Free Radic Biol Med 2003; 35:327–340.
95. Li N, Sioutas C, Cho A, et al. Ultrafine particulate pollutants induce oxidative stress and mitochondrial damage. Environ Health Perspect 2003; 111(4):455–460.
96. Fernádez-Urrusuno R, Fattal E, Rodrigues JM, et al. Effect of polymeric nanoparticle administration on the clearance activity of the mononuclear phagocyte system in mice. J Biomed Mater Res 1996; 31(3):401–408.
97. Fernádez-Urrusuno R, Fattal E, Féger J, et al. Evaluation of hepatic antioxidant systems after intravenous administration of polymeric nanoparticles. Biomaterials 1997; 18(6):511–517.
98. Rison L, Mollar P, Loft S. Oxidative stress induced DNA damage by particulate air pollution. Mutat Res 2005; 592:119–137.
99. Yamakoshi Y, Umezawa N, Ryu A, et al. Active oxygen species generated from photoexcited fullerene (C_{60}) as potential medicines: O_2^{-*} versus $^1O_2$. J Am Chem Soc 2003; 125:12,803–12,809.

100. Dotan Y, Lichtenberg D, Pinchuk I. Lipid peroxidation cannot be used as a universal criterion of oxidative stress. Prog Lipid Res 2004; 43(3):200–227.
101. Black MJ, Brandt RB. Spectrofluorometric analysis of hydrogen peroxide. Anal Biochem 1974; 58(1):246–254.
102. Wang H, Joseph JA. Quantitating cellular oxidative stress by dicholorofluorescein assay using microplate reader. Free Radic Biol Med 1999; 27:612–616.
103. Dubuisson MLN, de Wergifosse B, Trouet A, et al. Antioxidative properties of natural coelenterazine and synthetic methyl coelenterazine in rat hepatocytes subjected to *tert*-butyl hydroperoxide-induced oxidative stress. Biochem Pharmacol 2000; 60(4):471–478.
104. Shaik IH, Mehvar R. Rapid determination of reduced and oxidized glutathione levels using a new thiol-masking reagent and the enzymatic recycling method: Application to the rat liver and bile samples. Anal Bioanal Chem 2006; 385(1):105–113.
105. Van Cruchten S, Van den Broeck W. Morphological and biochemical aspects of apoptosis, oncosis and necrosis. Anat Histol Embryol 2002; 31:214–223.
106. Bottini M, Bruckner S, Nika K, et al. Multi-walled carbon nanotubes induce T lymphocyte apoptosis. Toxicol Lett 2006; 160:121–126.
107. Cui D, Tian F, Ozkan CS, et al. Effect of single wall carbon nanotubes on human HEK293 cells. Toxicol Lett 2005; 155:73–85.
108. Kuo JH, Jan MS, Chiu HW. Mechanism of cell death induced by cationic dendrimers in RAW 264.7 murine macrophage like cells. J Pharm Pharmacol 2005; 57:489–495.
109. Lovric J, Cho SJ, Winnik FM, et al. Unmodified cadmium telluride quantum dots induce reactive oxygen species formation leading to multiple organelle damage and cell death. Chem Biol 2005; 12:1227–1234.
110. Wang ZB, Liu YQ, Chi YF. Pathways to escape activation. Cell Biol Int 2005; 29:489–496.
111. Gurtu V, Kain SR, Zhang G. Fluorometric and colorimetric detection of caspase activity associated with apoptosis. Anal Biochem 1997; 251(1):98–102.
112. Loo DT, Rillema JR. Measurement of cell death. Methods Cell Biol 1998; 57:251–264.
113. Xia T, Korge P, Weiss JN, et al. Quinones and aromatic chemical compounds in particulate matter induce mitochondrial dysfunction: Implications for ultrafine particle toxicity. Environ Health Perspect 2004; 112(14):1347–1358.
114. Isakovic A, Markovic Z, Todorovic-Markovic B, et al. Distinct cytotoxic mechanisms of pristine versus hydroxylated fullerene. Toxicol Sci 2006; 91(1):173–183.
115. Qi LF, Xu ZR, Li Y, et al. In vitro effect of chitosan nanoparticles on proliferation of human gastric carcinoma cell line MGC803 cells. World J Gastroenterol 2005; 11:5136–5141.
116. Amacher DE. Drug associated mitochondrial toxicity and its detection. Curr Med Chem 2005; 12:1829–1839.
117. Gogvadze V, Orrenius S, Zhivotovsky B. Analysis of mitochondrial dysfunction during cell death. In: Current Protocols in Toxicology. New York: Wiley, 2004:10.1–2.10.27.
118. Guo W-X, Pye QN, Williamson KS, et al. Mitochondrial dysfunction in choline deficiency-induced apoptosis in cultured rat hepatocytes. Free Radic Biol Med 2005; 39(5):641–650.
119. Wu EY, Smith MT, Bellomo G, et al. Relationships and cytotoxicity in isolated rat hepatocytes. Arch Biochem Biophys 1990; 282:358–362.

13 In Vivo Evaluations of Solid Lipid Nanoparticles and Microemulsions

Maria Rosa Gasco
Nanovector s.r.l., Turin, Italy

Alessandro Mauro
Department of Neurosciences, University of Turin, Turin and IRCCS—Istituto Auxologico Italiano, Piancavallo (VB), Italy

Gian Paolo Zara
Department of Anatomy, Pharmacology and Forensic Medicine, University of Turin, Turin, Italy

SOLID LIPID NANOPARTICLES

Solid lipid nanoparticles (SLNs) are reported to be an alternative system to emulsions, liposomes, or polymeric nanoparticles (1–3). Different approaches can be employed to prepare SLNs, including (*i*) high-pressure homogenization at high or low temperatures, (*ii*) warm microemulsions, (*iii*) solvent emulsification–evaporation–diffusion, (*iv*) high-speed stirring, and/or (*v*) sonication (4). This chapter considers their in vivo application, discussing, in particular, SLNs obtained from warm microemulsions.

Muller and colleagues (5,6) studied the preferential adsorption of blood proteins onto intravenously injected particulate carriers of different origins; they found apolipoprotein E (Apo E) to be enriched on the surface of polysorbate 80–coated SLNs after their incubation in human citrate plasma, whereas no Apo E adsorption occurred after incubation with other surfactants. The same adsorption was observed on different kinds of nanoparticles. Apo E can play an important role in the transport of lipoprotein into the brain via the low-density lipoprotein (LDL) receptor present on the blood–brain barrier (BBB) (7). The hypothesis is that Apo E–adsorbing nanoparticles may mimic LDL particles, leading to their uptake through endocytic processes. Delivery to the brain by using nanoparticulate drug carriers in combination with the targeting principles of "differential protein adsorption" has therefore been proposed (8,9). The Pathfinder technology (10) exploits proteins present in the blood which absorb onto the surface of intravenously injected carriers for targeting nanoparticles to the brain. Apo E is one such targeting molecules for delivering nanoparticles to brain vessel endothelial cells of the BBB.

Atovaquone (11) is a drug that is poorly adsorbed after oral administration, showing low therapeutic efficacy against *Toxoplasma gondii*. Nanocrystals of the drug were produced and their surface was modified with Tween 80, leading to in vivo preferential adsorption of Apo E; the nanosuspension was administered intravenously in a murine model of *Toxoplasmic encephalitis*, leading to the disappearance of parasites and cysts at a dose 10 times smaller than that required if atovaquone was administered by the oral route.

SLNs carrying the lipophilic antipsychotic drug clozapine (12) were prepared by hot homogenization followed by ultrasonication; clozapine has very low bioavailability. The SLNs were administered by the intravenous and duodenal routes to Swiss albino mice. For intravenous administration, stearylamine was entrapped with clozapine in SLNs; the area under the curve (AUC) in the brain increased by up to 2.91-fold versus that of clozapine suspension. The same group (13) developed SLNs as carriers of the highly lipophilic cardiovascular drug nitrendipine (NDP), using different triglycerides for the lipid matrix, soy lecithin, and poloxamer 188. Positively and negatively charged NDP–SLN carriers were produced and were examined to explore the influence of charge on oral bioavailability. Different kinds of SLNs were administered to rats by the intravenous and intraduodenal routes. The pharmacokinetics of NDP–SLNs was examined, and tissue distribution versus that of an NDP suspension was studied in Swiss albino mice. Following intravenous administration, NDP-loaded SLNs were taken up to a greater extent than NDP suspension in the organs studied. The AUC and the mean residence time (MRT) of NDP-loaded SLNs were higher than those of NDP suspension, especially in the brain and the heart. Positively charged SLNs were taken up markedly by the brain and moderately by the heart. Uptake by reticuloendothelial system (RES) organs, such as the liver and the spleen, was compared with that occurring after the administration of NDP suspension. The higher levels of the drug were maintained for over 6 hours versus only 3 hours with NDP suspension.

SLNs were investigated for their ability to deliver quinine dihydrochloride for the management of cerebral malaria (14). Quinine was incorporated into SLNs and the SLNs were then coupled with transferrin via a cross-linker. Intravenous administration of transferrin-conjugated SLNs enhanced the uptake of quinine in the brain versus that offered by SLNs loaded with quinine alone.

To enhance the delivery of atazanavir, an HIV protease inhibitor, spherical SLNs carrying the drug were tested using a well-characterized human brain microvessel endothelial cell line (hCMEC/D3). Cell viability experiments showed that SLNs possess no toxicity against hCMEC/D3 cells up to a concentration corresponding to 200 nM of the drug. Delivery of ^{3}H-atazanavir by SLNs led to a significantly higher accumulation by the endothelial cell monolayer than by the drug in aqueous solution (15).

The in situ transport of lipid nanoparticles to the brain was evaluated by Koziara et al. (16): lipidic nanoparticles were prepared from warm microemulsion precursors, followed by the hot homogenization technique. The components used were emulsified wax (E wax) or Brij 72 as the matrix, water, and Brij 78 as the surfactant. The warm microemulsion was cooled under stirring, the SLNs were obtained and homogenized. The SLNs were labelled with ^{3}H-cetyl alcohol. Transport of the nanoparticles was measured by an "in situ" rat brain perfusion method; significant uptake of SLNs was observed, suggesting uptake by the central nervous system (CNS). The same group also studied the effect of charged nanoparticles on the integrity of the brain (17). They chose three surfactants, namely, neutral Brij 78, anionic SDS, and cationic (*N*-octadecylcholine), and evaluated the effect on the BBB integrity and nanoparticles' brain permeability by "in situ" rat brain perfusion. Neutral SLNs and low concentrations of anionic SLNs can be utilized as colloidal carriers to the brain; cationic SLNs give an immediate toxic effect on the BBB.

The anticancer drug camptothecin–SLN, injected intravenously, produced a significant prolonged drug residence time in the body compared with a plain drug solution: the AUC in the brain was 2.6-fold that of the solution (18).

Reddy and colleagues prepared tripalmitin nanoparticles incorporating the anticancer drug etoposide (19) by melt emulsification and high-pressure homogenization, followed by spray drying of the nanodispersed material. The resulting nanoparticles possessed either a negative (ETN) or a positive (ETP) charge. Radiolabelled etoposide nanoparticles of both types were injected into mice; the ETP nanoparticles produced a relatively high distribution in the bone and the brain (14-fold that of etoposide alone) 4 hours postinjection, which was much better than that by the ETN nanoparticles. The ETP nanoparticles possessed a long-circulating property, and their effectiveness for targeting drugs both to tumors and to the brain should be beneficial.

In another study, 3′,5′-dioctanoyl-5-fluoro-2′-deoxyuridine was incorporated into SLNs (20): drug-loaded SLNs and the drug solution were then administered intravenously, and the AUC level in the brain was double than that by injecting a plain drug solution.

Actarit is a poorly water-soluble drug used in the treatment of rheumatoid arthritis. SLNs carrying actarit (21) were produced by a modified solvent diffusion–evaporation method and administered intravenously to rabbits; the performance was compared with that of actarit 50% propylene glycol solution. The AUC of plasma concentration–time for actarit-loaded SLNs was 1.88-fold that of actarit in the solution; the MRT was 13.5 hours compared with 1.3 hours for the propylene glycol solution.

Different groups have studied insulin-loaded SLNs. Insulin was incorporated into SLNs by a modified solvent emulsification–evaporation method based on a w/o/w double emulsion (22); after oral administration of insulin-loaded SLNs to diabetic rats, a hypoglycemic effect was observed and lasted for 24 hours. A solvent-in-water emulsion–diffusion technique was devised and tested in rats (23). The insulin-loaded SLNs, prepared by a reverse-micelle, double-emulsion method, were studied for pulmonary administration as an alternative and noninvasive systemic delivery modality for therapeutic agents (24). During nebulization, the insulin-loaded SLNs remained stable. The study examined entrapment delivery, respirable fraction, and nebulization efficiency. Fluorescent-labelled insulin incorporated into SLNs showed them to be distributed in the lung alveoli. Dexamethasone acetate was also incorporated into SLNs (25).

SLNs Prepared from Warm Microemulsions

Warm microemulsions are prepared at temperatures ranging from 55°C to 80°C by using melted lipids (such as triglycerides/fatty acids) as oil, surfactants (such as lecithin), and cosurfactants (such as short-chain carboxylates or biliary salts); the warm microemulsions are subsequently dispersed in cold water. With this procedure, the nanodroplets of the warm microemulsion become SLNs; they are then washed by tangential-flow filtration. SLNs are spherical in shape and have a narrow size distribution. Their zeta-potential is normally high (30/40 mV) and can be either positive or negative depending on the starting formulation.

Hydrophilic and lipophilic molecules (drugs or diagnostics) can be incorporated into SLNs by using different methods. SLNs are able to carry drugs of different structure and lipophilicity, such as cyclosporine A (26), paclitaxel (27), doxorubicin

(28), tobramycin (29), steroids (30), peptides (31), antisense oligonucleotides (ODNs) (32), melatonin (33), apomorphine (34), and baclofen (35). Diagnostic compounds, such as Gadolinium(III) (GdIII) complexes (36) and iron oxides (37), have also been incorporated into SLNs.

SLNs are internalized within 2 to 3 minutes into all the tested cell lines (38,39) and, if administered by the duodenal route, they are targeted to the lymph (34). Stealth SLNs (SSLNs) can also be prepared to avoid their recognition by the RES, thus prolonging their residence time (40). SLNs, drug-loaded or otherwise, stealth or nonstealth, are transported through the BBB (41,42).

Unloaded SLN Biodistribution by the Intravenous and Duodenal Routes

To evaluate biodistribution in vivo, drug-free stealth and non-SSLNs were administered intravenously to rats to evaluate tissue distribution and transport across the BBB (41). Two types of experiments were performed using unlabelled and labelled SLNs. Rats were injected with labelled nonstealth or stealth nanoparticles and tissue distribution was monitored for 60 minutes. In another experiment, rats were injected with unlabelled and labelled SLNs and the cerebrospinal fluid was analyzed using transmission electron microscopy to confirm the presence of SLNs. Some biodistribution differences were found between labelled non-SSLNs and SSLNs. In particular, radioactivity in the liver and the lung was much lower with the stealth formulation than with the nonstealth counterpart, confirming that there is a difference in their uptake. Both types of SLNs were detected in the brain, and the electron microscopy images showed both types of SLNs in the rat cerebrospinal fluid (41).

The gastrointestinal uptake of SLNs was studied; unloaded labelled and unlabelled SLNs were administered duodenally to rats in two different amounts in equal volumes. Using electron microscopy, SLNs were observed in the lymph; the size of the particles was practically unchanged after administration. To evaluate the lymphatic uptake, labelled SLNs were used. The radioactivity data confirmed targeting of the particles to lymph and blood (40,43).

SLNs as Potential MRI Diagnostics

Superparamagnetic iron oxides are classified as contrast agents for magnetic resonance imaging (MRI). They affect water relaxation times T_1 and T_2; their ability to alter these properties is quantified through the parameter relaxivity. Iron oxides preferentially affect tissue T_2 relaxation times (and are called T_2-relaxing agents), while paramagnetic contrast agents, such as Gd complexes, chiefly affect T_1 and are known as T_1-relaxing agents.

Iron oxides are insoluble in water; therefore, to be used clinically, they must be transformed into modified colloids while their magnetic properties should remain unchanged. The surface of iron oxide nanoparticles can be modified, covering them with hydrophilic macromolecules, such as dextran in the case of Endorem®. Research (37) has examined whether SLNs can load iron oxides and whether, thus loaded, they reach the brain. Two kinds of SLNs containing iron oxides, SLN–Fe^A and SLN–Fe^B, were prepared from warm microemulsions and were initially studied in vitro; these preparations were compared with Endorem. Both SLN–Fe preparations showed in vitro relaxometric properties similar to those of Endorem. In view of the good T_2 relaxation time–enhancing property, an "in vivo" study of their distribution by using MRI was then performed. SLN–Fe^B, at a higher Fe concentration, was administered intravenously to rats; the comparison was again with Endorem. Images obtained after Endorem intravenous administration showed early modification but a rapid return to baseline; this is consistent with the short Endorem retention time in the blood. Results with the SLN–Fe^B were different: for each brain region studied, Signal Suppression was reached in the last images (135 minutes after administration) and it increased steadily from first to last acquisition. This shows that, after inclusion in SLNs, Endorem becomes a new type of contrast agent: Endorem is normally taken up by the liver and does not cross the BBB, whereas Endorem incorporated into SLNs–Fe^B is taken up by the CNS. This means that the SLN–Fe kinetics depends on the SLNs and not on their iron oxide content, as already seen with unloaded SLNs (40).

SLNs as Ocular Drug Delivery Systems

The rationale for the development of various nanoparticulate systems for sustained drug delivery in the ophthalmological field is based on their possible entrapment in the mucous layer covering the eye surface; this would increase precorneal residence time, extending absorption time. Another important challenge in the field of nanoparticulates is to deliver therapeutic doses of drugs to treat diseases involving the posterior part of the eye.

The poor ocular bioavailability of pilocarpine instilled from conventional preparations is well known. In an "in vitro" study (44), ion pairs of pilocarpine (Pilo-IP) were prepared to increase the drug's lipophilicity. Lipid nanoparticles containing Pilo-IP and an aqueous solution of Pilo-IP were examined. A biological study (44,45) was performed using two different formulations, administered topically to male New Zealand albino rabbits: an aqueous dispersion of Pilo-IP–SLNs and an aqueous solution of Pilo-IP. The formulations were tested by comparison with reference solutions. Miotic activity tests were achieved; each preparation was tested on at least six animals. The area under the miotic effect versus time curves was evaluated: AUCs increased 2.48 times for Pilo-IP in the aqueous solution and 2.84 times for Pilo-IP–SLN dispersion, both versus the reference solutions. Both formulations were biocompatible, and no irritation of the ocular tissues was observed.

SLNs as ocular delivery system for tobramycin (TOB–SLN) (46) were prepared, evaluated, and administered topically to rabbits. The SLNs were in the colloidal size range (average diameter < 100 nm, polydispersity index $= 0.2$). The SLN dispersion contained 2.5% of tobramycin as ion pair. The preocular retention of SLNs in rabbit eyes was examined using drug-free fluorescent SLNs (F-SLN); these were retained for longer times on the corneal surface and in the conjunctival sac than was a fluorescent aqueous solution. A dispersion of tobramycin (0.3% w/v)-loaded SLNs was administered topically to rabbits: the aqueous humor concentration of tobramycin was monitored for up to 6 hours. Compared with an equal dose of tobramycin administered in the form of standard commercial eye drops, TOB–SLN produced a significantly higher tobramycin bioavailability in the aqueous humor.

SLNs Administered Intravenously

Doxorubicin

In conscious rabbits, the pharmacokinetics and tissue distribution of doxorubicin enclosed in both non-SSLNs and SSLNs (three formulations at increasing

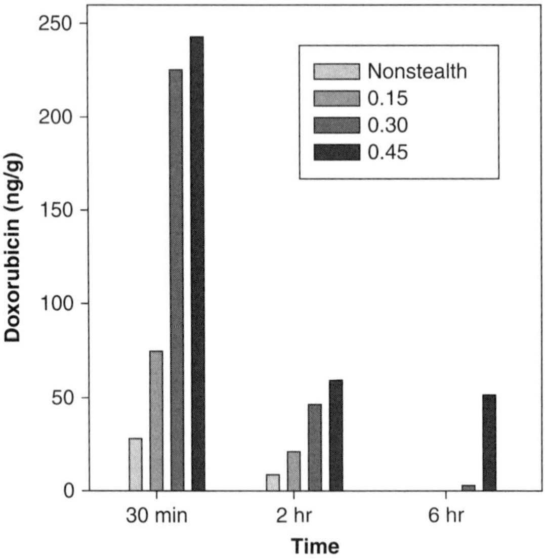

FIGURE 1 Doxorubicin concentration in brain after intravenous doxorubicin-loaded SLNs administration at different stealth concentrations.

concentrations of stealth agent) were evaluated after intravenous administration. The control was commercial doxorubicin solution. The experiments lasted 6 hours, and blood samples were collected at fixed times after the injection. In all samples, the concentrations of doxorubicin and its metabolite doxorubicinol were determined. Doxorubicin AUC increased as a function of the amount of stealth agent present in the SLNs; doxorubicin was still present in the blood 6 hours after the injection of SLNs or SSLNs, while none was detectable after the intravenous injection of doxorubicin solution. Tissue distribution of doxorubicin was determined 30 minutes, 2 hours, and 6 hours after the formulation administration; the drug was present in the brain after SLN administration. The amount of stealth agent increased the amount of doxorubicin transported to the brain; 6 hours after the injection of SSLNs, doxorubicin was detectable in the brain only with SSLNs at the highest amount of stealth agent (Fig. 1) (42).

The amount of doxorubicin present in other rabbit tissues (liver, lungs, spleen, heart, and kidneys) was lower after the injection of any of the four formulations of SSLNs than after the injection of the commercial solution. The pharmacokinetics and tissue distribution of doxorubicin in SLNs were also studied after intravenous administration to conscious rats and were compared with the commercial solution of doxorubicin. The same dose of each formulation (6 mg/kg) of doxorubicin was injected in the rat jugular vein. Blood samples were collected 1, 15, 30, 45, and 60 minutes and 2, 3, 6, 12, and 24 hours after the injection. Rats were killed after 30 minutes, 4 hours, or 24 hours, and samples of liver, spleen, heart, lung, kidneys, and brain were collected; the concentrations of doxorubicin and its metabolite doxorubicinol were determined in all tissue samples. Doxorubicin and doxorubicinol were still present in the blood 24 hours after the injection of SSLNs and non-SSLNs, while they were not detectable after the injection of the commercial solution.

The results confirmed the prolonged circulation time of SLNs compared with the doxorubicin solution. In all rat tissues except the brain, the amount of doxorubicin was lower after the injection of either type of SLNs than after the injection of the commercial solution. In particular, SLNs significantly decreased the concentration of doxorubicin in the heart (28).

Antisense ODNs

Particular interest deserves studies aimed to test nanoparticles as carriers for ODNs, which are molecules that are potentially useful for gene therapy. The delivery of ODNs to cells and tissues is highly limited by different factors, including their length, charge, and degradation by nucleases. In fact, unmodified ODNs show a very low bioavailability due to their fast degradation operated by exo- and endonucleases in both extra- and intracellular compartments. So, the incorporation of ODNs in suitable delivery systems could protect these molecules from nucleases, increase their cellular uptake, target them to specific tissues or cells, and reduce toxicity to other tissues.

Tondelli and colleagues (47) compared the efficacy to inhibit in vitro HL-60 leukemia cell proliferation of *c-myb* antisense ODNs administered as free solution or vehiculated by SLNs, and they found that the maximum dosage of ODN–SLN effective was fivefold less than the maximal nontoxic dose of unloaded SLNs. Furthermore, these authors found a more rapid (detectable 2 hours after treatment) and consistent cellular uptake of fluoresceinated ODNs vehiculated by SLNs, as well as a more prolonged effect (for up to 8 days in culture), than the free ODNs (47).

More recently, two publications addressed attention to the use of nanoparticles as delivery systems for antisense ODNs against the vascular endothelial growth factor (VEGF) (32,48). VEGF has been proposed as a target for antiangiogenetic therapy in different human diseases (including diabetic retinopathy and age-related macular degeneration) and tumor pathologies (including high-grade gliomas). In an in vitro study (48), the effectiveness of a VEGF antisense ODN vehiculated by a biodegradable nanoparticulate delivery system was evaluated using a human retinal epithelial cell line (ARPE-19). Nanoparticles of ODNs were prepared using poly(lactide-co-glycolide) copolymer with a double-emulsion solvent evaporation method. In this in vitro model, VEGF antisense ODNs demonstrated capability to inhibit VEGF expression of retinal cells if carried by nanoparticles or delivered by lipofectin but not in free solutions. Cellular uptake of antisense ODNs was increased by 4-fold for nanoparticles and by 13-fold for lipofectin.

Finally, the efficacy of SLNs carrying VEGF antisense ODNs to downregulate VEGF expression has been evaluated in rat glioma cells in vitro and in vivo (32). C6 rat glioma cells, maintained in hypoxic conditions to stimulate VEGF production, were treated in vitro with VEGF antisense ODNs, either free or carried by SLNs at different concentrations, for 24 and 48 hours. Western blot analysis showed that cellular VEGF expression was significantly reduced ($p < 0.01$) after 48 hours' treatment with SLNs carrying VEGF antisense ODNs, while expression remained stable after free VEGF antisense ODNs treatment. Moreover, in an in vivo experimental murine model of glioma (orthotopic intracerebral stereotactic implant of C6 glioma cells in Wistar rats), intravenous treatment for 3 days with VEGF antisense ODNs carried by SLNs produced a marked reduction of VEGF expression by tumor cells in both central and peripheral tumor areas. On the contrary, VEGF expression was unaffected by the treatment with free VEGF antisense ODNs. Thus, both in vitro and in vivo experiments in this glioma model demonstrated the effectiveness of SLNs carrying VEGF antisense ODNs in reducing VEGF expression, suggesting that SLNs can be regarded as a good carrier for the delivery of gene therapeutic agents to the CNS (32). However, this same study showed that after treatment with SLNs carrying VEGF antisense ODNs, the reduction of VEGF expression was evident not only in tumor cells but also in neuronal populations (especially in hippocampal neurons) known to express this growth factor. This finding indicates that SLNs can transport antisense ODNs throughout the BBB, not only where it is partially damaged (i.e., within and around the tumor) but also where it is intact, suggesting that the reduction of VEGF expression may occur not only where it is desired but also where it could be potentially dangerous. In particular, the reduction of VEGF production by neurons may reduce the ability of healthy cells to protect themselves from damage produced by edema-related hypoxia or by concomitant antineoplastic treatments.

SLNs by Intraperitoneal Administration

Baclofen

The intraperitoneal route of administration has been used very recently (35) to test the feasibility of incorporating baclofen in SLNs, aiming to obtain an efficacious new pharmaceutical preparation of this drug.

Baclofen, an analogue of the inhibitory neurotransmitter γ-aminobutyric acid, is traditionally the drug of choice for the treatment of spasticity of different origins. Although its precise mechanism of action in the CNS is not completely understood, it is known that the drug binds to γ-aminobutyric acid B receptors, inhibiting spinal cord reflexes. In fact, systemic administration of baclofen depresses mono- and polysynaptic spinal reflexes, however, its clinical use may be limited by important adverse central effects. Since oral baclofen is ill tolerated at higher doses, intrathecal (in the lumbar subarachnoid space) delivery of the drug through a pump system has become the standard of care for patients whose spasticity is not sufficiently managed by oral baclofen or for those who experience intolerable adverse effects. Localized, spinal application appears to reduce unwanted side effects, allowing higher baclofen concentrations at the site of action and reducing plasma levels.

However, surgical involvement, together with the risk of infection or catheter dysfunction, and the serious adverse effects caused by abrupt withdrawal, limit the number of potentially treatable patients.

Groups of Wistar rats were intraperitoneally injected with physiological solution, or with unloaded SLNs at 10 mL/kg (control groups), as well as with baclofen–SLN (baclofen concentration in water-reconstituted SLN suspension was 1.7 mg/mL) or with baclofen solution at increasing dosages (2.5, 5, 7.5, 8.5, and 10 mg/kg). Effects of the different treatments were tested at different times up to fourth hour by means of H-reflex analysis. Moreover, CNS adverse effects were evaluated by behavioral characterization with two scales validated and already used for motor symptoms due to spinal lesions and sedation in rat models (35).

Analysis of H-reflex following baclofen–SLN injection showed that H/M amplitude curve was characterized by a dose-dependent reduction at first and second hours, clearly confirming its efficacy; moreover, a rebound increase at fourth hour was observed, indicating an unexpected belated spinal hyperexcitability. Similarly, the effect of baclofen–SLN on the behavioral scales was stronger than that produced by baclofen-free solution, with the maximum effect at 1 hour. An important finding is that clinical effects were detectable after lower doses of baclofen–SLN (2.5 mg/kg) in comparison with the ones needed with baclofen solution (7.5 mg/kg). Four hours after the injection, only rats treated with the higher doses of baclofen-loaded SLNs still presented clinical signs consisting in sedation (8.5 mg/kg) or complete paralysis and piloerection (10 mg/kg). On the whole, these data suggest a dose-dependent modulation of spinal reflex excitability (associated with important cortical effects), which is not so evident after the administration of the standard formulation of baclofen. Clinical results showed a good correlation with plasma and tissue concentrations of baclofen: after 2 and 4 hours, only baclofen-loaded SLNs produced detectable baclofen plasma concentrations (with an almost linear decrease in the drug level throughout the 4 hours), while 2 hours after the administration of baclofen solution, the amount of baclofen in plasma was undetectable. Moreover, baclofen concentration in the brain 2 hours after SLN administration was almost double that after baclofen solution, suggesting that baclofen–SLN may pass the BBB (35). (Fig. 2)

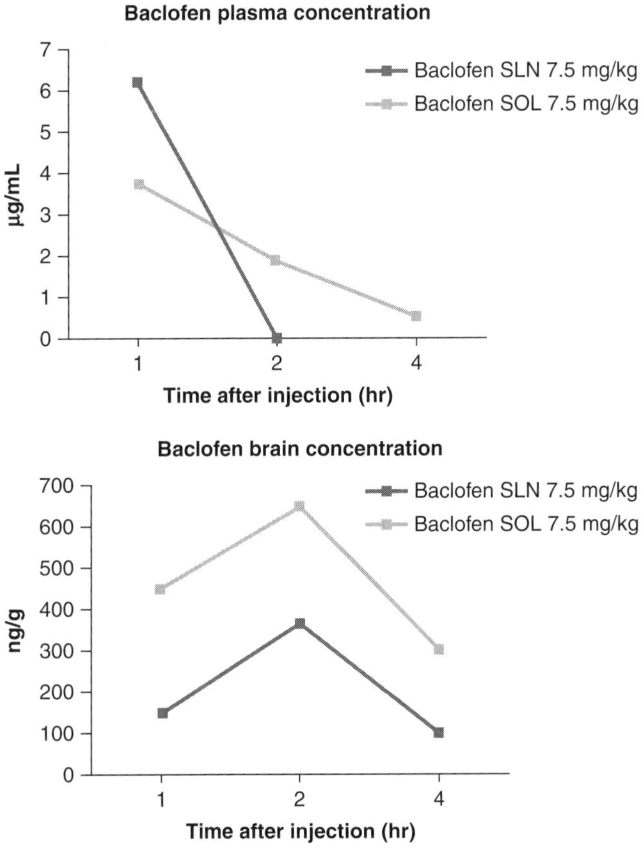

FIGURE 2 Plasma and brain baclofen concentrations after intraperitoneal injection of baclofen-loaded SLNs and baclofen solution.

Melatonin

Melatonin, the chief product secreted by the pineal gland, is a potent antioxidant molecule with a marked ability to protect tissues from damage caused by oxidative stress and from lipid peroxidation, as demonstrated in many experimental models. Moreover, it has been reported to reduce cyclosporine A (CsA) cardiotoxicity. Recently, an experimental study was designed to examine whether intraperitoneally administered melatonin is a useful tool for counteracting CsA-induced apoptosis in the rat heart and whether SLNs can be used as effective melatonin delivery devices. Oxidative stress in heart tissue was estimated by evaluating lipid peroxidation, and the expression of the isoform of inducible nitric oxide (iNOS) was studied. The antiapoptotic effect of melatonin was examined using the TUNEL technique and evaluating Bcl-2 protein family expression. Melatonin markedly reduced lipid peroxidation and normalized iNOS expression; it also restored cardiac morphology, blocking cell death. Its antiapoptotic efficacy was more marked when melatonin was loaded into SLNs (33).

SLNs as Delivery Systems by the Duodenal Route

The bioavailability of idarubicin (IDA) was studied after the administration of IDA–SLN duodenally to rats. Idarubicin and its main metabolite idarubicinol were determined in plasma and tissues by reverse-phase high-performance liquid chromatography. The pharmacokinetic parameters of idarubicin differed after duodenal administration of the two formulations: AUC versus time and elimination half-life were respectively ~21 times and 30 times higher after IDA–SLN administration than after solution administration. Tissue distribution also differed: idarubicin and idarubicinol concentrations were lower in the heart, lung, spleen, and kidneys after IDA–SLN administration than after solution administration. The drug and its metabolite were detected in the brain only after IDA–SLN administration, indicating that SLNs pass the BBB. The AUC of idarubicin was lower after intravenous IDA–SLN administration than after duodenal administration of the same formulation. Duodenal administration of IDA–SLN modifies the pharmacokinetics and tissue distribution of idarubicin. These data show that IDA–SLN act as a prolonged release system for the drug (49).

To evaluate gastrointestinal absorption of drugs incorporated in SLNs, tobramycin was used as model because it is not absorbed from the gastrointestinal tract, and thus is still administered by the parenteral route. Tobramycin-loaded SLNs were administered to rats by intravenous route and into the duodenum, and the outcome was compared with that after tobramycin administration. When administered intravenously, TOB–SLN showed a significantly higher AUC, associated with lower clearance, providing a sufficiently high level of the drug even after 24 hours. Furthermore, after duodenal administration, TOB–SLN produced a much higher AUC and decreased clearance. It is suggested that the very high blood level of the drug after duodenal administration of TOB–SLN may be due to the transmucosal transport of SLNs to the lymph and that TOB–SLN acts as a sustained release system when administered duodenally (50).

The time–concentration curve of SLNs containing three different percentages of tobramycin was evaluated after intraduodenal administration of the same dose of the drug in rats. The pharmacokinetic parameters varied considerably with the percentage of tobramycin administered. It is suggested that these differences can be due to differences among the three types of SLNs, in particular the number of SLNs administered, average diameter, total surface area, and drug concentration in each nanoparticle. The highest percentage of tobramycin in SLNs corresponded to the fastest release rate, whereas the lowest percentage produced the most prolonged release. Tobramycin was still present in lymph mesenteric nodes 21 hours after duodenal administration, confirming that SLNs can be considered as a sustained drug release carrier (29).

More recently, a comparison between intravenous and intraduodenal administration of SLNs was carried out in rats (34) by studying SLNs loaded with apomorphine (APO–SLN). Pharmacokinetic profile and biodistribution of APO–SLN following intraduodenal or intravenous administration into rat were compared with those obtained with apomorphine aqueous solution (APO). After intravenous administration, peak plasma concentration was higher using APO than using APO–SLN (1418.35 vs. 845.29 ng/mL); however, the total area under curve (AUC_{tot}) was higher following APO–SLN administration [18872.49 vs. 14274.90 ng/(mL min)]. The terminal half-life was significantly longer following APO–SLN administration (80.59 vs. 34.42). Instead, following intraduodenal administration, C_{max}, AUC_{tot}, and terminal half-life were significantly higher with APO–SLN than with APO, while clearance was shorter using APO–SLN. Moreover, concerning apomorphine brain concentrations after intravenous administration, at 30 minutes, they were significantly higher following APO–SLN administration than with APO administration, while at 4 hours, the drug was detectable only in the case of APO–SLN. Similarly, 30 minutes after duodenal administration, the drug was detectable exclusively following APO–SLN administration, whereas no apomorphine could be found 4 or 24 hours after the injection of either formulation (Fig. 3).

SLNs as Drug Delivery Systems of Melatonin in Humans by the Transdermal or Oral Routes

Melatonin is a hormone produced by the pineal gland at night, and is involved in the regulation of circadian rhythms. For clinical purposes (mainly disorders of the sleep–wake cycle and insomnia) in the elderly, exogenous melatonin administration should mimic the typical nocturnal endogenous melatonin levels, but its pharmacokinetics is not favourable due to its short half-life of elimination (51,52). Recently, the pharmacokinetics of melatonin incorporated in SLNs (MT–SLN) has been examined in humans after administration by oral and transdermal routes (53). Three kinds of freeze-dried MT–SLN containing different amounts of melatonin were prepared and characterized: (*i*) MT–SLN: MT = 1.8% for in vitro experiments (average diameter: 85 nm, polydispersity index: 0.135); (*ii*) MT–SLN: MT = 2% for transdermal application (average diameter = 91 nm, polydispersity index = 0.140); (*iii*) MT–SLN: MT = 4.13% for oral route (average diameter = 111 nm, polydispersity index = 0.189).

In vitro, MT–SLN produced a flux of 1 $\mu g/(h\ cm^2)$ of melatonin through hairless mice skin, following pseudo-zero-order kinetics (54). At the same time, in vivo study produced very interesting results, confirming in humans that SLNs can act as a reservoir that allows a constant and prolonged release of the included drugs (37). Melatonin (3 mg) incorporated in SLNs was orally administered at 8.30 a.m. to seven healthy subjects; for control purposes, 1 week later, the same subjects orally received a standard formulation of melatonin at the same dose (3 mg), again at 8.30 a.m. Compared with melatonin standard solution, T_{max} observed after MT–SLN administration was delayed by about 20 minutes while mean AUC and mean half-life of elimination were significantly higher [respectively 169944.7 ± 64954.4 pg/(mL h) vs. 85148.4 ± 50642.6 pg/(mL h), $p = 0.018$; and 93.1 ± 37.1 minutes vs.

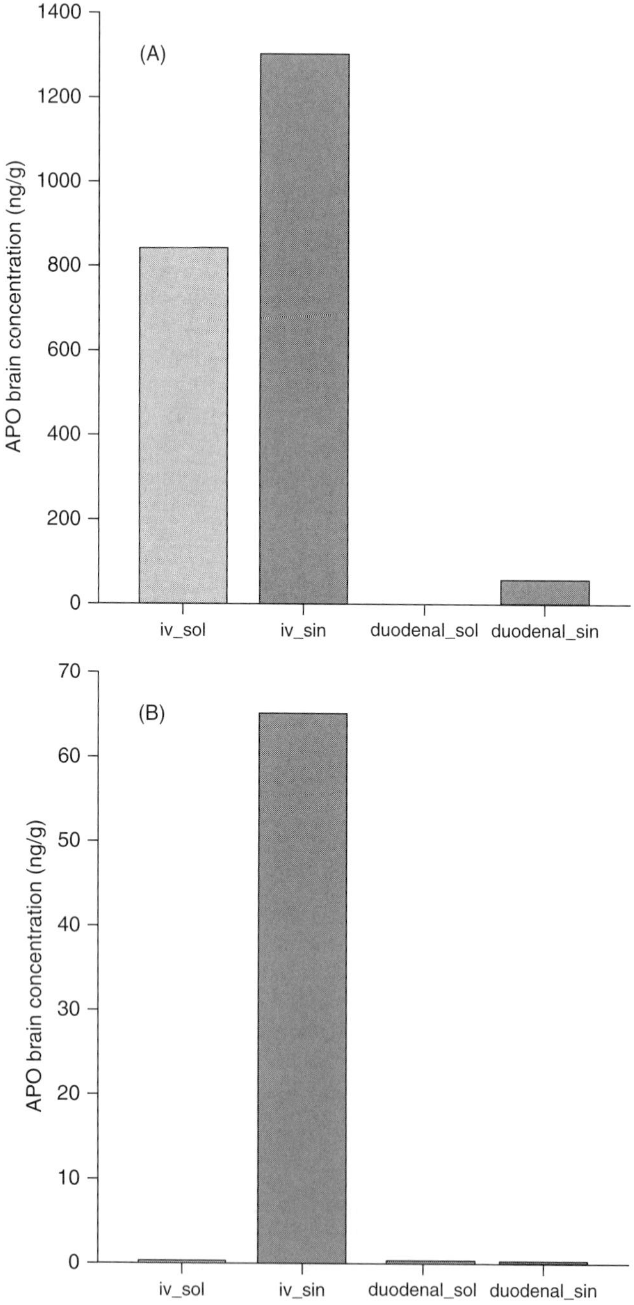

FIGURE 3 Apomorphine brain concentrations 30 min (A) and 4 hr (B) after i.v. (1,5 mg/Kg) or duodenal (4 mg/Kg) administration of apomorphine solution or apomorphine-loaded SLNs.

48.2 ± 8.9 minutes, $p = 0.009$]. Even more, standard formulation and MT–SLN after oral administration produced similar peak plasma levels of melatonin, even if delayed by about half an hour in the case of MT–SLN. More interestingly, detectable and clinically significant melatonin plasma levels after MT–SLN oral administration were maintained for a longer period of time, suggesting that SLNs orally administered to humans can yield a sustained release of the incorporated drug, a feature that could be particularly useful for molecules, such as melatonin, characterized by unfavourable kinetics (53). Previous studies in laboratory animals indicated a probable targeting of SLNs – either drug loaded or unloaded – to lymph, after duodenal administration (40). Similarly, the significantly longer half-life of melatonin observed in the study of Priano et al. (53) may suggest a targeting of MT–SLN to human lymph, although the capsules used to administer SLNs were not gastro resistant. In fact, melatonin half-life of elimination has been calculated in about 40 minutes after an intravenous bolus, and following oral administration, low bioavailability and rapid clearance from plasma have been shown, primarily due to a marked first-pass hepatic metabolism (2,3). In the above study (51), the mean half-life of melatonin elimination after oral MT–SLN was about double compared with the standard melatonin formulation used as control but also in comparison with data reported for intravenous administration (Fig. 4).

Moreover, pharmacokinetic analysis following transdermal administration of MT–SLN demonstrated that plasma levels of melatonin, similar to those produced by oral administration, may be achieved for more than 24 hours (53). In 10 healthy subjects, SLNs incorporating melatonin were administered transdermally by applying a patch at 8.30 a.m. and leaving it in place for 24 hours. In this delivery system, melatonin absorption and elimination were slow (mean half-life of absorption = 5.3 ± 1.3 hours; mean half-life of elimination = 24.6 ± 12.0 hours) so that melatonin plasma levels above 50 pg/mL were maintained for at least 24 hours (Fig. 5).

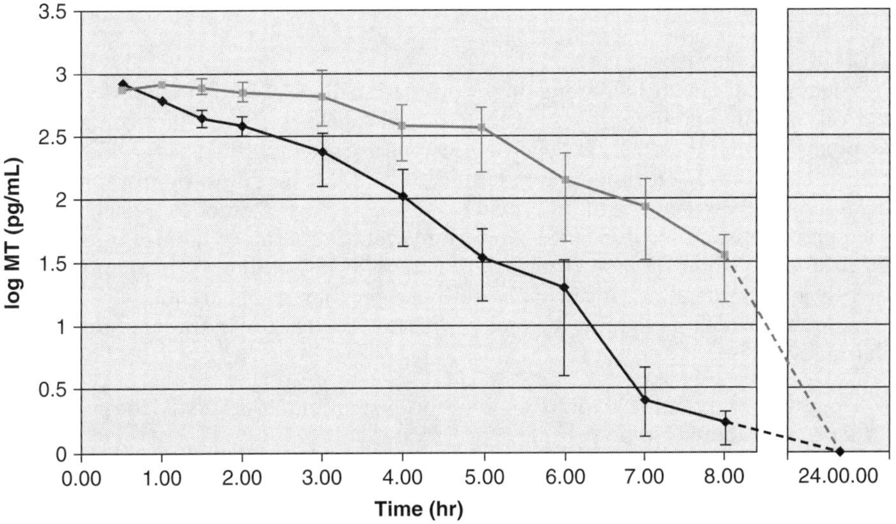

FIGURE 4 Melatonin (MT) plasma profile in humans after melatonin and MT-loaded SLN oral administration.

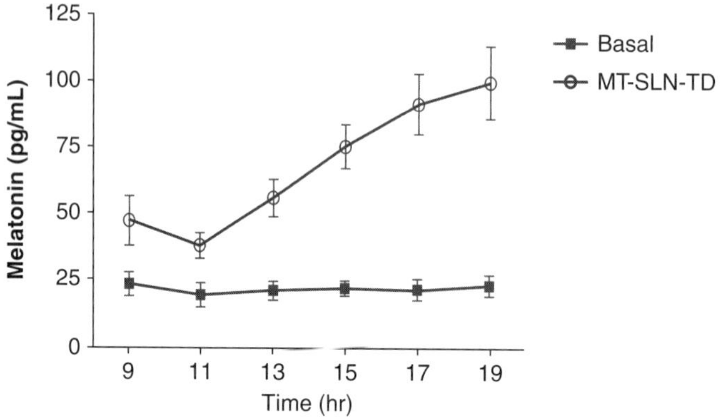

FIGURE 5 Melatonin plasma levels in humans at baseline and after transdermal administration of melatonin–loaded SLNs.

Tolerability of MT–SLN administered transdermally or by oral route was good and no adverse effect occurred, apart from a predictable mild somnolence and transient erythema after gel application. This means that, at least at the doses used in that study (54), SLN administration via the oral or transdermal route is safe.

These very favourable results, obtained in humans on administering melatonin-loaded SLNs, clearly suggest that SLNs can be considered effective in vivo delivery systems that could be suitably applied to different drugs, and in particular to those requiring prolonged high plasma levels but that have unfavourable pharmacokinetics. Finally, it must be stressed that, since doses and concentrations of drugs included in SLNs can be varied, different plasma level profiles could be obtained, thus disclosing new chances for sustained delivery systems adaptable to a variety of clinical conditions (53).

MICROEMULSIONS

Microemulsions are transparent, thermodynamically stable dispersions of water and oil, usually stabilized by a surfactant and a cosurfactant. They contain nanodroplets smaller than 0.1 μm. Microemulsions are often defined as thermodynamically stable liquid solutions; their stability is a consequence of the ultralow interfacial tension between the oil and water phases. A clear distinction exists between microemulsions and coarse emulsions. The latter are thermodynamically unstable, the droplets of their dispersed phase are generally larger than 0.1 μm and, consequently, their appearance is normally milky rather than transparent.

Microemulsions exhibit several properties that are of particular interest in pharmacy:

- Their thermodynamic stability enables the system to self-emulsify, the properties not being dependent on the followed process.
- The dispersed phase, lipophilic for oil-in-water (o/w) and hydrophilic for water-in-oil (w/o) microemulsions, can behave as a potential reservoir of lipophilic or hydrophilic drugs, respectively.
- The mean diameter of the droplets in microemulsions is below 100 nm. Such a small size yields a very large interfacial area, from which the drug can be quickly released into the external phase when in vitro or in vivo absorption takes place.

- The technology required to prepare microemulsions is simple, because their thermodynamic stability means that no significant energy contribution is required.
- Microemulsions can be sterilized by filtration, as the mean diameter of the droplets is below 0.22 μm.

The limits in the use of microemulsions in the pharmaceutical field derive, chiefly, from the need for all components to be acceptable, particularly surfactants and cosurfactants. The amounts of surfactants and cosurfactants required to form microemulsions are usually higher than those required for emulsions.

Microemulsions offer several advantages for pharmaceutical use, such as ease of preparation, long-term stability, high solubilization capacity for hydrophilic and lipophilic drugs, and improved drug delivery. They can also be used in oral and parenteral delivery (54), but this review is limited to in vivo studies by the transdermal route (55).

A microemulsion carrying methylnicotinate was prepared using lecithin, water, and isopropylmiristate (56) and was applied onto the skin of human volunteers; appreciable transport of the bioactive substance was obtained.

An o/w microemulsion and an amphiphilic cream, both carrying curcumin, were applied onto the skin of human volunteers; curcumin was chosen as model drug to compare the stratum corneum penetration of the two formulations. A deeper part of the stratum corneum was found to be accessible to the microemulsion than to the cream (57). Niflumic acid was incorporated in a sugar-based surfactant and tested in humans (58). It was found that the microemulsion formulation saturated with the drug (1%) was as efficient as a commercially available 3% o/w emulsion.

Good human skin tolerability of a lecithin-based o/w microemulsion compared with a conventional vehicle (o/w, w/o, and gel) was reported (59).

The transport of azelaic acid through mouse skin from a microemulsion (6.4%) and from a gel (15%) were compared in vitro (60); the lag time was determined for both systems and was rather longer for the colloidal vehicle than for the gel. However, the amount that emulsions permeated from the microemulsion was sevenfold that from the gel, although the concentration of azelaic acid in the microemulsion was less than half. The thickened microemulsion was then applied to lentigo maligna (61) and confirmed the efficacy of azelaic acid to treat this variety of melanoma; the microemulsion led to the regression of the lesions. Comparison between this treatment and treatment with a cream (20% azelaic acid and 3% salicylic acid) showed that the microemulsion led to regression earlier than did the cream. Ten cases were treated; the average time for the complete remission was halved compared with the times required with the cream.

Microemulsions for Transdermal Application of Apomorphin

Apomorphine, a potent, short-acting dopamine agonist at D_1 and D_2 dopamine receptors, potentially represents a very useful adjunctive medication for patients with Parkinson's disease. However, its clinical use is significantly limited by its pharmacokinetic profile characterized by a short half-life (approximately 30 minutes), rapid clearance from plasma, absence of storage or retention in brain regions, poor oral bioavailability (5%), and first-pass hepatic metabolism. Several, unsuccessful attempts have been made to overcome these limits by using other routes of administration, but at present, its use remains limited to few clinical conditions.

Recently, apomorphine was incorporated into microemulsions to study whether they are a feasible vehicle for transdermal transport of this drug. In the preparatory in vitro study (62), two different microemulsions whose components were all biocompatible were studied; the concentration of apomorphine was 3.9% in each. Since apomorphine is highly hydrophilic, apomorphine–octanoic acid ion pairs were synthesized to increase its lipophilicity. At pH 6.0, log P_{app} of apomorphine increased from 0.3 in the absence of octanoic acid to log $P_{app} = 2.77$ for a molar ratio 1:2.5 (apomorphine–octanoic acid). The flux of drug from the two thickened microemulsions through hairless mouse skin was respectively 100 $\mu g/(h\ cm^2)$ and 88 $\mu g/(h\ cm^2)$. The first formulation, having the higher flux, was chosen for in vivo administration in patients with Parkinson's disease.

For the in vivo study, 21 patients with idiopathic Parkinson's disease who presented long-term L-dopa syndrome, motor fluctuation, and prolonged "off" periods were selected (63). Ten grams of apomorphine hydrochloride (3.9%) included in the microemulsion for transdermal delivery (APO–MTD) was applied to a 100-cm^2 skin area on the chest; the area was delimited by 1-mm thick biocompatible foam tape and covered with a polyester-based membrane and an occlusive membrane to prevent evaporation. In these conditions, a single layer of microemulsion (1 mm thick) was directly in contact with the skin surface and acted as a reservoir of apomorphine. APO–MTD was applied at 8.00 a.m. and left for 12 hours. In all patients except two, apomorphine was detected in blood samples after a variable lag time. Pharmacokinetic analysis revealed that epicutaneous–transdermal apomorphine absorption was rapid (mean half-life of absorption = 1.03 hours) with a variability among patients (half-life of absorption SD = 1.39 hours). Mean C_{max} was above the therapeutic range (mean $C_{max} = 42.81 \pm 11.67$ ng/mL), with a mean T_{max} of 5.1 ± 2.24 hours. Therapeutic concentrations of apomorphine were reached after a mean latency of 45 minutes (range = 18–125), and stable concentrations, above the therapeutic range, continued for as long as APO–MTD was maintained in place. At hour 12, APO–MTD was removed, and the apomorphine plasma concentration then decreased at a rate comparable with that described for subcutaneous administration (mean half-life of elimination = 10.8 ± 1.93 hours). C_{max} and AUC showed good correlation with the clinical parameters studied ($r = 0.49$–0.56 and $p = 0.02$–0.04). Overall tolerability of APO–MTD was good; systemic adverse effects were similar to those caused by subcutaneous apomorphine injection (sleepiness, mild orthostatic hypotension, and transient nausea), and in the case of nausea, strictly related to the highest plasma level of apomorphine. Moreover, regarding local adverse effects, the large majority of patients (71.4%) presented a transient mild erythema at the site of APO–MTD application, with a complete regression within 48 hours, whereas in two cases, the erythema lasted more than 3 days and required local therapy. This study clearly demonstrated that in most patients with Parkinson's disease, APO–MTD is absorbed by the epicutaneous–transdermal route. This result is in contrast with other reports, in which the transdermal route did not produce detectable plasma levels of apomorphine, or in which no apomorphine was transported passively through the skin (64,65). Probably, this difference was mainly due to the peculiar pharmaceutical preparation used. Even if pharmacokinetic parameters are variable, APO–MTD demonstrated the feasibility of providing therapeutic apomorphine plasma levels for much longer periods of time than previously tested apomorphine preparations (several hours allowing more constant dopaminergic stimulation). These results are encouraging and APO–MTD might be of

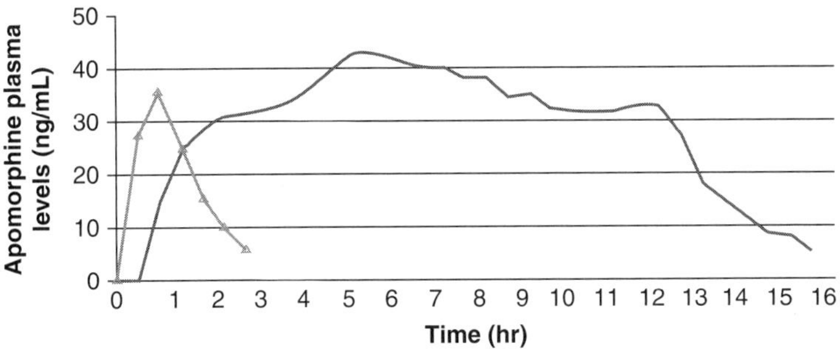

FIGURE 6 Comparison of apomorphine plasma levels after application of a preparation of apomorphine included in a microemulsion and administered transdermally (applied at time 0, removed at hour 12) and after 3 mg of apomorphine injected subcutaneously. Triangles, Apo-MTD; solid line, Apomorphine s.c.

clinical value in some patients with Parkinson's disease suffering from uncontrolled "wearing off" and prolonged "off" phenomena. On the contrary, because of the lag time of about 1 hour before therapeutic concentrations are reached, APO–MTD may not be the "ideal" preparation for rapid relief of "off" periods (Fig. 6).

Since APO–MTD was found to provide constant drug release over several hours, other studies have been addressed for its use for the nocturnal sleep disorders of patients with Parkinson's disease. Twelve patients with Parkinson's disease underwent standard polysomnography on basal condition and during one-night treatment with APO–MTD (50 mg applied to 100-cm^2 area from 10 p.m. until 8 a.m.) (66). Sleep analysis during APO–MTD treatment in comparison with basal condition showed very favourable findings: 16% increment of total sleep time, 12% increment of sleep efficiency, 16% increment of stage 3 and 4 nonrapid eye movement, 15% reduction of periodic limb movements index, 22% reduction of arousal index, and 23% reduction of cyclic alternating pattern rate (an objective measure of disruption and fragmentation of sleep). Pharmacokinetic analysis confirmed the absorption of apomorphine and the maintenance of therapeutic plasma levels for several hours (mean $C_{max} = 31.8 \pm 9.7$ ng/mL; mean $T_{max} = 3.1 \pm 1.6$ hours; mean half-life of absorption $= 1.2 \pm 1.4$ hours; mean half-life of elimination $= 8.8 \pm 1.9$ hours). On the whole, this study confirmed that APO–MTD in Parkinson's disease might be able to reduce nocturnal anomalous movements, akinesia, and rigidity and might be efficacious for reducing the instability of sleep maintenance typical of parkinsonian sleep.

CONCLUSION

This chapter has presented considerable evidence concerning the in vivo administration of drugs incorporated in SLNs; it has illustrated the special features of nanoparticles and how they could act as alternative carriers to deliver drugs in specific manners. Results of in vivo experiments in laboratory animals and humans are very encouraging: efficient drug protection, cell internalization, controlled release, and passage through biological anatomical barriers have been achieved.

It is to be hoped that in the future, novel biological and pharmacological approaches on the problem of associating drugs with SLNs will emerge and will be able to improve the pharmacological and clinical utility of these drug carriers.

REFERENCES

1. Uner M, Yener G. Importance of solid lipid nanoparticles (SLN) in various administration routes. Int J Nanomed 2007; 2:289–300.
2. Blasi P, Giovagnoli S, Schoubben A, et al. Solid lipid nanoparticles for targeted brain drug delivery. Adv Drug Deliv Rev 2007; 59:454–477.
3. Kaur IP, Bhandari R, Bhandari S, et al. Potential of solid lipid nanoparticles in brain targeting. J Control Release 2008; 127:97–109.
4. Muller RH, Kader K, Gohla S. Solid lipid nanoparticles (SLN) for controlled drug delivery. A review of the state of the art. Eur J Pharm Biopharm 2000; 50:161–177.
5. Harnish S, Muller RH. Plasma protein adsorption patterns on emulsions for parenteral administration: establishment of a protocol for two-dimensional polyacrylamide electrophoresis. Electrophoresis 1988; 19:345–354.
6. Luck M, Paulke BR, Schroder W, et al. Analysis of plasma protein adsorption on polymeric nanoparticles with different surface characteristics. Biomed Mater Res 1998; 39:478–485.
7. Dehouck R, Ferrari L, Dehouck MP, et al. A new function for the LDL receptor. Transcytosis of LDL across the blood brain barrier. J Cell Biol 1997; 38:877–889.
8. Muller RH, Schmidt S. Pathfinder technology for the delivery of drugs to the brain. New Drugs 2002; 2:38–42.
9. Radtke M, Muller RH. Nanostructured lipid drug carriers 2001; 2:48–52.
10. Muller RH, Keck CM. Challenges and solutions for the delivery of biotech. Drugs – a review of drug nanocrystal technology and lipid nanoparticle. J Biotechnol 2004; 113:151–170.
11. Scholer N, Krause K, Kayser O, et al. Atovaquone nanosuspensions show excellent therapeutic effect in a new murine model of reactivated toxoplasmosis. Antimicrob Agents Chemother 2001; 45:1771–1779.
12. Manjunath K, Venkateswarlu V. Pharmacokinetics, tissue distribution and bioavailability of clozapine solid lipid nanoparticles after intravenous and intraduodenal administration. J Control Release 2005; 107:215–228.
13. Manjunath K, Venkateswarlu V. Pharmacokinetics, tissue distribution and bioavailability of nitrendipine solid nanoparticles after intravenous and intraduodenal administration. J Drug Target 2006; 14:632–645.
14. Gupta Y, Jain A, Jain SK. Transferrin conjugate solid lipid nanoparticles for enhanced delivery of quinine dihydrochloride to the brain. J Pharm Pharmacol 2007; 59:935–940.
15. Chattopadhyay N, Zastre J, Wong HL, et al. Solid lipid nanoparticles enhance the delivery of the HIV protease inhibitor, atazanavir, by a human brain endothelial cell line. Pharm Res 2008; 25:2262–2271.
16. Koziara JM, Lockman PR, Allen DD, et al. In situ blood-brain barrier transport of nanoparticles. Pharm Res 2003; 20:1772–1778.
17. Lockman PR, Koziara JM, Mumper RJ, et al. Nanoparticle surface charges alter blood-brain barrier integrity and permeability. J Drug Target 2004; 12:635–641.
18. Yang S, Zhu J, Lu Y, et al. Body distribution of camptothecin solid lipid nanoparticles after oral administration. Pharm Res 1999; 16:751–757.
19. Reddy LH, Sharma RK, Chuttani K, et al. Etoposide -incorporated tripalmitin nanoparticles with different surface charge; formulation, characterization, radiolabeling, and biodistribution studies. AAPS J 2004; 6:e23.
20. Wang JX, Sun X, Zhang ZR. Enhanced brain targeting by synthesis of 3′,5′-dioctanoyl-5-fluoro-2′-deoxyuridine and incorporation into solid lipid nanoparticles. Eur J Pharm Biopharm 2002; 54:285–290.
21. Jieseng Y, Qun W, Xuefeng Z, et al. Injectable actarit loaded solid lipid nanoparticles as passive targeting therapeutic agents for rheumatoid arthritis. Int J Pharm 2008; 352:273–279.
22. Sarmento B, Martyins S, Ferreira D, et al. Oral insulin delivery by means of solid lipid nanoparticles. Int J Nanomed 2007; 2:743–749.
23. Battaglia L, Trotta M, Gallarate G, et al. Solid lipid nanoparticles formed by solvent in water emulsion technique: Development and influence on insulin stability. J Microencapsul 2007; 24:660–672.

24. Liu J, Gong T, Fu H, et al. Solid lipid nanoparticles for pulmonary delivery of insulin. Int J Pharm 2008; 356:333–344.
25. Xiang QY, Wang MT, Chen F, et al. Lung-targeting delivery of dexamethasone acetate loaded solid lipid nanoparticles. Arch Pharm Res 2007; 30(4):519–525.
26. Ugazio E, Cavalli R, Gasco MR. Incorporation of cyclosporin A in solid lipid nanoparticles in solid lipid nanoparticles. Int J Pharm 2002; 241:341–344.
27. Cavalli R, Caputo O, Gasco MR. Preparation and characterization of solid lipid nanospheres containing paclitaxel. Eur J Pharm Sci 2000; 10:305–308.
28. Fundarò A, Cavalli R, Bargoni A, et al. Stealth and non-stealth solid lipid nanoparticles (SLN) carrying doxorubicin: Pharmacokinetics and tissue distribution after IV administration to rats. Pharmacol Res 2000; 42:337–343.
29. Cavalli R, Bargoni A, Podio V, et al. Duodenal administration of solid lipid nanoparticles loaded with different percentages of tobramycin. J Pharm Sci 2003; 92:1085–1094.
30. Dianzani C, Cavalli R, Zara GP, et al. Cholesteryl butyrate solid lipid nanoparticles inhibit adhesion of human neutrophils to endothelial cells. Br J Pharmacol 2006; 148:648–656.
31. Morel S, Cavalli R. Thymopentin in solid lipid nanoparticles. Int J Pharm 1996; 132:259–262.
32. Brioschi A, Calderoni S, Pradotto LG, et al. Solid lipid nanoparticles carrying oligonucleotides inhibit vascular endothelial grow factor expression in rat glioma models. J Nanoneurosci 2009; 1:1–10.
33. Rezzani R, Fraschini F, Gasco MR, et al. Melatonin delivery in solid lipid nanoparticles: Prevention of cyclosporin A induced cardiac damage. J Pineal Res 2009; 46:255–261.
34. Mauro A, Pradotto L, Cattaldo S, et al. A new apomorphine formulation for oral administration. Neurol Sci 2007; 28(suppl):S6.
35. Priano L, El Assawy N, Gasco M, et al. Baclofen-loaded solid lipid nanoparticles: H-reflex modulation study, behavioural characterization and tissue distribution in rat after intraperitoneal administration. Neurol Sci 2008; 29(suppl):S443–S444.
36. Morel S, Terreno E, Ugazio E, et al. Relaxometric investigations of solid lipid nanoparticles (SLN) containing gadolinium (III) complexes. Eur J Pharm Biopharm 1998; 45:157–163.
37. Peira E, Marzola P, Podio V, et al. In vitro and in vivo study of solid lipid nanoparticles loaded with superparamagnetic iron oxide. J Drug Target 2003; 11:19–24.
38. Miglietta A, Cavalli R, Gasco MR, et al. Cellular uptake and cytotoxicity of solid lipid nanospheres (SLN) incorporating doxorubicin or paclitaxel. Int J Pharm 2000; 210: 61–67.
39. Serpe L, Cavalli R, Gasco MR, et al. Intracellular accumulation and cytotoxicity of doxorubicin with different pharmaceutical formulations in human cancer cells. J Nanosci Nanotechnol 2006; 6:3062–3069.
40. Bargoni A, Fundaro A, Zara GP, et al. Solid lipid nanoparticles in lymph and plasma after duodenal administration to rats. Pharm Res 1998; 15:745–750.
41. Podio V, Zara GP, Carazzone M, et al. Biodistribution of stealth and non-stealth solid lipid nanoparticles after intravenous administration to rats. J Pham Pharmacol 2001; 52:1057–1063.
42. Zara GP, Cavalli R, Gasco MR, et al. Intravenous administration to rabbits of non-stealth and stealth doxorubicin loaded solid lipid nanoparticles at increasing concentration of stealth agent: Pharmacokinetics and distribution of doxorubicin in brain and in other tissues. J Drug Target 2002; 10:327–335.
43. Gasco MR, Bargoni A, Cavalli R, et al. Transport in lymph and blood of solid lipid nanoparticles after oral administration in rats. Presented at the Proceedings of the 24th International Symposium on Controlled Release of Bioactive Materials, Stockholm; 1997:179–180.
44. Cavalli R, Morel S, Gasco MR, et al. Preparation and evaluation in vitro of colloidal lipospheres containing pilocarpine as ion-pair. Int J Pharm 1995; 117:2434–2246.
45. Cavalli R, Morel S, Gasco MR, et al. Evaluation in vitro/in vivo of colloidal lipospheres containing pilocarpine as ion-pair. Presented at the APGI 7th International Conference on Pharmaceutical Technology, Budapest, Hungary; 1995:801–802.
46. Cavalli R, Gasco MR, Chetoni P, et al. Solid lipid nanoparticles (SLN) as ocular delivery system for tobramycin. Int J Pharm 2002; 238:241–245.

47. Tondelli L, Ricca A, Laus M, et al. Highly efficient cellular uptake of c-myb antisense oligonucleotides through specifically designed polymeric nanospheres. Nucleic Acids Res 1998; 26:5425–5431.
48. Aukunuri JV, Ayalasomayajula SP, Kompella UB. Nanoparticle formulation enhances the delivery and activity of a vascular endothelial growth factor antisense oligonucleotide in human retinal pigment epithelial cells. J Pharm Pharmacol 2003; 22:1199–1206.
49. Zara GP, Bargoni A, Cavalli R, et al. Idarubicin solid lipid nanospheres administration to rats by duodenal route: Pharmacokinetics and tissues distribution. J Pharm Sci 2002; 91:1324–1333.
50. Cavalli R, Zara GP, Caputo O, et al. Transmucosal transport of tobramycin incorporated in SLN after duodenal administration; part I: Pharmacokinetic study. Pharmacol Res 2000; 42:541–545.
51. Mallo C, Zaidan R, Galy G, et al. Pharmacokinetics of melatonin in man after intravenous infusion and bolus injection. Eur J Clin Pharmacol 1990; 38:297–301.
52. DeMuro RL, Nafziger AN, Blask DE, et al. The absolute bioavailability of oral melatonin. J Clin Pharmacol 2000; 40:781–784.
53. Priano L, Esposti D, Esposti R, et al. Solid lipid nanoparticles incorporating melatonin as new model for sustained oral and transdermal delivery systems. J Nanosci Nanotechnol 2007; 7:3596–3601.
54. Gasco MR. Industrial Application of Microemulsions. New York: Marcel Dekker, 1997:97–123.
55. Heuschkel S, Goebel A, Neibert RHH. Microemulsions – Modern colloidal carrier for dermal and transdermal drug delivery. J Pharm Sci 2008; 97:603–631.
56. Bonina FP, Montenegro L, Scrofani L, et al. Effects of phospholipids based formulations on in vitro and in vivo percutaneous absorption of methyl nicotinate. J Control Release 1995; 34:53–63.
57. Teichmann A, Heuschkel S, Jacobi U, et al. Comparison of stratum corneum penetration and localization of a lipophilic model drug applied in an o/w microemulsion and an amphiphilic cream. Eur J Pharm Biopharm 2007; 54:176–181.
58. Bolzinger MA, Carduner TC, Poelman MC. Bicontinuous sucrose ester microemulsion: A new vehicle for topical delivery of niflumic acid. Int J Pharm 1998; 176:39–45.
59. Paolino D, Ventura CA, Nistico S, et al. Lecithin microemulsions for the topical administration of ketoprofen. Percutaneous adsorption through human skin and "in vivo" human skin tolerability. Int J Pharm 2002; 244:21–31.
60. Gasco MR, Gallarate M, Pattarino F. In vitro permeation of azelaic acid from viscosized microemulsions. Int J Pharm 1991; 69:193–196.
61. Gasco MR, Gallarate M, Pattarino F, et al. Effect of azelaic acid in microemulsion on Lentigo Maligna. Presented at the Proceedings of 17th International Symposium on Controlled Release of Bioactive Materials, Reno, NV; 1990:419–420.
62. Peira E, Scolari P, Gasco MR. Transdermal permeation of apomorphine through hairless mouse skin from microemulsions. Int J Pharm 2001; 226:47–51.
63. Priano L, Albani G, Brioschi A, et al. Transdermal apomorphine permeation from microemulsions: A new treatment in Parkinson's disease. Mov Disord 2004; 19:937–942.
64. Gancher ST, Nutt JG, Woodward WR. absorption of apomorphine by various routes in parkinsonism. Mov Disord 1991; 6:212–216.
65. van der Geest R, van Laar T, Gubbens-Stibbe JM, et al. Iontophoretic delivery of R-apomorphine; part II: An in vivo study in patients with Parkinson's disease. Pharm Res 1997; 14:1804–1810.
66. Priano L, Albani G, Brioschi A, et al. Nocturnal anomalous movement reduction and sleep microstructure analysis in parkinsonian patients during 1-night transdermal apomorphine treatment. Neurol Sci 2003; 24:207–208.

14 Microscopic and Spectroscopic Characterization of Nanoparticles

Jose E. Herrera and Nataphan Sakulchaicharoen

Department of Civil and Environmental Engineering, University of Western Ontario, London, Ontario, Canada

INTRODUCTION

Nanomaterials, specifically nanoparticles, are, without a doubt, key components in the development of new advanced technologies. Although nanoparticles are perhaps the simplest of nanostructures, nanoparticle-based technologies are broadly covering different fields, ranging from environmental remediation, energy generation, and storage all the way to applications in bioscience (1–5).

The need to fine-tune different nanoparticle properties to make them suitable for specific applications has sparked a large number of worldwide research efforts aimed at their tailoring. However, full use of these structures in these applications requires more detailed information and a feedback of data coming from reliable characterization techniques (6–8). Several methods have been applied to obtain this information and some of them are described in different chapters of this book. In general, most of these techniques comprise local probes, such as scanning electron microscopy (SEM), transmission electronic microscopy (TEM), electron diffraction, scanning tunneling microscopy, and atomic force microscopy, with bulk-sensitive probes such as optical absorption spectroscopy, infrared (IR) spectroscopy (Fourier transform IR), and Raman scattering, and X-ray–based techniques such as X-ray diffraction, X-ray photoelectron spectroscopy, and X-ray absorption (X-ray absorption near edge structure and extended X-ray absorption fine structure).

In this contribution, an overview of the recent progress in nanoparticle characterization is presented. Some of the aforementioned methods will be introduced and the kind of information that can be obtained from them will be discussed. However, a detailed account of a specific characterization method and its variations is outside the scope of this review.

ELECTRON MICROSCOPY

Since diffraction effects restrict the resolution of optical microscopy, structures smaller than 1 μm cannot be observed with light. Therefore, if imaging at considerably higher resolution is required, electromagnetic radiation of shorter wavelengths must be used. Electron beams present this possibility. The development of electron microscopes has resulted in instruments that are able to routinely achieve magnifications of the order of 1 million and that can disclose details with a resolution of up to about 0.1 nm.

When an electron beam interacts with a sample, many measurable signals are generated and electrons can be transmitted, backscattered, and diffracted. TEM uses the transmitted electrons to form a sample image, while SEM uses backscattered electrons and secondary electrons emitted from the sample. Depending on the sample thickness, transmitted electrons pass through it without suffering significant energy loss. Since the attenuation of the electrons depends mostly on the density and thickness of the sample, the transmitted electrons form a two-dimensional projection of the sample. This is the basis for TEM imaging. Electrons can also get diffracted by particles if these are favorably oriented toward the electron beam; the crystallographic information that can be obtained from these diffracted electrons is the basis for electron diffraction. Finally, the electrons in the primary beam can collide with atoms in the sample and be scattered back, or, in turn, remove more electrons from these atoms (secondary electrons). These two processes (backscattering and generation of secondary electrons) are more effective as the atomic number of the atom increases. If the primary electron beam is targeted over the sample surface and the yield of secondary or backscattered electrons is plotted as a function of the position of the primary electron beam, it is possible to obtain three-dimensional images of the samples analyzed; this method is the basis for SEM.

Transmission Electron Microscopy

The progress made in TEM has enabled the direct imaging of atomic structures in solids and surfaces. Nanometer-sized particles are commonly present in many different types of materials and the use of TEM allows for gathering information about particle size, shape, and any surface layers or absorbates (9,10). More recently, changes in nanoparticle structure as a result of interactions with gas-, liquid-, or solid-phase substrates can now be monitored by this technique (11).

In recent years, a large number of new and novel developments have been made in electron microscopy for nanotechnology. This includes new techniques such as in situ microscopy used for imaging dynamic processes, quantitative chemical mapping, holographic imaging of electric and magnetic fields, and ultrahigh-resolution imaging (12). For instance, the study of nanoparticles can be greatly improved with the use of aberration-corrected lenses, enabling image resolutions at levels sometimes lower than 1 Å (13,14). This level of image resolution yields a new level of understanding of the behavior of matter at the nanoscale.

An important precaution to be taken into consideration when performing TEM measurements on nanoparticle-containing samples is that they can be susceptible to the highly energetic electron beam of the TEM instrument (15). Beam susceptibility makes it very difficult sometimes to carry out electron diffraction studies on nanoparticles that are prone to beam damage. In this case, by using low electron beam currents, it is possible to obtain lattice fringe images and electron diffraction.

Figure 1 shows an example of a study using an aberration-corrected electron microscope to study the structure and morphology of AuPd bimetallic particles (16). The authors matched the experimental intensities of atomic columns with theoretical models of three-layered AuPd nanoparticles in different orientations. Based on this information, the authors indicated that the surface layer of the metallic nanoparticles contains kinks, terraces, and steps at the nanoscale. The effect of adding the second metal induced the formation of such defects. Figure 1(A) shows a dark-field micrograph of several AuPd nanoparticles. The inset indicates the authors' proposed sketch of element-rich locations in the layers. The intensity profile through a typical AuPd nanoparticle is displayed in Figure 1(B), and it depends on the atomic number and the column thickness. Figure 1(C) shows

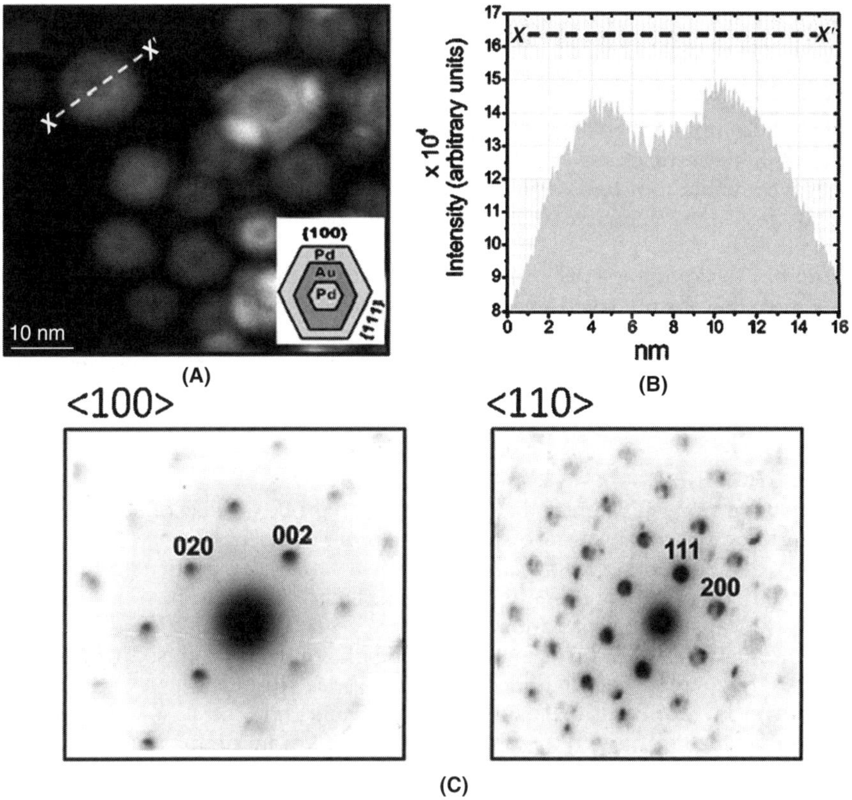

FIGURE 1 An aberration-corrected electron microscope to study the structure and morphology of AuPd bimetallic particles. (**A**) Transmission electronic microscopy image of AuPd nanoparticles. The contrast is due to a core–shell structure consisting of three layers as sketched in the inset. (**B**) The experimental intensity profile shows a lower magnitude on the central portion on the particle. (**C**) Nanodiffraction patterns of an individual AuPd nanoparticle showing a single-crystalline structure. Two orientations, <1 0 0> and <1 1 0>, are observed. *Source*: From Ref. 16.

the microdiffraction patterns obtained on two particles showing a single crystalline structure and two orientations (<1 0 0> and <1 1 0>).

In spite of all these advantages, TEM imaging still presents a series of challenges. For instance, image overlap is a typical problem during observation. When this occurs, the surrounding matrix usually tends to mask the supported nanoparticles. In some special cases, however, the existence of an epitaxial relationship between the nanoparticles and their support can be used to obtain size and shape information (17). Moreover, nanoparticles can be susceptible to damage under the electron beam irradiation conditions normally used for high-resolution imaging.

Scanning Electron Microscopy

SEM is, to a certain extent, a limited tool to characterize nanoparticles. The main problem with the application of SEM to nanoparticle characterization analysis is that sometimes it is not possible to clearly differentiate the nanoparticles from the substrate. Problems become even more exacerbated when the nanoparticles under study have tendency to adhere strongly to each other, forming agglomerates. In contrast to TEM, SEM cannot resolve the internal structure of these domains.

Nevertheless, SEM can yield valuable information regarding the purity of a nanoparticle sample as well as an insight on their degree of aggregation. Moreover, when nanoparticles are part of secondary and tertiary nanostructures, SEM becomes a valuable tool to assess their location (18). Furthermore, SEM imaging can clearly reveal the degree of dispersion and uniformity of metallic nanoparticles over substrates.

The big disadvantage of both SEM and TEM in this context is that one can never be sure that the observed image is truly representative of the bulk nanoparticle sample. Consequently, bulk-sensitive methods that provide information regarding the quality, size, and structural properties of a given sample must be employed. Among these methods, Raman spectroscopy and optical absorption deliver the most comprehensive results.

OPTICAL ABSORPTION

A very effective analysis method that can be used to probe the size of nanoparticles is through their optical absorption spectra (19–21). This technique is based on the well-known phenomenon of light absorption by a sample. In particular, the information obtained on the band energy gap is extremely useful to evaluate the dispersion and local structure of nanoparticles formed by d^0 transition metal oxides, sulfides, and selenides (22–26). Several methods have been proposed to estimate the band energy gap of these materials by using optical absorption spectroscopy. A general power law form has been suggested by Davis and Mott (27), which relates the absorption coefficient with the photon energy. The order of this power function is determined by the type of transition involved.

For instance, in the particular case of tungsten oxide nanoparticles, Barton et al. recommended using the formalism of an indirect allowed transition and therefore to use the square root of the Kubelka-Munk function multiplied by the photon energy (24). By plotting this new function versus the photon energy, the position of the absorption edge can then be determined by extrapolating the linear part of the rising curve to zero (24,26). The values thus obtained carry information about the average domain size of the oxide nanoparticles since, as the case of the particle in a box, the energy band gap decreases as the particle size increases (28). Based on the position of the absorption bands, a relative comparison can be made between the energies of the samples under investigation and those of references of a known particle size. Figure 2 shows one of such analysis; here, the authors compared absorption edge values obtained for several tungsten oxide species of known domain size to those of nine different tungsten oxide nanoparticle samples. These materials were obtained over a SBA-15 mesoporous substrate by using atomic layer deposition techniques at different temperatures (29). For the case of the samples obtained at 400°C, using different tungsten oxide loadings (Fig. 2(A)), the values obtained (3.0–3.8 eV) lie closer to the values corresponding to $(NH_4)_{10}W_{12}O_{41}$, indicating an apparent octahedral coordination environment for these samples. Moreover, the variation on the edge energy values clearly indicated that the average size of these tungsten oxide nanoparticles changes when the overall tungsten oxide loading in the substrate increases. In fact, previous studies by Barton et al. (24) have

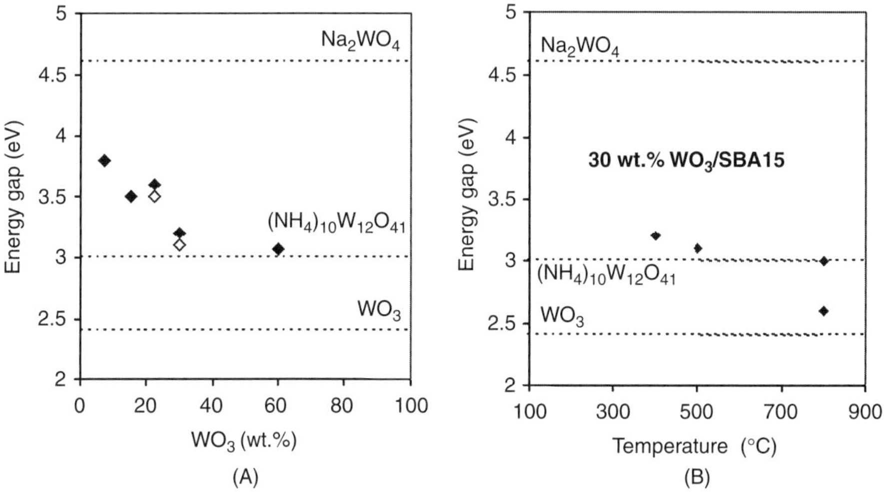

FIGURE 2 Edge energy values obtained for different WO_x nanoparticle samples. Edge energy values obtained for different WO_x nanoparticle samples supported on SBA15 as a function of (**A**) tungsten loading and (**B**) synthesis temperature. The values corresponding to analytical references of known domain size are also included for reference as dashed lines in the plot. *Source*: From Ref. 29.

suggested that for edge energy values above 3.5 eV, tungsten oxide nanoparticles do not interact with each other to form bridging W–O–W bonds and that these species exist in a distorted octahedral symmetry.

In a similar way, in Figure 2(B), the authors compared the edge energy values obtained from tungsten oxide nanoparticle samples prepared using a total tungsten oxide loading of 30% wt at different temperatures. First, the edge energy value (3.2 eV) for the sample prepared at 400°C corresponded to highly dispersed WO_x nanoparticles. In the case of the sample obtained at 500°C, a shift to lower energies is observed in the edge energy value. This is consistent with the formation of slightly larger WO_x domains, probably through condensation (formation of bridging W–O–W bonds) at the expense of W–OH sites as the authors observed in another study by using nuclear magnetic resonance spectroscopy (30). A contrasting behavior is observed when the same was prepared at 800°C. The authors reported that for this sample, the optical absorption spectra showed two different regions. A tail was observed at values below 3.0 eV, which is a clear indication of the formation of large agglomerates of WO_3 species, which agreed with X-ray diffraction results obtained in another study (31). However, part of the optical absorption spectra also showed edge energy values close to 3.0 eV, which the authors attributed to the formation of tungsten oxide nanoparticles, even at such a high temperature.

While for the case of tungsten oxide nanoparticles, an indirect allowed transition formalism yields the best way to obtain edge energy values, in the case of vanadium oxide nanoparticles, it seems that this choice is based mostly on the best linear fit of the energy gap curve (32–34). Wachs and colleagues have suggested the use of direct allowed transition formalism in the Davis and Mott correlation as

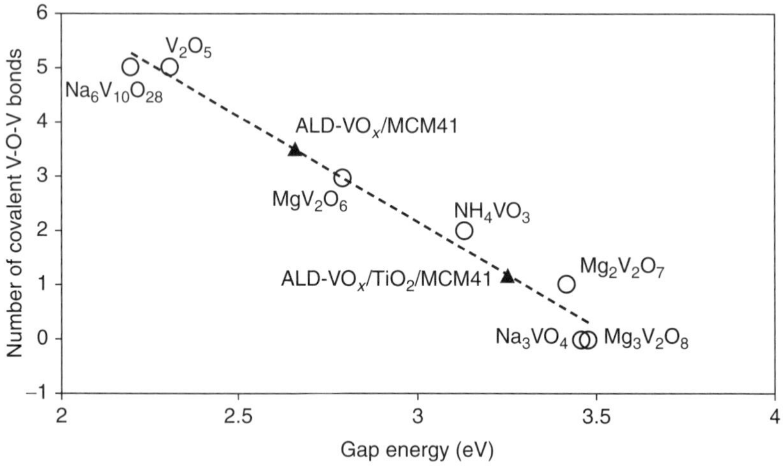

FIGURE 3 Average number of covalent V–O–V bonds present in different vanadium oxide nanoparticle samples. The values corresponding to references of vanadia are also included. *Source*: From Ref. 29.

the best way to obtain a linear fitting. The edge energy values obtained for three different supported vanadia nanoparticle–containing samples by using this correlation are shown in Figure 3. Here, the results are compared with the values for some reference samples of known domain size obtained by Gao and Wachs (33). This comparison supports the view of the formation of vanadia oxide nanoparticles with small sizes for a sample obtained using atomic layer deposition by using TiO_2/MCM41 as the substrate (ALD VO_x/TiO_2/MCM41). Indeed, if the correlation developed by Gao and Wachs (33) is used to obtain the average number of covalent V–O–V bonds (CVB), a value of 1.2 is obtained for this sample. In contrast, the energy band gap value obtained for vanadia nanoparticle samples prepared over a bare MCM41 substrate (ALD VO_x/MCM41) corresponds to an average number of CVB close to 3.5, clearly indicating how the average size of the vanadium oxide nanoparticle is affected by the existence of the titanium oxide phase.

While in the case of supported metal oxide nanoparticles, the information on particle size obtained from the optical absorption spectra is at best semiquantitative, recent reports indicated that for the case of metallic nanoparticles, optical absorption spectra can indeed provide accurate values for the quantification of size and clearly compete with well-established methods such as light scattering (35). Two recent studies report the development of experimental correlations between the size of gold nanoparticles and the concentration with its optical absorption spectra (36,37). However, these two methods seemed limited to particles with ideal spherical shapes. A more recent report provides quantitative relationships between Au nanoparticle size and concentration, accounting for the deviation of the particle size from ideal spheres (38). Figure 4 shows part of the results obtained in this study. The authors performed a long-term collection of experimental measurements of the extinction resonance position as a function of the mean equivolume particle

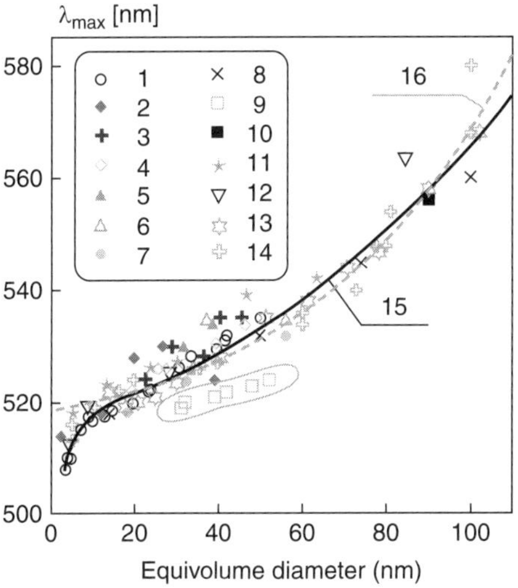

FIGURE 4 Visible absorption. Position of the maximum observed in the visible absorption as a function of the mean equivolume particle diameter for gold nanoparticles dispersed in water. *Source*: From Ref. 38.

diameter. The solid line presents an averaged calibration curve, described by the following equation (all quantities in nanometers):

$$d = \begin{cases} 3 + 7.5 \times 10^{-5} X^4, \ X < 23 \\ \left[\sqrt{X - 17} - 1\right]/0.06, \ X \geq 23 \end{cases} \qquad X = \lambda_{max} - 500$$

In spite of the evident distribution of experimental points around this curve, the authors claimed that there is a general agreement between most of the data. This remarkable result clearly showed that not only nanoparticle size but also nanoparticle concentration in liquid phases can be accurately determined from optical absorption spectra, provided that shape effects are taken into consideration.

RAMAN SPECTROSCOPY

Among the several techniques used to characterize nanomaterials, Raman spectroscopy is perhaps the most powerful tool to get information on their vibrational and electronic structures (39). Raman spectroscopy is based on the inelastic scattering of visible light by matter. Light scattering may be elastic or inelastic. Elastic scattering is the most common phenomenon and occurs without loss of photon energy (i.e., without any change in the frequency of the original wave). In contrast, a very small fraction of the incoming radiation undergoes inelastic scattering, in which the scattered wave compared with the incoming wave results in a different frequency. This frequency difference is called the Raman shift, which can be positive or negative. If, upon collision, the photon loses some of its energy, the resulting radiation has a positive Raman shift (Stokes radiation). In contrast, when the incoming photons gain energy, the resulting radiation has higher frequencies (anti-Stokes radiation) and a negative Raman shift is observed.

If an instrument with adequate bandwidth is used to measure the energy of both the incoming and outgoing light, it is observed that both the Stokes and anti-Stokes radiations are composed of discrete bands, which are intimately related to molecular vibrations of the substance under investigation. The information obtained by measuring the Raman shift is therefore most valuable since direct information at the molecular level of the nature of the chemical bonds and symmetry on the material under investigation can be obtained. In the case of solids, the most elementary processes are associated with degrees of freedom of ions and electrons in crystalline and amorphous solids, with the only exception of long acoustic phonons and acoustic magnons, which are associated with Brillouin scattering (40).

A limitation of Raman spectroscopy is the extremely low quantum efficiencies associated with the process (almost 1 photon is inelastically scattered for every 10^6 photons that interact with the sample). Thus, a very intense light source must be used to get a signal strong enough to measure satisfactorily the Raman shift. Moreover, since a very precise measure of the frequency of the incoming light is needed to calculate the Raman shift, the use of a monochromatic excitation light is preferred. A laser light source satisfies both conditions: monochromaticity and high intensity; it is thus the obvious choice for excitation light source to perform Raman spectroscopy.

The low efficiency of Raman scattering can be counterbalanced by an intensity enhancement due to the so-called resonant Raman effect. This resonant effect occurs when the photon energy of the exciting or scattered light beam matches the energy of an allowed optical electronic transition of a chromophoric group within the sample. Excitation within the absorption band of the sample then results in the selective enhancement of just those vibrational modes on the sample that selective couple with the oscillating dipole moment induced by the excitation electric field (41). The intensities of the resonance Raman-enhanced bands can increase as much as 10^8-fold. Thus, it is possible to selectively study the vibrational spectra of very dilute samples or chromophores in solids by choosing excitation wavelengths in resonance with a particular analyte chromophore. However, just the intensity of Raman bands associated to this resonant process will be amplified while all other bands will fade away on the spectrum background.

Another mechanism for the intensity enhancement of Raman signals is through the excitation of localized surface plasmons (SPs) (42). Similarly in this case, it is necessary that the incident laser light used to obtain the spectra and the Raman signal are in or near resonance with the plasmon frequency (43). This situation is satisfied normally for the case of organic molecules adsorbed on metals. The technique is then generally known as surface enhanced Raman spectroscopy (SERS). In general the choice of surface metal is dictated by the substrate's plasmon resonance frequency. Silver and gold are typical metals for SERS experiments because their plasmon resonance frequencies fall within the wavelength ranges of typical lasers used to perform Raman experiments, providing maximal enhancement for visible and near-IR light. Copper is another metal whose absorption spectrum falls within the range acceptable for SERS experiments (7).

While in general Raman, resonant Raman, and SERS vibrational spectra of metallic nanoparticles are widely used for chemical, bioanalytical, and biomedical applications (44–48), the study of the correlation between the characterization of the nanoparticle itself with these spectra is rather limited. Recent studies indicate, however, that there is a correlation between the SP resonance properties of gold nanoparticles immobilized on gold substrates and the resonance-enhanced Raman

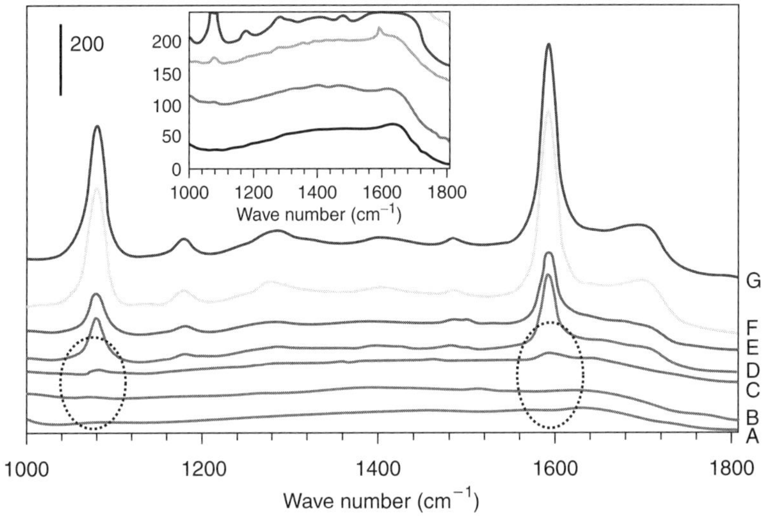

FIGURE 5 Surface enhanced Raman spectra of 4-mercaptobenzoic acid adsorbed on gold nanoparticles of different sizes in aqueous solution (**A**) 30 nm, (**B**) 40 nm, (**C**) 50 nm, (**D**) 60 nm, (**E**) 70 nm, (**F**) 80 nm, and (**G**) 90 nm. (*Inset*): Magnified view of circled curves. *Source*: From Ref. 36.

scattering spectroscopic properties (49). Indeed, it has been recently reported that the SP band undergoes a red shift as the particle size increases and the substrate particle distance decreases. Moreover, the intensity of this Raman signal was shown to be tunable by varying the particle size, particle–substrate separation, and substrate material.

These results can be used to link nanoparticle diameters with the values obtained from the Raman spectra. Since the vibrational and electronic structures can be probed at the same time, direct information of the nanoparticle diameter can be obtained from the Raman spectra. Based on this effect, Raman spectroscopy now provides a particularly valuable tool to examine not just vibrational structure of nanonsystems but also the merits of theoretical modes for the phonon dispersion relations in this systems (50).

Figure 5 shows an example of this approach – the results obtained by Njoki and colleagues for a set of SERS spectra for 4-mercaptobenzoic acid adsorbed on gold nanoparticles of different sizes in aqueous solution are presented (36). The authors indicated that the size correlation with the intensity of SERS revealed that this intensity increases with particle size. Although there are many reports in the literature on SERS spectra for metallic nanoparticles, these focus on particles deposited on solid substrates (51,52). In this case, the observation of the SERS signals in the solution is noteworthy. The authors also analyzed the SERS intensity for the 1078 and 1594 cm^{-1} bands observed in Figure 5. Figure 6 shows this analysis; here the authors plotted the intensity against the particle size. These results are compared with the ones obtained for the values of the absorption maximum on the optical absorption spectra of the gold nanoparticles, again against particle size. The authors indicated that in both cases, it is possible to correlate particle size either

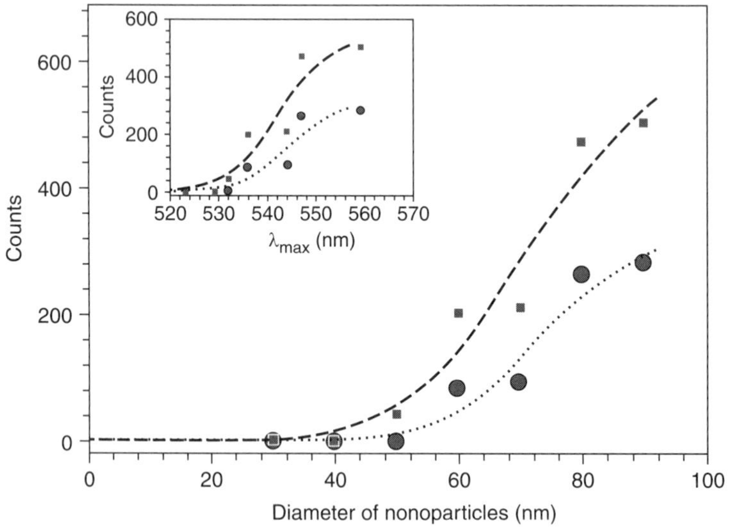

FIGURE 6 Plots of the intensity of the surface enhanced Raman spectra. Plots of the intensity of the surface enhanced Raman spectra for 4-mercaptobenzoic acid adsorbed on gold nanoparticles from 1078 cm^{-1} (filled circles) and 1594 cm^{-1} bands (filled squares) as observed in Figure 6 against the particle size. The inset shows the results obtained by the authors when a similar correlation is performed using the maximum on the gold nanoparticles absorption spectra instead of the surface enhanced Raman intensity. *Source*: From Ref. 36.

with the absorption maximum in the optical absorption spectra or with the SERS intensity. This size correlation of the SERS demonstrates the validity of determining the wavelength of the SP resonance band probed by SERS as a measure of the particle size.

OUTLOOK

The examples presented herein show how different techniques can be used to characterize different types of nanoparticles. Although electron microscopy can be used not only to obtain direct images of nanoparticles but also to monitor and understand dynamic effects, it is extremely difficult, if not impossible, to extrapolate the information obtained to the full collection of nanoparticles in the same sample. Although electron microscopy provides essential information for the development of nanoscale materials and devices for various applications, it not feasible to warranty that the observed images truly represent the bulk nanoparticle sample. Complementary spectroscopic techniques such as optical absorption and Raman spectroscopy, as well as others that were not mentioned in this contribution (e.g., near-IR and IR and electron diffraction – see contributions by Kim and Moeck, respectively, in this issue), must also be incorporated to properly characterize this structure, offering many additional insights into their phenomena. Thus, it is anticipated that the role and capability of all these techniques and their combinations will only continue to increase in the future.

REFERENCES

1. Navalakhe RM, Nandedkar TD. Application of nanotechnology in biomedicine. Indian J Exp Biol 2007; 45(2):160–165.
2. Liu WT. Nanoparticles and their biological and environmental applications. J Biosci Bioeng 2006; 102(1):1–7.
3. Baker CC, Pradhan A, Shah SI. Metal nanoparticles. In: Nalwa HS, ed. Encyclopedia of Nanoscience and Nanotechnology, vol. 5. Stevenson Ranch, CA: American Scientific Publishers 2004; 5:449–473.
4. Yu W, Su S. The artificial cell design: nanoparticles. Artif Cell Cell Eng Ther 2007; 103–114.
5. Raimondi F, Scherer GG, Kötz R, et al. Nanoparticles in energy technology: Examples from electrochemistry and catalysis. Angew Chem Int Ed 2005; 44:2190–2209.
6. Hall JB, Dobrovolskaia MA, Patri AK, et al. Characterization of nanoparticles for therapeutics. Nanomedicine 2007; 2(6):789–803.
7. Sequeira M, Dixit SG. Nanoparticle characterization: Analytical techniques & instrumentation. Chem Ind Digest 1999; 12(1):60–64, 66–69.
8. Lechner MD, Machtle W. Characterization of nanoparticles. Mater Sci Forum 2000; 352:87–90.
9. Henry CR. Morphology of supported nanoparticles. Prog Surf Sci 2005; 80(3–4):92–116.
10. Brydson R, Brown A. An investigation of the surface structure of nanoparticulate systems using analytical electron microscopes corrected for spherical aberration. In: Kenneth D, Harris M, Edwards P, eds. Turning Points in Solid-State, Material and Surface Science. London: RSC Publishing, 2008:778–791.
11. Howe JM, Mori H, Wang ZL. In situ high-resolution transmission electron microscopy in the study of nanomaterials and properties. MRS Bull 2008; 33(2):115–121.
12. Wang ZL. New developments in transmission electron microscopy for nanotechnology. Adv Mater 2003; 15(18):1497–1514.
13. Hutchison JL, Titchmarsh JM, Cockayne DJH, et al. A C_S corrected HRTEM: Initial applications in materials science. JEOL News 2002; 2:37E.
14. Yamasaki J, Kawai T, Kondo Y, et al. A practical solution for eliminating artificial image contrast in aberration-corrected TEM. Microsc Microanal 2008; 14:27–35.
15. Bentley J, Gilliss SR, Carter CB, et al. Nanoscale EELS analysis of oxides: composition mapping, valence determination and beam damage. J Phys Conf Ser 2005; 26:69–72.
16. Ferrer D, Blom DA, Allard LF, et al. Atomic structure of three-layer Au/Pd nanoparticles revealed by aberration-corrected scanning transmission electron microscopy. J Mater Chem 2008; 18:2442–2446.
17. Smith DJ. Characterisation of nanomaterials using transmission electron microscopy. In: Hutchison J, Kirkland A, eds. Nanocharacterisation. Cambridge, England: The Royal Society of Chemistry, 2007:1–27.
18. Debe MK. Novel catalysts, catalyst support and catalyst coated membrane methods. In: Vielstich W, Gasteiger HA, Lamm A, eds. Handbook of Fuel Cells, Vol. 3. New York: John Wiley & Sons, Inc., 2003:576–589.
19. Roy D, Fendler J. Reflection and absorption techniques for optical characterization of chemically assembled nanomaterials. Adv Mater 2004; 16(6):479–508.
20. Norman TJ, Grant CD, Zhang JZ. Optical and dynamic properties of gold metal nanomaterials: From isolated nanoparticles to assemblies. In: Kotov N, ed. Nanoparticle Assemblies and Superstructures. Boca Raton, FL: CRC Press/Taylor & Francis, 2006:193–206.
21. Van Dijk MA, Tchebotareva AL, Orrit M, et al. Absorption and scattering microscopy of single metal nanoparticles. Phys Chem Chem Phys 2006; 8(30):3486–3495.
22. Singha A, Satpati B, Satyam PV, et al. Electron and phonon confinement and surface phonon modes in CdSe-CdS core-shell nanocrystals. J Phys Condens Mater 2005; 17(37):5697–5708.
23. Badr Y, Mahmoud MA. Effect of PbS shell on the optical and electrical properties of PbSe core nanoparticles doped in PVA. Physica B 2007; 388(1–2):134–138.
24. Barton DG, Shtein M, Wilson RD, et al. Structure and electronic properties of solid acids based on Tungsten oxide nanostructures. J Phys Chem B 1999; 103(4):630–640.

25. Gutierrez-Alejandre A, Castillo P, Ramirez J, et al. Redox and acid reactivity of Wolframyl centers on oxide carriers: Bronsted, Lewis and redox sites. Appl Catal A Gen 2001; 216(1–2):181–194.
26. Khodakov A, Yang J, Su S, et al. Structure and properties of vanadium oxide-zirconia catalysts for propane oxidative dehydrogenation. J Catal 1998; 177(2):343–351.
27. Davis EA, Mott NF. Conduction in noncrystalline systems; part V; Conductivity, optical absorption, and photoconductivity in amorphous semiconductors. Philos Mag 1970; 22(179):903–922.
28. Weber RS. Effect of local structure on the UV-visible absorption edges of molybdenum oxide clusters and supported molybdenum oxides. J Catal 1995; 151(2):470–474.
29. Herrera JE, Kwak JH, Hu JZ, et al. Synthesis of nanodispersed oxides of vanadium, titanium, molybdenum, and tungsten on mesoporous silica using atomic layer deposition. Top Catal 2006; 39(3–4):245–255.
30. Hu JZ, Kwak JH, Herrera JE, et al. Line narrowing in ^{1}H MAS spectrum of mesoporous silica by removing adsorbed H_2O using N_2. Solid State Nucl Mag 2005; 27(3): 200–205.
31. Herrera JE, Kwak JH, Hu JZ, et al. Synthesis, characterization, and catalytic function of novel highly dispersed tungsten oxide catalysts on mesoporous silica. J Catal 2006; 239(1):200–211.
32. Hossein AA, Hogarth CA, Beynon J. Optical absorption in CeO_2-V_2O_5 evaporated thin films. J Mater Sci Lett 1994; 13(15):1144–1145.
33. Gao X, Wachs IE. Investigation of surface structures of supported vanadium oxide catalysts by UV-vis-NIR diffuse reflectance spectroscopy. J Phys Chem B 2000; 104(6):1261–1268.
34. Khan GA, Hogarth CA. Optical absorption spectra of evaporated vanadium oxide (V_2O_5) and co-evaporated V_2O_5/boron oxide thin films. J Mater Sci 1991; 26(2):412–416.
35. Khlebtsov NG, Bogatyrev VA, Melnikov AG, et al. Differential light-scattering spectroscopy: a new approach to studying of colloidal gold nanosensors. J Quant Spectrosc Radiat Transfer 2004; 89(1–4):133–142.
36. Njoki PN, Lim IS, Mott D, et al. Size correlation of optical and spectroscopic properties for gold nanoparticles. J Phys Chem C 2007; 111(40):14664–14669.
37. Haiss W, Thanh NTK, Aveyard J, et al. Determination of size and concentration of gold nanoparticles from UV-vis spectra. Anal Chem 2007; 79(11):4215–4221.
38. Khlebtsov NG, Determination of size and concentration of gold nanoparticles from extinction spectra. Anal Chem 2008; 80(17):6620–6625.
39. Weinstein BA. Raman spectroscopy under pressure in semiconductor nanoparticles. Phys Status Solidi B 2007; 244(1):368–379.
40. Merlin R, Pinczuk A, Weber WH. Overview of phonon Raman scattering on solids. In: Weber WH, Merlin R, eds. Raman Scattering in Material Science. Berlin: Springer Scientific, 2000:1–219.
41. Armstrong RS, Gallagher SH, Noviandri I, et al. Solvent effects on the resonance Raman spectroscopy, electronic absorption spectroscopy and electrochemistry of fullerenes and fullerides. Fullerene Sci Technol 1999; 7(6):1003–1028.
42. Willets KA, Van Duyne RP. Localized surface plasmon resonance spectroscopy and sensing. Annu Rev Phys Chem 2007; 58:267–297.
43. Campion A, Kambhampati P. Surface-enhanced Raman scattering. Chem Soc Rev 1998; 27:241–250.
44. Cheng MM, Cuda G, Bunimovich YL, et al. Nanotechnologies for biomolecular detection and medical diagnostics. Curr Opin Chem Biol 2006; 10(1):11–19.
45. Kreibig U, Vollmer M. Spectroscopy of nanoparticles. Praxis der Naturwissenschaften, Physik in der Schule 2003; 52(4):24–28.
46. Martins R, Baptista P, Silva L, et al. Identification of unamplified genomic DNA sequences using gold nanoparticle probes and a novel thin film photodetector. J Non-Cryst Solids 2008; 354(19–25):2580–2584.
47. Rosi NL, Mirkin CA. Nanostructures in biodiagnostics. Chem Rev 2005; 105(4):1547–1562.

48. Stewart ME, Anderton CR, Thompson LB, et al. Nanostructured plasmonic sensors. Chem Rev 2008; 108(2):494–521.
49. Driskell JD, Lipert RJ, Porter MD. Labeled gold nanoparticles immobilized at smooth metallic substrates: Systematic investigation of surface plasmon resonance and surface-enhanced Raman scattering. J Phys Chem B 2006; 110(35):17444–17451.
50. Arora AK, Rajalakshmi M, Ravindran TR, et al. Raman spectroscopy of optical phonon confinement in nanostructured materials. J Raman Spectrosc 2007; 38(6):604–617.
51. Wadayama T, Oishi M. Surface-enhanced Raman spectral study of Au nano-particles/alkanethiol self-assembled monolayers/Au(1 1 1) heterostructures. Surf Sci 2006; 600(18):4352–4356.
52. Qian XM, Nie SM. Single-molecule and single-nanoparticle SERS: From fundamental mechanisms to biomedical applications. Chem Soc Rev 2008; 37(5):912–920.

15 Introduction to Analytical Scanning Transmission Electron Microscopy and Nanoparticle Characterization

Zhiqiang Chen
Institute for Advanced Materials and Renewable Energy, University of Louisville, Louisville, Kentucky, U.S.A.

Jinsong Wu
Department of Materials Science and Engineering, Northwestern University, Evanston, Illinois, U.S.A.

Yashwant Pathak
Department of Pharmaceutical Sciences, Sullivan University College of Pharmacy, Louisville, Kentucky, U.S.A.

INTRODUCTION

Scanning transmission electron microscopy (STEM) was invented by Baron Manfred von Ardenne at Siemens in 1938 (1), 6 years after the invention of the first transmission electron microscope by Knoll and Ruska in 1932 (2), and its resolution was lower than that of transmission electron microscope at that time. The STEM technique did not well develop until Albert Crewe developed the field emission gun (FEG) at the University of Chicago. Crewe and coworkers demonstrated the ability to visualize single, heavy atoms with their scanning transmission electron microscope in 1970 (3). However, atomic resolution chemical analysis by using the STEM was not reported until 1993 (4,5). Now, STEM has become an essential tool for material characterization, with the highest spatial resolution among all analytical techniques.

In STEM, images are formed with a focused electron beam scanning over and passing through an ultrathin specimen. The interaction between the electron beam and the ultrathin specimen involves elastic and inelastic scattering, which carry not only structural but also chemical information. The usefulness of STEM in the field of nanocharacterization was quickly recognized, and the technique is currently among the physical tools commonly applied in nanomaterial research (6,7), especially in nanoparticle characterization (7–9). Practical nanoparticle characterization with STEM has been focused not only on the particle shapes and size distribution but also on the structural and chemical information.

In this chapter, basic STEM instrumentation, its related techniques, as well as some experimental considerations are overviewed, followed by an introduction of the theoretical background. Examples of applications in the recurrent themes of nanoparticles are given.

BASIC THEORY

STEM is based on the interaction between high-energy electrons and a thin specimen. The electrons are a low-mass, negatively charged particle with both particle and wave characteristics. In the electron microscope, the wavelength of electrons, λ, is determined by the accelerating voltage, V (10,11):

$$\lambda = \frac{h}{\left[2m_0 eV\left(1 + \frac{eV}{2m_0c^2}\right)\right]^{\frac{1}{2}}} \tag{1}$$

where m_0 is the electron mass, e the electron charge, c the speed of light in vacuum, and h the Planck constant.

Two scattering phenomena may occur when high-energy electrons interact with the atoms of a thin specimen: elastic scattering and inelastic scattering. In elastic scattering, an incident electron beam does not suffer the transfer of energy to the specimen. In inelastic scattering, the beam loses part of energy by transferring to the specimen. When high-energy electrons interact with atoms of the specimen inelastically, the lost energy may excite the specimen atoms to generate a wide range of signals. These signals may include secondary electrons, Auger electrons, X ray, visible light, etc. (Fig. 1). Meanwhile, backscattered as well as transmitted electrons (could be inelastic or elastic) are formed by the incident electrons due to the interaction.

The elastic scattering process can be quantitatively described by the time-independent Schrödinger equation for a fast electron accelerated by a potential E and traveling through a crystal potential $V(r)$ (10,11),

$$\nabla^2\Psi(r) + \frac{8\pi^2 me}{h^2\left[E + V(r)\Psi(r)\right]} = 0 \tag{2}$$

where $\Psi(r)$ is the wave function of the electrons, m the mass of the electrons, e the charge of the electrons, and E the accelerating voltage of the electrons.

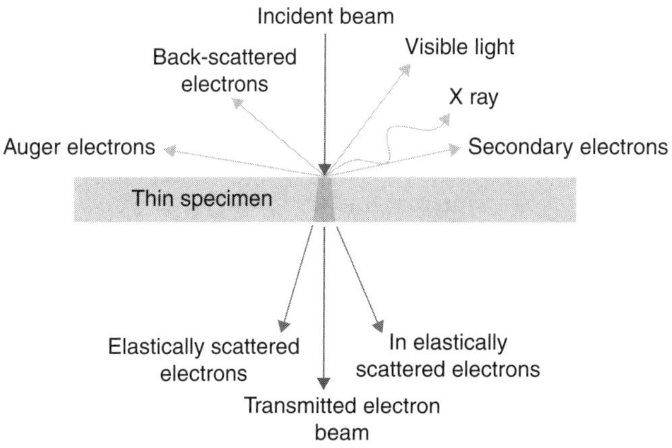

FIGURE 1 Schematic of interaction between electrons and a thin specimen.

The propagation of the incident beam in the microscope can be given by a plane wave with amplitude $|\Psi|$ and phase $2\pi kr$ (10,11),

$$\Psi = \Psi_0 e^{2\pi kr} \tag{3}$$

where k is the wave vector, $|k| = 1/\lambda$.

The elastically scattered wave is expressed by (10,11),

$$\Psi = \Psi_0 \left[e^{2\pi k_1 z} + if(\theta)\ e^{2\pi kz} \right] \tag{4}$$

where $f(\theta)$ is the atomic scattering factor.

The wavefunction in the image plane of the microscope is the convolution of the scattered wave and a point spread function $g(r)$ determined by the microscope parameters. The wavefunction at each point in the specimen is $\Psi(r)$ and each point in the image is defined as $\Phi(r)$. Each point in the image has the contributions from many points from the specimen. It can be expressed by (10,11)

$$\Phi(r) = \int \Psi(r) g(r - r')\, dr' = \Psi(r) \otimes g(r) \tag{4a}$$

where $g(r - r')$ is the weighting factor of each point in the specimen.

When absorption is taken into account by including a function $\mu(r)$, the specimen function is given by (10,11)

$$\Psi(r) = \exp\left[-i\sigma V_t(r) - \mu(r)\right] \tag{5}$$

If the specimen is thin enough, $V_t(r)$ is $\ll 1$ and the weak-phase object approximation is valid. Then, through the expansion of exponential function and neglecting $\mu(r)$ and higher order terms, $\Psi(r)$ becomes

$$\Psi(r) = 1 - i\sigma V_t(r) \tag{6}$$

So, $\Phi(r)$ becomes

$$\Phi(r) = [1 - i\sigma V_t(r)] \otimes g(r) \tag{7}$$

In STEM, images are formed with a transmitted electron beam and $g(r)$ is a probe formation function. The transmitted signals are detected by a detector. If we assume that the detector inner diameter is large enough to neglect the effect of the hole in the center of the detector, the image formation in the STEM almost corresponds to incoherent imaging conditions, and for a weak-phase object, the image intensity is (12–17)

$$I = \Psi^2(r) \left|g(r)\right|^2 \tag{8}$$

The probe formation function is an instrument function. The size and shape of the probe are dependent on the electron source, beam-defining apertures, the electron energy, and probe-forming lenses. In reciprocal space, the Fourier transformations of $\Psi(r)$ and $g(r)$ are $F(u)$ and $G(u)$. The probe amplitude $G(u)$ is given by (12–17)

$$G(u) = \int A(K) e^{+}(i\gamma(K)) e^{+}(i(Ku))dK \tag{9}$$

where u and K are two-dimensional vectors of real and reciprocal space, respectively, $A(K)$ the amplitude of the objective lens back focal plane, and $\gamma(K)$ the objective lens transfer function factor. $\gamma(K)$ is given by a function of the defocus, Δf, and the lens spherical aberration, C_s, in a system without C_s correctors,

$$\gamma(K) = \frac{\pi}{\lambda}\left(\Delta f \theta^2 \frac{1}{2} C_s \theta^4\right) = \frac{1}{2}\left(\Delta f K^2 + \frac{1}{2} C_s \frac{K^4}{\chi^2}\right) \tag{10}$$

The entire probe is coherent since the probe can be thought as a coherent superposition of plane waves but not as comprising those plane waves individually. The dynamic diffraction effect might be canceled with a proper detector geometry. The contribution of thermal diffuse scattering to the detector makes the high-angle annular dark-field (HAADF)-STEM imaging an incoherent imaging with phase flip-over (18). The contrast of the image might be contributed from the atomic number Z.

INSTRUMENTATION

There are two kinds of scanning transmission electron microscopes in use, the dedicated scanning transmission electron microscope and the commercial scanning transmission electron microscope coupled with the transmission electron microscope. Although most dedicated scanning transmission electron microscopes have a gun under the specimen and detectors above the specimen, the commercial scanning transmission electron microscope is inverse; however, the optics are quite similar. A schematic optics of an analytical scanning transmission electron microscope is shown in Figure 2. A scanning transmission electron microscope consists of an illumination system, a specimen stage, an imaging system, related attachments such

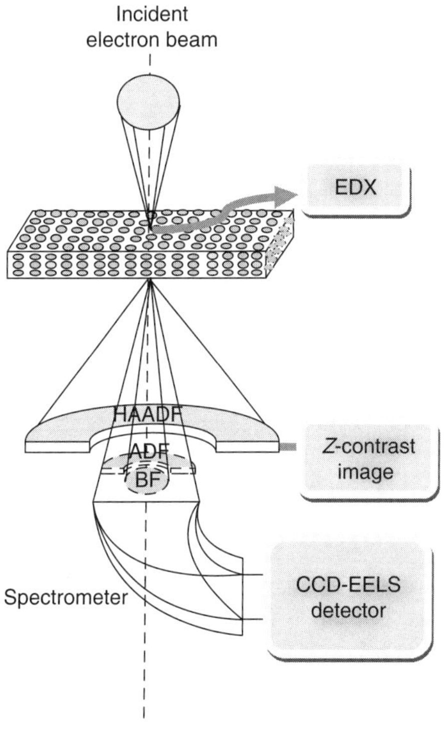

FIGURE 2 Schematic of the commercial STEM showing the geometry of bright field (BF), annular dark field (ADF), and high-angle dark-field (HAADF) detectors. Energy-dispersive X-ray (EDX) composition analysis can be performed with an EDX detector above the specimen. Electron energy loss spectrometry may also be performed with the high spatial resolution with BF and ADF detectors retracted and HAADF detectors inserted.

as energy-dispersive X-ray (EDX) spectrometer, and an electron energy loss spectrometer. These allow for the observation and analysis of materials down to the atomic scale.

Currently, two types of electron sources are commonly used in an electron microscope: thermionic and field emissions. Thermionic sources are either tungsten or lanthanum hexaboride (LaB_6) crystals. Their coherence and brightness are limited. The field emitters are ultrafine tungsten needles. The electrons generated by the FEG are of much higher coherence and brightness than those generated by thermionic sources. It is worth mentioning that STEM would not have been widely used without the invention of FEG. In FEG, electrons are generated from a very small field emission needle tip so that a small overall demagnification factor is allowed. These have made sure that a ultrasmall probe can be formed with enough beam intensity, so that signal-to-noise ratios in EDX spectroscopy and electron energy loss spectrometry (EELS) are high enough for the analytical applications of materials. Typically, the tip size is 5 nm in diameter and a demagnification of about 25 is sufficient to produce a probe size of 2 Å (19). However, FEGs require high vacuum in comparison with thermionic sources. Especially, cold FEGs operating at room temperature require extremely high vacuum so that gas absorption on the tip surface can be avoided and the electron emission barrier is low. High vacuum systems are expensive. On heating to moderate temperature, FEGs can emit electrons at relative lower vacuum. This type of thermally emitted FEGs have been widely used as the electron source in commercial FEG TEM-STEM.

An STEM image is an integration of different scattered or transmitted electrons on the detectors. Three types of detectors are usually installed on an STEM system as shown in Figure 2, namely, bright field (BF), annular dark field (ADF), HAADF detectors. The BF detector collects transmitted and scattered electrons within a semiangle of less than 10 mrad at the center of optical axis of the microscope. The correspondent BF images are relatively noisier and have lower resolution than do ADF and HAADF images (10). Its resolution is lower than that of the BF TEM image, too, in general. But the BF STEM image can be used to visualize Ronchigram shadow image, which is an essential technique for STEM optics fine alignment (20).

An ADF STEM detector is a small ring disk gathering most scattered electrons with a semiangle between 10 and 50 mrad (10). It is less noisy than a TEM DF image. As the hit of scattered electrons on the detector is adjustable by varying the camera length, the image contrast is easily improved by adjusting the camera length.

TABLE 1 Important Characteristics of the Electron Guns (10,19)

	Field emission gun			
	Cold	Thermal	LaB_6	Tungsten
Work function (eV)	4.5	4.5	2.4	4.5
Operation temperature (K)	300	2000	1700	2700
Vacuum (mbar)	10^{-11}	10^{-9}	10^{-7}	10^{-6}
Current density (A/m^2)	10^{10}	10^{10}	10^{6}	5×10^{4}
Brightness ($A/(cm^2\ sr)$)	10^{9}	10^{8}	10^{6}	10^{5}
Energy spread (eV)	0.3	0.5	1.5	3
Crossover size (μm)	<0.01	<0.01	10	50
Life time (hr)	>1000	>1000	500	100

An HAADF STEM detector is a ring disk with a hole collecting only electrons scattered through a semiangle of more than 50 mrad (10). It is larger than the ADF detector. High-angle scattered electrons are mainly caused by incoherent Rutherford scattering from a thin specimen. The scattered intensity is proportional to the square of atomic number, that is, Z^2. It is, therefore, also known as Z-contrast imaging. As shown in Figure 2, the HAADF detector allows most electrons to go through the detector. With an electron energy loss spectrometer underneath, Z-contrast imaging and chemical analysis can be performed simultaneously. More details are discussed in the next section.

STEM-RELATED TECHNIQUES

Z-Contrast Imaging

Z-contrast images are formed by the high-angle scattered (50–100 mrad at 200 kV) incoherent electrons on an annular detector (Fig. 2). At such a high angle, Rutherford scattered electrons are detected without Bragg diffraction effect (12,18). The Rutherford scattering intensity is proportional to Z^2. It is thus termed as Z-contrast imaging. At such a high angle, the contribution of thermal diffuse scattering becomes dominant. Phonon wave vectors in thermal scattering are significant in magnitude but have random phases. Each scattering event leads to a scattered wave with a slightly different wave vector and phase. The interaction of 1s state scattering with thermal diffuse scattering results in a fraction of 1s state intensity loss. The coherence effects between neighboring atomic columns are effectively averaged. Therefore, Z-contrast images show incoherent imaging characteristics. With a probe as small as 1s state atom column, each atom column can be considered as an independent scatter. The phase oscillation problem associated with the interpretation of conventional high-resolution TEM images is therefore eliminated (12–18). In thin specimens, the image intensity is dependent on the specimen composition. Heavy atoms have stronger scattering; they appear brighter in the image. Light elemental atoms have weaker scattering; they appear darker in the images.

In practice, a high-resolution Z-contrast imaging requires the electron beam probe on the exact zone axis of the crystalline specimen so that channeling effect at two-beam condition can be avoided and projected spacing of atomic column is smaller than the probe size. At the same time, the microscope should be aligned accurately. The Ronchigram, alternatively known as a shadow image or microdiffraction pattern, is the most useful method to optimize the electron probe and accurately align the microscope optics (18,20). The slight change in optical components results in apparent translations in the pattern from circular symmetry or the presence or absence of interference fringes in the pattern (18,20). The Ronchigram can be directly observed either on the microscope phosphor screen or on a TV rate-CCD camera. Camera length and positioning can be controlled with the projector lenses and shift coils. In addition, the STEM small objective aperture must be inserted to cut out high-angle rays to form Z-contrast images.

Energy-Dispersive X-Ray Spectroscopy

The incident high-energy electrons may suffer inelastic scattering and generate a wide range of secondary signals. One of the most important secondary signals is X ray. The incident electrons have enough energy to penetrate the outer shell of specimen atoms and interact with inner-shell electrons. It may result in a vacant site

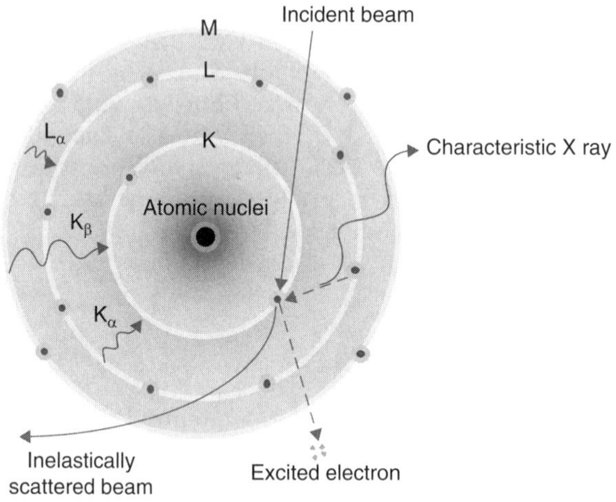

FIGURE 3 Schematic of X-ray origin. The incident electron beam causes inner-shell electron to be excited and escape the attraction of atomic nuclei. Electron hole is generated on the inner shell. The outer shell refills the inner-shell electron hole and releases X-ray photon.

at the inner shell via the excitation of an inner-shell electron to the valence or conduction band, or out-of-atom nucleus attractive field. Also, subsequently, an outer-shell electron may refill the vacant site through the release of a X-ray photon or an Auger electron with energy equal to the energy difference between the excited and final atomic states so that the ionized atom can keep its lowest energy state. As shown in Figure 3, this energy difference is unique to the atom. When the electron from the L shell fills the vacant site on the K shell, K_α X ray is released. When the electron from the M shell fills the vacant site on the K shell, K_β X ray is released (Fig. 3) (10,19).

The energy-dispersive spectrometer is an X-ray photon detection system with three main parts: the detector, the processing electronics, and the multichannel analyzer (MCA) display (Fig. 4). The detector is a reverse-biased p-i-n diode composed of silicon or germanium. Silicon is implanted with Li ions to avoid electrical breakdown. Electron–hole pairs are generated inside the detector when incoming X ray bombards on the detector. The number of electrons or holes is proportional to the energy of the incoming X ray. The electrons or holes are finally converted to a

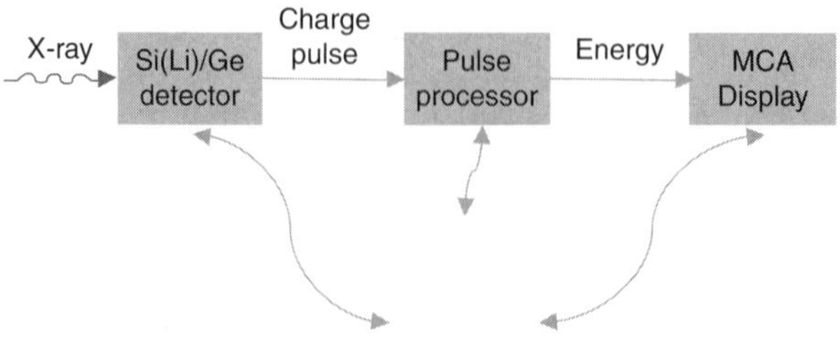

FIGURE 4 Schematic of the energy-dispersive X-ray system. X ray is collected by a Si/Ge semiconductor detector and transferred into charge pulses. Finally, charge pulses are processed and displayed by multichannel display. All the processes are controlled by a computer.

charge voltage pulse with a charge-sensitive preamplifier. Thermal energy also activates electron–hole pairs in the semiconductor detector. Hence, the detector requires liquid nitrogen to cool down the detector surface to about 90K so that noise level is low enough and detector will not be destroyed by the diffusion of Li atoms (10). Contaminations such as hydrocarbon and ice accumulation on the cold surface detector lead to an absorption of low-energy X rays. To solve this problem, a window is required to isolate the detector and the microscope chamber. This window is made from beryllium or nonberyllium materials such as polymer, diamond, boronitride, silicon nitride, or composite Al/polymer. Beryllium window hampers the passage of light elemental characteristic X ray and subsequently affects the microanalysis of light elements such as C, N, and O. Ultrathin window made from nonberyllium materials allows most of the elemental characteristic X ray to reach the detector except H, He, Li, Be, etc. (10).

After the generation of a charge pulse, the pulse is converted into a voltage pulse and amplified by the pulse processor. The final spectrum is displayed as a function of X-ray energy via the appropriate channels in the MCA display controlled by a computer.

A typical EDX spectrum consists of elemental characteristic X-ray peaks and continuous Bremsstrahlung X-ray background as shown in Figure 5. Several artifacts may appear on the EDX spectrum generated from the EDX system: escape peak, internal fluorescence peak, and sum peak. The escape peak is generated from the EDX detector. It is a satellite peak with 1.74 eV lower energy than the characteristic X-ray peaks of elements. The escape peak is recognizable in the quantitative analysis software. The software can automatically remove any escape peak during the analysis. The internal fluorescence peak is also produced from the EDX detector. After long counting times, the K_a of Si occurs in the spectrum. The intensity is so small that we could neglect it during the quantification analysis. The sum peak arises from the pulse-processing electronics. The sum peak intensity is dependent on the count rate. Reasonable count rate can avoid the appearance of a sum peak.

Energy resolution of the EDX detector is dependent on the detector electronic quality, leakage current, incomplete charge collection, and electron–hole pair fluctuations. The spectrum energy resolution varies from high to low as an increase of characteristic X-ray energy. IEEE standard energy resolution is defined as the full-width half-maximum (FWHM) of the Mn K_a. The energy resolution of the EDX detector is much lower than that of the electron energy loss spectrometer. It is generally between 130 and 140 eV.

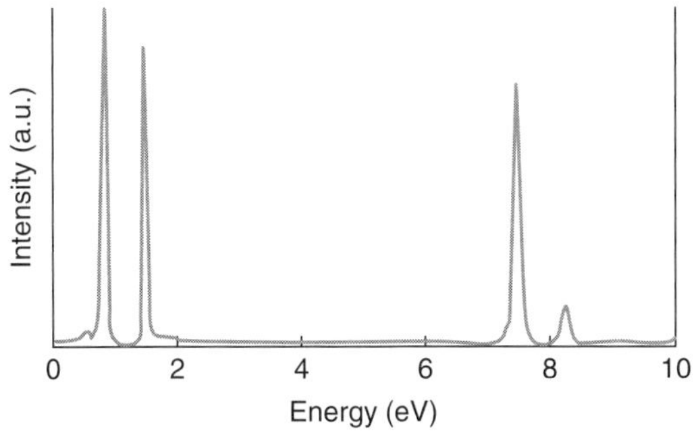

FIGURE 5 Schematic of energy-dispersive X-ray spectrum showing the characteristic X-ray peaks.

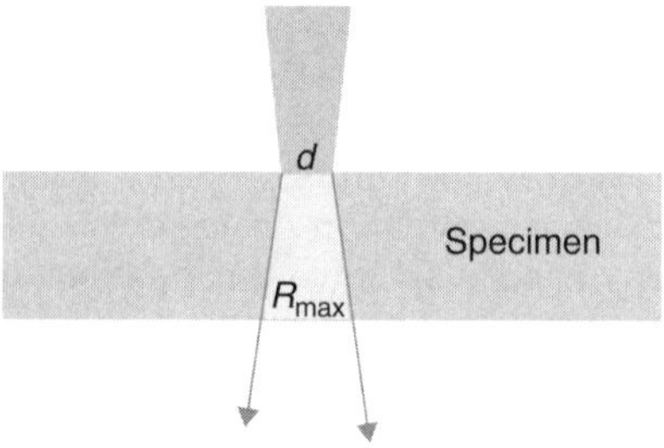

FIGURE 6 Definition of spatial resolution in a thin specimen.

Microanalysis with EDX-STEM gives a very high spatial resolution, particularly with an FEG electron source. The spatial resolution R of EDX is defined as (10)

$$R = \frac{d + R_{max}}{2} \tag{11}$$

where d is the incident beam size and R_{max} the maximum beam spread inside the thin specimen (Fig. 6).

Clearly, the spatial resolution of EDX-STEM is determined by the incident electron beam size and the beam spread. The beam spread is dependent on the specimen thickness. As the specimen in STEM is a thin foil, the excited volume observed in the bulk specimen is eliminated. The spatial resolution is thus high in STEM. Although the probe size can be as small as subangstrom in modern STEM systems with C_c and C_s correctors, relatively large probe size is required so that a reasonable signal-to-noise ratio can be achieved.

EDX QUANTIFICATION

Electron beam bombardment induces the generation of elemental characteristic X ray. It is used not only to identify the presence of elements but also to quantify the element content in the local area with electron beam interaction. In principle, quantitative analysis by using EDX-STEM is a most straightforward technique. The specimen for STEM is thin enough so that X-ray absorption and fluorescence can be neglected. According to the Cliff-Lorimer ratio method, the weight percentage of two elements A and B can be related to the integration intensities of the two-element peaks (10,19)

$$\frac{C_A}{C_B} = k_{AB}\frac{I_A}{I_B} \tag{12}$$

where C_A and C_B are the concentrations of element A and B, respectively, k_{AB} the Cliff-Lorimer k factor; and I_A and I_B the characteristic X-ray counts of elements A and B, respectively.

The k_{AB} factor of different elements can be determined experimentally with known composition of the specimen or calculated theoretically (10). These factors have been saved in the database provided by the EDX manufacturer.

Electron Energy Loss Spectroscopy

As the fast incident electrons interact with the sample, they may cause excitations of electrons in the conduction band, or discrete transitions between atomic energy levels, for example, 1s → 2p transitions. It leads the scattered electron beam to lose energy. The lost energy is related to the electron excitation energy of a given

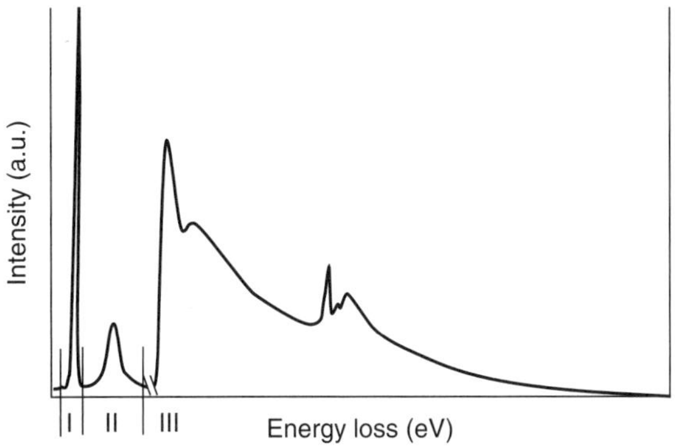

FIGURE 7 Schematic of electron energy loss spectrum. Three regions can be divided: zero-loss peak (I), low-loss region (II), and high-loss (ionization) region (III).

element. The excitation energy can be used for the composition, chemical bonding, and electron structure analysis of materials.

Due to the inelastic scattering, the beam passes through the specimen as a beam with various wavelength electrons, analogous to the visible light. Just as a glass prism can be used to separate the different colors of visible light, a magnetic prism is used to disperse various wavelength electrons (Fig. 2). The dispersed electrons, namely, EELS diffraction, are collected by a CCD camera or single or an array of diode scintillators.

Typical electron energy loss spectrum is shown in Figure 7. The spectrum can be divided into three regions (10,21,22). Region I is an intense peak composed of both unscattered and elastically scattered electrons, namely, zero-loss peak. The EELS resolution is defined as FWHM of a zero-loss peak. Generally, the zero-loss peak is used for the electron energy loss spectrometer alignment and focus. In addition, energy spread of an electron gun can be measured with the zero-loss peak.

Region II extends from the edge of the zero-loss peak out to about 50 eV. It results from the plasma excitation. The peak is thus called a plasmon peak. The electron beam energy loss, E_p, is a function of frequency, ω_p, of the generated plasma. The plasmon peak contains valuable information about the electronic structure of the valence or conduction bands (21,22).

$$E = h\omega_p = h\sqrt{\frac{4\pi e^2 \rho}{m_e}} \tag{13}$$

where ρ is the electron density and m_e the electron mass.

Plasmon peak intensity is related to the specimen thickness. It provides a practical way to measure the specimen thickness t (21,22) as follows

$$t = \lambda ln\left(\frac{I_p}{I_0}\right) \tag{14}$$

where λ is the average mean free path of electrons, I_p the intensity of the plasmon peak, and I_0 the intensity of the zero-loss peak.

Region III is the high-loss portion above 50 eV. The spectrum in this region contains information on inner- or core-shell excitation or ionization. The spectrum in this region has a smoothly decreasing background with superimposition of abrupt edges. These edges are identical to the X-ray absorption edges. These edges contain the most important information on the element binding or ionization energy and the ionization cross section. The binding energy in the electron energy loss spectrum identifies the element of interest: the intensity in the edge is proportional to the differential scattering cross section, which is given by Fermi's golden rule as (10,21,22)

$$\frac{\partial^2 I}{\partial \Omega \partial E} = \frac{4\gamma^2}{a_0^2 q^2} \left| \langle f | e^{iq-r} | i \rangle \right|^2 \rho(E) \tag{15}$$

where γ is the relativistic correction, q the momentum transfer, $\rho(E)$ the density of the final state, a_0 the Bohr radius, $|i\rangle$ the initial state of wave functions, $\langle f|$ the final state of wave functions, E the energy loss, and Ω the solid angle.

Similarly, we used the X-ray intensity to determine the relative quantities of the elemental constituents. We can determine the elemental ratio by using electron energy loss spectrum, as these characteristic edges are normally well separated and unique for atomic number Z. EELS quantifications give us an absolute value of the atomic content of the specimen. The absolute quantities N_i of element atom i in the analyzed area A is given by (10,21,22)

$$N_i = \frac{I_K(\beta, \Delta)}{I_l(\beta, \Delta)\, \sigma_K^i(\beta, \Delta)} \tag{16}$$

where β is the EELS collection angle defined as in Figure 8, Δ the energy window in the electron energy loss spectrum, $I_K(\beta, \Delta)$ the intensity integration of the element

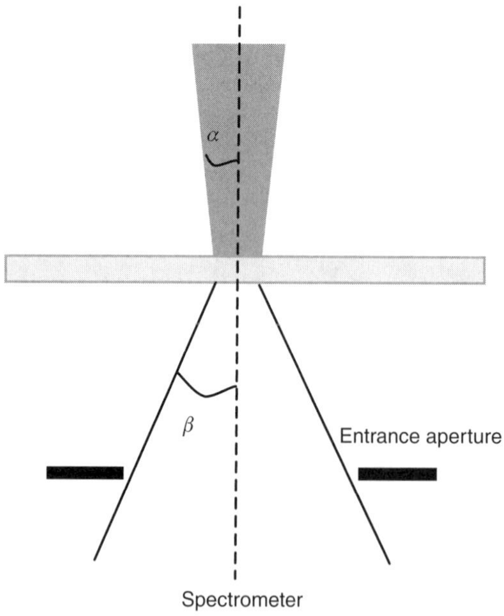

FIGURE 8 Schematic of electron energy loss spectrometry collection angle and beam convergence angle.

i's ionization edge, $I_l(\beta, \Delta)$ the intensity integration of EELS low-loss region, and $\sigma_K(\beta, \Delta)$ the partial ionization cross section.

Obviously, the ratio of two elements A and B in the specimen can be determined by (10,21,22)

$$\frac{C_A}{C_B} = \frac{I_K^A(\beta, \Delta)\,\sigma_K^B(\beta, \Delta)}{I_K^B(\beta, \Delta)\,\sigma_K^A(\beta, \Delta)} \tag{17}$$

Elemental ionization edge in EELS can provide not only information on the composition but also fingerprints of atomic bonding, coordination, or nearest neighbor distances which reflect on the ionization edge shape in the electron energy loss spectrum. The fine structure near the threshold within 30 eV is known as the energy loss near-edge fine structure (ELNES). The theory for ELNES is not well defined. But it is of great interest in practical applications. Usually, it is used to determine the chemical bonding of elements in the specimen. In addition, the extending several hundred electron volts edge spectrum known as extended energy loss fine structure provides information about the atomic positions. Details on fine structure analysis in EELS are described elsewhere (10,21,22).

Although EELS techniques appear not limited to STEM, EELS-STEM is unique because it allows information about the composition and chemical bonding of elements in the specimen to directly couple with images. As can be seen from Figure 2, the annular detector used for *Z*-contrast imaging allows the electron energy loss spectrometer to directly detect the low-angle scattering electrons used for EELS. The *Z*-contrast image can be used to position the electron probe over a particular structural feature for the acquisition of a spectrum. This ability is of great importance for nanoparticle characterization. With this technique, the surface and internal chemical bonding information of the nanoparticles can be clarified.

PRACTICE IN NANOPARTICLE CHARACTERIZATION

In STEM imaging mode, a highly focused electron probe is scanned through a thin sample (sample thickness range = 1–100 nm), while the transmitted electrons are collected by the detectors. A variety of detectors are arranged around the sample. Depending on the positions of the detectors, the STEM imaging can have BF mode, ADF mode, and HAADF mode (10,19,23). The HAADF mode is the most important one since the image contrast shows *Z*-contrast: the higher the atomic number, the brighter is the image. Meanwhile, it can collect electrons at high angles for image formation, while allowing small-angle scattering to pass through a hole in the detector to an electron energy loss spectrometer. This is one of the great strengths of the technique. Additional detectors may be arranged around the probe, such as an EDX detector, a topic that has been discussed in detail in the previous sections. Figure 9(A) shows a TEM image of Au nanoparticles on a carbon supporting film. The grid used to support the nanoparticles was purchased from Ted Palle Inc., CA, USA. The supporting film imposes a background to the image. Figure 9(B) shows an HAADF-STEM image of the same sample (from a different area). Since the element number of Au ($Z = 79$) is much higher than that of carbon ($Z = 6$), the electrons scattered to high angles from Au atoms are much more than those by carbon atoms. Thus, the image shows *Z*-contrast: while all the Au nanoparticles are lighten up, the background becomes very dim. Thus, a *Z*-contrast HAADF-STEM image can provide us a better chance to accurately measure the size of nanoparticles. In modern

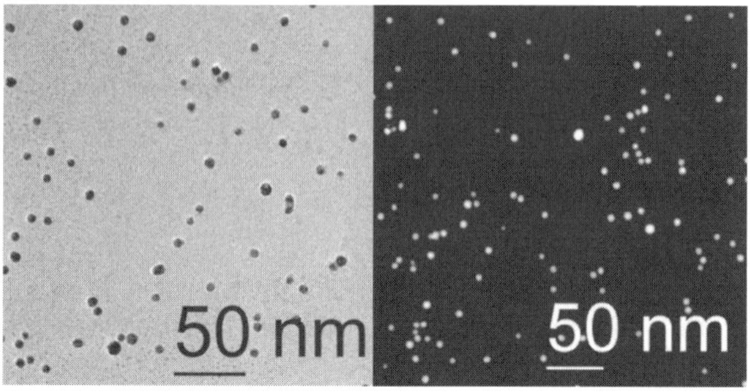

(A) (B)

FIGURE 9 (**A**) Transmission electron microscopy image of Au nanoparticle and (**B**) high-angle annular dark-field scanning transmission electron microscopy (STEM) image of the Au nanoparticles, which shows *Z*-contrast. It is clear that in the STEM image, the background due to amorphous carbon (supporting film) has quite lower intensity.

scanning electron microscopes, there is the possibility to install STEM detectors too. Due to the lower voltage and relatively simpler lens system used in a scanning electron microscope, usually it has a lower spatial resolution than a transmission electron microscope. The situation has improved recently, and it is routine to obtain a nanometer resolution with a modern scanning electron microscope, such as Hitachi S-5500. Figure 10(A) shows a BF STEM image of PbTe nanoparticles, while Figure 10(B) shows an HAADF-STEM image of the same sample that shows Z-contrast.

The ultimate resolution of the scanning electron microscope is used to determine the size of the electron probe. The invention of FEG by Crewe and coworkers provided a high-density electron beam that can be focused into a small size (i.e., subnanometer scale) (3). They used the bright electron source in the first scanning electron microscope to obtain images of individual heavy atoms lying on a thin carbon film, which are the first atomic resolution STEM images (3). It is well recognized that the spatial resolution of an electron microscope is mainly restricted by lens aberrations. Recently, there is big progress in the development of aberration

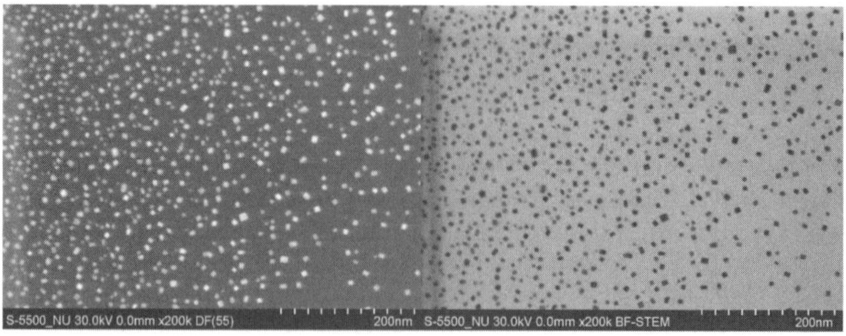

(B) (A)

FIGURE 10 (**A**) Bright-field scanning transmission electron microscopy (STEM) image of PbTe nanoparticle and (**B**) high-angle annular dark-field scanning transmission electron microscopy (STEM) image of the PbTe nanoparticle, which shows *Z*-contrast. Both images were obtained from a Hitachi S-5500 scanning electron microscope with STEM detectors.

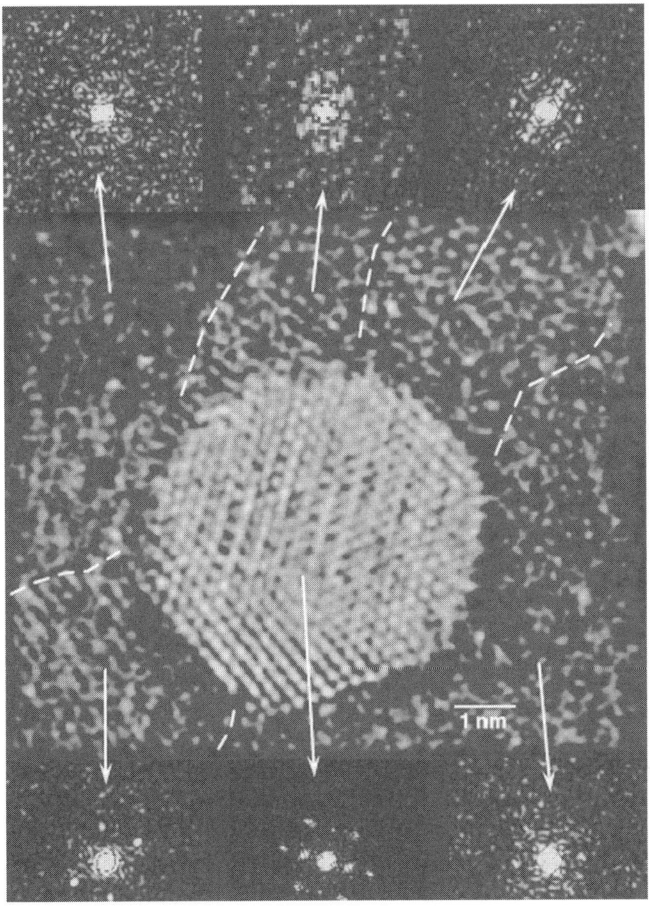

FIGURE 11 Atomic resolution image of an Au island on an amorphous carbon substrate. Surrounding the island are "rafts" of single atomic layers of Au. Further away, small clusters and single atoms of Au are present. Diffraction patterns from various regions surrounding the island show that the rafts are ordered in various structures adjacent to the built-up islands. *Source*: From Figure 1 of Ref. 5.

correctors. A practical aberration corrector was developed by Haider et al. (24), being the first one to demonstrate a resolution improvement in a particular instrument. For example, by implementation of an aberration correction system in the scanning electron microscope, Batson et al. (25) could achieve an electron probe smaller than 0.1 nm. Figure 11 shows the atomic resolution STEM image of the 10-nm Au nanoparticles on amorphous carbon obtained in their system (VG microscope HB501 with a quadrupole–octupole aberration corrector). Nowadays, the correctors of lens aberration have gradually become the standard component in modern transmission electron microscopes which makes the atomic resolution STEM routine. Meanwhile, the pursuit of ultimate performance with the aberration-corrected STEM places severe demands on the environment of the STEM in terms of vibration and electrical fields (26).

EELS is a useful technique to probe empty states, which contain the composition and charge state information of the nanoparticles. In an electron microscope, the spectra are collected by passing the high-energy (100–1000 keV) electron beam, focused down to a small probe, through a thin film and recording the transmitted energy loss spectrum. If the probe is scanned over the sample, the result may be presented as a map of chemical composition. STEM imaging mode becomes unique when its atomic resolution imaging capability is combined with EELS. In STEM mode, an electron energy loss spectrum can be recorded for each pixel. This thus forms a powerful tool in composition mapping, which can be achieved at atomic resolution. It is the highest resolution that can be achieved by all the available techniques of structural analysis. Atomic column sensitivity in EELS was first demonstrated by Batson (27). Figure 12(A) shows an STEM image of 5-nm Au nanoparticles used to perform EELS-STEM line profile. The yellow square shows the area used for sample drifting correction: to correct the sample drift during data acquisitions. The green line defines the line used for EELS imaging: the electron probe will scan across the line at a given spacing and an electron energy loss spectrum is collected at every stop. A typical electron energy loss spectrum (raw data without any process) is shown in Figure 12(B), in which an edge can be found for O atoms at 532 eV (the core-loss edge of oxygen). For N atoms, the core-loss characteristic energy loss is at 401 eV. It is hard to observe such an edge for N atoms in the spectrum. A line profile for O and N atoms across the line by using their EELS signal is shown in Figure 12(C). A combination of imaging and spectroscopic techniques provides us a powerful tool for the nanostructural analysis.

In an electron microscope, a variety of detectors are arranged around the sample. A most commonly installed one is the energy-dispersive spectroscopy detector. By detecting the characteristic X ray emitted from the sample activated by high-energy electrons, the composition can be identified. In the STEM imaging mode, while the electron probe is scanned across the sample, an energy-dispersive spectrum can be collected for each pixel. By doing so, the sample composition can be mapped similar to EELS-STEM mapping. Figure 13(A) shows an STEM image of CoFe core and SiO_2 shell nanoparticles. The area in the yellow rectangle is defined to correct sample drift, while the green line defines the line for which energy-dispersive spectrum is collected for each point. Figure 13(B) shows a typical energy-dispersive spectrum of the core–shell nanoparticle when the beam is scanned through both the core and the shell. In the pattern, the Fe and Si peaks can easily be seen. Figure 13(C) shows the line profile of both Si and Fe peaks by using their integrated energy-dispersive intensity at each stop. The Fe core and Si shell properties of the nanoparticle are clearly observed.

CONCLUSIONS

The importance of STEM and related techniques for the characterization of nanoparticles is justified by the fact that STEM is involved in the investigations of many different features: nature of the particle phase, size, dispersion, chemical composition, and bonding. The aim of this chapter was to provide the reader with the basis to understand the conceptions and techniques of STEM.

In summary, STEM presently allows the observation of the structural and composition details at the atomic scale. It is a powerful experimental tool for nanoscience. However, we need to be always aware of radiation damage and contamination.

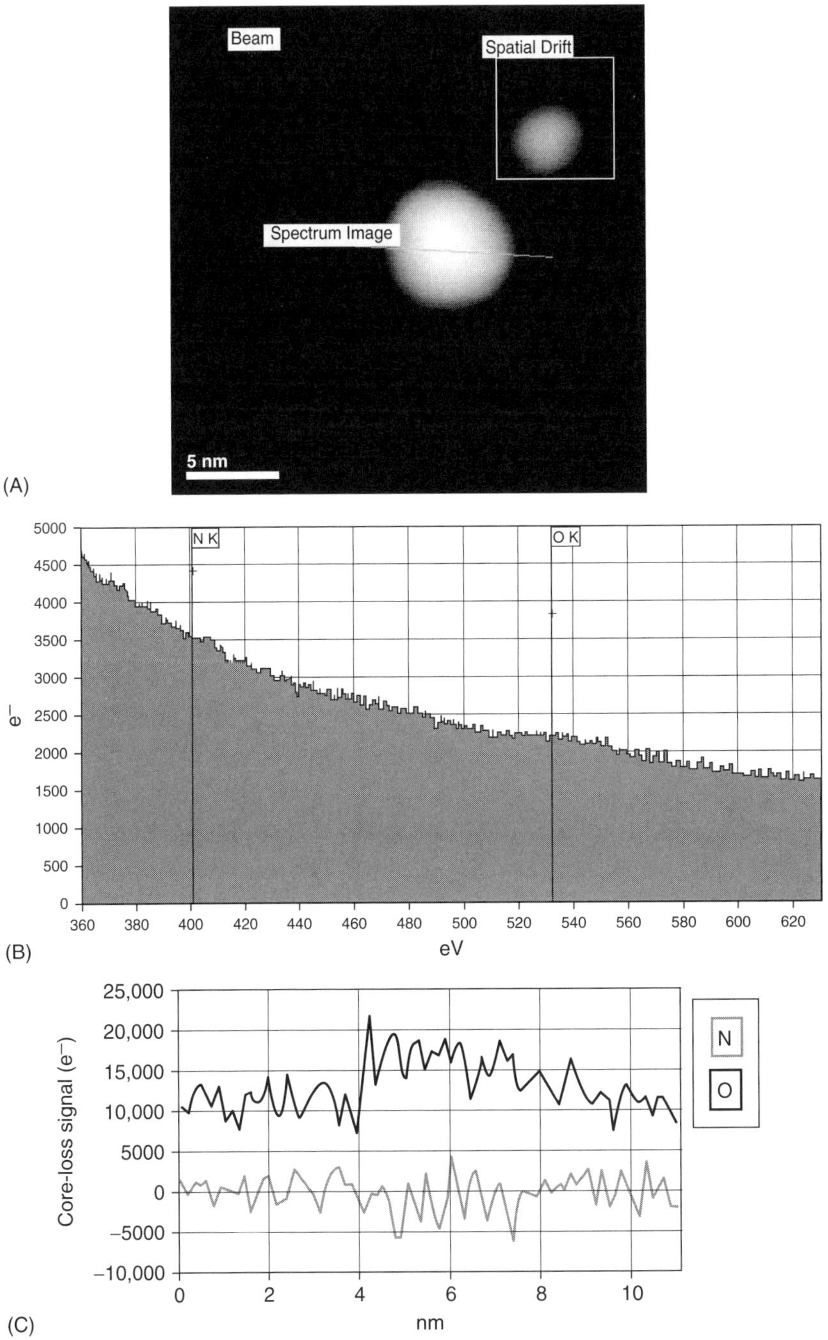

FIGURE 12 (**A**) A scanning transmission electron microscopy image of 5-nm Au nanoparticles used to perform electron energy loss spectrometry (EELS) line profile. (**B**) A typical and raw electron energy loss spectrum is obtained. The N signal is barely detected. (**C**) Line profile of O and N signals by using an EELS signal.

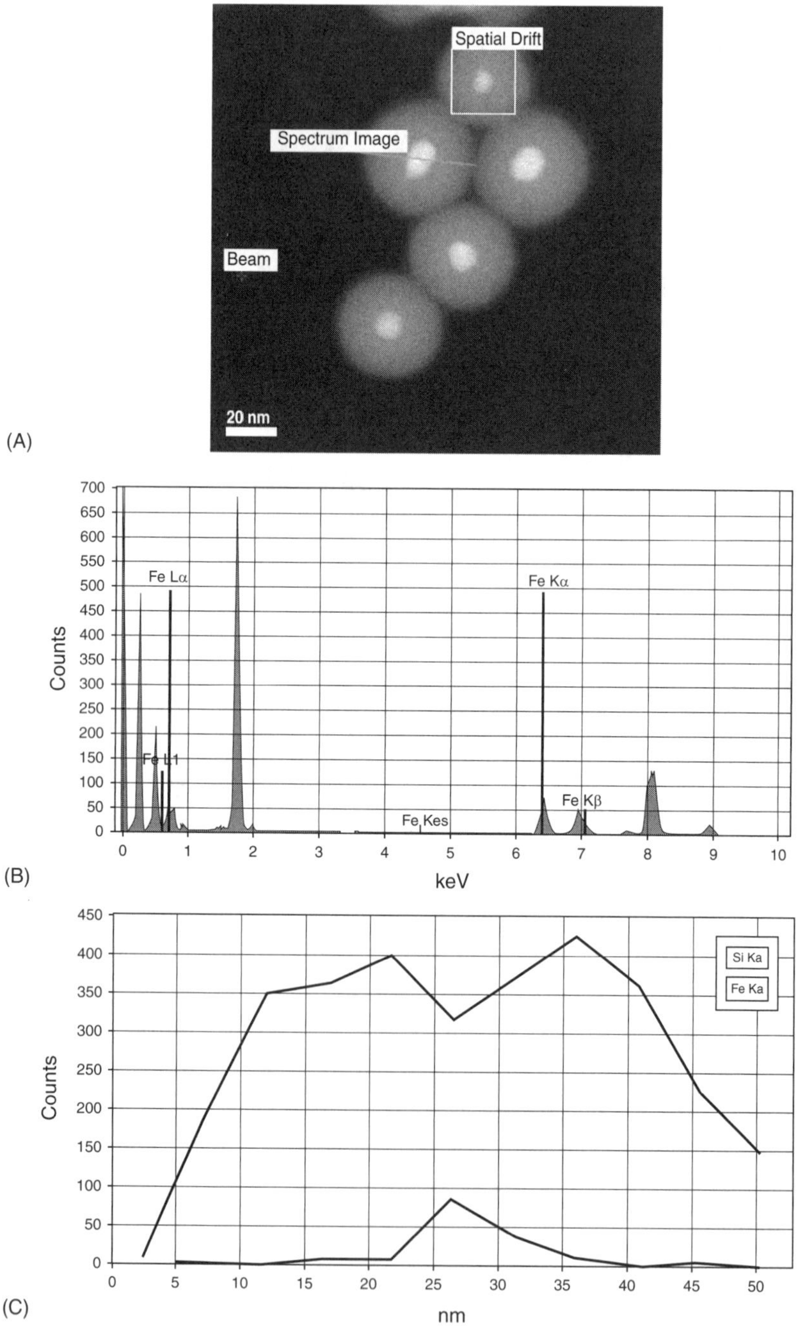

FIGURE 13 (**A**) A scanning transmission electron microscopy image of CoFe core and SiO_2 shell nanoparticles used to perform energy-dispersive X-ray (EDX) line profile. (**B**) A typical and raw energy-dispersive spectrum, in which the Fe peaks are labeled. (**C**) Line profile of Si and Fe by using their EDX signal across the line. The nature of the core–shell structure is clearly observed.

REFERENCES

1. von Ardenne M. Das Elektronen-Rastermikroskop: Theoretische Grundlagen. Z Phys 1938; 109:553.
2. Knoll M, Ruska E. Das Elektronenmikroskop. Z Phys 1932; 78:318.
3. Crewe AV, Wall J, Langmore J. Visibility of a single atom. Science 1970; 168:1338.
4. Browning ND, Chisholm MF, Pennycook SJ. Atomic-resolution chemical analysis using a scanning transmission electron microscope. Nature 1993; 366:143.
5. Browning, ND, Chisholm MF, Pennycook SJ. Corrigendum: Atomic-resolution chemical analysis using a scanning transmission electron microscope. Nature 2006; 444:235.
6. Mao C. Nanomaterials characterization: structures, compositions, and properties. Microsc Res and Tech 2006; 69:519.
7. Wang ZL. Characterization of Nanophase Materials. Weinheim, Germany: Wiley VCH Verlag GmbH, 2000.
8. McBride JR, Kippeny TC, Pennycook SJ, et al. Aberration-corrected Z-contrast scanning transmission electron microscopy of CdSe nanocrystals. Nano Lett 2004; 4:1279.
9. Li ZY, Yuan J, Chen Y, et al. Direct imaging of core-shell structure in silver-gold bimetallic nanoparticles. Appl Phys Lett 2005; 87:243103.
10. Williams DB, Carter CB. Transmission Electron Microscopy. New York: Plenum Press, 1996.
11. Spence JCH. High Resolution Electron Microscopy, 3rd ed. New York: Oxford University Press, 2002.
12. Nellist PD, Pennycook SJ. Incoherent imaging using dynamically scattered coherent electrons. Ultramicroscopy 1999; 78:111.
13. Nellist PD, Pennycook SJ. The principles and interpretation of annular dark-field Z-contrast imaging. Adv Imaging Electron Phys 2000; 113:147.
14. Nellist P, Pennycook S. Subangstrom resolution imaging using annular dark-field STEM. Adv Imaging Electron Phys 2000; 113:148.
15. Pennycook SJ, Rafferty B, Nellist PD. Z-contrast imaging in an aberration-corrected scanning transmission electron microscope. Microsc Microanal 2000; 6:343.
16. Rafferty B, Nellist PD, Pennycook SJ. On the origin of transverse incoherence in Z-contrast STEM. J Electron Microsc (Tokyo) 2001; 50:227.
17. Rafferty B, Pennycook SJ. Towards atomic column-by-column spectroscopy. Ultramicroscopy 1999; 78:141.
18. Browning ND, James EM, Kishida K, et al. Investigating atomic scale phenomena at materials interfaces with correlated techniques in STEM/TEM. Res Adv Mater Sci 2006; 1:1.
19. Keyse RJ, Garratt-Reed AJ, Goodhew PJ, et al. Introduction to Scanning Transmission Electron Microscopy. Oxford: BIOS Scientific Publishers, 1998.
20. James EM, Browning ND. Practical aspects of atomic resolution imaging and analysis in STEM. Ultramicroscopy 1999; 78:125.
21. Ahn CC. Transmission Electron Energy Loss Spectrometry in Materials Science and the EELS Atlas, 2nd ed. Grunstadt, Germany: Willey-VCH Verlag Gmbh & Co. KGaA, 2004.
22. Egerton RF. Electron Energy Loss Spectroscopy in the Electron Microscope, 2nd ed. New York: Plenum Press, 1996.
23. Cowley J. Scanning-transmission electron-microscopy of thin specimens. Ultramicroscopy 1976; 2:3.
24. Haider M, Uhlemann S, Schwan E, et al. Electron microscopy image enhanced. Nature 1998; 392:768.
25. Batson PE, Dellby N, Krivanek OL. Sub-angstrom resolution using aberration corrected electron optics. Nature 2002; 418:617.
26. Muller D, Grazul J. Optimizing the environment for sub-0.2 nm scanning transmission electron microscopy. J Electron Microsc 2001; 50(3):219.
27. Batson P. Simultaneous stem imaging and electron-energy-loss spectroscopy with atomic-column sensitivity. Nature 1993; 366:727.

16 Structural Fingerprinting of Nanocrystals in the Transmission Electron Microscope: Utilizing Information on Projected Reciprocal Lattice Geometry, 2D Symmetry, and Structure Factors

Peter Moeck and Sergei Rouvimov

Laboratory for Structural Fingerprinting and Electron Crystallography, Department of Physics, Portland State University, Portland, Oregon, U.S.A.

INTRODUCTION

The goal of this chapter is to outline two novel strategies for fingerprinting nanocrystals from structural information that is contained either in a precession electron diffraction (PED) pattern of a single nanocrystal (1,2) or in a fine-grained crystal powder electron diffraction ring (CPEDR) pattern with random orientation of the nanocrystallites (3–5). While structural fingerprinting from CPEDR patterns can be conveniently performed with a large primary beam diameter so that experimental data may be recorded for many nanocrystals at once, PED of individual nanocrystals requires the application of nanobeam diffraction techniques.

The main emphasis of this chapter is on an outline of the theoretical foundation of these two structural fingerprinting strategies. It complements our earlier studies that had an experimental emphasis (6–9) and concentrated on advanced structural fingerprinting based on the Fourier transform of high-resolution transmission electron microscopy (HRTEM) images. As there are already several publications (6–8) and an MSc thesis (9) (in open access) on the latter subject, this kind of advanced structural fingerprinting is dealt with here only very briefly. Because there is very little published research on advanced structural fingerprinting based on PED patterns, this chapter also provides a brief introduction to this comparably new diffraction technique. Some of our recent experimental results with PED on silicon crystals are shown here for the first time.

Note that we published recently an extensive review of structural fingerprinting strategies in the transmission electron microscope (10). There is, thus, no need to discuss and quote any of the "traditional" structural fingerprinting strategies in the transmission electron microscope (TEM). The term "traditional" refers here to strategies that combine information on the projected reciprocal lattice geometry with either spectroscopic information as obtainable from an analytical transmission electron microscope from the same sample area or prior knowledge on the chemical composition of the sample.

In contrast, the novelty of the advanced strategies that are briefly described here is due to the combination of information on the projected reciprocal lattice geometry with information on the two-dimensional (2D) symmetry, and structure factor moduli or phase angles. These strategies are applicable only to nanocrystals, since they rely on kinematic or quasi-kinematic scattering approximations. [Quasi-kinematic means that the electron scattering is of an intermediate nature (3), which is dealt with by utilizing approximate correction factors to kinematic predictions.] Other limitations of these strategies in connection with the nature of nanocrystals and the currently existing crystallographic databases are also mentioned. A brief discussion of powder X-ray fingerprinting and its limitations is also given.

Note that kinematic and quasi-kinematic approximations to the scattering of fast electrons do allow for the successful solving and refining of unknown crystal structures [see Ref. (11) for a recent review, which is also in open access]. Although with much more data processing, this is achieved on the basis of the same kind of data that are employed in our advanced structural fingerprinting strategies. This whole field is now known as "structural electron crystallography," while its predecessor and a part of its theoretical foundation are referred to as "electron diffraction structure analysis" (EDSA).

Boris Konstantinovich Vainshtein, the 1990 P. P. Ewald Prize laureate, concluded about EDSA almost 50 years ago that there "is no doubt now that electron diffraction may be used for the complete analysis of crystals whose structure is unknown" (3). Structural electron crystallography from diffraction patterns and HRTEM images (either alone or in combination with other diffraction techniques) has so far led to several hundreds of solved and refined crystal structures.[a] (*See* NOTES page 310.)

One may straightforwardly argue on the basis of the undeniable success of EDSA and structural electron crystallography that if the solving of crystal structures from electron scattering data is feasible in kinematic or quasi-kinematic scattering approximations for sufficiently thin inorganic crystals, the much less sophisticated structural fingerprinting based on the same scattering theories must be feasible as well (10). In order for "2D symmetry" and "structure factor fingerprinting" to yield a more discriminatory identification of nanocrystals that cannot already be distinguished on the basis of their projected reciprocal lattice geometry, one may need only qualitative or semiquantitative information while structural electron crystallography needs to employ fully quantitative information.

The corollary that it "is always possible, even easy, to collect intensity data that *cannot* be analyzed by conventional phasing methods or to record high-resolution images where the resemblance to any known structure is not at all obvious" (12) by Douglas L. Dorset, the 1999 A. Lindo Patterson Award laureate, is to be taken very seriously in the structural fingerprinting of nanocrystals. In order to allow the reader to appreciate the limits that dynamical electron scattering effects set on the application of our novel structural fingerprinting strategies, theory sections on kinematic and quasi-kinematic approximations to the scattering of fast electrons are given.

Since there are simple relationships between structure factor moduli and diffracted intensities (and structure factor phases and the phases of Fourier coefficients of the HRTEM image intensity distribution) for kinematic diffraction conditions only, the standard procedure is to utilize quasi-kinematic approximations, and when necessary, to correct the experimental data for dynamical scattering effects. This approach is analogous to the one typically taken in structural electron crystallography and constitutes the first pillar of structural fingerprinting. Because model structures from a comprehensive database form the second pillar of structural fingerprinting and semiquantitative structure factor information suffices, the task is reduced to finding the one model structure that fits a certain set of experimental data best. A short section on criteria for deciding what is the best fit to the model data is presented in the following text. The typical hierarchy of dynamical diffraction corrections to quasi-kinematic data and the "principle of minimal corrections" are also illustrated.

The term "phase" has different meanings in different branches of the natural sciences. Because we will occasionally mention the extraction of structure factor phase angles (structure factor phases) from the Fourier transform of HRTEM images, we will never refer to the identification of "crystal phases" from their characteristic "crystal phase fingerprints" in this chapter. Genuine crystal phases in the thermodynamic sense (i.e., regions in space with homogenous physical and chemical properties) are referred to in this chapter simply as "crystal structures." This should help avoid confusion between genuine crystal phases and structure factor phases, as both are entirely different concepts. To avoid confusion, the reader needs to be also aware of the conceptual difference between crystal structure phases and the phases of electron waves in the TEM (11,13). While the latter are (real space) properties of the electron wave (with picometer dimensions), which are modified both by scattering on the electrostatic potential of a crystal and by the objective lens, and are finally lost in the process of recording diffraction patterns or HRTEM images, the former can be reliably extracted from the Fourier transform of HRTEM images (11,13,14). This is because the structure factor phases are (reciprocal space) entities that are directly related to the Fourier coefficient phases of that electrostatic potential.

Finally, the mainly inorganic subset (15) (with some 20,000 entries) of the Crystallography Open Database (16–18) (with currently more than 70,000 entries overall) needs to be mentioned here because we are in the process of interfacing open-access search-match capabilities to this database. We also provide visualizations of so-called "lattice-fringe fingerprint plots" [i.e., one of the key concepts of our new strategies (10,19)] at our Web server (20). [Note that an early version of Ref. (19) is in open access.] Interactive visualizations of the atomic arrangement of these entries in three dimensions are also provided at our Web server (20).

NANOCRYSTALS CANNOT BE FINGERPRINTED STRUCTURALLY BY POWDER X-RAY DIFFRACTOMETRY

In powder X-ray diffraction fingerprinting, the three-dimensional (3D) crystal structure information is collapsed into a one-dimensional intensity profile plotted over the angles between the primary and scattered beams (Fig. 1). This ensures that the relative large abundance of structural 3D information can be utilized for the fingerprinting (at just one orientation of the sample in a diffractometer). The angular position and relative heights of Bragg peaks in X-ray diffractograms constitute the information that is principally employable for structural fingerprinting. Since there is no simple experimental test as to the presence of textures in the crystalline powder when the very popular Bragg-Brentano parafocusing diffractometer geometry is employed, the information on the relative peaks heights is often not utilized in structural fingerprinting. [Note that textures may result in significant deviations of the experimental Bragg peak heights from their counterparts in the database. Advanced structural fingerprinting strategies in powder X-ray diffractometry do, however, utilize fitting procedures to the whole pattern (21).]

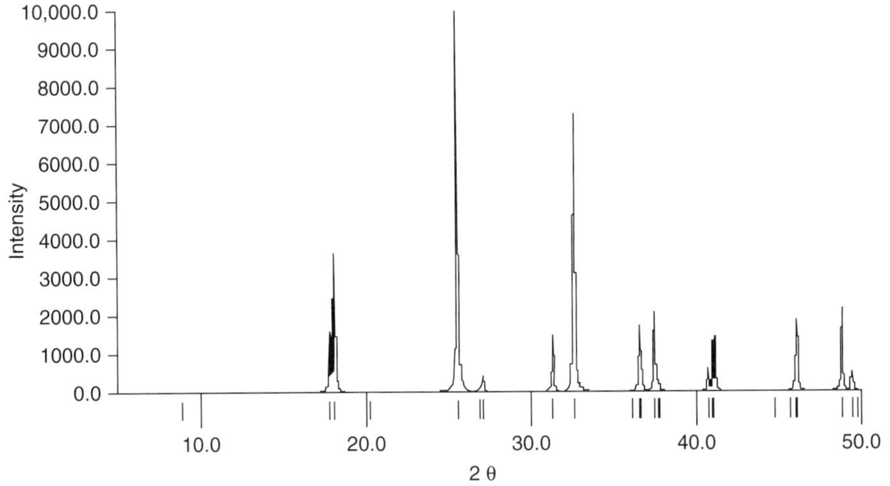

FIGURE 1 Calculated X-ray powder diffractogram of the mineral pseudo-brookite, $Fe_2(TiO_3)O_2$, utilizing the characteristic K_α radiation from a Cu target. The theoretical positions of the Bragg peaks are marked at the abscissa. There are several strong peaks so that both a Hanawalt search (22) and a Fink search would work well for the identification of this crystalline material. This powder diffraction pattern was simulated with the freeware program Mercury, Cambridge Crystallographic Data Centre, downloadable at http://www.ccdc.cam.ac.uk/products/mercury/.

While a "Hanawalt search" (22) employs the angular positions (reciprocal lattice vectors) of the three most intense X-ray powder diffraction peaks, a "Fink search" utilizes the eight (or 10) shortest reciprocal lattice vectors with reasonably high peak intensities (Fig. 1). Utilizing either or both of these classical search strategies leads, usually together with some prior knowledge of chemical information, to an identification of an unknown by comparison with the entries of a comprehensive database such as the well-known Powder Diffraction File (23).

The powder X-ray method works best for crystal sizes in the micrometer range, in which kinematic X-ray diffraction on otherwise almost perfect crystal lattices results essentially in delta functions for the line profiles of the individual reflections. The convolution of these delta functions with the instrumental broadening function of a diffractometer determines the shape and width of Bragg peaks in a powder X-ray diffractogram. For smaller crystals, the situation becomes rather complex and the Bragg peaks may get simultaneously as well as asymmetrically shifted or even anisotropically broadened (24). All of these small crystal size and morphology effects are detrimental to an unambiguous identification of a crystalline material from its powder X-ray diffractogram.

For nanocrystals with a relative large unit cell and low symmetry (e.g., Ta_2O_5), the powder X-ray diffraction pattern gets less and less characteristic with nanocrystal size because more and more Bragg peaks overlap due to their broadening (25). As these peaks broaden, their intensity also diminishes until they become difficult to distinguish from the background. This has been demonstrated by simulations of Ta_2O_5 diffraction patterns (25) utilizing the Debye equation, which assumes only

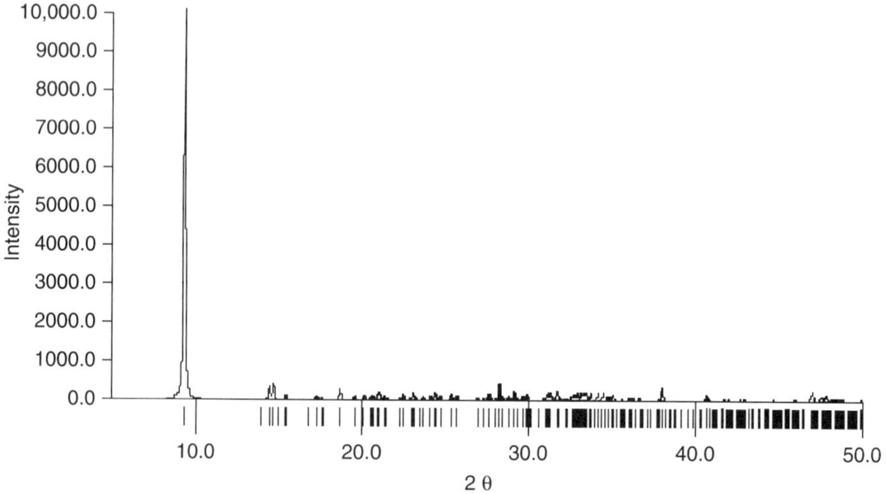

FIGURE 2 Calculated X-ray powder diffractogram of vanadium oxide nanotubes (29) utilizing the characteristic K_α radiation from a Cu target. The theoretical positions of the Bragg peaks are marked at the abscissa. There is only one strong peak (with several higher order peaks) so that a classical Hanawalt search (22) would not work for the identification of this nanocrystalline material with tubular morphology. It is highly questionable if a Fink search would lead to an unambiguous identification either. In addition, it is known that the angular position of the strong (002) peak depends sensitively on the growth and processing conditions since small cations, such as H^+ or Li^+, may get intercalated in this material. The freeware program Mercury (see caption of Fig. 1) has been used for the simulation.

atomicity but not regular arrangements of atoms. Further complications arise from size and shape distributions in the nanocrystal population (26).

Nanocrystals may also possess surface and near-surface regions that are highly distorted or relaxed with respect to the bulk crystal structure. Such distinct surface structures, in turn, result in X-ray powder diffraction patterns that are no longer characteristic of the crystalline bulk core (27). Anatase (TiO_2) nanocrystals of size less than about 2 nm may, for example, not possess a core region that corresponds to the bulk lattice structure at all (28).

Finally, certain technologically important materials (e.g., carbon or vanadium oxide nanotubes (29)] do *not* give characteristic powder X-ray diffraction fingerprints by which the crystal structure may be identified out of a range of candidate structures from a comprehensive database (Fig. 2). Such nanomaterials will, therefore, most likely not become part of general purpose X-ray powder diffraction databases. It is, therefore, fair to conclude that the otherwise very powerful powder X-ray diffraction technique becomes quite useless for crystal structure identifications in the nanometer size range.

ADVANTAGES OF UTILIZING FAST ELECTRONS FOR STRUCTURAL FINGERPRINTING OF NANOCRYSTALS

Nanometer crystal sizes have, on the other hand, exactly the opposite effect on the feasibility of our novel strategies of lattice-fringe fingerprinting (with either partial or complete structure factor extraction) from a single HRTEM image or a single PED pattern (10). This is because as the crystals become smaller, more lattice fringes become visible over a wider angular range in an HRTEM image (19). Correspondingly, the shape function of a nanocrystal becomes more extended in reciprocal space and more diffraction spots appear in a PED pattern (1,2). The combined weak-phase object/kinematic diffraction or phase object/quasi-kinematic diffraction approximations of TEM will also be reasonably well adhered to during the recording of the experimental data when crystal sizes are "very small." (Some approximate quantifications of such "smallness" are given in the following text.)

The atomic scattering factors for fast electrons of the elements are about three orders of magnitudes larger than for X rays. (The term "fast" refers here to some 50% to 80% of the speed of light, corresponding to electron wavelengths in the picometer range.) This, on the one hand, ensures that there will be sufficient diffracted intensity so that structural information is conveyed for fingerprinting purposes in the TEM even for the smallest of nanocrystals. On the other hand, this strong scattering of electrons by matter may complicate the analysis. The section on electron scattering theories in the following text clarifies how the most prominent dynamical diffraction effects can be taken into account and corrected for in our novel structural fingerprinting strategies.

Fast electrons can be focused by electromagnetic fields and lenses that act as natural Fourier transformers, besides providing magnification in a transmission electron microscope. Structural fingerprinting in the TEM is, therefore, not just limited to the analysis of electron diffraction patterns but also works on the basis of so-called "structure"[b] (*see* NOTES page 310) HRTEM images that were recorded close to the Scherzer (de)focus.[b] The structure factor phase angles that are lost in the recording of a diffraction pattern can be extracted by Crystallographic Image Processing (11,13) from HRTEM images (14).

KINEMATIC AND QUASI-KINEMATIC APPROXIMATIONS TO THE SCATTERING OF FAST ELECTRONS

The scattering of electron waves by atoms and the periodic electrostatic potential of a crystal is governed by the Schrödinger equation. The solutions to this equation provide the basis of the multiple-beam dynamical theory of electron scattering, which is the only strictly correct description of the scattering of electrons by matter. The predictions of the dynamical theory for crystals depend very sensitively on the exact crystal orientation, morphology, and thickness so that various approximations are used under different circumstances. Inelastic scattering may be treated as an absorption effect and, for small crystals, is typically neglected altogether.

For many purposes, the two-beam dynamical scattering theory (also known as the first Bethe approximation) suffices. This approximation is an exact solution of the Schrödinger equation for the special case of only one strong diffracted beam in the diffraction pattern. As is shown in the following text, for vanishing crystal thickness, the predictions of the two-beam dynamical scattering theory closely approach the predictions of the kinematic theory.

The conceptual basis of the kinematic theory is the single scattering of electrons by the electrostatic potential out of the primary beam into the diffracted beams while the former is negligibly attenuated. This is an idealized case for the scattering of fast electrons [while it typically suffices for the scattering of X rays by crystals that are composed of mosaic blocks in the millimeter to centimeter range (3)].

For nanometer-sized crystals, one can, however, reliably base crystallographic analyses by means of electron scattering on quasi-kinematic approximations and correct for primary extinction effects. (The utilized primary extinction correction is conceptually very similar to that employed in X-ray crystallography.) Secondary scattering effects can approximately be corrected for on the basis of the ratios of reflection intensities to the intensities of kinematically forbidden reflections (3).

The physical process of electron diffraction can be described mathematically by a Fourier transform. Electrons are scattered at the electrostatic potential energy distribution within the unit cell. This distribution peaks strongly at the positions of atoms. Following Ref. (3) and its notation closely, the Fourier coefficients, Φ_{hkl}, of the electrostatic potential $\phi(x, y, z)$ are given by the relation

$$\Phi_{hkl} = \int_{\Omega} \phi(x, y, x) \exp\left\{2\pi i \left(h\frac{x}{a} + k\frac{y}{b} + l\frac{z}{c}\right)\right\} \mathrm{d}x\mathrm{d}y\mathrm{d}z \tag{1}$$

where the dimension of Φ_{hkl} is volts times cube of length; integers h, k, l the Miller indices, that is, labels of the reflecting net plane (with reciprocal spacing $|\vec{d}^*_{hkl}| = |nh\vec{a}^* + nk\vec{b}^* + nl\vec{c}^*| = |^1/_{d_{hkl}}| = d^*_{hkl}$, where $\vec{a}^*, \vec{b}^*, \vec{c}^*$ are the basis vectors of the reciprocal crystal lattice with $\vec{a}^* = \vec{b} \times \vec{c}/_{\Omega}$, $\vec{b}^* = \vec{c} \times \vec{a}/_{\Omega}$, and $\vec{c}^* = \vec{a} \times \vec{b}/_{\Omega}$, $\Omega = \vec{a}(\vec{b} \times \vec{c})$ the volume of the unit cell, and $\vec{a}, \vec{b}, \vec{c}$ the basis vectors of the (direct) crystal lattice, with a, b, c as their respective magnitudes); n an integer that described the order of the reflection); and x, y, z the coordinates of atoms in the unit cell. [Following the practice of EDSA, these Fourier coefficients are in relation (1) not normalized by the volume of the unit cell (3).]

These (mathematical) Fourier coefficients physically represent electron waves that are scattered by the electrostatic potential of the crystal in directions that are defined by Bragg's law ($\lambda = 2d_{hkl} \sin\Theta$, where Θ is half of the angle between the transmitted and scattered electron beams) and recorded a large distance away from the crystal. [An alternative name for these Fourier coefficients is "structure amplitudes" (3), whereby other authors frequently use a definition according to relation (1) with an additional normalization to the unit cell volume.]

The structure factor, F_{hkl}, with dimension of length, is a key concept in structural crystallography (14) and is in the case of electron scattering given by

$$F_{hkl} = \frac{\sigma}{\lambda}\Phi_{hkl} \tag{2}$$

where $\sigma = 2\pi me\lambda/h^2$ is the "interaction parameter," with m as the (relativistic) mass of the electron, e the elemental charge (i.e., modulus of the electronic charge), λ the (relativistic) electron wavelength, and h the Planck constant. A so-called "reflection" (i.e., a spot of diffracted intensity in a diffraction pattern) may be referred to simply by its respective label (hkl).

Relations (1) and (2) can be interpreted in words as "knowing and *identifying a crystal structure in direct space is equivalent to* knowing and *identifying its structure factors in reciprocal space.*" To obtain experimental data on a crystal structure (in reciprocal space), one "*needs to perform an electron scattering experiment*, which can mathematically be *described by a Fourier transform.*" Novel and highly discriminatory structural fingerprinting strategies can, thus, be based on combining partial or complete structure factor information with information on the projected reciprocal lattice geometry and 2D symmetry (10).

In the first Born approximation (to the solution of the Schrödinger equation for the scattering of an electron by the electrostatic potential of an atom), that is, in the kinematic electron scattering theory, the structure factors are given by the relation

$$F_{hkl} = \sum_j f_j f_T^j \exp 2\pi i(hx_j + ky_j + lz_j) \tag{3a}$$

where f_T^j are the atomic scattering factors for electrons and $f_T\, j$ the respective temperature factors for all j atoms in the unit cell. (Since atomic scattering factors are typically tabulated in Å, structure factors are also typically given in Å.) Note that temperature factors are much less important for electrons than for X rays. This is because the atomic scattering factors for electrons fall off with $(\sin\Theta/_\lambda)^2$, that is, much more rapidly than their counterparts for X rays. When the unit cell volume Ω is given in nm^3, the relation $|\Phi_{hkl}| = 0.047875\,|F_{hkl}|$ is valid for structure factor moduli in Å. While the first Born approximation ensures that the atomic scattering factors are real numbers, the structure factors (and the corresponding Fourier coefficients of the electrostatic potential) are, however, complex numbers with a modulus and phase (angle)

$$F_{hkl} = |F_{hkl}|\,\cos 2\pi(hx_i + ky_i + lz_i) + i\,\sin 2\pi(hx_i + ky_i + lz_i) = |F_{hkl}|\,e^{i\alpha_{hkl}} \tag{3b}$$

$$F_{hkl} = A(h,k,l) + i\,B(h,k,l) = \sqrt{A^2(h,k,l) + B^2(h,k,l)}\,e^{i\,\arctan\{B(h,k,l)/A(h,k,l)\}} \tag{3c}$$

For an *ideal* single crystal, the two-beam dynamical diffraction theory gives the intensity of a diffracted beam I_{hkl} (i.e., of a reflection) as a function of a (reciprocal) distance h_3, which represents the dimension of the (reciprocal 3D) shape transform of a (parallelepiped shaped) crystal that is measured from the exact reciprocal lattice point position, $h_3 = 0$, parallel to the primary beam direction, by the relation

$$I_{hkl}(h_3) = I_0 S Q^2 \frac{\sin^2\{A_3[(\pi h_3)^2 + Q^2]^{0.5}\}}{(\pi h_3)^2 + Q^2} \tag{4a}$$

where I_0 is the intensity of the primary electron beam; S the area of the crystal that is illuminated by the primary electron beam; $Q = \lambda|^{F_{hkl}}/_{\Omega}|$ an entity that is proportional to a particular structure factor; and A_3 the (direct space) dimension of a crystal in the direction of the transmitted beam (i.e., what is usually understood in TEM as crystal thickness). For comparison with treatments by other authors (30,31), note that $Q = \pi\cos\Theta/_{\xi_{hkl}}$, where ξ_{hkl} is known as the "extinction distance." [Since $\cos\Theta \approx 1$ for the diffraction of fast electrons, it is dropped in many of the formulae of Ref. (3).]

The central maximum and subsidiary maxima of the Fourier transform of the shape function in the direction of the primary beam are given by the relation

$$h_3 = \left[\left(\frac{n}{2A_3}\right)^2 - \frac{Q^2}{\pi^2}\right]^{0.5} \tag{4b}$$

In relation (4b), n is an integer that is even (and >2) for all zero values of (4a) that separate the central interference maximum and all attenuating subsidiary maxima of this function.

The respective predictions of the kinematic theory for an *ideal* single crystal are as follows:

$$I_{hkl}(h_3) = I_0 S Q^2 \frac{\sin^2 \pi A_3 h_3}{(\pi h_3)^2} \tag{5a}$$

where

$$h_3(n) = \frac{n}{2A_3} \tag{5b}$$

If Q is much smaller than πh_3, relations (4a) and (4b) can be approximated by relations (5a) and (5b), respectively. Because h_3 is inversely proportional to the size of the crystal, it becomes larger the smaller the nanocrystal gets. In other words, for sufficiently thin crystals, the two-beam dynamical diffraction theory is well approximated by the kinematic theory. For $h_3 = 0$, relation (4a) becomes

$$I_{hkl}^{\mathrm{MAX}}(h_3 = 0) = I_0 S\ \sin^2(QA_3) \tag{6a}$$

and relation (5a) becomes

$$I_{hkl}^{\mathrm{MAX}}(h_3 = 0) = I_0 S(QA_3)^2 \tag{6b}$$

The square of the sine function in (6a) may be replaced by the square of the argument for small QA_3 so that relation (6b) becomes a good approximation to the former relation. Note that the nature of the electron scattering phenomena is revealed in relations (4a) to (6b), but one does not base structural electron crystallography or structural fingerprinting strategies that employ structure factor information directly on them. For this, these relations need to be modified by Lorentz factors, as is discussed next.

The ratio of the integrated scattered beam intensity to the initial beam intensity received by a *real* crystalline sample from the primary beam in a *real* electron scattering experiment should be called "integrated coefficient of reflection" (3), and it is in the kinematic theory given by the relation

$$\frac{I_{hkl}}{I_0 S} = \lambda^2 \left| \frac{F_{hkl}}{\Omega} \right|^2 A_3 L \tag{7a}$$

where L is a Lorentz factor and possesses the unit of length. Analogously to their counterparts in X-ray diffraction, Lorentz factors account for the physical particulars (including the relative time intervals) of the intersections of the Ewald sphere, with the shape transform of the nanocrystals at the accessible reciprocal lattice points.

Note that no Lorentz factor was given in relations (4a), (5a), (6a), and (6b) because these relations refer to nonintegrated intensities for an *ideal* single crystal. For such a crystal, L is unity (without a dimension), as there is no time dependency of the intensity for all reflections. Similarly, relations (8a) to (9f) in the following refer to theoretical concepts of a general nature so that no Lorentz factor needs to be considered.

The nature of the Lorentz factor differs from experimental setup to setup, that is, with both the diffraction technique and the crystalline sample type. Within a certain diffraction technique and crystalline sample type, the Lorentz factor varies only quantitatively (3,4). For now, it may suffice that making L smaller than A_3 and/or Q^{-1} by choice of certain parameters of a diffraction technique or by choice of the selection of a certain crystalline sample reduces the integrated coefficient of reflection so that structural fingerprinting may proceed within the frameworks of the kinematic or quasi-kinematic theories.

Since a particular Q is proportional to a particular Fourier coefficient of the electrostatic potential, which is a parameter of a crystal structure, relations (5a) and (5b) will, for different reflections (*hkl*) of the same nanocrystal with a fixed size, be better or worse approximations to relations (4a) and (4b). The electron wavelength, size/thickness, and structure of the nanocrystal as well as the volume of its unit cell are fixed in a typical experiment, but they are also parameters that determine how well the two-beam dynamical diffraction theory will be approximated by the kinematic theory.

It is, therefore, quite appropriate to introduce a "range of crystal sizes/thicknesses, electrostatic potential values, electron wavelengths, and unit cell volumes" in which a nanocrystal diffracts quasi-kinematically. In general, for the same scattering angle, the "electron scattering centers" (i.e., atoms and ions) with higher atomic number possess higher scattering factors than their lower atomic number counterparts. The mutual arrangement of the "electron scattering centers" also determines the electrostatic potential. While for face-centered cubic structures of elements such as aluminum, silver, and gold all atoms scatter in phase, that is, their individual contributions to the scattered waves add up, there will be constructive and destructive interferences in more complex structures. Also, there are typically more reflections for structures with large unit cell volumes than there are for structures with small unit cell volumes. In addition, the reflections from large unit cells tend to be weaker than their counterparts from structures with small unit cells. The crystal orientation determines through Bragg's law which reflections will be activated in a given experiment and, therefore, also affects the "range" in which a nanocrystal diffracts quasi-kinematically.

As no definitive crystal size/thickness limit for the quasi-kinematic diffraction range can be given that would apply to all nanocrystals and all experiments, one may employ the relation

$$\lambda \left| \frac{F_{hkl}}{\Omega} \right| A'_3 \approx 1 \tag{8a}$$

where A'_3 has the meaning of Vainshtein's "critical thickness range" (3–5), as an evaluation criterion for the gradual transition from the kinematic theory to the dynamical two-beam theory. The relation

$$\lambda \left| \frac{F_{\text{ave}}}{\Omega} \right| A'_3 \approx 1 \tag{8b}$$

where F_{ave}, the average over the structure factor of a certain structure, is also used as such as an evaluation criterion (5,14,32). [Note that relations analogous to equations (8a) and (8b) apply to X-ray scattering as well and there are also relations analogous to those given further above for the kinematic and two-beam dynamical scattering theories of X rays (14).]

Somewhat arbitrary, one may define another "critical thickness" by the relation

$$A''_3 \leq 0.5 \frac{\Omega}{\lambda F_{hkl}} \tag{9a}$$

which corresponds to "nearly kinematical" scattering, as there will be, at most, an 8% intensity reduction with respect to a "truly kinematical" scattered beam by means of primary two-beam extinction (33).

A gradual transition from the kinematic theory to the dynamical two-beam theory will occur in the range

$$0.5 < \lambda \left| \frac{F_{hkl}}{\Omega} \right| A_3 \approx 1 \tag{9b}$$

At the upper range of relation (9b), a primary extinction correction after Blackman (34) will become a necessity. Such corrections can be employed advantageously when integration over the excitation errors is achieved by the specifics of the employed diffraction technique (3,4) and are described in more detail in the following text.

Note that

$$\lambda \left| \frac{F_{hkl}}{\Omega} \right| A_3 \approx \frac{\pi}{2} \tag{10}$$

implies that the two-beam dynamical diffraction theory becomes gradually valid (3,34) (with an A_3 of approximately half of the extinction distance). The range

$$1 < \lambda \left| \frac{F_{hkl}}{\Omega} \right| A_3 \approx \frac{\pi}{2} \tag{9c}$$

is, therefore, subject to gradually increasing primary extinction effects (4). From practical experience with mosaic nanocrystals and polycrystals with either random orientation or textures (3), it was recommended that a Blackman primary extinction correction should be employed within the whole range

$$0.7 \leq \lambda \left| \frac{F_{hkl}}{\Omega} \right| A_3 \leq 2 \tag{9d}$$

Reducing somewhat arbitrarily the lower bound in relation (9d) to 0.5 in order to connect to relation (9a) seamlessly, one may define an "extended quasi-kinematic region" by the full range

$$0.5 \leq \lambda \left| \frac{F_{hkl}}{\Omega} \right| A_3 \leq 2 \tag{9e}$$

for all electron scattering techniques that provide an effective integration over the excitation errors.

Individual reflections of the same nanocrystal (i.e., for a fixed A_3, λ, and Ω) may well behave differently. The reflections that possess small structure factor moduli may behave nearly kinetically and the ones with intermediate and large structure factor moduli may behave quasi-kinematically.

Note that Blackman primary extinction corrections are in principle applicable to all of the ranges of relations (9a) to (9e) as long as the diffraction technique provides an effective integration over the excitation errors (3). For low values of $QA_3 = \lambda |F_{hkl}/_{\Omega}| A_3$, the corrections will be small. They may, therefore, frequently not be justified when microphotometry of stacks of photographic films with varying exposure times is employed in order to obtain the integrated coefficient of reflections. Since the accuracy of this method of dealing with the raw data is about 10% to 15% (4), primary extinction corrections are typically not necessary for QA_3 values that obey relation (9a).

This illustrates a problem that structural electron crystallography (of unknowns) has learned to circumvent in an iterative manner: the approximate identification of which reflection needs to be dealt with by what dynamical correction (4). The so-called "crystallographic reliability" values, R values for short, and model structures are used for this purpose. (These R values represent the relative deviation of an experimentally obtained crystal structure from the calculated model structure that fits the totality of the experimental data best. This representation is typically made on the basis of the Fourier coefficients of the electrostatic potential (i.e., in reciprocal space) and is given in percentage. R values are to be discussed briefly further down next.)

While the R values are without dynamical corrections for sufficiently small crystals usually in the 20% range, a successful correction of the data set for dynamical effect reduces these values to 3% to 10% (4). Appropriate correction for dynamical scattering effects of the most important reflections can thus be identified by their effect on the R value. Note that structural fingerprinting at the structure factor level works on the basis of a range of preidentified model structures that correspond within the experimental error bars to the projected reciprocal lattice geometry and 2D symmetry. Since these model structures are obtained from a comprehensive database, there is no shortage of model structures with which the experimentally obtained (and appropriately corrected) structure factor data can be compared. Overall, in order to avoid structural misidentifications, one must try to minimize the total amount of corrections necessary to obtain a minimal R value.

Within the ranges of relations (9b) to (9e), where the scattering of electrons is of an intermediate nature, it is also possible to approximate experimentally obtained intensities $I_{\text{ave}}{}^{\text{EXP}}$ that are averaged over certain ranges of $\sin\Theta/_{\lambda}$ (and also normalized to the number of electrons that transmit the nanocrystal elastically, I_0S) as

$$\frac{I_{\text{ave}}^{\text{EXP}}}{I_0 SL} \approx c_{\text{kin}} Q_{\text{ave}}^2 + c_{\text{dyn}} Q_{\text{ave}} \tag{11a}$$

where c_{kin} and c_{dyn} are kinematic and two-beam dynamic fitting coefficients; $Q_{\text{ave}} = \lambda|F_{\text{ave}/\Omega}|$; and $c_{\text{kin}} + c_{\text{dyn}} = 1$ (32). When model structures are available, there is no need for averaging, that is,

$$\frac{I_{hkl}^{\text{EXP}}}{I_0 SL} \approx c_{\text{kin}} Q^2 + c_{\text{dyn}} Q \tag{11b}$$

Provided that the chemical composition of a crystal and the number of chemical formula units per unit cell (but not the crystallographic structure itself) are known, the degree of "scattering dynamicity" (32) can be estimated from (11a) by comparing curves of averaged integrated intensities over certain ranges of $\sin\Theta/_{\lambda}$ with the respective curve for the sum of the atomic scattering factors of all atoms in the unit cell in the same $\sin\Theta/_{\lambda}$ range, the so-called "sum of the f-curves." Analogously, a comparison of the former curves with the respective curve for the sum of squares of the atomic scattering factors of all atoms in the unit cell in the same $\sin\Theta/_{\lambda}$ range, the so-called "sum of the f^2-curves," allows for an estimation of the degree of "scattering kinematicity." It has also been suggested to use averaged atomic scattering factors for such comparisons (35).

Note that these comparisons can be done without the benefit of a known model structure as a result of the averaging over certain $\sin\Theta/\lambda$ ranges. Such estimations work because the mean square value of the phase factors of relations (3a) to (3c) is unity when the atoms are uniformly distributed throughout the unit cell (36). Structure factor modulus information may, thus, be extracted for structural fingerprinting purposes very pragmatically either with or without the benefit of a structure model.

The usage of relations (11a) and (11b) is especially recommended if there is some dispersion in the nanocrystallite size distribution in a polycrystalline sample, that is, in which some small crystals diffract kinematically ($\sim Q_{ave}{}^2$ or $\sim Q^2$) and others diffract dynamically ($\sim Q_{ave}$ or $\sim Q$) because they are of a larger size (3). If the falloff of averaged integrated intensities at small values of $\sin\Theta/\lambda$ corresponds to the sum of the f-curves and at large values of $\sin\Theta/\lambda$ to the sum of the f^2-curves, the crystal thickness will be in the quasi-kinematic range and a Blackman correction may advantageously be employed to the small-angle Bragg reflections (3). A somewhat related pragmatic approach to extracting relative structure factor moduli from measured relative intensities that is also applicable in the quasi-kinematic range is to determine an exponent of Q_{ave} that is intermediate between 2 and unity by a fitting and averaging procedure (3).

The phase grating approximation to dynamical multiple-beam scattering can also be used to extract quasi-kinematic structure factors from electron diffraction intensities on the basis of two experimental data sets that were recorded for the same kind of crystals at a highest voltage (e.g., 1000 kV) transmission electron microscope and an intermediate voltage (e.g., 100 kV) transmission electron microscope (37). This approach is highly advantageous, as the reflections that need to be corrected can be identified directly. (Note that the phase grating approximation neglects the curvature of the Ewald sphere but accounts well for dynamical scattering. Data sets that are recorded from the same kind of crystals at highest and intermediate voltages will, therefore, be in better or worse agreement with the predictions of this theory so that dynamical and kinematical scattering effects can to some extent be separated.)

The above-mentioned falloff of the atomic scattering factors of electrons with $(\sin\Theta/\lambda)^2$ results in small-angle Bragg reflections, with low (hkl) indices being typically the first to cease following the equations of the kinematic theory with increasing crystal size, while large-angle Bragg reflections may do so later. The former reflections are found in a diffraction pattern close to the primary beam and are typically the only ones that contribute to an HRTEM image that is recorded in a non–aberration-corrected transmission electron microscope. Correspondingly, crystal sizes are for structural fingerprinting from non–aberration-corrected HRTEM images restricted to thicknesses of some 5 to 10 nm only. PED patterns, on the other hand, also show large-angle Bragg reflections and can (for this and other reasons that are discussed in the following text) be employed to fingerprint nanocrystals structurally in the thickness range of about 10 to 50 nm. Because PED avoids the excitation of more than one strong diffracted beam (as much as this is possible with current technology[c] (see NOTES page 310) in a transmission electron microscope) and integrates over the excitation errors, Blackman corrections are frequently applicable to strong small-angle Bragg reflections. These corrections are discussed next.

BLACKMAN CORRECTIONS OF ELECTRON DIFFRACTION INTENSITIES

On the basis of Blackman's seminal paper (34) and careful experimental verifications of its conclusions, a practical method for the correction of the intensities of fine-grained CPEDR patterns for primary extinction effects was developed (3). While Ref. (47) provides, for example, experimental tests for aluminum polycrystals with negligible textures for the range

$$0.05 \leq \lambda \left| \frac{F_{hkl}}{\Omega} \right| A_3 \leq 2.5 \tag{9f}$$

the precision and accuracy of some of the respective measurements were significantly improved in Ref. (48).

For a fine-grained crystal powder with a random distribution of nanocrystal orientations, relations (4a) and (5a) can be integrated over all (reciprocal) distances h_3 (which represent the shape transform of the crystal parallel to the primary beam direction). For the two-beam dynamical diffraction case, this leads relation (4a) to

$$\frac{I_{hkl}}{I_0 S Q^2} = \frac{1}{Q} \int_0^{QA_3} J_0(2x)\mathrm{d}x = R(A_3, Q) \tag{12a}$$

where J_0 is the zero-order Bessel function and R a function of A_3 and Q. The corresponding integration of the kinematical counterpart relation (5a) leads to

$$\frac{I_{hkl}^{\mathrm{KIN}}}{I_0 S Q^2} = A_3 \tag{12b}$$

For small thicknesses A_3, where the integral over J_0 is of the order of unity, $R(A_3, Q) \approx A_3$ and relation (12a) becomes relation (12b) (i.e., the scattering will be kinematic). For large thicknesses, the upper limit QA_3 of the integral in (12a) may be approximated by infinity, and with $\int_0^{\infty} J_0(2x)\mathrm{d}x = {}^1/_2$, one obtains $R = {}^1/_2 Q$ (with the dependence on A_3 lost).

(Note that for finite QA_3 of the order of about 2 and larger, the value of this integral oscillates with ever-decaying amplitude around the value of 0.5. For crystal parallelepipeds with some thickness variations, these oscillations will be damped out.) The integrated coefficient of a reflection, thus, approximates for large thicknesses in the two-beam dynamical case to

$$\frac{I_{hkl}^{\mathrm{DYN}}}{I_0 S Q^2} \approx \frac{1}{2Q} \tag{12c}$$

that is, it is proportional to the first power of the structure factor. In the kinematical case (12b), this proportionality is, however, to the square of the structure factor.

The "kinematic correction function"

$$K_{\mathrm{kin}}(A_3, Q) = \frac{R(A_3, Q)}{A_3} = \frac{1}{A_3 Q} \int_0^{QA_3} J_0(2x)\mathrm{d}x \tag{13}$$

can, therefore, be employed for a correction of primary extinction effects in the two-beam approximation so that one obtains for the experimentally obtained integrated coefficient of a reflection of a real crystal in a diffraction experiment that integrates effectively over the excitation error the following relation:

$$\frac{I_{hkl}^{\mathrm{EXP}}}{I_0 S} = \lambda^2 \left| \frac{F_{hkl}}{\Omega} \right|^2 A_3 L K_{\mathrm{kin}}(A_3, Q) \tag{14}$$

For nearly kinematic scattering, one would have $K_{\mathrm{kin}} \approx 1$ (but a little smaller than unity). With increasing degree of dynamical scattering (i.e., increasing A_3Q), K_{kin} will decrease. For $A_3Q < 2$, K_{kin} may be approximated (3) by the relation

$$K_{\mathrm{kin}}(A_3, Q) \cong \exp\left(-\frac{1}{3A_3^2Q^2}\right) \tag{15a}$$

Since the first four factors on the right-hand side of relation (14) are the kinematic theory expression for an experimentally obtained integrated coefficient of a reflection from a real crystal, that is,

$$\lambda^2 \left| \frac{F_{hkl}}{\Omega} \right|^2 A_3 L = \frac{I_{hkl}^{\mathrm{KIN}}}{I_0 S} \tag{7b}$$

one can interpret relation (14) also as

$$\frac{I_{hkl}^{\mathrm{EXP}}}{I_{hkl}^{\mathrm{KIN}}} = K_{\mathrm{kin}}(A_3, Q) \tag{16a}$$

that is, as giving the ratios of the normalized intensity of experimentally obtained reflections to the respective normalized intensity as predicted by the kinematic theory.

At the practical level, it is customary to fit the experimental normalized intensity data to the function

$$I_{hkl}^{\mathrm{EXP}} = QLI_0S \int_0^{QA_3} J_0(2x)\mathrm{d}x \tag{16b}$$

and also to obtain as a result of the fitting an average value for A_3 (3).

In the spirit of a transition from kinematic scattering to two-beam dynamic scattering, one may alternatively use a "dynamic correction function"

$$K_{\mathrm{dyn}}(A_3, Q) = \frac{I_{hkl}^{\mathrm{EXP}}}{I_{hkl}^{\mathrm{DYN}}} \cong \exp\left(\frac{1}{6A_3^2Q^2}\right) \tag{15b}$$

for values of $A_3Q < 2.5$ that are close to the upper bounds of relations (9c) to (9e) in order to obtain integrated intensities that are proportional to the first power of the structure factor moduli (49).

In summary, Blackman corrections deliver a thickness–structure-dependent Lorentz factor. This Lorentz factor, in turn [i.e., relation (7)], ensures that structural fingerprinting can be performed within the validity range of quasi-kinematic approximations set by relations (9b) to (9e). [Relation (9f), on the other hand, gives the range for which this procedure has been experimentally verified.] As already mentioned, these corrections are applicable whenever the particulars of the diffraction experiments allow for the extraction of reflection intensities that were effectively integrated over the excitation errors. This is obviously the case for fine-grained CPEDR patterns (3,5,14) and also (to some extent in dependency of the precession angle) for PED patterns. Structural fingerprinting from both types of diffraction patterns is discussed next.

STRUCTURAL FINGERPRINTING FROM FINE-GRAINED CPEDR PATTERNS

Blackman corrections as well as relations (11a) and (11b) became the foundation of the structural electron crystallography work of the "EDSA school" around Pinsker, Vainshtein, and others (3–5,14,32,49–51). Electron diffraction cameras rather than transmission electron microscopes were initially (3,50) used and primary electron beam sizes were up to several hundreds of micrometers. These large beam sizes and crystallite sizes in the nanometer range ensured that the scattering of electrons was kinematic or quasi-kinematic, that is, could be accounted for by utilizing relations (7b) and (16b).

Normalized integrated intensities of mosaic nanocrystals, oblique textures of nanocrystallites, and fine-grained crystal powder with random orientation of the nanocrystallites have been utilized by the EDSA school in the kinematic and quasi-kinematic approximations

$$\frac{I_{hkl}^{\mathrm{EXP}}}{(I_0 S)^{\mathrm{EXP}}} = \lambda^2 \left| \frac{F_{hkl}}{\Omega} \right|^2 A_3 L \tag{17a}$$

by means of Blackman corrections for primary extinction effects. With the Lorentz factor

$$L = L_{\mathrm{ring}} = \frac{p d_{hkl}}{2} \tag{17b}$$

where p is the multiplicity factor of the reflection *hkl* and relations (7b) and (17a) are applicable to the normalized integrated intensity of rings in CPEDR patterns (3,5,14,51). [Lorentz factors for mosaic nanocrystals and oblique textures are also given in Refs. (3,5,51) but are not further discussed here. Since dynamical effects are typically more pronounced for mosaic nanocrystals, a correction for primary extinction following Blackman is more often required for them than for samples with textures or randomly oriented nanocrystallites (3,14).]

For a small (direct space) segment, Δ, of a Debye-Scherrer ring in a CPEDR pattern, the Lorentz factor of relations (7b) and (17a) becomes

$$L = L_{\mathrm{ring\,segment}} = \frac{p d_{hkl}^2 \Delta}{4\pi D \lambda} \tag{17c}$$

where D is the effective distance from the specimen to the detector, the so-called "camera length" of a transmission electron microscope (3–5). For per unit length of a Debye-Scherrer ring in a CPEDR pattern (52), the Lorentz factor of relations (7b) and (17a) becomes

$$L = L_{\mathrm{ring\,per\,unit\,length}} = \frac{p d_{hkl}^2}{8\pi^2 D \lambda} \tag{17d}$$

The only serious objection that the 1987 P. P. Ewald Prize laureate John M. Cowley, FAA, FRS, raised to this body of work was that it did not address "... the question of how to deal with the 'systematic' n-beam interactions which will inevitably affect some reflections strongly"(53). This statement is entirely correct and shall serve as a severe warning *not to augment structural fingerprinting of nanocrystals in the TEM with extracted structure factor information if the crystals are simply too thick* (and contain atoms that are rather heavy)! The route to success is obvious; keep the crystals as thin as possible to avoid systematic n-beam and other multiple dynamical interactions.

If this is for some reason not a viable option, one may deal with systematic n-beam interactions of selected systematic rows, for example, for (h00), (hh0), and (hhh) reflections that are higher orders ($n = 2, 3, \ldots$) of strong reflections with $h =$ 1 or 2, by means of the second Bethe approximation, the so-called "Bethe dynamic potentials" (4,14,32,48,54,55). As in the Blackman primary extinction correction, no knowledge of either the crystal thickness or orientation is needed for the application of this correction.

John M. Cowley repeatedly emphasized that the approximations of the EDSA school work best when the primary beam is large so that averaging over a large number of nanocrystals takes place, for example: "The averaging over crystal orientation reduced the dynamical diffraction effects to an extent that practical structure analysis was feasible" (52). CPEDR patterns should, thus, be recorded for structural fingerprinting purposes with a very wide beam and large selected area apertures should be utilized. These very wide beams also reduce the effects of structural damage of the individual nanocrystals by energy that is deposited by the primary beam. Assessing many nanocrystals at once also alleviates problems that are typically associated with collecting (energy-dispersive X-ray) spectroscopic information from individual nanocrystals. While the collection of such information from individual nanocrystals requires a focused, that is, high energy density primary electron beam that may damage a nanocrystal structurally, possible structural damage to the individual nanocrystal is minimized by collecting an energy-dispersive X-ray spectroscopic signal from an ensemble of many nanocrystals.

There are also many statements by John M. Cowley that emphasize the importance of both the correct usage of Lorentz factors and the dynamical scattering effect corrections, for example:

> Criteria in quantitative form are available to determine whether the intensities might be modified by such factors as crystal size, morphology or lattice defects. Provided that such criteria are applied with sufficient care, there can be very little objection to the use of the intensities from such material for purposes for structure analysis. (56)

The main objection of Douglas L. Dorset to the body of work of the EDSA school is that corrections based on a fitting and averaging procedure to find an exponent of Q_{ave} that is intermediate between 2 and unity (35) are not specific to the reflections (37). His correction scheme on the basis of the phase grating approximation and two experimental data sets that were recorded with highest and intermediate voltage transmission electron microscopes indeed offers specificity as to which reflections need to be corrected (37), albeit at the price of a higher experimental effort.

For structures with a wide range of chemical compositions, the analyses of the CPEDR patterns by the EDSA school resulted in crystallographic R values in the range of 5% to 8% (50). These are some of the lowest reported R values for structural electron crystallography and quite comparable with what is routinely obtained by means of X-ray crystallography on much larger crystals from large data sets. In direct space (i.e., the space of the electrostatic potential), these very good R values of the EDSA school correspond to positional accuracies of 1% to 2% for medium-weight atoms and up to 10% for light atoms (50).

Fine-grained crystal powders with randomly oriented nanocrystallites can be straightforwardly distinguished from textures of nanocrystallites because the intensity distribution in the diffraction patterns of the former does not change when the whole specimen is tilted with respect to the primary beam by means of the specimen goniometer (57). (Note that, as mentioned in the section "Nanocrystals cannot be fingerprinted structurally by powder X-ray diffractometry," a comparably simple experimental test for the existence of preferred orientations in a powder sample is not readily available for X rays so that the experimentally obtained peaks heights are often discarded in powder X-ray diffraction fingerprinting.)

Zero-loss energy filtering is highly recommended for the accurate extraction of structure factor moduli from CPEDR patterns (58). New software tools for the more accurate and precise analysis of CPEDR patterns have been developed (59–61) recently and are available as free downloads at the Web site of the Digital Micrograph™ Script Database (62). Some of these software tools provide the correction for a posteriori distortions to the electron diffraction ring geometry by the projector lenses of the transmission electron microscope.

STRUCTURAL FINGERPRINTING FROM PED PATTERNS OF INDIVIDUAL NANOCRYSTALS

The PED method, also referred to as Vincent-Midgley technique due to the seminal paper (33) of these authors, is formally analogous to the well-known X-ray (Buerger) precession technique, but it utilizes a precession movement of the primary electron beam around the microscope's optical axis rather than that of a single crystal around a fixed primary X-ray beam direction. The primary and diffracted electron beams are descanned in such a manner that stationary diffraction patterns are obtained. The primary electron beam can be either parallel (57) or slightly convergent (33), and its precession creates a hollow illumination cone[c] with its vertex on the crystalline sample.

Figure 3 shows experimental PED patterns from a silicon crystal in both the "just-precessed" and "properly descanned" modes (for 200-kV electrons). In the latter mode [Figs. 3(B) and 3(D)], all of the fine arcs and circles of intensity of Figures 3(A) and 3(C) are integrated into sharp diffraction spots, which remain motionless on the screen of the transmission electron microscope. The strongest circles in Figures 3(A) and 3(C) are due to the primary electron beam and its proper descanning

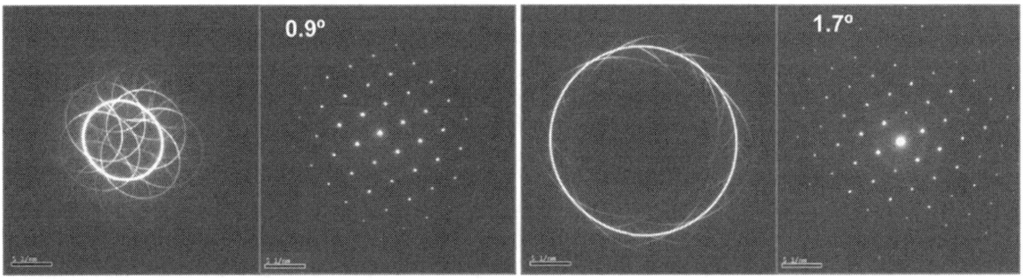

(A) (B) (C) (D)

FIGURE 3 Experimental precession electron diffraction patterns of a silicon crystal: [110] zone axis, 40-nm approximate thickness, and 200 kV. (**A**) and (**C**) "Just-precessed" mode and (**B**) and (**D**) "properly descanned" mode so that stationary spot diffraction patterns result. Smaller and larger[d,e] (*see* NOTES page 312) precession angles of either 0.9° (**A**) and (**B**) or 1.7° (**C**) and (**D**) were utilized.

results in the central 000 spot in the PED patterns of Figures 3(B) and 3(D). (All of the following experimental PED patterns were taken at 200 kV as well and are shown only in the properly descanned mode.)

Figure 4 illustrates the PED geometry with sketches for a [110] oriented cubic nanocrystal. The precession movement of the primary electron beam around the center of the screen of the transmission electron microscope (Figs. 3(A) and 3(C)] in direct space can be visualized in reciprocal space by the rotation of the so-called "Laue circle" (Fig. 4(A)] around the central 000 spot in the stationary diffraction patterns of Figures 3(B) and 3(D). The radius of the Laue circle is determined by the precession angle, that is, the half angle of the hollow illumination cone of the precessing primary electron beam. The precession angle can be calibrated on the basis of the radius of the primary electron beam circle in "just-precessed" mode recordings as Figures 3(A) and 3(C).

The Ewald sphere will be intersected sequentially at positions that are close to the circumference of the Laue circle [Fig. 4(A)]. Note how individual rows of reflections are excited sequentially (as much as this is possible with current technology in a transmission electron microscope[c]) in Figure 4(A) for a precession angle of 2.8°. This sequential excitement of reflections and rows of reflections reduces the number of viable multiple diffraction scattering paths between different reflections and rows of reflections at any one time and, thus, reduces nonsystematic multiple

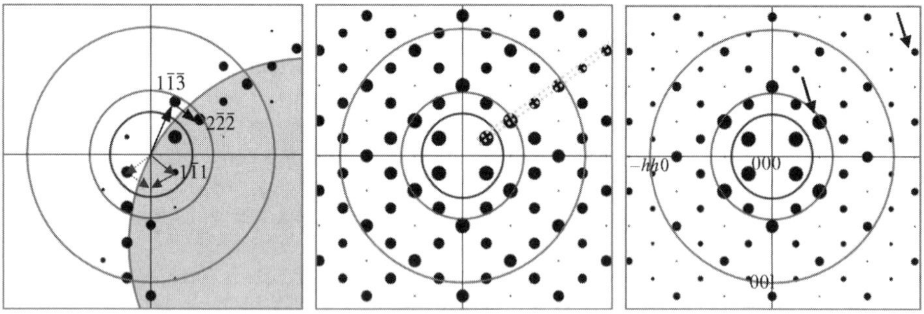

(A) (B) (C)

FIGURE 4 Sketches to illustrate the precession electron diffraction (PED) geometry, [110] zone axis of a 3.3-nm thin crystal with space group $Fd\bar{3}m$ (no. 227, origin choice 2, one atom in asymmetric unit at position 000), 200 kV, 2.8° precession angle. The concentric rings represent direct-lattice spacings of 0.225, 0.15, and 0.075 nm, respectively. **(A)** Snapshot of the formation of a PED pattern, intensity distribution (size of the reflection disks) proportional to the structure factor, $\sim F_{hkl}$. The partly shown shaded circle represents the (rotating) Laue circle. The short full and dotted arrows represent the double scattering paths that are mainly responsible for the intensity of the kinematically forbidden (002) reflection. (**B**) Two-beam dynamical intensity distribution, $\sim F_{hkl}$. The double-dashed line represents one of the $\{hhh\}$ systematic rows. (**C**) Kinematical intensity distribution, $\sim F_{hkl}{}^2$. The arrows points to two $\{hhh\}$ reflections with indices that are all even but not a modulus of 4, that is, they are kinematically forbidden for this crystal. These reflections are also present in the snapshot of (**A**) where a double-diffraction path that leads to the $(2\bar{2}\bar{2})$ reflection over $(1\bar{1}\bar{3})$ and $(1\bar{1}1)$ is marked by arrows. Note that the overall intensity distribution in (**B**) and (**C**) is far from the "nearly leveled out" intensity distribution that may be caused by strong multiple scattering effects in experimental electron diffraction patterns. In addition, both figures have essentially the same intensity ordering between the strongest and weakest reflections. The program "eMap" (Version 1.0, 2007) of the AnaliTEX company[f] has been used for the creation of these sketches.

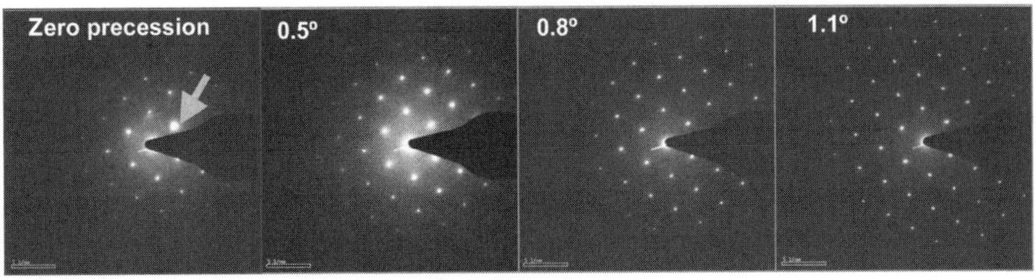

(A) (B) (C) (D)

FIGURE 5 Experimental diffraction patterns from a crushed silicon crystal, approximate thickness 60 nm, orientation close to the [110] zone axis. (**A**) Selected area electron diffraction (SAED) pattern (zero precession). (**B**) to (**D**) Precession electron diffraction (PED) patterns from the same sample area with increasing precession angle.[d] Note that while the intensity of the $(1\bar{1}\bar{1})$ reflection, marked by an arrow, is much higher than that of its Friedel pair $(\bar{1}11)$ and that of the other two symmetry equivalent $\pm(1\bar{1}1)$ reflections in the SAED pattern (**A**), the intensities of all four symmetry equivalent $\{1\bar{1}1\}$ reflections are very similar for the PED patterns (**B**) to (**D**). To appreciate a beneficial side effect of the precessing primary electron beam, note that all electron diffraction patterns arise from the same nanocrystal area in the same crystallographic orientation.

scattering effects significantly. [The long, double-dashed line in Fig. 4(B) represents a systematic row of reflections. Systematic dynamical interactions along such rows are not suppressed by the precession movement of the primary electron beam.]

A comparison of the conventional selected area electron diffraction (SAED) pattern [Fig. 5(A)] with its PED pattern counterparts [Figs. 5(B) to 5(D)] demonstrates that a nanocrystal does not need to be oriented with a low-indexed zone axis exactly parallel to the optical axis of the transmission electron microscope in order to support advanced structural fingerprinting in the TEM. The primary electron beam may also be slightly tilted with respect to the optical axis of the transmission electron microscope, as demonstrated by the comparison of the SAED pattern [Fig. 6(A)] with its PED counterpart patterns [Figs. 6(B) and 6(C)]. These tolerances to crystal misorientations and beam tilt misalignments lessen the experimental efforts for effective structural fingerprinting based on PED patterns.

Compared with the SAED patterns from the respective three silicon crystals [Figs. 5(A), 6(A), and 7(A)], there are frequently many more reflections in the PED patterns [Figs. 5(B) to 5(D), 6(B), 6(C), 7(B), and 7(C)] from nanocrystals with low defect content. This is especially true for higher precession angles [Figs. 5(D), 6(B), 6(C), 7(B), and 7(C)]. More reflections allow for least-squares fits to larger systems of inhomogeneous linear equations. This results in more precise determinations of projected reciprocal lattice geometries. The initial projected reciprocal lattice geometry–based identification step of our structural fingerprinting procedure can, thus, be improved on the basis of PED patterns of high-quality nanocrystals.

For thin nanocrystals with high defect content and layer structures, for example, zeolites (63) and organic crystals, in which there can be a large amount of defect-mediated secondary scattering (64), the number of reflections may decrease when PED is utilized. For thick nanocrystals in which a large amount of the intensity of the diffraction spots might be due to multiple scattering, there may also be a reduced number of reflections in PED patterns. This is because PED effectively reduces the number of viable multiple scattering paths.

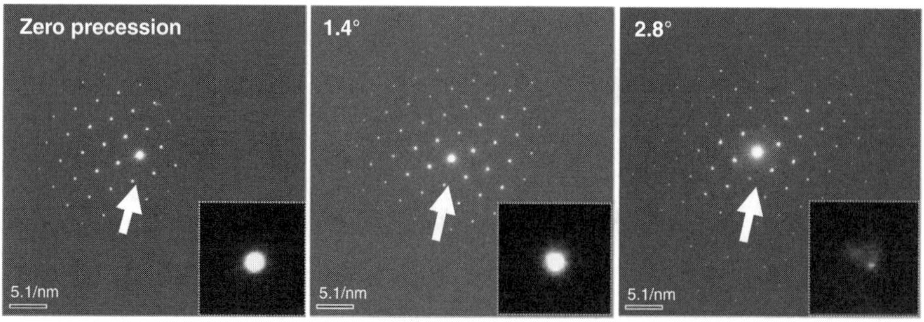

(A) (B) (C)

FIGURE 6 Experimental diffraction patterns from a thicker part of a wedge-shaped silicon crystal that was prepared in a focused ion-beam microscope. The thickness is approximately 56 nm. (**A**) Selected area electron diffraction pattern (zero precession) close to the [110] zone axis. Note the slight misalignment of the primary beam. (**B**) and (**C**) Precession electron diffraction patterns from the same sample area with increasing precession angle.[d] One member of the kinematically forbidden ±(002) reflections is marked by an arrow in each of the diffraction patterns and also shown magnified in the insets. To appreciate a side effect of the precessing primary beam, note that all of the electron diffraction patterns of this figure arise from the same nanocrystal area with the same initial beam tilt misalignment with respect to the optical axis of the transmission electron microscope.

Most important for advanced structural fingerprinting purposes in the TEM (10), single-crystal PED patterns deliver integrated diffraction spot intensities that can be treated as either kinematical or quasi-kinematical for crystals that are up to several tens of nanometers thick (33,40,42–46,54,55,57,63,65–78).

The {222} reflections are kinematically forbidden for silicon, but they are rather strong in all diffraction patterns [Figs. 3(B), 3(D), 5, 6, 7(B), and 7(C)] [except in the SAED pattern of the thinnest crystal, Fig. 7(A)]. The special term "perturbation reflections" has been suggested (79) for such reflections and their intensity is due to an electron diffraction equivalent of the Renninger ("Umweganregungs") effect of X-ray diffraction. The subsequent diffraction of the $(1\bar{1}\bar{3})$ beam on the $(1\bar{1}1)$ net plane results, for example, in the $(2\bar{2}\bar{2})$ reflection.

The {222} and {666} reflections of silicon are located in densely populated systematic rows. These reflections have a very low intensity in the SAED pattern of the thinnest crystal [Fig. 7(A)], but they are strong in the corresponding PED patterns [Figs. 7(B) and 7(C)]. This can partly be explained by the precession geometry, which tends to excite whole systematic rows at once. As has been noted earlier (33), PED cannot suppress multiple-beam dynamic scattering within a systematic row of reflections.

As demonstrated in Figures 5 to 7, kinematically forbidden reflections, for example, ±{002} and {222} or {666} reflections of silicon, are frequently present in electron diffraction patterns as a result of multiple dynamical scattering. PED, does, however reduce the intensity of ±(002) reflections significantly, especially at large precession angles[d] and for thicker crystals [Figs. 6(B) and 6(C)].

The ±(002) reflections in the [110] orientation of silicon arise mainly from the double scattering by the $\pm(\bar{1}1\bar{1})$ and $\pm(\bar{1}11)$ reflections. Since these $\pm(\bar{1}1\bar{1})$ and $\pm(\bar{1}11)$ reflections possess the largest net plane spacing, the effect of the geometrical part of the Lorentz factor[e] on their intensities will be rather significant for low

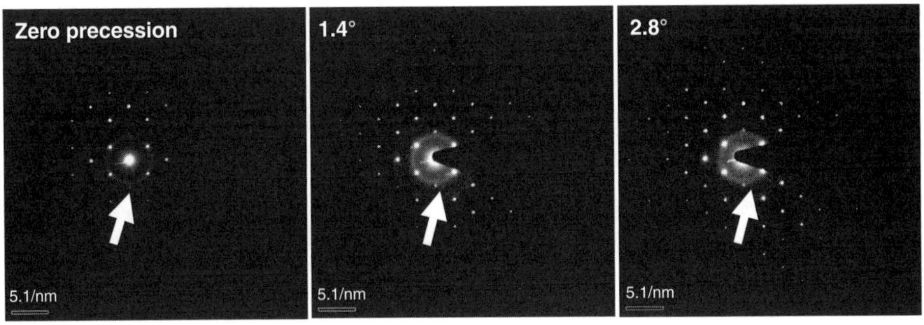

(A) (B) (C)

FIGURE 7 Experimental diffraction patterns from the thinnest part of a wedge-shaped silicon crystal that was prepared in a focused ion-beam microscope and is oriented close to the [110] zone axis. The thickness is approximately 6 nm. (**A**) Selected area electron diffraction (SAED) pattern (zero precession). (**B**) and (**C**) Precession electron diffraction patterns from the same sample area with increasing precession angle. All diffraction patterns were recorded close to the amorphized edge region of the sample that borders on the vacuum region within the microscope. This explains the relatively strong asymmetry of the SAED pattern (**A**) with respect to its center. The concentric rings in all diffraction patterns arise from the above-mentioned amorphized edge region. The effect of the geometrical part of the Lorentz factor[e] seems to dominate over its structure–thickness-dependent part for this thin crystal and may explain the initial absolute increase of the intensity of the kinematically forbidden ±(002) reflections with precession angle. One member of this pair of reflections is marked by an arrow in each diffraction pattern.

precession angles and gradually decline as this angle increased. (The literature so far disagrees about the exact formulae for both parts of the Lorentz factor and under which conditions they need to be applied or can be igonred.[e]) (*See* NOTES page 312.)

An exponential decay of peak intensities with precession angle was observed for the kinematically forbidden ±(002) reflections of silicon for four thicknesses in the range of approximately 22 to 56 nm. (These results are not shown explicitly and will be published elsewhere.) As observed earlier by other authors (42), this decay appeared to be independent of the nanocrystal thickness for the experimentally tested thickness range.

Figure 8 provides a comparison of the effect of the precession angle on the intensity of the ±(002), ±(111), and ±(111) reflections for the 6-nm thin silicon nanocrystal of Figure 7 with the corresponding dependency of the thickest silicon nanocrystal of that range, that is, the 56-nm thick silicon nanocrystal of Figure 6. Principally different dependencies of the integrated intensities of kinematically forbidden and "allowed" reflections on the precession angle for both thicknesses are revealed in Figure 8. While there is an exponential decay of the intensities for the ±(002) reflections of the thick crystal and an analogous decay with nearly the same slope between 1.1° and 2.2° precession angle for the (002) reflection of the thin crystal, the {111} reflection intensities decrease much more slowly with precession angle for both crystals and settle to a certain value that is mainly determined by thickness-dependent primary extinction effects.

These principally different dependencies may allow for a quite unambiguous identification of some of the kinematically forbidden reflections and could be utilized for advanced structural fingerprinting. If a crystal projects very well and

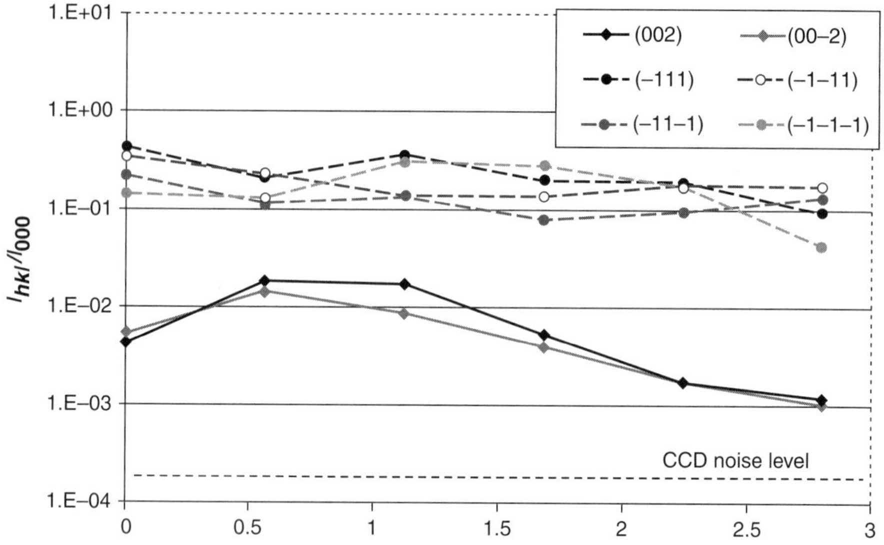

(A)

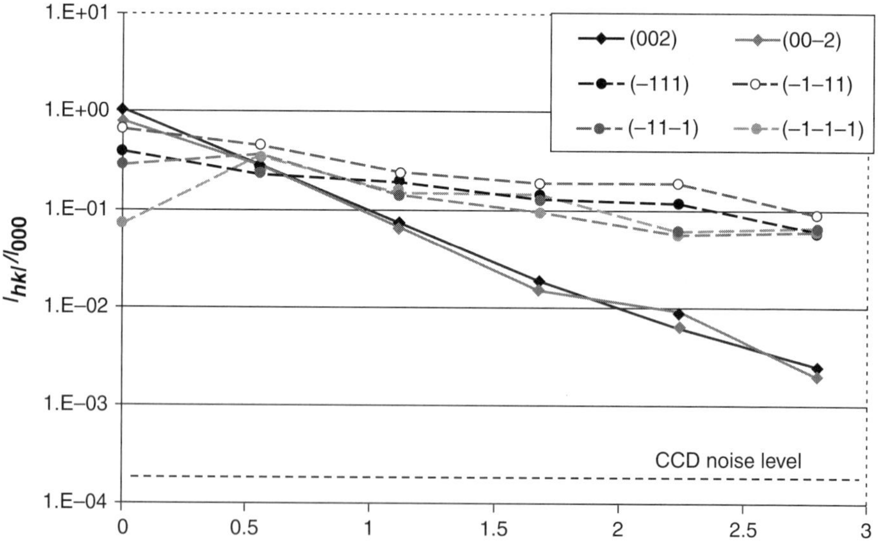

(B)

FIGURE 8 Effects of the precession angle on the normalized peak intensities of the kinematically forbidden ±(002) and the "allowed" {111} reflections that mainly produce them by double diffraction in the [110] zone axis orientation for two silicon nanocrystals with thickness of approximately (**A**) 6 nm and (**B**) 56 nm. The normalization was performed by dividing the maximal peak intensity of the reflections by the maximal peak intensity of the primary beam. The relative large difference in the intensities of members of the two {111} Friedel pairs in (**A**) is due to the recording of the diffraction patterns close to the amorphized edge region of the sample, bordering on the vacuum region in the microscope (see caption of Fig. 7). Note that for a precession angle between about 1.1° and 2.2°, the intensity of the (002) reflection of the 6-nm thick crystals declines at nearly the same rate as the ±(002) reflections of the 56-nm thick crystal. *Abbreviation*: CCD, charge-coupled device.

reflections from higher order Laue zones are present in a PED pattern, the space group may be determined from a few crystal projections (80,81). The program "Space Group Determinator" from the Calidris company[f] supports such identifications (82).

As already noted [by a comparison of the SAED pattern, Fig. 5(A), with its PED pattern counterparts, Figs. 5(B) to 5(D)], a nanocrystal does not need to be oriented with a low-indexed zone axis exactly parallel to the optical axis of the transmission electron microscope in order to support advanced structural fingerprinting in the TEM. This effect is also demonstrated by the integrated peak intensities of the ±(111) Friedel pair reflections in Figure 8(B). While there is a noticeable difference between the intensity of the members of this Friedel pair in the SAED pattern, there is a small difference between these peak intensities in the PED patterns. (If normalized integrated intensities were compared, there would probably be an even smaller difference in the latter case.)

PED allows for the collection of integrated reflection intensity data at smaller and larger precession angles from the same crystalline sample area. A suitable modification of Douglas L. Dorset's correction scheme (37) may, therefore, be developed on the basis of two experimental data sets that differ with respect to their "effective curvature" of the Ewald sphere but are recorded successively from the same crystalline sample area.

Automated "crystal orientations and structures" mapping[c] in the TEM is enhanced significantly by PED (38). The tendencies of PED patterns to show more reflections with kinematic or quasi-kinematic intensity and the suppression of real structure–mediated double-diffraction effects are the causes of this enhancement.

REPRESENTING PROJECTED RECIPROCAL LATTICE GEOMETRIES IN LATTICE-FRINGE FINGERPRINT PLOTS

The extraction of the projected reciprocal lattice geometry from PED patterns or HRTEM images is the first step of structural fingerprinting. The latter have to show at least two sets of crossing lattice fringes in real space, resulting (in reciprocal space) in at least two (nonzero) Fourier coefficient (plus their respective Friedel pairs) in the Fourier transform of the HRTEM image intensity. The crystal structure, the orientation of the crystal, and the Scherzer[b] resolution of a non–aberration-corrected transmission electron microscope determine the number of lattice fringes in the HRTEM image and, thus, the number of Fourier coefficients in the transform of the image intensity. (The following section on advanced instrumentation illustrates how the number of Fourier coefficients in transforms of the HRTEM intensity increases with increasing resolution of the transmission electron microscope.) Throughout the following section, it is assumed that the Fourier transform of the HRTEM image intensity contains at least two Friedel pairs, that is, allows for the assignment of two independent (noncoplanar) reciprocal lattice vectors that define the projected reciprocal lattice geometry.

The extraction of information on the projected reciprocal lattice geometry is very similar for both sources of structural data. Due to the large radius of the Ewald sphere in electron diffraction (i.e., λ^{-1}), one will obtain a reciprocal 2D lattice as a projection of any reciprocal 3D lattice. The "position" of each reflection in this lattice (or each Fourier coefficient of the HRTEM image intensity) is characterized by either a single set or several sets of three parameters. One of these parameters is the distance of the reflection to the reflection 000, in other words, the length of the reciprocal lattice vector of this reflection. The other parameter is the acute angle this reflection makes with another reflection. The remaining parameter is the length of the other reciprocal lattice vector that was used in order to define the (acute) "interfringe angle."

This definition of the "position parameters" of reflections has several advantages. The most obvious being that the position of a reflection does not depend on the orientation of the projected reciprocal lattice with respect to the edges of the medium on which the PED patterns or HRTEM images were recorded. Experimental plots of projected reciprocal lattice geometry are thus independent of this orientation. Another advantage of this definition of the position parameters of reflections is connected to the ways in which lattice centering and space group symmetry elements with glide component that result in kinematically forbidden reflections are dealt with in such plots. This is described in more detail in the following text. For now, it suffices to say that the experimental plots will represent the whole projected reciprocal lattice geometry in a consistent manner.

For the initial part of advanced structural fingerprinting in TEMs, such experimental plots can be straightforwardly compared with their theoretical counterparts, which we call "lattice-fringe fingerprint plots" (Figs. 9 and 10). The latter plots can be calculated "on the fly" over the Internet from our mainly inorganic subset (15) of the Crystallography Open Database (16–18) and contain all of the data points for all of the zone axes of a crystalline material up to a predefined resolution in reciprocal space. Identifying a crystal from its projected reciprocal lattice geometry is, thus, frequently equivalent to finding the 2D data points of the experimental plot within the theoretical lattice-fringe fingerprint plot.

Figure 9 shows the theoretical lattice-fringe fingerprint plot for the mineral rutile for a 0.19-nm Scherzer[b] resolution in both the kinematic and (two-beam) dynamical diffraction limits. Screw axes and glide planes result in systematic absences of reflections in 2D projections of the reciprocal lattice geometry and are revealed in "kinematic lattice-fringe fingerprint plots" by missing rows [compare Figs. 9(A) and 9(B)]. The so-called "Gjønnes and Moodie dynamically forbidden reflections" (83) are shown in dynamical lattice-fringe fingerprint plots [Fig. 9(B)], because small beam tilts and somewhat larger crystal tilts may cause these fringes to be present in HRTEM images (84). Analogously, these reflections should be present in PED patterns when the hollow cone illumination is not perfectly symmetric about the zone axis and the crystals are not exactly oriented with their zone axis parallel to the optical axis of the transmission electron microscope. The other type of systematic absences of reflections in 2D projections of reciprocal lattice geometries, which are due to 3D Bravais lattice centerings, results in systematic absences of entire rows in lattice-fringe fingerprint plots independent of the "kinematic" or "dynamic" type of these plots.

While there are two data points in lattice-fringe fingerprint plots for reflections with different spacings, the crossing of two symmetrically reflections results in just one data point (because the latter possess by symmetry the same spacing). These plots extend in reciprocal space out to the resolution of a PED pattern or either to the Scherzer resolution or to the instrumental resolution[b] of the transmission electron microscope in the imaging mode. All of the resolvable lattice fringes and reflections up to the appropriate resolution will be included for a certain crystal

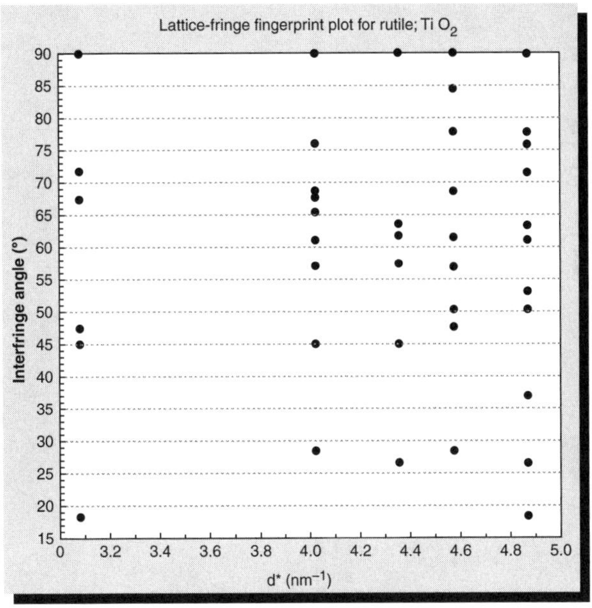

(A)

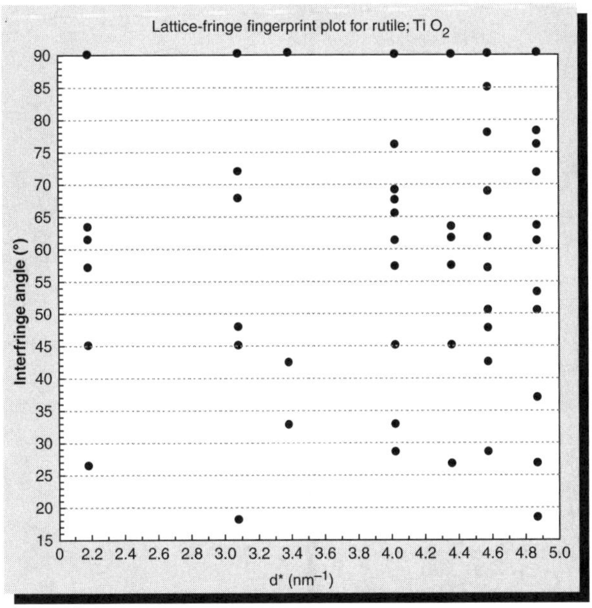

(B)

FIGURE 9 Theoretical lattice-fringe fingerprint plots for rutile (TiO_2) for a microscope with a Scherzer[b] resolution of 0.19 nm. (**A**) Kinematic diffraction limit. (**B**) Two-beam dynamic diffraction limit. Note the (Gjønnes and Moodie) dynamically forbidden reflections, for example, {100} in the case of rutile in the exact [001] orientation, are included in the (two-beam) dynamic diffraction limit plot.

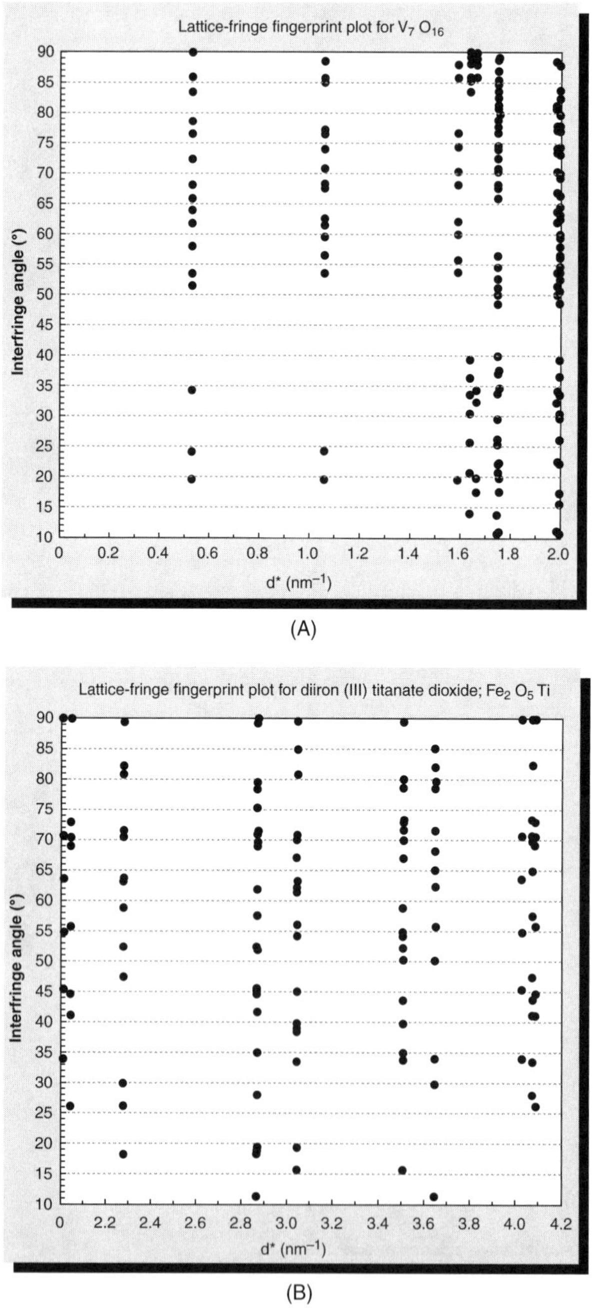

FIGURE 10 Theoretical lattice-fringe fingerprint plots for (**A**) vanadium oxide nanotubes (chemical composition: V_7O_{16}), structural details in Ref. (29), and (**B**) the mineral pseudo-brookite, $Fe_2(TiO_3)O_2$. Note the characteristically different distribution of the two-dimensional data points in both plots and that the abscissas are on different length scales.

structure into these plots. The appearance of lattice-fringe fingerprint plots is, thus, both crystalline material and reciprocal space resolution specific (Figs. 9 and 10), while there is no essential conceptional difference at the projected reciprocal lattice geometry level for both types of source data.

Figure 10(A) shows a theoretical lattice-fringe fingerprint plot that has been calculated for vanadium oxide nanotubes, a crystalline material that did *not* give a characteristic X-ray powder diffraction fingerprint (Fig. 2). Due to the rather large unit cell dimensions of the vanadium oxide nanotubes (29), an older transmission electron microscope with a very modest Scherzer[b] resolution of 0.5 nm is already sufficient to fingerprint this material structurally from the Fourier transform of an HRTEM image. The abscissa in Figure 10(A) is accordingly restricted to 2 nm^{-1}. A modern analytical transmission electron microscope with a Scherzer[b] resolution of 0.24 nm is, on the other hand, required to structurally fingerprint the higher symmetric (space group *Cmcm*) mineral pseudo-brookite with a unit cell that is approximately half the size of that of the vanadium nanotubes. The abscissa extends to 4.2 nm^{-1} in Figure 10(B) accordingly. Whenever PED patterns are used as experimental data source, there are many more data points in the respective experimental lattice-fringe fingerprint plots, as the resolution restriction of data in reciprocal space is determined by the nanocrystals themselves.

Figure 10(B) shows a theoretical lattice-fringe fingerprint plot for the mineral pseudo-brookite, for which a characteristic X-ray powder diffraction fingerprint was shown as Figure 1. From the comparison of Figures 10(A) and 10(B), one can conclude that lattice-fringe fingerprinting works for both types of crystalline materials, those that do *not* (Fig. 2) and those that do (Fig. 1) give characteristic X-ray powder diffraction fingerprints.

An initial search in a database of theoretical lattice-fringe fingerprints that is based on the 2D positions of data points in lattice-fringe fingerprint plots alone may result in several candidate structures. In the following steps, the search can be made more discriminatory both by trying to match crystallographic indices to the 2D positions and by determining the projected symmetry. We now deal with the former.

Because one will always project along one zone axis, all indices of the reflections must be consistent with a certain family of zone axes. As far as the lattice-fringe fingerprint plots are concerned, this follow-up search is equivalent to assigning crystallographic indices to the 2D data points. This can be done on the basis of Weiss' zone law for the zero-order Laue zone. Each (vertical) column of data points in a lattice-fringe fingerprint plot corresponds to one family of reflections (net planes). Discrete points on a second x-axis in a lattice-fringe fingerprint plot can, therefore, be labeled with the respective Miller indices, $\{hkl\}$, of a family of reflections. Each (horizontal) row of data points in a plot such as Figures 9 and 10, on the other hand, belongs to a family of zone axes. Discrete points on a second y-axis of such a plot can, thus, be labeled with the respective Miller indices, $<uvw>$, of a family of zone axes. The cross product of the Miller indices of two data points from two different columns (representing two different reciprocal spacings) that are also located within the same row (representing one interfringe angle) gives the zone axis symbol, $<uvw> = \{h_1k_1l_1\} \times \{h_2k_2l_2\}$. While each family of reflections will show up only once on such a second x-axis, the same family of zone axis symbols may be showing up multiple times on such a second y-axis. Guided by the added

Miller indices for columns and rows on such a lattice-fringe fingerprint plot, kinematically forbidden reflections can be easily identified (in kinematical diffraction limit plots). Higher orders of a family of net planes $\{nh,nk,nl\}$ possess an (n times) integral multiple of the spatial frequency of that family. Such higher orders of families are also easily spotted in a lattice-fringe fingerprint plot because their "columns" look identical. This is because the respective higher order net planes will intersect other net planes at precisely the same interfringe angles as those net planes from a lower order.

Within the error bars and especially in lattice-fringe fingerprint plots for a very high microscope resolution, it is possible that families of net planes or zone axes coincide on the second x- or y-axis. There are also cases in which the net plane spacings of two different families are exactly the same, for example, $\{122\}$ and $\{003\}$ or $\{553\}$ and $\{731\}$ in the cubic system. While the respective data points will be located in a lattice-fringe fingerprint plot in the same column, they will most likely possess different interfringe angles, that is, will be located in different rows. Because all interfringe angles between identically indexed reflections are the same in the cubic system, space group information can be extracted straightforwardly from lattice-fringe fingerprint plots (in the kinematic diffraction limit) of cubic crystals.

More elaborate lattice-fringe fingerprint plots may contain in the third and fourth dimension information on structure factor phases and amplitudes. Possibly, in a fifth dimension, histograms of the probability of seeing crossed lattice fringes in an ensemble of nanocrystals may be added to both types of lattice-fringe fingerprint plots and may facilitate the structural fingerprinting of an ensemble of nanocrystals. The equations for calculating such probabilities for an ensemble of randomly oriented nanocrystals are given in Ref. (19) (of which an early version is in open access). Instead of employing higher dimensional spaces, one could also stick to 2D displays such as Figures 9 and 10 and simply add to selected data points sets of numbers that represent additional information (e.g., structure factor phases and amplitudes), expected intensities in the kinematic or two-beam dynamic limits, all with their respective error bars.

Similarly to the classical Hanawalt (22) search strategy of powder X-ray diffraction databases (23), one could divide lattice-fringe fingerprint plots into 2D geometric data sectors of experimental condition–specific average precisions and accuracies and also allow for some overlap between the sectors. Larger reciprocal spacings and interfringe angles can be measured inherently more accurately and precisely than smaller reciprocal spacings and interfringe angles. The location of the respectively more precise and accurate data points will be in the upper right-hand corners of lattice-fringe fingerprint plots.

If there are many 2D data points in theoretical lattice-fringe fingerprint plots, as in the case of crystals with large lattice constants and for data from PED patterns with high precession angle, the initial search in an advanced structural fingerprinting proceeding may be based just on the reciprocal lattice spacings and interfringe angles of the two reflections with the shortest reciprocal lattice length. These three data point position parameters are a minimalistic characteristic of a certain zone axis of a crystalline material. Such search strategies are in the process of being implemented under the name "reduced lattice-fringe fingerprint plots" in both the kinematic and (two-beam) dynamic diffraction limits at our Web site (20) on the basis of data of the mainly inorganic subset (15) of the Crystallography Open Database (16–18).

2D SYMMETRIES AND THEIR UTILIZATION IN STRUCTURAL FINGERPRINTING

Because the diffraction spot intensities in PED patterns are integrated by the precession movement of the primary electron beam, they must possess the "2D diffraction symmetry of the projected electrostatic potential" of the nanocrystal. Fourier coefficients of the image intensity of an HRTEM structure image,[b] on the other hand, are complex entities and must possess the "full 2D projection symmetry of the electrostatic potential."

Both symmetries can be straightforwardly extracted by computer programs such as CRISP/ELD and Space Group Determinator from Calidris[f] from the experimental data so that this information can be utilized for advanced structural fingerprinting of nanocrystals in the TEM. One must, however, be aware that symmetry is to some extent "in the eye of the beholder," as it refers strictly only to mathematical entities. The programs CRISP/ELD and Space Group Determinator provide, however, crystallographic reliability values so that theoretical and experimental symmetries can be meaningfully compared.

Plane groups are the 2D projections of the crystallographic (3D) space groups. The former are sometimes referred to as the "wallpaper groups," because any wallpaper can be classified as belonging to one of these groups. Since there is a simple linear relationship between the Fourier coefficients of the image intensity of an HRTEM structure[b] image and the structure factors of a crystal (11,13), the former must be related to each other by the symmetry elements of the respective plane group that results from the projection of the 3D symmetry elements of the electrostatic potential. A "beginner's guide" to the application of symmetry in 2D electron crystallography (which concentrates on TEM images but does not cover PED patterns) has recently been published (85).

While there are 230 space groups in total, their projections in two dimensions in any direction results in just one of the 17 plane groups. Volume I of the *International Tables for X-ray Crystallography* (86) and volume A of the updated versions of these well-known encyclopedic reference books, the *International Tables for Crystallography* (*IT-A*) (87), describe both groups comprehensively and give rules on how to project symmetry elements from 3D onto any net plane. There is also a "teaching edition" that gives a comprehensive description of the 17 plane groups (88) and the rules on how to obtain plane groups from space groups. While the earlier versions of these texts (86) gave the plane symmetry group of the projections along the crystal axes for the triclinic, monoclinic, and orthorhombic space groups only, *IT-A* gives the plane symmetry groups for three low-indexed projections for each space group.

Since the symmetry element projection rules are somewhat cumbersome to apply, we are in the process of developing a universal space group projector program that will be later on interfaced to the mainly inorganic subset (15) of the Crystallography Open Database (16–18) and accessible openly at our Web server (20). We utilize calculation procedures that are suggested in Ref. (89).

In short, the projected 2D coordinates (r, s) of the 3D fractional atomic coordinates (x, y, z) (also representing 3D direct space vectors from the 3D origin to the respective atoms) along any axis $[uvw]$ are obtained by multiplication with the projection matrix P_{ij}

$$\begin{bmatrix} r \\ s \end{bmatrix} = \begin{bmatrix} P_{11} & P_{12} & P_{13} \\ P_{21} & P_{22} & P_{23} \end{bmatrix} \cdot \begin{bmatrix} x \\ y \\ z \end{bmatrix} \quad (18)$$

The projection of $[uvw]$ is $[0, 0]$ = origin of 2D mesh and the projections of (the direct space 3D lattice) vectors $\mathbf{p}$ and $\mathbf{q}$ will be the new (2D) unit mesh vectors = $(1, 0)$ and $(0, 1)$ so that one has six equations to solve for the six components of P_{ij}

$$\begin{bmatrix} 010 \\ 001 \end{bmatrix} = \begin{bmatrix} P_{11} & P_{12} & P_{13} \\ P_{21} & P_{22} & P_{23} \end{bmatrix} \cdot \begin{bmatrix} u & p_1 & q_1 \\ v & p_2 & q_2 \\ w & p_3 & q_3 \end{bmatrix} \quad (19)$$

with vectors $\mathbf{p} = p_1\mathbf{a} + p_2\mathbf{b} + p_3\mathbf{c}$ and $\mathbf{q} = q_1\mathbf{a} + q_2\mathbf{b} + q_3\mathbf{c}$. The simplest matrices P_{ij} are obtained in cases when $\mathbf{p}$ and $\mathbf{q}$ are both chosen to be unit cell vectors ($\mathbf{a}$, $\mathbf{b}$, or $\mathbf{c}$) of the respective 3D lattice. These matrices are as follows:

$$\left[P_{ij}\right]^{\vec{a},\vec{b}} \begin{bmatrix} 1 & 0 & -\left({}^{u}/_{w}\right) \\ 0 & 1 & -\left({}^{v}/_{w}\right) \end{bmatrix} \mathbf{p} = \mathbf{a} = (100),\ \mathbf{q} = \mathbf{b} = (010) \quad (20a)$$

$$\left[P_{ij}\right]^{\vec{a},\vec{c}} \begin{bmatrix} 1 & -\left({}^{u}/_{v}\right) & 0 \\ 0 & -\left({}^{w}/_{v}\right) & 1 \end{bmatrix} \mathbf{p} = \mathbf{a} = (100),\ \mathbf{q} = \mathbf{c} = (001) \quad (20b)$$

$$\left[P_{ij}\right]^{\vec{b},\vec{c}} \begin{bmatrix} -\left({}^{v}/_{u}\right) & 1 & 0 \\ -\left({}^{w}/_{u}\right) & 0 & 1 \end{bmatrix} \mathbf{p} = \mathbf{b} = (010),\ \mathbf{q} = \mathbf{c} = (001) \quad (20c)$$

For the determination of the projected 2D symmetry (plane group) for any space group, one needs to take all symmetry equivalent positions (x, y, z), (x', y', z'), ..., for the space group [from Ref. (86) or *IT-A*] and choose P_{ij} according to $[uvw]$ and multiply. The multiplicity of the general position (x, y, z) [i.e., a number in Ref. (86) or *IT-A*] will be the number of projected 2D positions (X, Y). Between certain (X, Y) and (X', Y'), ..., there will be plane group symmetry relations that are conveniently listed for each of the plane groups in Ref. (86), *IT-A*, and Ref. (88). Since the multiplicity of the general position of a space group is generally higher (i.e., 192 for $Fd\bar{3}m$, no. 227) than the multiplicity of the general position of a plane group (i.e., 12 for *p6mm*, no. 17, which results from a projection of no. 227 down [111]), there are usually several sets of symmetry-related 2D positions (X, Y) and (X', Y'). Finally, one needs to identify the correct plane group by the fulfillment of the condition that all of its symmetry relations for the general position are obeyed.

Note that for projections of 3D symmetry elements, the 2D projection mesh axes do not need to be perpendicular to $[uvw]$. One may, thus, utilize one of the simplest P_{ij} [relations (20a) to (20c)]. Obviously, this does not work for projections along the crystal axes, but these projections are given for each (nonrhombohedral) space group in *IT-A*.

Friedel's law, that is, $|F_{hkl}| = |F_{-h-k-l}|$ and $\alpha_{hkl} = -\alpha_{-h-k-l}$, applies for kinematic and quasi-kinematic scattering of fast electrons so that there is always at least a twofold rotation axis present in the zero-order Laue zone of any PED pattern. As a consequence, only those six 2D diffraction symmetry groups that contain a twofold rotation axis can be distinguished on the basis of the reflections of the zero-order

Laue zone. For the higher order Laue zones of a PED pattern, reflections that are related to each other geometrically by a twofold rotation axis in projection are, however, not necessarily Friedel pairs because $|F_{hkl+n}| \neq |F_{-h-k-l+n}|$, where n is an integer equal or larger than unity. Those reflections can, therefore, possess different intensities and the projected 2D diffraction symmetry of the electrostatic potential is for PED reflections in higher order Laue zones' one of the ten 2D point groups. This results in a higher level of structural discrimination.

USAGE OF CRYSTALLOGRAPHIC RELIABILITY (*R*) VALUES

Experimentally obtained projected reciprocal lattice geometry, 2D symmetry, structure factor moduli, and/or structure factor phases (depending on the data source) all need to be compared with their theoretical counterparts for candidate structures from a database. For each of these "search-match entities," we suggest the usage of a crystallographic R value, as it is standard practice for structure factor moduli and reflection intensities in structural electron and X-ray crystallography. The generalized R value is, given by

$$R_X = \frac{\sum \left| X_{\text{exp}} - X_{\text{theory/candidate}} \right|}{\sum X_{\text{exp}}} \tag{22}$$

and the best "overall fit" between the experiment and the candidate structure is obtained by giving each of the individual R values an appropriate weight and adding them all up in some appropriate fashion. The lowest weighted sum of all R values shall then indicate a quite unambiguous structural identification.

Obviously, all experimental search-match entities possess random and systematic errors that will determine their respective relative weight. For example, the accuracy and precision of the extracted structure factor phase angles and moduli will depend on how accurately and precisely the contrast transfer function of the objective lens can be determined at every point of interest and on how well the kinematic or quasi-kinematic approximations were obeyed by the electron scattering in the case of HRTEM images as data source.

For PED patterns as data source, the situation is simpler. The accuracy and precision of the extracted structure factor moduli will depend on how accurately and precisely the integrated intensities of the reflections can be measured, how well they are integrated by the precession movement of the primary electron beam, and how well they are described by the kinematic or quasi-kinematic scattering approximations.

If it is expected that some of the experimentally obtainable pieces of structural information possess particularly large random and/or systematic errors, they may simply be excluded from the respective R value in order not to bias the overall fit unduly. For example, for PED reflection intensities from thick crystals, it may make sense to exclude all low-angle Bragg reflections from the R value of the squares of the structure factors.

A comparatively minor problem is that the theoretical values of the search-match entities are not precisely known either. The accuracy of theoretical structure factors depends on the (not precisely known) accuracy of the atomic scattering factors, which might be for heavier atoms up to 10% (66). The atomic scattering factors for larger scattering angles are known to be more accurate than their counterparts for smaller scattering angles (3). The theoretical structure factors for larger scattering angles will, therefore, be more accurate than their counterparts for smaller scattering angles.

Finally, there is also the possibility that a certain structure may not be in the respective database. With so much experimentally extractable structural fingerprinting information that can be combined in different ways for searches and matches with low individual *R* values, it seems highly impractical to try to predict what the more and most successful identification strategies might be. We, therefore, propose to simply test a range of strategies on different sets of candidate structure data in order to see pragmatically what works well.

ADVANCED INSTRUMENTATION

The solving of materials science problems by means of TEM is currently undergoing the lens-aberration correction revolution (90–92). Reliable spatial information down to the sub-Å length scale can nowadays be obtained in both the parallel illumination and the scanning probe (scanning transmission electron microscopic) mode [when there is an effective correction for scan distortions (93) in the latter mode]. Objective lens aberration–corrected transmission electron microscopes and condenser lens aberration–corrected scanning transmission electron microscopes in the bright-field mode allow for sufficiently thin crystals the retrieval of Fourier coefficients of the projected electrostatic potential down to the sub-Å length scale and, thus, represent a novel type of crystallographic instrument.

The higher the directly interpretable[b] resolution in an aberration-corrected transmission electron microscope is, the lower will, in principle, be the lateral overlap of the electrostatic potentials from adjacent atomic columns and the more zone axes will be revealed by crossed lattice fringes in structureal[b] images. Note that the relationship between directly interpretable image resolution and visibility of zone axes is strongly superlinear. This is, for example, demonstrated in Table 1 for a densely packed model crystal with a very small unit cell.

TABLE 1 Relationship Between Directly Interpretable HRTEM Image Resolution and the Visibility of Zone Axes*

Directly interpretable image resolution (nm)	Number and type of visible net plane families	Number and type of visible zone axes (lattice-fringe crossings)
0.2	**2**, i.e., {111}, {200}	**2**, i.e., [001], [011]
0.15	**3**, i.e., {111}, {200}, {220}	$\mathbf{2^2}$, i.e., [001], [011], [111], [112]
0.1	**4**, i.e., {111}, {200}, {220}, {311}	$\mathbf{2^3}$, i.e., [001], [011], [111], [112], [013], [114], [125], [233]
$\leq$**0.05**	$\geq$**18**, i.e. {111}, {200}, {220}, {311}, {331}, {420}, {422}, {511}, {531}, {442}, {620}, {622}, {551}, {711}, {640}, {642}, {731}, {820}	$>\mathbf{2^5}$, e.g., [001], [011], [111], [012], [112], [013], [122], [113], [114], [123], [015], [133], [125], [233], [116], [134], [035], ...

*Relationship between directly interpretable HRTEM image resolution and the visibility of net plane families and zone axes within one stereographic triangle [001]–[011]–[111] for a hypothetical cubic AB compound with 0.425-nm lattice constant and space group $Fm\bar{3}m$. This corresponds to the halite structural prototype, which has a face-centered cubic packing of the A element with all the octahedral intersites filled by the B element. This hypothetical material is very densely packed, as 8 atoms occupy one unit cell. It is assumed that both hypothetical atoms have similar atomic scattering factors.

Aberration-corrected transmission electron microscopes and scanning transmission electron microscopes are also expected to have significant impact on the feasibility of structural electron crystallography based on PED patterns (94). A complementary integrated diffraction spot–based technique that utilizes a large-angle defocused incident beam and a spherical aberration corrector (and which will be especially useful for beam-sensitive crystals) has recently been developed (95).

NANOCRYSTAL-SPECIFIC LIMITATIONS TO STRUCTURAL FINGERPRINTING

The lattice constants of inorganic nanocrystals with diameters between 10 and 1 nm may be contracted or expanded by a fraction of a tenth of a percent (96), a fraction of a percent (97), or up to a few percentages at worst (for the very smallest crystals) either due to free-energy minimization effects in the presence of large surface areas (98) or due to reduced lattice cohesion (96). For noncubic nanocrystals, these lattice constant changes may result in changes of the angles between net planes (and, therefore, affect the position of data points in lattice-fringe fingerprint plots) on the order of a tenth of a degree to a few degrees for some combinations of reciprocal lattice vectors while other combinations may be negligibly affected.

There are also "crystallographically challenged materials" and "intercalated mesoporous materials," that is, tens of nanometer-sized entities with a well-defined atomic structure over small length scales that can be described by a relatively large crystallographic unit cell with a low symmetry (99,100). The above-mentioned vanadium oxide nanotubes [Figs. 2 and 10(A)] are examples of the former (29). Both of these materials lack long-range crystallographic order. Significant structural distortions that might be considered as classical defects or nanocrystal-specific defects to the average structure may be present to such an extent that it may make little sense to consider the disorder as a defect away from an ideal structure. In short, the deviations from the perfect atomic structure might be rather severe in these materials but remnants of the crystallinity might still be present. We are quite confident that both of these materials can be fingerprinted structurally in the TEM, while this is at least for crystallographically challenged materials not possible by X-ray diffractometry (Fig. 2).

CRYSTALLOGRAPHY DATABASES FOR ADVANCED STRUCTURAL FINGERPRINTING IN THE TRANSMISSION ELECTRON MICROSCOPE

The authors of Ref. (99) appeal to individual researchers who deal with structural problems of nanocrystals to communicate more frequently and openly so that progress can be made more rapidly within an emerging scientific community. Structural fingerprinting of nanocrystals in the TEM has a role to play here because it makes sense to solve only those structures that are really new to science. For those structures that are really new to science, there will be no entries in the existing databases. Because crystallographically challenged materials do not possess a characteristic powder X-ray diffraction fingerprint (Fig. 2), they may not become part of the classical comprehensive fingerprinting databases. Other structural fingerprinting databases and/or crystallographic reference databases should, therefore, be erected and such structural nanocrystal fingerprint information and/or whole nanocrystal structures should be collected for reference purposes.

The emerging community of "nanocrystallographers" may decide to start an open-access database all by themselves because database development has traditionally been done by active research scientists. In addition, the required computing infrastructure for such developments is better than ever before with ubiquitous desktop computing and quite universal Internet access.

Perceived copyright issues should not delay better communications within the emerging community of nanocrystallographers. As determined by the World Intellectual Property Organization (101) in the "Berne Convention for the Protection of Literary and Artistic Works" (93), crystallographic data are not copyrightable because they are "data" (102) and not "original, tangible forms of expression"[g] (*see* NOTE page 313) (103).

Utilizing the Crystallographic Information File[g] (CIF) format (104,105), entries for nanocrystal structures are currently being collected at Portland State University's Nano-Crystallography Database (NCD) Web site (105). The readers of this chapter are encouraged to send their published (or preliminary) nanocrystal structure results per e-mail attachment to the first author of this chapter so that they can be expressed in CIFs and made openly accessible at the NCD. Ready-made CIFs can be uploaded and edited directly by members of the nanocrystallography community. The only condition for inclusion in the NCD is that crystal size information should be explicitly or implicitly contained in the respective CIF. As all structural electron crystallography and EDSA results were obtained from nanometer-thick crystals, they have a natural home in the NCD.

Note that at the NCD (106) and other open-access databases that are housed at our Web servers (20), we provide interactive 3D visualization of the atomic arrangement of a structure and allow for "on the fly" calculations of lattice-fringe fingerprint plots as well as direct searches of the entries by keywords, lattice parameters, chemical composition, reduced lattice-fringe information, etc. We recently added the capability to display crystal morphologies interactively in 3D at one of these databases (107) because the morphology of nanocrystal has been shown to be crucial to catalytic properties (108).

None of the databases that have traditionally been employed for structural fingerprinting in the TEM[h] (*see* NOTE page 313) (based on the projected reciprocal lattice geometry combined with either energy-dispersive X-ray spectroscopy or prior information on the chemical content of a sample) contain sufficient crystallographic information to support our novel structural fingerprinting strategies (which combine information on the projected reciprocal lattice geometry, 2D symmetry, and structure factors).

Other databases such as the Inorganic Crystal Structure Database (ICSD) of the Fachinformationszentrum (FIZ) Karlsruhe (109), the Powder Diffraction File, Version 4 (PDF-4), of the International Center for Diffraction Data (ICDD) (21), the Pearson's Crystal Data database of Material Phases Data System (MPDS) (Vitznau, Switzerland), and ASM International (the Materials Information Society, Materials Park, OH), which is distributed by Crystal Impact (Bonn, Germany) (110), or the two open-access databases Linus Pauling File of the Japan Science and Technology Agency (JST) (Tokyo, Japan) and MPDS (111) and Crystallography Open Database (16–18) or its mainly inorganic subset (15), will, therefore, need to be employed for such advanced structural fingerprinting in the TEM. A review on crystallographic databases was published recently (112) and an entry on the same topic has also been placed at the wikipedia (113).

SUMMARY

An industrial-scale need for structural fingerprinting of nanocrystals is emerging and laboratory-based X-ray diffraction methods cannot satisfy this need. Two novel strategies for structural fingerprinting of nanocrystals by electron diffraction in a transmission electron microscope (or in an electron diffraction camera) were, therefore, outlined. Another rather novel structural fingerprinting method that relies on the analysis of HRTEM images was briefly mentioned.

B. K. Vainshtein, B. B. Zvyagin, and A. S. Avilov wrote in 1992: "For quite a long time the majority of electron diffraction patterns obtained, sometimes striking in diversity and even grandeur, appeared as a store of concealed structural information and as collector's items" (4). With quite ubiquitous mid-voltage transmission electron microscopes and the increased availability of PED add-ons, image plates, slow-scan, charge-coupled device detectors, and precision electrometers, the time is now to extract this information and utilize it not only for structural electron crystallography but also for structural fingerprinting of nanocrystals.

ACKNOWLEDGMENTS

This research was supported by grants from the Office of Naval Research to the Oregon Nanoscience and Microtechnologies Institute. Additional support was provided by the NorthWest Academic Computing Consortium as well as by Faculty Development, Faculty Enhancement, and Internationalization Awards by Portland State University. We thank Boris Dušek, Jan Zahornadský, Jan "Irigi" Olšina, Hynek Hanke, and Ondrej Certik, students of the Charles University of Prague, Czech Republic, for the creation of Portland State University's open-access crystallography databases including the associated 3D structure and lattice-fringe fingerprint plot visualizations and search capabilities. Dr. Peter Sondergeld is acknowledged for his creation of a wikipedia entry on crystallographic databases. We also thank selected members of the International Center for Diffraction Data (and especially members of its Electron Diffraction Subcommittee) for their encouragement of our developments of novel strategies for structural fingerprinting of nanocrystals.

REFERENCES

1. Nicolopoulos S, Moeck P, Maniette, et al. Identification/fingerprinting of nanocrystals by precession electron diffraction. In: Luysberg M, Tillmann K, Weirich T, eds. EMC 2008, Vol. 1: Instrumentation and Methods. Berlin: Springer, 2008:111–112.
2. Moeck P, Bjoerge R, Maniette Y, et al. Structural fingerprinting in the transmission electron microscope. Powder Diffract 2008; 23(2):155–162. Presented at the 2008 Annual Meeting of the Members of the International Center for Diffraction Data; March 10–14, 2008. Available at: http://www.icdd.com/profile/march08files/Moeck-Nicolopoulos-ab.pdf.
3. Vainshtein B K, Structure Analysis by Electron Diffraction. Oxford, U.K.: Pergamon Press, 1964.
4. Vainshtein BK; Zvyagin BB, Avilov AS. Electron diffraction structure analysis. In: Cowley JM, ed. Electron Diffraction Techniques. Oxford, U.K.: Oxford University Press, 1992:216–312.
5. Vainshtein BK, Zvyagin BB. Electron-diffraction structure analysis. In: Shmueli U, ed. International Tables for Crystallography, Vol. B, Reciprocal Space, 2nd ed. Dordrecht, The Netherlands: Kluwer Academic, 2001:306–320.
6. Moeck P, Rouvimov S, Nicolopoulos S, et al. Structural fingerprinting of a cubic iron-oxide nanocrystal mixture: A case study. NSTI-Nanotech 2008; 1:912–915. Available at: www.nsti.org. ISBN 978-1-4200-8503-7. Open access version of a similar conference paper: arXiv:0804.0063.
7. Moeck P, Bjorge R, Mandell E, et al. Lattice-fringe fingerprinting of an iron-oxide nanocrystal supported by an open-access database. Proc NSTI-Nanotech 2007; 4:93–96. ISBN 1-4200637-6-6. Available at: www.nsti.org.

8. Moeck P, Bjorge R. Lattice-fringe fingerprinting: structural identification of nanocrystals by HRTEM. In: Snoeck E, Dunin-Borkowski R, Verbeeck J, Dahmen U, eds. Quantitative Electron Microscopy for Materials Science. Mater Res Soc Symp Proc 2007:1026E: paper 1026-C17-10.
9. Bjorge R. Lattice-Fringe Fingerprinting: Structural Identification of Nanocrystals Employing High-Resolution Transmission Electron Microscopy [MSc thesis]. Portland State University, Portland, OR, May 9, 2007. Published on the Internet after peer review: J Dissertation 2007; 1(1). Available at: http://www.scientificjournals.org/journals2007/j_of_dissertation.htm.
10. Moeck P, Fraundorf P. Structural fingerprinting in the transmission electron microscope: Overview and opportunities to implement enhanced strategies for nanocrystal identification. Z Kristallogr 2007; 222:634–645.
11. Zou XD, Hovmöller S. Electron crystallography: Imaging and single-crystal diffraction from powders. Acta Crystallogr A 2008; 64:149–160. Available at: http://journals.iucr.org/a/issues/2008/01/00/issconts.html.
12. Dorset DL. Correlations, convolutions and the validity of electron crystallography. Z Kristallogr 2003; 218:237–246.
13. Zou XD. On the phase problem in electron microscopy: The relationship between structure factors, exit waves, and HREM images. Microsc Res Tech 1999; 46:202–219.
14. Vainshtein BK. Modern Crystallography, Vol. I, Symmetry of Crystals, Methods of Structural Crystallography. New York: Springer, 1981. Springer Series in Solid-State Sciences 15.
15. Available at: http://nanocrystallography.research.pdx.edu/CIF-searchable/cod.php *and* http://nanocrystallography.research.pdx.edu/search.py/index.
16. Leslie M. Free the crystals. Science 2005; 310:597.
17. Gražulis S, Chateigner D, Downs RT, et al. Crystallography Open Database – An open access collection of crystal structures. A tribute to Michael Berndt. J Appl Cryst. In press.
18. Available at: http://crystallography.net *mirrored* at http://cod.ibt.lt/ (in Lithuania), http://cod.ensicaen.fr/ (in France), *and* http://nanocrystallography.org (in Oregon, U.S.A.).
19. Fraundorf P, Qin W, Moeck P, et al. Making sense of nanocrystal lattice fringes. J Appl Phys 2005; 98:114308-1–114308-10. Open access: arXiv:cond-mat/0212281 v2. Virtual J Nanoscale Sci Technol 2005; 12(25).
20. Available at: http://nanocrystallography.research.pdx.edu *and* http://nanocrystallography.org.
21. Faber J, Blanton J. Full pattern comparison of experimental and calculated powder patterns using the integral index method in PDF-4+. Powder Diffract 2008; 23:141–149.
22. Hanawalt JD, Rinn HW. Identification of crystalline materials: Classification and use of X-ray diffraction patterns. Ind Eng Chem Anal Ed 1936; 8:244–247.
23. Faber J, Fawcett T. The Powder Diffraction File: Present and future. Acta Cryst B 2002; 58:325–332.
24. Ungár T. Microstructure of nanocrystalline materials studied by powder diffraction. Z Kristallogr Suppl 2006; 23:313–318.
25. Pinna N. X-ray diffraction from nanocrystals. Prog Colloid Polym Sci 2005; 130: 29–32.
26. Kaszkur Z. Test of applicability of some powder diffraction tools to nanocrystals. Z Kristallogr Suppl 2006; 23:147–154.
27. Pielaszek R, Łojkowski W, Matysiak H, et al. In: Lojkowski W, Turan R, Proykova A, Daniszewska A, eds. 8th Nanoforum Report: Nanometrology. 2006:79–80. Available at: www.nanoforum.org.
28. Banfield JF, Zhang H. Nanoparticles in the Environment. In: Banfield JF, Navrotskyed A, eds. Nanoparticles and the Environment. Ribbe PH, series ed. Reviews in Mineralogy & Geochemistry, Vol. 44. Mineralogical Society of America, 2001:1–58.
29. Petkov V, Zavalij PY, Lutta S, et al. Structure beyond Bragg: Study of V_2O_5 nanotubes. Phys Rev B 2004; 69:85410-1–85410-6.
30. Spence CH. High-Resolution Electron Microscopy, 3rd ed. New York: Oxford University Press, 2004:391, appendix 4.

31. Zou XD. PhD thesis, Electron crystallography of Inorganic structures–theory and practice, Stockholm University, Stockholm, Sweden, 1995.
32. Klechkovskaya VV, Imamov RM. Electron crystallography of inorganic structures – theory and practice. Crystallogr Rep 2001; 46:534–549.
33. Vincent R, Midgley P. Double conical beam-rocking system for measurement of integrated electron diffraction intensities. Ultramicroscopy 1994; 53:271–282.
34. Blackman M. On the intensities of electron diffraction rings. Proc R Soc (Lond) A 1939; 173:68–82.
35. Vainshtein BK, Lobachev AN. Differentiation of dynamic and kinematic types of scattering in electron diffraction. Sov Phys Crystallogr 1956; 1:370–371.
36. Giacovazzo C, Capitelli F, Cuocci C, et al. Direct methods and applications to electron crystallography. In: Hawkes PW, Merli PG, Calestani G, Vittori-Antisari M, eds. Microscopy, Spectroscopy, Holography and Crystallography with Electrons, Vol. 123. New York: Academic Press, 2002:291–310.
37. Dorset DL. Structural Electron Crystallography, New York: Plenum Press, 1995.
38. Rauch EF, Véron M, Portillo J, et al. Automatic crystal orientation and phase mapping in TEM by precession diffraction. Microsc Anal 2008; 22(93):S5–S8.
39. Hoppe W. Three-dimensional low dose reconstruction of periodical aggregates. In: Hoppe W, Mason R, eds. Unconventional Electron Microscopy for Molecular Structure Determination. Advances in Structure Research by Diffraction, Vol. 7. Braunschweig, Wiesbaden: Friedrich Vieweg & Sohn, 1979:191–220.
40. Gemmi M, Zou XD, Hovmöller S, et al. Structure of Ti_2P solved by three-dimensional electron diffraction data collected with precession technique and high resolution electron microscopy. Acta Crystallogr A 2003; 59:117–126.
41. Own CS, Marks LD, Sinkler W. Electron precession: A guide for implementation. Rev Sci Instr 2005; 76:033703
42. Ciston J, Deng B, Marks LD, et al. A quantitative analysis of the cone-angle dependence in precession electron diffraction. Ultramicroscopy 2008; 108:514–522.
43. Sinkler W, Own CS, Marks L. Application of a 2-beam model for improving the structure factors from precession electron diffraction intensities. Ultramicroscopy 2007; 107:543–550.
44. Own CS, Sinkler W, Marks LD. Rapid structure determination of a metal oxide from pseudo-kinematical electron diffraction data. Ultramicroscopy 2006; 106:114–122.
45. Own CS. System Design and Verification of the Precession Electron Diffraction Technique [PhD thesis]. Evanston, IL: Northwestern University, 2005. Available at: http://www.numis.northwestern.edu/Research/Current/precession.shtml.
46. Own CS, Subramanian AK, Marks LD. Quantitative analyses of precession diffraction data for a large cell oxide. Microsc Microanal 2004; 10:96–104.
47. Horstmann M, Meyer G. Messung der elastischen Elektronenbeugungsintensitäten polykristalliner Aluminium-Schichten. Acta Crystallogr 1962; 15:271–281.
48. Horstmann M, Meyer G. Messung der Elektronenbeugungsintensitäten polykristalliner Aluminiumschichten bei tiefer Temperatur und Vergleich mit der dynamischen Theorie. Z Phys 1965; 182:380–397.
49. Vainshtein BK, Lobachev AN. Dynamic scattering and its use in structural electron diffraction studies. Sov Phys Crystallogr 1962; 6:609–611.
50. Pinsker ZG. Development of electron diffraction structure analysis in the USSR. In: Goodman P, ed. Fifty Years of Electron Diffraction. Dordrecht, The Netherlands: D. Reidel, 1981:155–163.
51. Vainshtein BK. Kinematic theory of intensities in electron-diffraction patterns; part 2: Patterns from textures and polycrystalline aggregates. Sov Phys Crystallogr 1956; 1:117–122.
52. Cowley JM. Electron diffraction and electron microscopy. In: Shmueli U, ed. International Tables for Crystallography, Vol. B, Reciprocal Space, 2nd ed. Dordrecht, The Netherlands: Kluwer Academic, 2001:277–284.
53. Cowley JM. crystal structure determination by electron diffraction. In: Chalmers B, Hume-Rothery W, eds. Progress in Materials Science, Vol. 13. Oxford, U.K.: Pergamon Press, 1967:267–321.

54. Gjønnes J. The dynamic potentials in electron diffraction. Acta Crystallogr 1962; 15:703–707.
55. Gjønnes K, Cheng Y, Berg BS, et al. Corrections for multiple scattering in integrated electron diffraction intensities; application to determination of structure factors in the [001] projection of Al_mFe. Acta Crystallogr A 1998; 54:102–119.
56. Cowley JM, Rees ALG. Fourier methods in structure analysis by electron diffraction. Rep Prog Phys 1958; 12:165–225.
57. Avilov A, Kuligin K, Nicolopoulos S, et al. Precession technique and electron diffractometry as new tools for crystal structure analysis and chemical bonding determination. Ultramicroscopy 2007; 107:431–444.
58. Mayer J, Deinigner C, Reimer L. Electron spectroscopic diffraction. In: Reimer L, ed. Energy-Filtering Transmission Electron Microscopy, Berlin: Springer, 1995. Springer Series in Optical Sciences, Vol. 71.
59. Mitchell DRG, Schaffer B. Scripting-customized microscopy tools for Digital Micrograph™. Ultramicroscopy 2005; 103:319–332.
60. Mitchell DRG. Circular Hough transform diffraction analysis: A software tool for automated measurement of selected area electron diffraction patterns within Digital Micrograph™. Ultramicroscopy 2008; 108:367–374.
61. Mitchell DRG. DiffTools: Software tools for electron diffraction in Digital Micrograph™. Microsc Res Tech 2008; 71:588–593.
62. Digital Micrograph™ Script Database. Available at: http://www.felmi-zfe.tugraz.at/dm_scripts.
63. Dorset LD, Gilmore CJ, Jorda JL, et al. Direct electron crystallographic determination of zeolites zonal structures. Ultramicroscopy 2007; 107:462–473.
64. Cowley JM, Rees AL, Spink JA. Secondary elastic scattering in electron diffraction. Proc Phys Soc A 1951; 64:609–619.
65. White TA, Eggeman AS, Midgley PA. "Phase-scrambling" multislice simulations of precession electron diffraction. In: Luysberg M, Tillmann K, Weirich T, eds. EMC 2008, Vol. 1: Instrumentation and Methods. Berlin: Springer, 2008:237–238.
66. Oleynikov P, Hovmöller S, Zou XD. Precession electron diffraction: Observed and calculated intensities. Ultramicroscopy 2007; 107:523–533.
67. Gemmi M, Nicolopoulos S. Structure solutions with three-dimensional sets of precessed electron diffraction intensities. Ultramicroscopy 2007; 107:483–494.
68. Dudka AP, Avilov AS, Nicolopoulos S. Crystal structure refinement using Bloch-wave method for precession electron diffraction. Ultramicroscopy 2007; 107:474–482.
69. Boulahya K, Ruiz-González L, Parras M, et al. Ab initio determination of heavy oxide perovskite related structures from precession electron diffraction data. Ultramicroscopy 2007; 107:445–452.
70. Nicolopoulos S, Morniroli JP, Gemmi M. From powder diffraction to structure resolution of nanocrystals by precession electron diffraction. Z Kristallogr Suppl 2007; 26:183–188.
71. Kverneland A, Hansen V, Vincent R, et al. Structure analysis of embedded nano-sized particles by precession electron diffraction. η'-precipitate in an Al-Zn-Mg alloy as example. Ultramicroscopy 2006; 106:492–502.
72. Weirich TE, Portillo J, Cox G, et al. Ab initio determination of the framework structure of the heavy-metal oxide $Cs_xNb_{2.54}W_{2.46}O_{14}$ from 100 kV precession electron diffraction data. Ultramicroscopy 2006; 106:164–175.
73. Gjønnes J, Hansen V, Kverneland A. The precession technique in electron diffraction and its application to structure determination of nano-size precipitates in alloys. Microsc Microanal 2004; 10:16–20.
74. Gemmi M, Calestani G, Migliori A. Strategies in electron diffraction data collection. In: Hawkes PW, Merli PG, Calestani G, Vittori-Antisari M, eds. Advances in Imaging and Electron Physics, Vol. 123, Microscopy, Spectroscopy, Holography and Crystallography with Electrons. New York: Academic Press, 2002:311–325.
75. Midgley PA, Sleight ME, Saunders M, et al. Measurement of Debye-Waller factors by electron precession. Ultramicroscopy 1998; 75:61–67.

76. Gjønnes J, Hansen V, Berg BS, et al. Structure model for the phase Al_mFe derived from three-dimensional electron diffraction intensity data collected by a precession technique. Comparison with convergent-beam diffraction. Acta Crystallogr A 1998; 54:306–319.
77. Berg BS, Hansen V, Midgley PA, et al. Measurement of three-dimensional intensity data in electron diffraction by the precession technique. Ultramicroscopy 1998; 74:147–157.
78. Gjønnes K. On the integration of electron diffraction intensities in the Vincent-Midgley precession technique. Ultramicroscopy 1997; 69:1–11.
79. Heidenreich RD. Theory of the "forbidden" (222) electron reflection in the diamond structure. Phys Rev 1950; 77:271–283.
80. Morniroli JP, Redjaïmia A, Nicolopoulos S. Contribution of electron precession to the identification of the space group from microdiffraction patterns. Ultramicroscopy 2007; 107:514–522.
81. Morniroli JP, Redjaïmia A. Electron precession microdiffraction as a useful tool for the identification of the space group. J Microsc 2007; 227:157–171.
82. Oleynikov P, Hovmöller S, Zou XD. Automatic space group determination using precession electron diffraction patterns. In: Luysberg M, Tillmann K, Weirich T, eds. EMC 2008, Vol. 1: Instrumentation and Methods. Berlin: Springer, 2008:221–222.
83. Gjønnes J, Moodie AF. Extinction conditions in the dynamic theory of electron diffraction. Acta Crystallogr 1965; 19:65–67.
84. Cowley JM, Smith DJ. The present and future of high-resolution electron microscopy. Acta Crystallogr A 1987; 43:737–751.
85. Landsberg MJ, Hankamer B. Symmetry: A guide to its applications in 2D electron crystallography. J Struct Biol 2007; 160:332–343.
86. Henry NFM, Lonsdale K, eds. International Tables for X-ray Crystallography, Vol. 1, 3rd ed. Chester, England: International Union of Crystallography, 1969.
87. Hahn T, ed. International Tables for Crystallography, Vol. A, Space-Group Symmetry, 5th ed. Chester, England: International Union of Crystallography, 2005:111–717, 91–109.
88. Hahn T, ed. Brief Teaching Edition of Volume A, Space-Group Symmetry, International Tables for Crystallography, 5th rev. ed. Chester, England: International Union of Crystallography, 2005.
89. Biedl A. The projection of a crystal structure. Z Kristallogr 1966; 123:21–26.
90. Krivanek OL, Corbin GJ, Dellby N, et al. An electron microscope for the aberration-corrected era. Ultramicroscopy 2008; 108:179–195.
91. Pennycook SJ, Varela M, Hetherington CJD, et al. Materials advances through aberration-corrected electron microscopy. MRS Bull 2006; 31:36–43.
92. Bleloch AL, Falke U, Goodhew PJ. Geometric aspects of lattice contrast visibility in nanocrystalline materials using HAADF STEM. Ultramicroscopy 2006; 106:277–283.
93. Sanchez AM, Galindo PL, et al. An approach to the systematic distortion correction in aberration-corrected HAADF images. J Microsc 2006; 221:1–7.
94. Own CS, Sinkler W, Marks LD. Prospects for aberration corrected electron precession. Ultramicroscopy 2007; 107:534–542.
95. Morniroli JP, Houdellier F, Roucau C, et al. LACDIF, a new electron diffraction technique obtained with the LACBED configuration and a C_S corrector: Comparison with electron precession. Ultramicroscopy 2008; 108:100–115.
96. Lu K, Zhao YH. Experimental evidences of lattice distortion in nanocrystalline materials. Nanostruct Mater 1999; 12:559–562.
97. Sui ML, Lu K. Variations in lattice parameter with grain size of a nanophase Ni_3P compound. Mater Sci Eng A 1994; 179/180:541–544.
98. Roduner E. Nanoscopic Materials, Size-Dependent Phenomena. Cambridge, U.K.: Royal Society of Chemistry, 2006.
99. Billinge SJL, Levin I. The problem with determining atomic structure at the nanoscale. Science 2007; 316:561–565.
100. Billinge SJL, Kanatzidis MG. Beyond crystallography: The study of disorder, nanocrystallinity, and crystallographically challenged materials with pair distribution functions. Chem Commun 2004; 749:749–760.

101. Available at: http://www.wipo.int/treaties/en/ip/berne/trtdocs_wo001.html.
102. Available at: http://www.wipo.int/treaties/en/ip/berne/pdf/trtdocs_wo001.pdf.
103. Brodsky MH. Fair and useful copyright, ACA RefleXions. Newslett Am Crystallogr Assoc 2006; (4):15–16. ISSN 1958-9945. Available at: http://aca.hwi.buffalo.edu//newsletterpg_list/Newsletters_PDF/Winter06.pdf.
104. Hall S, McMahon B, eds. International Tables for Crystallography, Vol. G: Definition and Exchange of Crystallographic Data. Chester, England: International Union of Crystallography, 2005. Available at: http://www.iucr.org/iucr-top/cif/index.html.
105. Brown ID, McMahon B. CIF: The computer language of crystallography. Acta Crystallogr B 2002; 58:317–324.
106. Available at: http://nanocrystallography.research.pdx.edu/CIF-searchable/ncd.php.
107. Available at: http://nanocrystallography.research.pdx.edu/CIF-searchable/cmd.php.
108. Tian N, Zhou Z-Y, Sun S-G, et al. Synthesis of tetrahexahedral platinum nanocrystals with high-index facets and high electro-oxidation activity. Science 2007; 316:732–735.
109. Available at: http://icsdweb.fiz-karlsruhe.de/index.php *and* http://icsd.ill.fr/icsd/index.html *and* http://www.stn-international.de/stndatabases/databases/icsd.html.
110. Available at: http://www.crystalimpact.com/ *and* http://www.crystalimpact.com/pcd/download.htm.
111. Villars P, Onodera N, Iwata S. The Linus Pauling File (LPF) and its applications to materials design. J Alloys Compd 1998; 279:1–7. Available at: http://crystdb.nims.go.jp.
112. Moeck P. Crystallographic Databases: past, present, and development trends. Proc Mater Sci Technol 2006; 1:529–540.
113. Available at: http://en.wikipedia.org/wiki/Crystallographic_database.

NOTES

a. An analysis of the content of the 2006 edition of the Inorganic Crystal Structure Database by Thomas Weirich resulted in 522 structures that were solved partially or completely by structural electron crystallography (http://www.gfe.rwth-aachen.de/sig4/index.htm).

b. The Scherzer (or point-to-point or directly interpretable) resolution of a transmission electron microscope is obtained at a special underfocus that leads to a "structure image" for a weak-phase object (30). The term "structure image" has been proposed by John M. Cowley to describe a member of the restricted set of lattice images in TEM that can be directly interpreted (to some limited resolution) in terms of a crystal's projected atomistic structure. Structure images need to be obtained under instrumental conditions that are independent of the crystal structure. It has been pointed out that the imaging of phase objects at the Scherzer (de)focus maximizes contributions from the "linear or first-order image," in which there is a linear relationship between the projected electrostatic potential and the image intensity, and minimizes contribution from the "second-order or quadratic image," in which the proportionality is quadratic (31). Structure images can, thus, be obtained at the Scherzer (de)focus for crystals that are slightly thicker than weak-phase objects. The instrumental resolution of a transmission electron microscope is for field-emission gun–fitted microscopes without a spherical aberration corrector much higher than the Scherzer resolution.

c. Electron precession add-ons to older and newer mid-voltage transmission electron microscopes have been commercialized by Stavros Nicolopoulos and coworkers at NanoMEGAS SPRL (http://www.nanomegas.com). Portland

State University's "Laboratory for Structural Fingerprinting and Electron Crystallography" serves as the first demonstration site for the NanoMEGAS company in the Americas. A first-generation PED device "Spinning Star" is interfaced to an analytical FEI Tecnai G^2 F20 field-emission gun transmission electron microscope and can be demonstrated on request. The whole suite of electron crystallography software from Calidris and AnaliTEX (footnote f) can also be demonstrated at this laboratory.

A second-generation "Spinning Star" is at the core of the ASTAR system (38) of the NanoMEGAS company. This PED device allows for fast and highly reliable "crystal orientations and structures" map acquisitions with a transmission electron microscope. The ASTAR system is superior to the complementary electron backscatter diffraction technique in scanning electron microscopy, because it is based on precessed transmission diffraction spot patterns rather than "near-surface backscattered" Kikuchi diffraction patterns. The former patterns are much less sensitive to the plastic deformation state of the crystals and their real structure content than Kikuchi patterns. In addition, the orientations and crystal structures of smaller nanocrystals can be mapped in a transmission electron microscope [or scanning transmission electron microscope (STEM)] due to the transmission geometry and higher acceleration voltages. For the automated "crystal orientations and structures" mapping of metals and minerals at approximately 30-nm spatial resolution, relative small precession angles of less that 0.5° were shown to result in significant enhancements of the "map reliability index" (when 300-kV electrons emitted by a LaB_6 electron gun were utilized) (38). Larger precession angles should result in an even better map reliability index, but there may then be a need to include reflections from higher order Laue zones in the analysis. (See also footnote d for instrumental limits on precession angles and sizes of precessing electron probes.)

The advantages of a precessing primary electron beam for structural electron crystallography in both the diffraction and high-resolution imaging modes of TEMs have been already realized in the mid-1970s of the last century (39). Many older transmission electron microscopes do allow for a hollow cone illumination but not for a "proper descanning," [Figs. 3(B) and 3(D)] of the resulting "just-precessed" PED patterns [Figs. 3(A) and 3(C)]. Stationary (or proper descanned) PED patterns have, on the other hand, been recorded from nanocrystals in a Philips CM30 T microscope that was operated in its selected area channeling pattern STEM mode (40). Laurence D. Mark's group at Northwestern University recently created (41) and utilized (42–46) three PED devices. Copies of these devices have been installed at the University of Illinois at Urbana-Champaign, Arizona State University, and UOP LLC (formerly known as Universal Oil Products at Des Plaines). It was, however, the recent creation of the above-mentioned electron precession add-ons to older and newer mid-voltage transmission electron microscopes by the NanoMEGAS company that resulted in the formation of a "PED community" that currently comprises about 40 research groups worldwide.

d. While large precession angles reduce multiple scattering more effectively, they may also lead to the excitement of reflections from higher order Laue zones for certain zone axes of crystals with large lattice constants. The (reciprocal space) radius in a PED pattern with no overlap of reflections from the zero-order Laue zone with their counterparts from higher order Laue zones may be estimated by the relation

$$R_{\text{no overlap}} = k \sin\left(\arccos\left\{\frac{k\cos\varepsilon - |uvw|^{-1}}{k}\right\} - \varepsilon\right)$$

where k is the electron-wave number $= \lambda^{-1}$, ε the precession angle, and $|uvw|$ the magnitude of the (direct space) zone axis vector (67). In a PED pattern, the maximal (reciprocal space) radial distance of reflections from the central 000 reflection can be estimated by the relation $R_0 = 2k \sin\varepsilon$ (66).

For silicon in the [110] orientation, 200-kV electrons, and a precession angle of 2.8°, one obtains 18.2 nm^{-1} for the "no-overlap radius" of a PED pattern. All reflections of the first-order Laue zone are kinematically forbidden by the space group symmetry of silicon. For the same precession angle and electron wavelength, the PED pattern may, in theory, extend out to 39 nm^{-1}. Because the atomic scattering factors fall off quickly for electrons with increasing scattering angle, there is typically no measurable reflection intensity for reflections with net plane spacings of a few tens of an Å. The reflections with the largest reciprocal lattice vectors (and appreciable intensity) that we observed in our experiments for Si, [110], 200 kV, and 2.8° precession angle, were of the {11,11,1} type, that is, at 28.7 nm^{-1}.

The maximally obtainable precession angle can for any transmission electron microscope be estimated from the maximally obtainable dark-field tilt angle (57). For our FEI Tecnai G2 F20 ST microscope, these maximally obtainable angles are approximately 4°. The spherical aberration coefficient of the objective lens, C_S, sets minimal limits to the electron probe size in PED experiments according to the relation $P_{\min} \approx 4C_s\delta\varepsilon^2$, where δ is the beam-convergence semiangle, ε the precession angle, and $\delta \ll \varepsilon$ (33,57). Precession and descanning distortions will increase the obtainable electron probe sizes. For our analytical field-emission transmission electron microscope with a C_s of 1.2 mm and a first-generation "Spinning Star" from NanoMEGAS, the minimal electron probe size can in the nanobeam mode be adjusted to a few tens of nanometers while utilizing rather large precession angles of up to 3°.

e. One part of the Lorentz factor should account for the precession diffraction geometry and the other part should depend on the structure and thickness of the crystal(s) in some "Blackman-type" fashion. Different research groups used different approximate formulae for the geometric part of the Lorentz factor, for example, both $[d^*(1 - \{d^*/_{R_0}\}^2)^{0.5}]^{-1}$ (40) and $[d^*(1 - d^*/_{R_0})^{0.5}]^{-1}$ (45) were utilized, with R_0 as given in footnote d and ε as the precession angle. Depending on the illumination conditions, an extra term that corrects for the primary beam divergence may be included in the geometric part of the Lorentz factor (78).

The structure- and thickness-dependent Lorentz factor should be analogous to relation (13) when full integration of the reflections can be achieved. When less than full integration is achieved (because the nanocrystal is, for example, very small so that its shape transform is widely spread out in reciprocal space and contains several subsidiary maxima or the precession angle is insufficiently small), the mathematical limits of integration of the structure- and thickness-dependent Lorentz factor part need to be modified accordingly. It may become necessary to include an additional Lorentz factor that accounts for subsequent elastic scattering of electrons that underwent an initial small energy loss.

Note that there are successful structure analyses on the basis of PED data for both, utilizing a Lorentz factor (40,44,45,71,77) or ignoring (63,72) it. Structures have also been solved successfully on the basis of PED data by means of direct methods, utilizing either the square root or the first power of integrated intensities. These successes of complementary strategies seem to depend on the particular type of sample. This suggests that more complex forms of Lorentz factors may be appropriate for certain sample types and also illustrates a long-known fact about quasi-kinematic diffraction theory–based electron crystallography:

> "As shown in practice, for any formula of transition from I to $|\Phi|$ the main features of the structure are revealed on the Fourier synthesis. However, the peaks corresponding to heavy and medium atoms of a given structure in the incorrect transition formula are displaced from the true positions...." [Vainshtein and Lobachev (49)]

f. The programs CRISP/ELD and Space Group Determinator run on IBM-compatible personal computers. These programs are part of a comprehensive software suite for electron crystallography, have been developed by Xiaodong Zou, Sven Hovmöller, and coworkers, and can be ordered at http://www.calidris-em.com. The program eMap by Peter Oleynikov (AnaliTEX) complements this electron crystallography suite (also runs on IBM-compatible personal computers) and can also be ordered over the Calidris Web site. With the latter program, one can, for example, simulate PED patterns and calculate structure factors from standard crystallographic information files (CIFs) (see footnote g).

g. The CIF format [and its underlying Self-defining Text Archive and Retrieval (STAR) structure] is owned by the International Union of Crystallography (IUCr), which "will not permit any other organization to 'capture' STAR or CIF and try to ransom it back to the community" (http://www.iucr.org/iucr-top/cif/faq/). Note that the IUCr asserts that "if you are putting out your CIF- or STAR-compliant application to the world for free, we are not going to ask you to start charging money for it so that you can pay the IUCr a license fee" (http://www.iucr.org/iucr-top/cif/faq/). In CIF-based open-access databases, the above-mentioned copyrightable "original, tangible forms of expression" (http://aca.hwi.buffalo.edu//newsletterpg_list/Newsletters_PDF/Winter06.pdf) are expressed in the CIF format so that, under the above-stated conditions, no license fee need to be paid to the IUCr or anybody else.

h. These databases are the "Powder Diffraction File, Version 2" (PDF-2), of the International Center for Diffraction Data (ICDD), the "Crystal Data" Database of the National Institute of Standards and Technology (NIST, formerly known as the National Bureau of Standards), and the NIST Standard Reference Database 15. The latter resulted from collaborations between the NIST, Sandia National Laboratories, and the ICDD, and is sometimes also referred to as the "NIST/Sandia/ICDD Electron Diffraction Database."

17 Mechanical Properties of Nanostructures

Vladimir Dobrokhotov
Department of Physics and Astronomy, Western Kentucky University, Bowling Green, Kentucky, U.S.A.

INTRODUCTION

The second part of the 20th century has seen significant developments in our understanding of fundamental material science, and thus also of the mechanical performance of materials. This understanding has generated profound changes in the field, leading to new families of materials, new concepts, and wide-ranging improvements in the mechanical behavior and in all other properties of materials. In our energy-conscious society, materials and structures are required to be more performant, lightweight, and cheap. The best answer to these requirements is often provided through the powerful concept of reinforcement of a "matrix" material with second-phase dispersion (clusters, fibers). It is an interesting fact that many natural forms of reinforcement possess a nanometric dimension, whereas most current synthetic composites include fibers in the micrometer range. Expected benefits of such "miniaturization" would range from a higher intrinsic strength of the reinforcing phase (and thus of the composite) to more efficient stress transfer, to possible new and more flexible ways of designing the mechanical properties of yet even more advanced composites (1). Presently, reinforcement of common materials (alloys, polymers) with nanostructures is one of the most promising areas of study. As one of the major factors that determine the quality of reinforcement is the mechanical strength of nanostructures, the studies of elastic properties of nanomaterials are of significant importance. Besides reinforcement, investigation of the mechanical properties of nanowires is essential to determine the material strength for practical implementation as electronic or optical interconnects, as components in microelectromechanics, and as active or passive parts in nanosensors. Mechanical failure of those interconnects or building blocks may lead to malfunction, or even fatal failure of the entire device. Mechanical reliability, to some extent, will determine the long-term stability and performance for many of the nanodevices currently being designed and fabricated. When nanowire properties have been adequately explored and understood, their incorporation into solutions of practical problems will become evident more quickly and feasible for active and concerted pursuit. Nanomechanical measurements are a challenge, but remain essential to the fabrication, manipulation, and development of nanomaterials and perhaps even more so to our fundamental understanding of nanostructures. For this purpose, various experimental techniques, or methods, have been developed in the last several years, including tensile, resonance, nanoindentation, and bending tests. Traditional optical microscopy lacks the resolution to investigate phenomena of colloidal dimensions adequately, and electron and X-ray techniques are greatly limited either by environmental (e.g., liquids) or material property (e.g., conductivity, cross-section for energy beam interactions) restrictions. Today, very few electron microscopes are capable of the true atomic resolution required for fundamental studies on intermolecular and colloidal behavior of two- or three-body interactions, for example. But scanning electron microscopes (SEMs) have played a crucial role in the study of mechanics with nanowires and nanosprings, and high-resolution optical microscopy is useful for locating these so-called one-dimensional objects on test substrates. In many cases, this is feasible because of lengths exceeding a few microns that scatter enough light for adequate contrast. The advent of atomic force microscopy (AFM) marked the beginning of significant advancement toward more routine molecular-scale imaging in three quantified dimensions with the simultaneous measurement of additional (one or more) physical properties. Researchers employ the AFM in various ways to determine sample mechanical properties, especially the elastic, or Young's, modulus. Bending tests with AFM are common for mechanical characterization of nanowire-like systems owing to high spatial resolution and direct force measuring sensitivity.

ATOMIC FORCE MICROSCOPY FOR BEND TEST

Atomic force microscopes are themselves nanomechanical instruments. AFM employs a sharp, cantilever-mounted probe to raster scan surfaces. Image resolution can be very high—scientists have observed subatomic-scale features—but depends on various factors including tip sharpness, acoustic isolation of the instrument, sampling medium, AFM controller precision, etc. The tip and sample positions are manipulated relative to each other with piezoelectric or other (e.g., electromagnetic coils) actuators. The AFM precisely controls the tip location on the sample by managing the voltage applied to the scanners. These are arranged either with three independent, orthogonal piezoelectric blocks or in a tube configuration. Piezoelectric scanner performance can be limited because of nonlinearities in the scanner material creep, noise, and drift in the high voltage supply, or thermal drift of the AFM apparatus itself. Various attractive and repulsive forces act between the tip and sample, such as van der Waals, electrostatic, and capillary forces. To some extent, such forces can be controlled by altering the sampling medium—for example, sampling under water can eliminate the effect of capillary forces. Typically, a diode laser reflects off the back of the cantilever onto a quadrant photodetector, which senses cantilever bending and twisting. If the cantilever spring constant is known, the cantilever deflections may be converted to quantitative force data. Cantilever calibration is not a trivial procedure, however. Cantilevers are sold with typical force constants that may, in fact, vary by an order of magnitude from reported average values. Alternately, the values can be calculated from the cantilever's spring geometry. For a uniform, rectangular cross-section, the cantilever's spring constant is given by $k_c = Ewt^3/4l^3$, where w is the width of the cantilever, l is its length, t is its thickness, and E is the elastic modulus. This, however, is seldom accurate enough for the required precision of quantitative AFM studies. Most cantilever probes are rectangular or triangular with a "two-beam" geometry connecting at the tip. Many AFM cantilevers are also coated with one or more layers of metal for reflectivity and other surface modifications. From the deflection of the cantilever, we calculate tip-sample force data by using Hooke's law $F = k_c z$, where F is the magnitude of the force acting between the tip and sample, k_c is the cantilever spring constant, and z is the cantilever deflection at its free end. AFM may be known best for its ability to generate high-resolution topographical images. In most imaging modes, a feedback system senses instantaneous cantilever deflection and adjusts scanner elements to maintain a constant interaction between the tip and sample. This instrument records and plots scanner adjustments as surface topography. In fact, the AFM may operate in any of various modes, depending on the interaction energies or forces of interest. For example, selective chemical functional groups may be attached to the probe, generating force data reflecting sample composition. The image produced will thus be a convolution map of chemical makeup, not merely surface topography. Also, the AFM can record twisting movements of the cantilever, which represent frictional forces acting between the tip and the sample. Force spectroscopy generates a force–distance curve for a single location on the sample. This is a plot of the magnitude of the force acting between tip and sample versus the position of the scanner in the direction normal to the substrate. Force–distance curves hold a wealth of information about the sample's mechanical properties. Points of discontinuity, the slopes of the approach, and retract curves, as well as any observed hystereses all cede hints to surface behavior. The difficulty arises in interpretation and deconvolution of multiple phenomena. Hysteresis, for example, is the result of adhesion, surface deformation, and/or nonlinear performance of the instrument, such as piezoelectric or other transducers for scanning and the probe detection sensor (e.g., photodiode). Another consideration is that the AFM does not directly measure the actual tip-sample separation distance. Rather, it controls and/or measures the vertical scanner position, the cantilever deflection, and the sample deformation. We often collect an array of force–distance curves at discrete sites over an area of the sample surface to determine effects of sample heterogeneity. This enables the AFM to produce spatially resolved maps of both topography and other sample material properties near the surface that are gleaned from force or energy profiles. Commercially available tools, such as pulsed-force mode (PFM) and force–volume imaging, accomplish this mapping with improved automation. In these cases, force, adhesion, and stiffness data are readily collected and made available for offline interpretation and analysis. Other physical interactions between the probe and sample may also be mapped with more difficulty, such as energy loss or long-range forces. It is primarily the stiffness data that are of interest to nanomechanical studies. Carbon nanotubes (CNT), nanowires, and nanosprings have been highly studied for their sometimes remarkable electrical and mechanical properties. Some nanowires exhibit surprisingly high tensile strength, possibly due to the presence of fewer mechanical defects per unit length than in their macroscopic analog. Being able to synthesize these structures reliably and understand their behavior will be vital to their ultimate application as future building blocks in everything from circuits to microelectromechanical or nanoelectromechanical systems (MEMS or NEMS), environmental sensors to biomimetic implants, hydrogen production and storage to smart composite materials, just to name the more popular examples. Deformation tests for determining the mechanical properties of edge-supported films are similar to bending nanowires and are plagued with similar difficulties of experimental setup and data interpretation at the nanoscale. While a significant literature base exists on bulge tests, the "two-dimensional" analog of beam bending, there is, surprisingly, little in the way of true nanomechanical investigations with AFM, nanoindentation, or other techniques, that is, nanoscopic in both morphology and stress (e.g., nanoNewton point loads). Edge-supported film deflections may be interpreted from expressions derived for the classic centrally loaded plate deformations or from extended models for membranes and shape memory materials. The films subjected to point loading with AFM were modeled as beams but are closer to rectangular plates, even though significant stretching of the polymeric (membrane) material may be expected to skew the results significantly from classic behavior. Local deformations in AFM bend tests cannot be neglected as in some classic experiments, but a satisfactory solution for nanomechanics is yet to be developed. We expect to observe discrepancies between actual deformations and predicted values based only on global bending model predictions at this stage in development for both suspended films and nanowires.

MECHANICAL PROPERTIES OF CARBON NANOTUBES

Rolling up a graphene sheet on a nanometer scale has dramatic consequences on the electrical properties. The small diameter of a carbon nanotube (CNT) also has an important effect on the mechanical properties, compared with traditional micron-size graphitic fibers. Perhaps the most striking effect is the opportunity to associate high flexibility and high strength with high stiffness, a property that is absent in graphite fibers. These properties of CNTs open the way for a new generation of high-performance composites. Theoretical studies on the mechanical properties of CNTs are more numerous and more advanced than those in experimental measurements, mainly due to the technological challenges involved in the production of nanotubes and in the manipulation of nanometer-sized objects. However, recent developments in instrumentation [particularly high-resolution transmission electron microscopy (HRTEM) and atomic force microscopy (AFM)], production, processing, and manipulation techniques for CNTs have given remarkable experimental results. The mechanical properties are strongly dependent on the structure of the nanotubes, which is due to the high anisotropy of graphene (2).

Knowledge of the Young's modulus (E) of a material is the first step towards its use as a structural element for various applications. The Young's modulus is directly related to the cohesion of the solid and, therefore, to the chemical bonding of the constituent atoms. For a thin rod of isotropic material of length l_0 and cross-sectional area A_0, the Young's modulus is then $E = \text{stress/strain} = (F/A_0)/(\delta l/l_0)$. A molecular solid has a low modulus (usually less than 10 GPa) because van der Waals bonds are weak (typically 0.1 eV), whereas a covalently bonded one (such as graphite, diamond, SiC, BN) has a high modulus (higher than 100 GPa). Moreover, in each class of solids (defined by the nature of the bonding), experiments show that elastic constants follow a simple inverse fourth power law with the lattice parameter. Small variations of the lattice parameter of a crystal may induce important variations of its elastic constants. For example, C33 of graphite (corresponding to the Young's modulus parallel to the hexagonal c-axis) depends strongly on the temperature due to interlayer thermal expansion. The Young's modulus of a CNT is therefore related to the sp^2 bond strength and should equal that of a graphene sheet when the diameter is not too small to distort the C–C bonds significantly.

It is interesting to compare the different theoretical results concerning the Young's modulus and its dependence on the nanotube diameter and helicity. The results are found to vary with the type of method and the potentials used to describe the interatomic bonding. The Young's modulus can be written as the second derivative of the strain energy divided by the equilibrium volume. Continuum elastic theory predicts a $1/R^2$ variation of the strain energy, with an elastic constant equal to C11 of graphite (which corresponds to the Young's modulus parallel to the basal plane), independent of the tube diameter. Therefore, in the classical approximation, the Young's modulus is not expected to vary when wrapping a graphene sheet into a cylinder. This is not surprising, as the atomic structure is not taken into account, so the elastic constants are the same as in a planar geometry. This classical approximation is expected to be valid for large-diameter CNTs. The question now is what happens in very small diameter tubes for which the atomic structure and bonding arrangement must be included in a realistic model. Both ab initio and empirical potential-based methods have been used to calculate the strain energy as a function of the tube diameter (and helicity). They all agree that only small corrections to the $1/R^2$ behavior are to be expected. As a consequence, only small deviations of the elastic constant along the axis (C33 in standard notation) are observed. It is worth noting that the dependence of the elastic constants on the nanotube diameter is found to be different for each model. For example, two different empirical potentials give different values for the elastic constant and show a different trend as a function of diameter. A decrease of C33 when the radius decreases is sometimes predicted; in other cases, the inverse behavior is observed (2).

The first measurement of the Young's modulus of MWNTs came from Treacy and colleagues (3). TEM was used to measure the mean-square vibration amplitudes of arc-grown MWNTs over a temperature range from RT to 800°C. The average value of the Young's modulus derived from this technique for 11 tubes was 1.8 TPa—0.40 TPa being the lowest and 4.15 TPa the highest. The authors suggest a trend for higher moduli with smaller tube diameters. The method of measuring thermal vibration amplitudes by TEM has been extended to measure SWNTs at room temperature (4). The average of 27 tubes yielded a value of $E = 1.3 + 0.6 - 0.4$ TPa, but there are two systematic errors in measuring the temperature and the nanotube length, which lead to an underestimate of E. Given these uncertainties in the method, it was not possible to state whether single-walled tubes are stiffer than multiwalled tubes. All measured values of E for nanotubes indicate that it may be higher than the currently accepted value of the in-plane modulus of graphite. The authors point out that either the cylindrical structure of the tubes imparts greater strength or the modulus of graphite has been underestimated. The latter statement is less likely considering the high precision of macroscopic methods and the variety of concordant experiments on single-crystal graphite and fibers. Salvetat et al. developed a method to measure the elastic modulus of CNTs deposited on a well-polished alumina ultrafiltration membrane (5). On such a substrate, nanotubes occasionally lie over the pores, either with most of the tube in contact with the membrane surface or with the tube suspended over a succession of pores (Fig. 1).

Attractive interactions between the nanotubes and the membrane clamp the tubes to the substrate. A Si_3N_4 AFM tip is then used to apply a force and to measure the resulting deflection of the CNT. Once a suspended nanotube is located with the AFM, its diameter, suspended length, and deflection midway along the suspended length *from a series of images* taken at different loads were determined. The apparent tube width is a convolution of the tube diameter and the tip radius, but the height remains a reliable measure of the tube diameter. The deflection, d, of a beam as a function of the applied load is known from small-deformation theory to be $d = FL^3/aEI$, where F is the applied force, L is the suspended length, E is the Young's modulus, I is the moment of inertia of the beam, and $a = 192$ for a clamped beam. The reversibility of the tube deflection and the linearity of the F–D curve show that the nanotube response is linear and elastic, at least at low load and deflection [Fig. 1(B)]. The slope of the curve gives directly the Young's modulus of

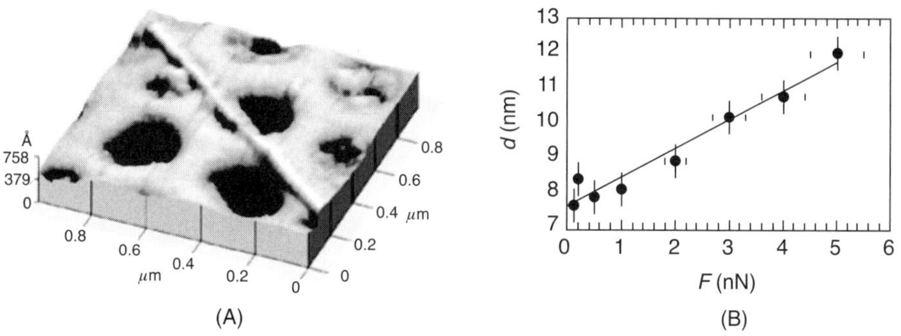

FIGURE 1 (**A**) 3D rendering of an AFM contact-mode image showing a SWNT bundle lying over a pore on the surface of a polished alumina membrane. (**B**) Typical deflection–force curve on a SWNT bundle. *Source*: From Ref. 2.

the CNT. The average value of the Young's modulus for 11 individual arc-grown carbon nanotubes was found to be between 400 and 1220 GPa. The most recent results of the CNT bend test were demonstrated in the work of Lee et al. (6). To measure the stiffness of a suspended nanotube, it was pushed down at its midpoint with an AFM cantilever, acquiring force–displacement data during the loading and unloading processes. Figure 2(B) shows a typical pair of loading and unloading *F–D* curves.

The curves appear generally linear, confirming that the small deformation model is still valid for this experiment. The Young's modulus of the CNTs varies significantly, depending on the two major factors: nanotube diameter and the method

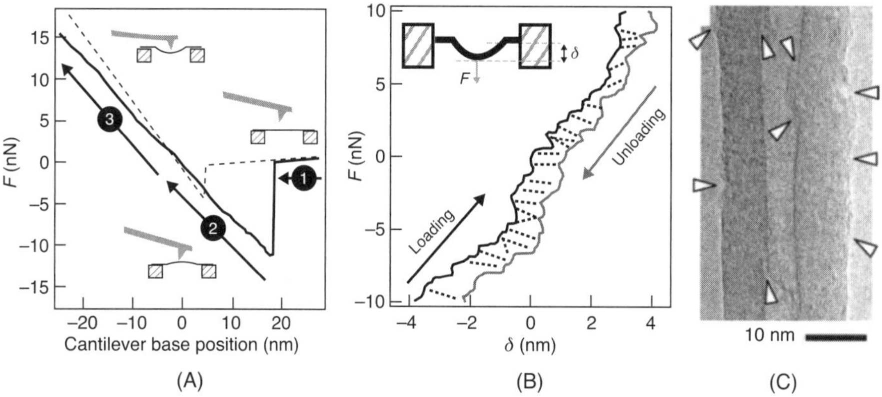

FIGURE 2 (**A**) In AFM force–displacement curve technique, the tip is first lowered into contact with the sample and then it is pulled off contact. Cantilever deflection (proportional to the loading force *F*) and cantilever base (*z* scanner) position are measured during the process. A loading curve on a suspended CNT (*solid line*) and that on a flat reference surface (*dashed line*) have different slopes due to the CNT deflection. (**B**) By comparing the loading and unloading curves on a CNT with those on a flat surface, a *F*–δ graph can be obtained. Some matching kinks in the loading and unloading curves are indicated by dashed lines. (**C**) TEM image of a CVD-grown MWCNT. Markers indicate structural defects. *Source*: From Ref. 6.

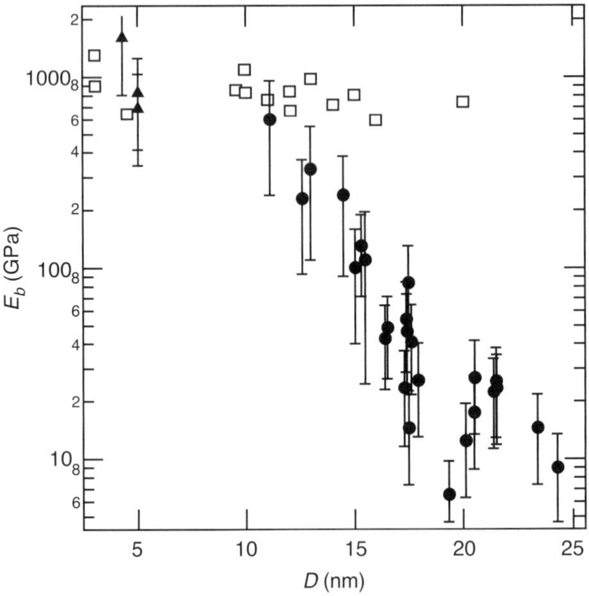

FIGURE 3 (*Circles*): Results on CVD-grown MWCNTs. (*Triangles*): Bending modulus of small-diameter CVD-grown doubel-walled CNT ropes. (*Squares*): Nanotubes grown by arc-discharge evaporation. *Source*: From Ref. 6.

of synthesis. Figure 3 summarizes the results obtained by Lee et al. (6). The elastic modulus varies dramatically in the 10 to 20 nm diameter range, increasing almost exponentially with decreasing diameter. Data for small-diameter double- to quadruple-walled CNT ropes have been included to show the upper limit possible with CVD-grown MWCNTs (triangles in Fig. 3). Previous results on CNTs and small-diameter CNT ropes grown by arc-discharge evaporation are also included in the graph for comparison (squares in Fig. 3). Unlike the CVD grown MWCNTs, the arc-discharge–grown MWCNTs showed fairly constant elastic modulus over the same diameter range. A number of theoretical works have actually predicted the CNT elastic modulus to drop with decreasing diameter because of the excessive strain imposed on the graphene shells at small diameters. However, Lee et al. (6) were the first, who found the dependence of Young's modulus upon the method of synthesis and concluded that the observed diameter dependence is a reflection of the diameter-dependent material quality; when grown in CVD, thinner MWCNTs are structurally superior to the thicker tubes. The observed diameter dependence is a strong evidence for the metastable-catalyst growth model proposed by Kukovitsky and colleagues (7,8). This model emphasizes the role of partially molten catalysts that have liquid *skins* covering solid cores. The liquid skin, which may be metastable, is extremely important because carbon diffusion would occur predominantly through this layer, and any instabilities at the layer would disturb the CNT growth. In CVD, the catalyst is continuously agitated by exothermic precursor dissociation and endothermic carbon precipitation, the processes that may appear to the catalyst as discrete events on the time scale at which changes can occur to its liquid skin. Therefore, it is realistic to assume that instabilities *do* exist and that the catalyst–nanotube interface suffers from numerous perturbations during CNT growth. TEM movies of CNT growth in CVD have shown catalysts undergoing numerous morphological changes. Perturbations at the catalyst–nanotube interface would cause structural defects in the produced MWCNT, degrading its quality. The frequency and the gravity of perturbation would depend on the catalyst size. Under the same conditions, the smaller catalyst would have a thicker liquid skin due to its larger surface-to-volume ratio. The thicker skin would lead to a less frequent fluctuation at the nanotube catalyst interface as well as allowing more stable carbon diffusion, despite some perturbations. So a MWCNT with a better structure would grow from the smaller catalyst. The observed *transition* in the elastic modulus over the 10 to 20 nm diameter range confirms this conclusion. Considering that 30 nm Co particles have been observed to melt at 600 (50) degrees Centigrade in methane atmosphere on silica support, our 10 to 20 nm transition range is situated slightly lower than expectation. The difference could be due to a number of reasons: a catalyst often produces a MWCNT thinner than its size, and the carbon precursor and the support were different in our case. Nonetheless, the diameter-dependent variation of the elastic modulus is a strong evidence for the metastable catalyst growth model. In summary, we have measured the elastic modulus of individual MWCNTs grown from a single CVD process. The data show the elastic modulus changing dramatically in a narrow diameter range. The diameter dependence in elastic modulus is a strong evidence for the metastable-catalyst growth of MWCNTs in CVD. It is difficult to arrive at the near-exponential diameter dependence starting from the growth model. Successfully modeling the liquid-skin instabilities and accurately predicting the type and frequency of induced defects are daunting tasks. Calculating how the different structural defects influence a MWCNT's mechanical strength is also very challenging. A possible solution is to model a metastable catalyst to be oscillating between liquid and solid forms, with the time constants determined by the environment and the catalyst size. The produced MWCNTs should then be modeled as a series of high-quality segments joined by poor-quality nodes, the lengths of these parts being related to the time constants of liquid and solid forms.

MECHANICAL PROPERTIES OF NANOWIRES

Researchers employ various deformation tests to study the moduli of elasticity of nanowires. A wide variety of different nanowires were studied by Shanmugham et al. (9), Chen et al. (10,11), and Withers and Aston (12) from the University of Idaho (9–12). Compared to the previously described bend-test of carbon nanotubes, where the CNT was bent by the AFM probe at the midpoint, the bend test presented by Aston shows a complete elastic profile of the nanowire by applying the AFM-induced force at several points along the entire length of nanowire (9–12). In general, the expressions used to relate elastic modulus to observed deflection and applied force come from classic texts on strength of macroscopic materials. From the deflection of the cantilever, we calculate tip–sample force data by using Hooke's law:

$$F = kz, \tag{1}$$

where F is the magnitude of the force acting between the tip and sample, k is the cantilever spring constant, and z is the cantilever deflection at its free end. This approach is valid when the beams follow linear elastic theory of isotropic materials

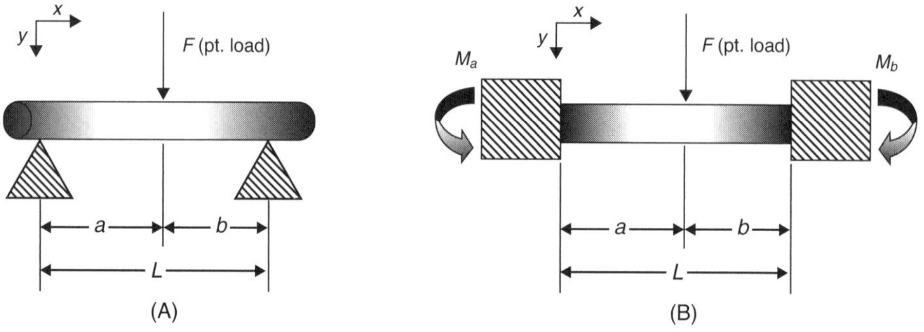

FIGURE 4 Schematic representation of beam support models: (**A**) simple and (**B**) clamped. *Source*: From Ref. 12.

and have high length-to-thickness ratios. A beam can be supported in various ways, resulting in mathematical description via boundary conditions of the type of supported ends. The two extremes in behavior are set by having both ends either free or built-in (that is, clamped or fixed) to resist both torque and slip. The first is the simply supported beam [Fig. 4(A)] and the latter is the clamped end configuration [Fig. 4(B)]. Other beam bending behavior may be described by mixed boundary conditions, or by modifications thereof, for example, a limited-slip or limited-torque end that might be related to adhesion and tribology.

The general bending equation expresses beam deflection, z, in the plane of applied stress as a function of the beam's moment:

$$EI = \frac{d^2z}{dx^2} = -M. \tag{2}$$

The first moment of inertia, I, depends on the beam's cross-sectional shape and physical dimensions. If the beam shape changes during measurement, as may be expected for nanomechanical studies, the in situ determination of I may be intractable for nanowires—or even unsolvable under AFM test designs—and will at worst need to be considered a second variable parameter for interpretive analysis. The moment M depends on the magnitude of the concentrated point load, F, the load location measured from the origin down the long axis (i.e., the x-coordinate), a, and the length of the beam, L. M is negative when the z-axis is defined as positive in the downward direction, convenient for AFM bending studies. In the case of built-in or clamped ends, M is expressed as

$$M = \frac{Fb}{L}x - \frac{Fab^2}{L^3}(L - x) - \frac{Fa^2b}{L^3}x, \quad x \leq a \tag{3}$$

and

$$M = \frac{Fb}{L}x - \frac{Fab^2}{L^3}(x - a) - \frac{Fab^2}{L^3}(L - x) - \frac{Fa^2b}{L^3}x, \quad x \geq a. \tag{4}$$

Substitution of Eq. (3) or Eq. (4) into Eq. (2) followed by double integration with respect to x provides the standard forms for the bent beam profiles as a function of applied load, representative of AFM force–distance profiling methods:

$$EIz = \frac{Fb^2x^3}{6L^3}(2a + L) - \frac{Fab^2x^2}{2L^2}, \quad x \le a \tag{5}$$

$$EIz = \frac{Fb^2x^3}{6L^3}(2a + L) - \frac{Fab^2x^2}{2L^2} + \frac{F}{6}(x - a)^3, \quad x \ge a \tag{6}$$

Relationships of these and similar forms can be used to calculate Young's modulus directly, assuming elastic behavior for classic mechanical comparisons. For nanomechanics, values calculated for E are currently interpreted as "apparent" elasticity because we have yet to investigate nanoscale bending in full depth. Boundary conditions are determined by the effect of the support scheme on the shape of the profile constructed from deflection data versus location on the nanowire. Letting $a = b = x = L/2$ yields $E = FL^3/192zI$, which corresponds to the previously considered midpoint bend test for carbon nanotubes. Alternately, we can substitute a known spring constant of a beam, for example, nanowire, and calculate its elasticity directly:

$$E = \frac{k_{\mathrm{w}} L^3}{192I}. \tag{7}$$

The slope of an AFM force–distance curve taken against a surface is the observable spring constant of the mechanical system, k_{sys}, composed of the cantilever probe being used and the material under test, which is ideally characterized by the cantilever and the (nanowire) sample being springs in series:

$$\frac{1}{k_{\mathrm{sys}}} = \frac{1}{k_{\mathrm{c}}} + \frac{1}{k_{\mathrm{w}}}. \tag{8}$$

We obtain k_{c} from a calibration cycle, and we obtain k_{sys} from the slope of the force–distance curve over a segment of constant compliance. Young's modulus can then be computed from Eq. (7) with k_{w} and the geometry of the nanowire setup (from AFM and/or SEM measurements). We find that for beams with identical material properties and physical dimensions, where the only difference is the support scheme (boundary conditions), the midpoint deflection for the simply supported beam will be larger by a factor of 4, as shown in Figure 5.

Knowing the boundary conditions for bending profiles is extremely important for measuring the elastic modulus of nanowires accurately. A fixed-beam boundary shows a shallow slope in the deflection profile near the ends and has more curvature throughout the profile, while a simple beam boundary gives a steeper slope at the ends, and the profile is more parabolic in shape rather than the subtle curve of the fixed beam. Profiles for mixed boundaries (one simply supported and one fixed) would show intermediate values of deformation and be asymmetric about the midpoint. Bending profiles of typical nanowires have been plotted as a function of relative tip position x/L (as shown in Fig. 6) to illustrate the experimental boundary conditions observed for GaN nanowires. The maxima in deflection at the middle and profile symmetry indicate comparable boundary conditions at each supported end. Theoretical profiles for fixed and simple beams are also shown in Figure 6. The bending profile for a 57-nm nanowire shows a smoother transition in slope from the edge toward the midpoint, which is the main feature of fixed ends. All other tested nanowires, with diameters ranging from 89.3 to 135.0 nm, show sharper and

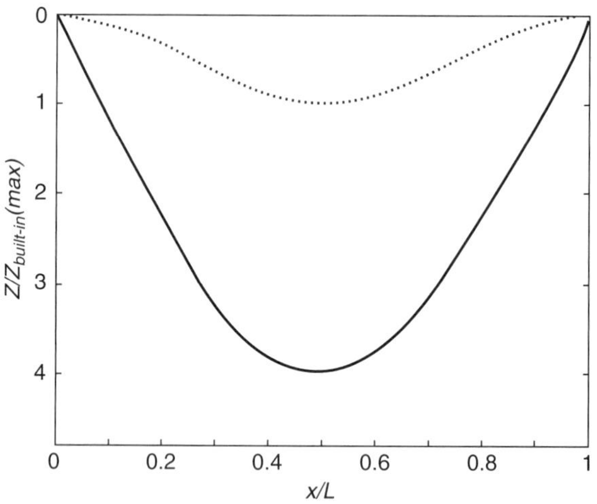

FIGURE 5 Comparison of deflection profiles for simply supported beam (*solid*) and beam with clamped or fixed ends (*dashed*). Units are normalized. *Source*: From Ref. 12.

steeper sloping for edge deflections, best fitted by the simple-beam boundary condition. All experimental deflection profiles exhibit a reasonable fit, though not perfect, with theoretical values from fixed or simple-beam models. These findings are similar to DPFM data for silver nanowires but different from experiments on polymer nanowires (silver and polymer nanowires did not require a polymer adhesion layer) (9–12). The deflection profiles of polymer nanowires were best fitted with a fixed beam model even though they were quite large (170–200 nm). Silver nanowires showed fixed-end behavior for smaller diameters under low loading conditions and resulted in simple supports for larger diameters, which consequently required high loading for similar deflections. Large silver nanowires under low applied force exhibited intermediate conditions, probably due to transitions from fixed to simple ends with increasing stress (Fig. 7). Higher work of adhesion with the substrate was suspected for small nanowires due to increased capillary effects, or meniscus forces. However, under relatively large applied force, the work of adhesion is not always sufficient to resist the resultant stresses. Compared to silver, GaN nanowires show relatively weak adhesion to the substrate, demonstrated from initial AFM images on the silicon test gratings without a polymer adhesion layer. Though the polymer thin film improved adhesion, it was still insufficient in most cases for larger nanowires. Aside from nanowire size and experimental loading, this suggests that the intrinsic surface energy of GaN nanowires is quite low, as is well known of boron nitride. (Contact angle measurements with the water on substrates of random GaN nanowire coatings also exhibit complete nonwetting.) After confirming the boundary conditions, GaN nanowire elastic moduli were computed by using the appropriate model. Elastic modulus for GaN nanowires decreases from 400.1 ± 14.9 GPa (standard deviation errors) to 195.6 ± 19.7 GPa as diameter increases from 57 to 135 nm. The value of 400 GPa for 57 nm nanowire is from the fixed-beam model, while the simple beam model was found to be valid for the 89.3–135.0 nm range. Moduli trends are similar to the reported results for many other nanowires, such as

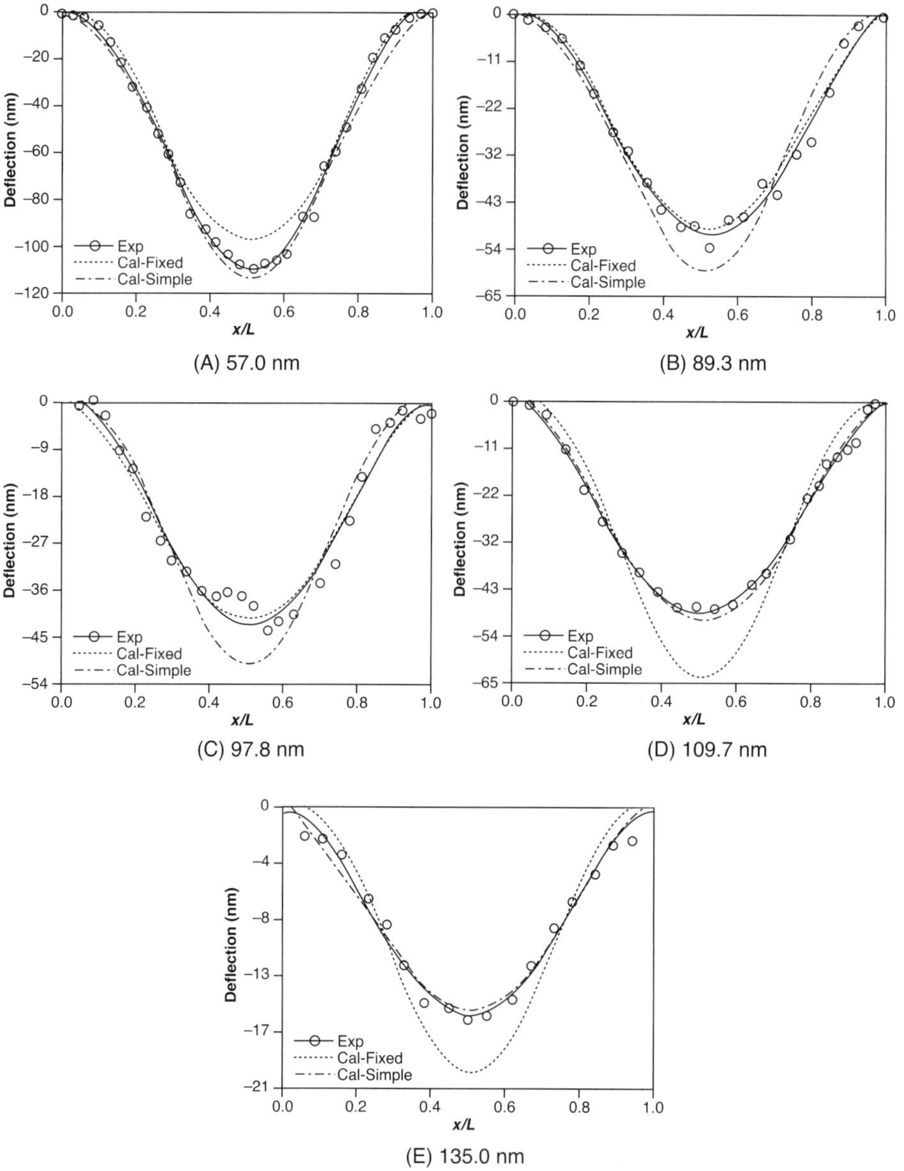

FIGURE 6 Comparison of experimental and model deflections for tested GaN nanowires. *Source*: From Ref. 11.

ZnO, SiC, and TiO_2. The elastic modulus for ZnO nanowires with diameters smaller than 120 nm increases dramatically with decreasing diameters and is significantly higher than larger nanowires or bulk ZnO. SiC nanorods were reported to approach the theoretical value for E when the diameter was reduced to ~20 nm. Generally, physical properties such as elastic modulus are believed to be directly related to the

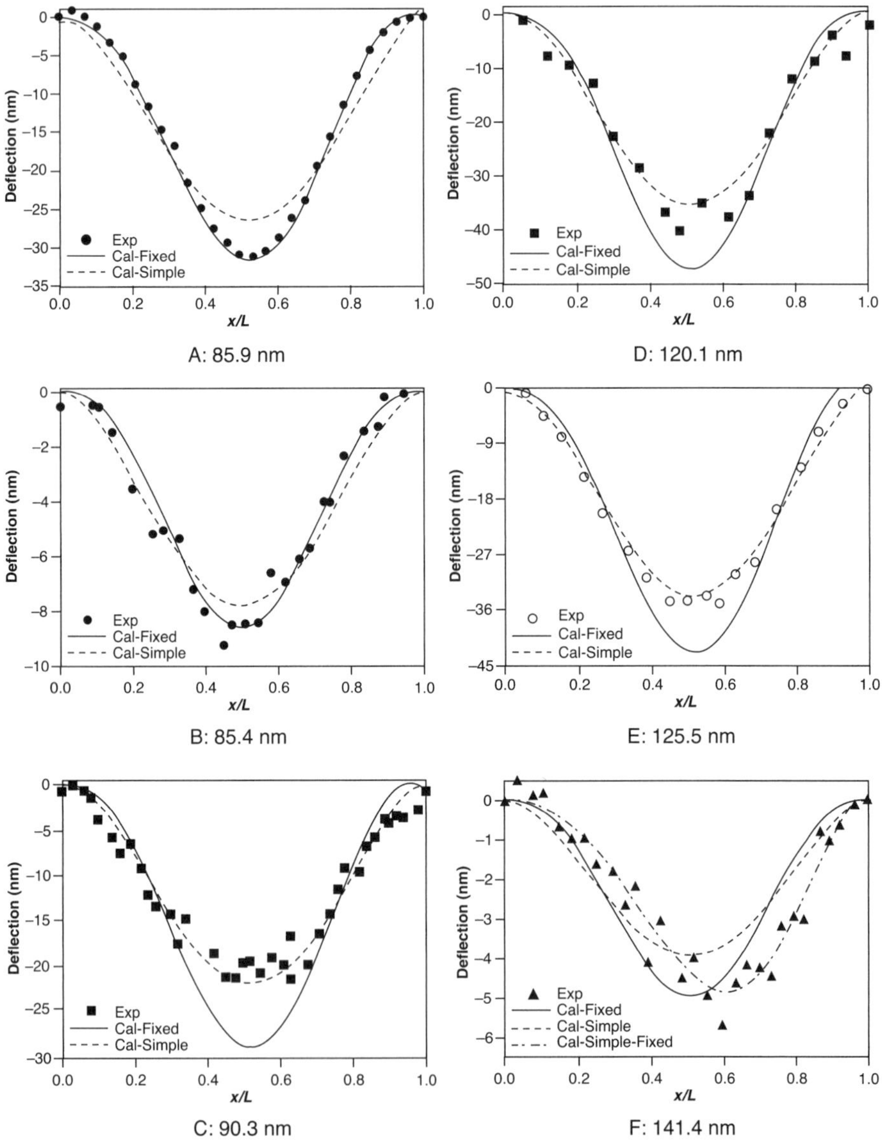

FIGURE 7 Comparison of experimental and model deflections for tested silver nanowires. *Source*: From Ref. 10.

material structural perfection. The elastic stiffening for smaller nanowires can be explained by the lower probability of finding a defect in smaller volumes. However, these findings are different from the results of silver and triangular GaN nanowires. The opposite trend was observed in E for triangular GaN nanowires within the diameter range of 36 to 84 nm. The authors suggested that this converse behavior was owing to the increase of surface-to-volume ratio (S/V) with decreasing d. They believed that the atomic coordination and cohesion near the surface were "poor" relative to that of bulk, and the increasing dominance of the surface would decrease the rigidity of the structure. Furthermore, nanowire shape and, microstructure are also determinate contributions to the difference in observed moduli trends: the GaN nanowires we produced have a hexagonal cross section grown along the [0001] direction, while those having a triangular cross-sectional shape were grown along the [120] direction. For most nanowires with diameters of 89 to 110 nm, the simple-beam model gave $E = 218.1–316.9$ GPa, which approaches or meets the literature values of ~295 GPa for bulk single crystal. Literature values for GaN nanowires with triangular cross sections were similar: 227 to 305 GPa for 36 to 84 nm diameter. But for the smallest nanowire with diameter of 57.0 nm, E is as high as 400.1 ± 14.9 GPa from the fixed-beam model, which far exceeds the literature values. This may be due to variations in nanowire shape.

MECHANICAL PROPERITES OF NANOSPRINGS

In 2001, research group of McIlroy at the University of Idaho synthesized the first coiled nanosprings with both coil and wire diameters of the order of tens of nanometers (Fig. 8) (13). Currently, AFM mechanical investigations of silicon oxide nanosprings are underway; although the measurements have become almost routine, their interpretation with respect to boundary condition uncertainty remains a difficult issue (12). Various other one-dimensional objects, similar to these nanosprings, have come under study. Silicon springs of much larger pitch and

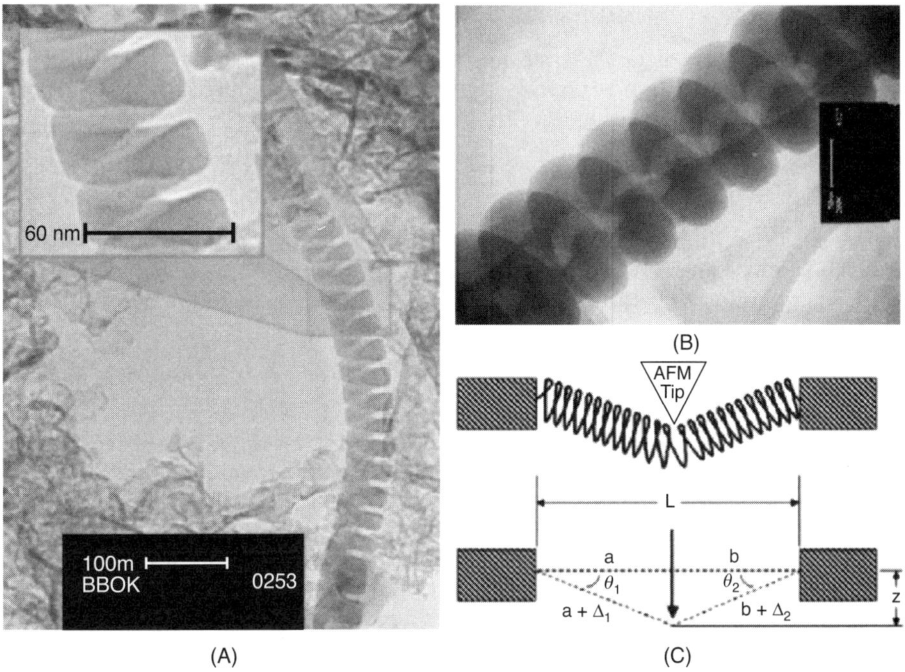

FIGURE 8 (**A**) BC nanospring, (**B**) SiC nanospring, and (**C**) lateral view of nanospring AFM bend test. *Sources*: From Refs. 11, 12.

coil diameter were reported to exhibit an electromechanical response to conducting mode AFM; these long-axis measurements are prone to simultaneous compression and bending, which can lead to indeterminate conclusions about material properties. One of the methods of nanospring analysis is when a nanocoil is clamped between two cantilevered AFM tips—one with a very compliant cantilever, another of an order of magnitude higher—by electron beam–induced deposited (EBID) residual hydrocarbons in a SEM environment. In this case, the force applied to the spring is determined from the observed deflection of the more compliant cantilever, and the nanocoil elongation is determined, both by using SEM (12). A spring constant for the coil is then found by dividing the applied load by the total elongation. It is assumed that the hydrocarbon "glue" does not deform (boundary conditions of the fixed ends). Bend tests may be conducted for nanosprings, as with the nanowires, to find Young's modulus. As long as the AFM tip is not located at either end of the spring, the force applied can be expressed with simple geometric relationships: $F = k(\Delta_1 \sin\theta_1 + \Delta_2 \sin\theta_2)$, where F is the applied load, k is the nanospring constant, Δ_1 and Δ_2 are the elongations of the spring segments to either side of the applied load beyond their relaxed states, and θ is the angular displacement of the spring segments. If we let the tip act at the spring's midpoint, employ appropriate trigonometric identities, and assume uniform geometries and material properties for the length of the spring, the equation may be written explicitly as a function of vertical displacement only:

$$F = 2kz\left(1 - \frac{L/2}{\sqrt{(L/2)^2 + z^2}}\right), \tag{9}$$

where L is the linear distance separating the spring's two anchored ends. The bending spring constant, k, can be expressed in terms of the physical geometry and properties of the spring material as

$$k = \frac{Gd^4}{8D^3N}, \tag{10}$$

where d is the wire diameter, D is the nanospring coil diameter, N is the number of involved nanospring coils, and G is the shear modulus of the material ideally related to elastic modulus. Substituting Eq. (10) into Eq. (9) yields the following result:

$$F = \frac{Gd^4}{4D^3N}\left(1 - \frac{L/2}{\sqrt{(L/2)^2 + z^2}}\right)z. \tag{11}$$

All values may be observed with AFM and/or SEM with the exception of the shear modulus, which may be estimated from the ideal relationship $E = 2(1 + \nu)G$, where ν is Poisson's ratio for the spring material. For transverse loading on the spring, the relationship between F and z is not expected to be linear. Under load within the elastic limit, the nanospring wire is twisted rather than stretched axially in the more usual testing geometry for a coiled spring. It is not subjected to the same effective loading in the way its nanowire analog would be in the three-point bend test. Another complication is the determination of Poisson's ratio. Values are tabulated for common bulk materials, but may in fact vary over a sample surface and are completely unknown when working with new materials, adding another dimension of uncertainty to the mechanics of nanosprings.

CONCLUDING REMARKS

Even though, the mechanical properties of nanostructures obey the classical laws of mechanics, the values of elastic moduli strongly deviate from the ones of a bulk material. The fundamental reasons for increased apparent elastic modulus in nanostructures remain unproved. There are multiple possible causes for these effects, including internal strain and specific atomic organization of the material. The mechanical properties of materials at nanoscale are determined not just by chemical compounds but also by the size and morphology of nanostructures. It is common to say that nanotechnology opens extra dimensions in the periodic table of elements. The advent of AFM was of great importance to nanomechanics. Phase contrast, pulsed-force, and other intermittent contact modes provide high spatial resolution of surfaces, highlighting inhomogeneities and relative surface property differences. Instrumentation and modeling have matured—and are constantly evolving—to where they offer a useful look into nanoscale mechanical performance. Although significant obstacles remain, the collection of recent advancements in nanomechanical measurements of material strength are laying a strong foundation to improve our understanding of basic material behavior such as beam bending and plastic deformation.

REFERENCES

1. Wagner HD. Reinforcement. Weizmann Institute of Science, Rehovot, Israel. Encyclopedia of Polymer Science and Technology 2002 by John Wiley & Sons, Inc.
2. Salvetat J-P, Bonard J-M, Thomson NH, et al. Mechanical properties of carbon nanotubes. Appl Phys A 1999; 69:255–260.
3. Treacy MMJ, Ebbesen TW, Gibson JM. Exceptionally high Young's modulus observed for individual carbon nanotubes. Nature 1996; 381:678.
4. Krishnan A, Dujardin E, Ebbesen TW, et al. Measurement of the Young's modulus of single-shell nanotubes using a TEM. Phys Rev B 1998; 58:14013.
5. Salvetat J-P, Kulik AJ, Bonard J-M, et al. Mechanical properties of carbon nanotubes. Adv Mater 1999; 11:161.
6. Lee K, Lukić B, Magrez A, et al. Diameter-dependent elastic modulus supports the metastable-catalyst growth of carbon nanotubes. Nano Lett 2007; 7(6):1598–1602.
7. Kukovitsky EF, L'vov SG, Sainov NA. VLS-growth of carbon nanotubes from the vapor. Chem Phys Lett 2000; 317;65–70.
8. Kukovitsky EF, L'vov SG, Sainov NA, et al. Correlation between metal catalyst particle size and carbon nanotube growth. Chem Phys Lett 2002; 355:497–503.
9. Shanmugham S, Jeong J, Alkhateeb A, et al. Polymer nanowire elastic moduli measured with digital pulsed force mode AFM. Langmuir 2005; 21;10214–10218.
10. Chen Y, Dorga BL, McIlroy DN, et al. On the importance of boundary conditions on nanomechanical bending behavior and elastic modulus determination of silver nanowires. J Appl Phy 2006; 100:104301.
11. Chen Y, Stevenson I, Pouy R, et al. Mechanical elasticity of vapour–liquid–solid grown GaN nanowires. Nanotechnology 2007; 18:135708.
12. Withers JF, Aston DE. Nanomechanical measurements with AFM in the elastic limit advances in colloid and interface. Science 2006; 120;57–67.
13. McIlroy DN, Zhang D, Kranov Y. Nanosprings. Appl Phys Lett 2001; 79:1540.

18 Fullerene-Based Nanostructures: A Novel High-Performance Platform Technology for Magnetic Resonance Imaging (MRI)

***Krishan Kumar, Darren K. MacFarland, Zhiguo Zhou, Christpher L. Kepley, Ken L. Walker, Stephen R. Wilson, and Robert P. Lenk**

Luna a nanoWorks (A Division of Luna Innovations, Inc.), Danville, Virginia, U.S.A.

INTRODUCTION

Diagnostic modalities, including computed tomography (CT), scintigraphic, or nuclear medicine (SPECT, single photon computed emission tomography and PET, positron emission tomography), magnetic resonance imaging (MRI), and ultrasound (US), are routinely used to investigate the architecture and physiological functions of the human body. The relatively low resolution of PET/SPECT requires that these methods be combined with CT scans for interpretation, and there is concern about the long-term effects of exposure to ionizing radiation (1). Magnetic resonance imaging (MRI), a noninvasive technique, is widely available, highly translatable, and can provide high contrast, especially for the study of soft tissue. The technique is based on the relaxation properties of water protons, the most abundant nuclei in the human body.

The signal intensity of MR images is a complex function of T_1 (spin:lattice or longitudinal), T_2 (spin:spin or transverse), T_E (spin echo delay time), and T_R (Pulse Repetition Time). Although MRI is used to detect the distribution of water molecules in different types of tissues, the image contrast is dependent on the relaxation characteristics of the protons in tissues, proteins, and lipids. Although lesions can be marked clearly in the images, there remains the issue of the insufficient contrast between healthy and diseased tissues due to relatively small differences, and hence a correspondingly low apparent MRI sensitivity for identifying the associated abnormalities. To overcome this deficiency of the technique, contrast agents are used to enhance the signal of the target by influencing the relaxation rates of local protons. It is estimated that 35% of the 60 million MRI procedures use contrast agents to improve imaging characteristics.

CONTRAST AGENTS IN MAGNETIC RESONANCE IMAGING (MRI)

Contrast agents catalyze relaxation rates or shorten T_1 and T_2 relaxation times of water protons and can be divided into two classes depending on whether they most influence T_1 or T_2. Superparamagnetic agents, such as iron oxide, influence the signal intensity by shortening T_2 relaxation time predominantly; thereby, producing darker images as compared to surroundings and are consequently called negative

*Contact at kumar070154@yahoo.com.

contrast agents. Paramagnetic metal ions (e.g. Gd^{3+}, Dy^{3+}, Mn^{2+}, $Fe^{2+/3+}$, Cr^{3+}) are known to mainly shorten T_1 relaxation time of water protons, and consequently produce brighter images.

The requirements of any metal ion to be a contrast agent for MRI are high paramagnetism, at least one labile coordinated water, and fixed oxidation state. Gadolinium (Gd) has the highest spin only magnetic moment ($\mu^2 = 63$ BM), exhibits labile coordinated water ($k_{ex} > 10^6$ s^{-1}), and possesses a relatively long electronic relaxation time (~0.1 ns at high fields), making it nearly an ideal catalyst for reduction of T_1 relaxation time of water through the Gd(dipole)–H(dipole) interaction.

Gadolinium (Gd^{3+}) hydrolyzes under physiological conditions and precipitates in the presence of inorganic phosphate, carbonate, and hydroxide. Free or unbound gadolinium ions have been found to be toxic in both in vitro and in vivo studies, with the LD_{50} in mice being low (i.e., 0.1 mmol/kg). To prevent these toxic side effects, gadolinium chelates with high thermodynamic stability and kinetic inertia are required to keep gadolinium in solution and to increase the tolerance. In addition, low osmolality and viscosity along with rapid clearance are also needed.

GADOLINIUM CHELATES–BASED MRI CONTRAST AGENTS

Based on this principle, and to prevent the toxicity effects of free gadolinium, significant research and development work has investigated this area over the past 25 years. Linear and macrocyclic polyaminocarboxylate chelating agents are used to form ionic, nonionic, kinetically inert, and thermodynamically stable chelates. These agents were found very safe and efficacious in the preclinical and clinical settings. The first gadolinium-based contrast agent, Magnevist®, was registered in the United States by Schering AG (Germany) more than 20 years ago. Since then, a series of contrast agents were approved and used as MRI contrast agents (MRI-CAs) for various applications (Fig. 1; Table 1).

Extracellular, low-molecular-weight gadolinium chelate–based MRI-CAs dominate the diagnostics market. However, a few novel and more specific MRI-CAs, containing metals other than Gd, Mn, and iron (Mn-DPDP as TESLASCAN and coated superparamagnetic iron-based nanoparticles as FERIDEX/ENDOREM), were also developed and marketed.

The physicochemical and biological properties of these products were reported previously, and a summary thereof is provided here: (*i*) The macrocyclic

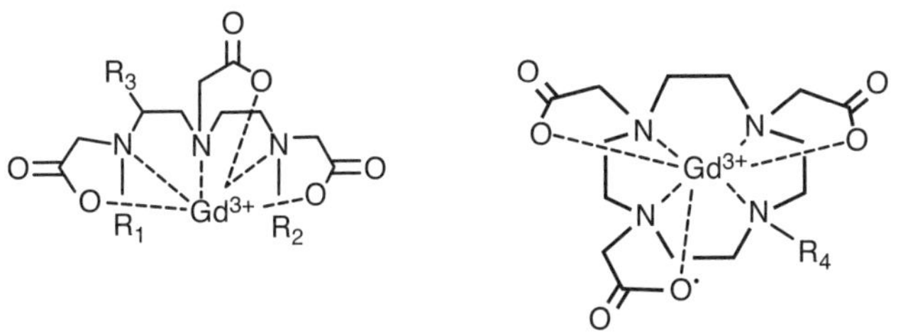

FIGURE 1 Structures of actives in commercial MRI contrast agents, where R_1 and R_2 = $-CH_2COO$, R_3 = H, (DTPA); R_1 and R_2 = $-CH_2CONHCH_3$, R_3 = H, (DTPA-BMA); R_1 and R_2 = $CH_2CONHCH_2CH_2OCH_3$, R_3 = H, (DTPA-BMEA); R_1 = $-CH_2COO$, R_2 = $-CH(COO)CH_2OCH_2C_6H_5$, R_3 = H, (BOPTA); R_1 and R_2 = $-CH_2COO$, R_3 = $-CH_2(C_6H_5)OC_2H_5$, (EOB-DTPA); R_4 = $CH_2CH(OH)CH_3$, (HP-DO3A); R_4 = CH_2COO, (DOTA); R_4 = $-CH(CH_2OH)CH(OH)CH_2OH$, (DO3A-butrol).

TABLE 1 Commercially Available Gadolinium-Based MRI Contrast Agents

Brand name	Generic name	Molecular formula	Company name	Indications
Magnevist	Gadopentate dimegluamine	Gd(DTPA)(H_2O)$^{2-}$	Schering AG (Germany) Now Bayer Health Care	Neuro, whole body
Ominiscan	Gadodiamide	Gd(DTPA-BMA)(H_2O)	Nycomed (Norway) Now GE Health Care	Neuro, whole body
OptiMARK	Gadoversetamide	Gd(DTPA-BMEA)(H_2O)	Mallinckrodt (US) Now Covidien	Neuro, whole body
MultiHance	Gadobenate dimeglumine	Gd(BOPTA)(H_2O)$^{2-}$	Bracco (Italy)	CNS, liver
Eovist	Gadooxate disodium	Gd(EOB-DTPA)(H_2O)$^{2-}$	Schering AG (Germany) Now Bayer Health Care	Liver
ProHance	Gadoteridol	Gd(HP-DO3A)(H_2O)	Bracco (Italy)	Neuro, whole body
Dotarem	Gadoterate megluamine	Gd(DOTA)(H_2O)$^{-}$	Guerbet (France)	Neuro, whole body
Gadovist	Gadobutrol	Gd(DO3A-butrol)(H_2O)	Schering AG (Germany) Now Bayer Health Care	CNS, whole body

polyaminocarboxylates form thermodynamically more stable and kinetically inert Gd^{3+} complexes than the chelates of linear polyaminocarboxylates; (*ii*) the relaxivity ($mM^{-1} s^{-1}$) of these chelates is in the range of 3.1–4.2 (0.47 T), 2.9–4.0 (1.5 T), 2.8–4.0 (3 T), and 2.8–4.0 (4.7 T); (*iii*) nonionic chelates have lower osmolality than ionic chelates and are more blood compatible, making them more tolerable than the ionic chelates; (*iv*) in vitro, in vivo biodistribution studies suggest that the macrocyclic polyaminocarboxylate chelates dissociate less than the chelates of linear polyaminocarboxylates; (*v*) the chelate-based MRI contrast agents distribute in extracellular spaces and excrete renally except MultiHance and Eovist, which are hepatobilliary agents.

LIMITATIONS OF CURRENT GADOLINIUM CHELATE–BASED MRI CONTRAST AGENT

The number of MRI scans that uses gadolinium-based MRI contrast agents has increased tremendously over the past 20 years; however, these agents have several limitations. These limitations are as follows: (*i*) High concentrations are needed to produce effective contrast owing to low relaxivity and diffusion effects experienced postinjection. For example, the proposed dose of the current MRI contrast agents is 0.1 to 0.3 mmol/kg. (*ii*) Low sensitivity and poor targeting ability severely limit the potential use of these agents in diagnosis. (*iii*) These agents have been associated with serious side effects recently, Nephrogenic Systemic Fibrosis (NSF), in patients with impaired renal function (i.e., with glomerular filtration rates <30) (2,3). These findings have been the subject of a black-box warning from the US FDA for this class of products (4). Presumably, the pathology is a result of greater gadolinium release from the chelates because of slower clearance in these patients. The susceptibility of gadolinium chelates to dissociate in vivo and in vitro was recognized some time ago (5).

Concerns related to the safety of these commercial products and the future development of chelate-based targeted MRI contrast agents are being raised. Therefore, it is crucial to develop a new generation of MRI contrast agents with greater efficiency (high relaxivity) and increased safety (low toxicity). A significant amount of work has been conducted in this area and is the subject of numerous review articles (6–8). This chapter will focus on the use of metallofullerenes and their derivatives that are used as MRI contrast agents.

FULLERENES AND FULLERENE-BASED NANOMATERIALS FOR BIOMEDICAL APPLICATIONS

The discovery of a spherical crystal form of carbon, bound by single and double bonds that form a three-dimensional geodesic spheroidal crystal, named fullerenes or buckminsterfullerene, by Kroto et al. in 1985 (9), opened a new field of science—nanomaterials. Fullerenes are created spontaneously when carbon is heated under low pressure in atmospheric conditions where there is little oxygen. Interestingly, researchers have found that they exist in nature and can be produced in small quantities simply by burning candles in a room with limited air. Similar conditions were used during antiquity to collect carbon soot to produce India ink, and fullerenes have been detected in the ink of Japanese manuscripts that are thousands of years old.

Since the discovery of C_{60}, other fullerenes, including C_{70}, C_{72}, C_{74}, C_{76}, C_{78}, and C_{84}, are also produced, some in smaller quantities. Fullerene structures have a number of unusual properties that make them unique. They are carbon crystals that are insoluble in water and in many organic solvents. Fullerene π electrons do not completely saturate the available orbitals, so they have an intrinsic affinity for absorbing available electrons. Since their report and wide-scale availability, dissemination through the academic research community was rapid; since then, numerous applications have emerged. Derivatized water-soluble fullerenes are novel nanomaterials, which are showing significant potential applications in biology and life sciences (10–14). For example, C_{60} derivatives can be used to block the allergic response (15), to inhibit HIV replication (16,17), and can also be used as neuroprotective agents (18).

It is important, at this point, to define fullerene and fullerene derivatives as nanoparticles. That is, their diameter is in the nanometer range (i.e., 1–100 nm). The icosahedral cage of C_{60} is approximately 1 nm in diameter, and other higher carbon fullerenes (C_{70}, C_{72}, C_{74}, C_{76}, C_{78}, and C_{84}) are only slightly larger. Thus, despite their molecular weight, fullerenes are similar in size to most small molecule therapeutics (unless these are aggregated)—an important property from the drug design and development perspective. Because of their spherical shape and tendency to aggregate, fullerenes are denoted commonly as nanoparticles. [On the contrary, the smaller nano size of the fullerenes and fullerene derivatives is a very useful property for biomedical application, as the smaller size of nanoparticles is less likely to be taken up by the Reticular-Endothelial System (RES) compared to the larger classical nanoparticles.]

NEW FRONTIERS IN MRI CONTRAST AGENTS

In the recent years, significant efforts have been invested in developing new generation MRI contrast agents. Some recent approaches have been focused with nanoparticulate platforms including gadolinium–silica, gadolinium–liposomes, perfluorocarbons, dendrimers, solid lipid, and gadolinium oxides. An overview of these studies is provided in an excellent recent review (19).

Fullerenes have hollow interiors inside, where other atoms and ions can be entrapped. Those materials that encapsulate metal atoms are called endohedral metallofullerens. In these cases, positive charge on the metal is balanced by the negative charge on the fullerene cage. For these applications, the most important property of the endohedral metallofullerene is highly stable encapsulation of metal ions. There is no covalent or coordinate bond formation, but rather the metal atoms are physically trapped inside the cage. The absence of any reactive site on the cage often makes the cage inert under physiological conditions. The metal inside the cage is not available by endogenously available ions and is not subject to any exchange or transmetallation, making it far more stable and inert than traditional metal chelates. This property has been exploited by several groups recently in encapsulating gadolinium for biomedical applications, such as MRI-CA (20). Recent work involves using derivatized gadofullerenes, $Gd@C_{60}$, $Gd@C_{82}$, and our own work by using $Gd_3N@C_{80}$. An account of the recent work is provided in the sections below.

TRIMETASPHERES®-BASED MRI CONTRAST AGENTS

Trimetaspheres® were discovered by serendipity at Virginia Tech when a graduate student (21) was trying to produce Fullerenes with a single metal atom entrapped within the cage. The result was the discovery of an unknown peak ($m/z = 1109$) in the mass spectrum, which was eventually shown to be an 80-carbon Fullerene

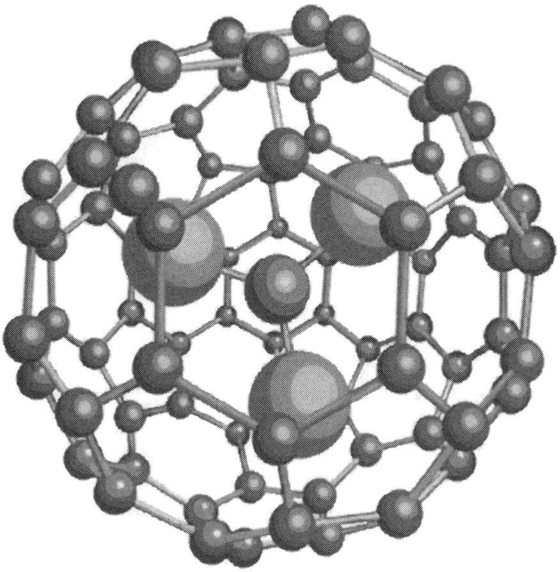

FIGURE 2 General structure of trimetasphere®.

sphere with a trimetal nitride, Sc_3N, complex entrapped within (Fig. 2). The Sc_3N moiety is encapsulated in a highly symmetric, icosahedral C_{80} cage, which is stabilized as a result of charge transfer between nitride cluster and fullerene cage. This metal nitride only exists briefly in nature, but the carbon shell prevents the unstable molecule from decomposing. In the nomenclature that has evolved, this compound is described as $Sc_3N@C_{80}$. Trimetal nitrides themselves are not stable; however, ironically their presence greatly stabilizes the Trimetasphere®. Luna Innovations Incorporated has exclusive rights to the issued US patent covering this unique class of endohedral metalofullerenes and has received a registered Trademark from the US Patent and Trademark Office on the name Trimetasphere®.

Because of their entrapped high-energy metal complex, Trimetaspheres® are truly a novel composition of matter. The entrapped metals provide unique physical properties that differentiate them from other carbon nanomaterials. For example, the carbon sphere in the Trimetasphere® is stable at temperatures hundreds of degrees above that at which the C_{60} sphere is destroyed. The nitride compound has a net charge of $+6$, and the fullerene cage has a compensating -6 charge. Thus, while the overall charge is zero, the surface of the sphere presents the electronegative moiety. This charge distribution affects the magnetic properties, chemical reactivity of the sphere, its ability to transfer electrons, and its biological behavior. For many applications, a key to using Trimetaspheres® successfully will be decoration of its surface with enabling chemical derivatives. That is, the carbon sphere is modular; it is possible to attach one or more side chains to functionalize the Trimetasphere® for specific applications.

When Trimetaspheres® were discovered, a variety of different trimetal nitride fullerenes could be produced. These include many of the lanthanides, including gadolinium as well as Sc and Y. Moreover, it is possible to manufacture combinations of various metals, such as $Sc_2GdN@C_{80}$ and $ScGd_2N@C_{80}$. Gd_3N is of particular interest for MRI because it contains 21 unpaired f-orbital electrons, which

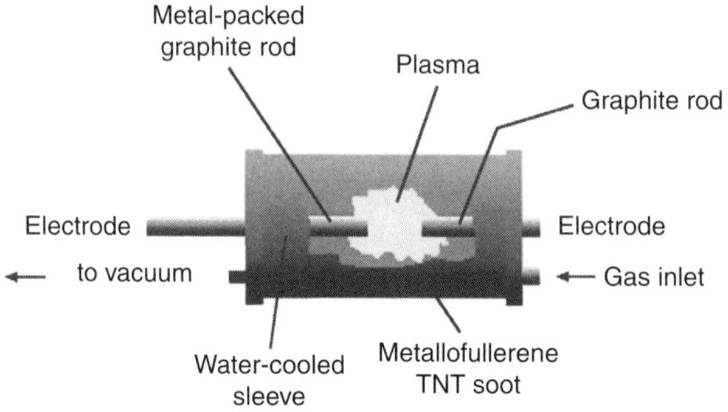

FIGURE 3 General description of the arc process for production of trimetasphere®.

makes it paramagnetic. Luna Innovations has developed a proprietary manufacturing process to produce larger quantities of the $Gd_3N@C_{80}$ by the arc process (Fig. 3). Graphite rods packed with desired metallic oxide or mixture thereof in a dynamic helium atmosphere containing a small amount of nitrogen is used in this process. Fullerene cages and endohedral metallofullerenes are extracted from the soot by various solvent systems, and high-performance liquid chromatography (HPLC) purification gives the desired product.

Fullerenes and endohedral metallofullerenes are insoluble in water and in some organic solvents. The water insolubility of these materials is a result of graphite-like surfaces having carbon–carbon single and double bonds. Consequently, surface modification is required to solubilize them in water and in other mixed solvents for biomedical applications. For example, $Gd@C_{60}$ has been water solubilized by polyderivatization with hydroxyl and carboxylic groups (21–26). Similarly, multiple hydroxyl groups (27,28), combination of hydroxyl groups and β-alanine (29,30), and organophosphates (31) were attached to the surface of the cage in $Gd@C_{82}$.

Higher stability, increased circulation, slower clearance, and high relaxivity are necessary for development of next generation contrast agents for safety of patients and for targeted MRI. It was considered that the attachment of hydroxyl, PEG, and sulfonate groups would address two fundamental issues: (*i*) solubility and (*ii*) biodistribution. Consequently, a series of $Gd_3N@C_{80}$ derivatives were prepared (32). This includes $Gd_3N@C_{80}$(di-PEG5000)$(OH)_x$, $Gd_3N@C_{80}(OH)_x$, and $Gd_3N@C_{80}(OSO_3H)_x(OH)_y$. These materials were characterized by various techniques, such as SQUID (superconducting quantum interference devices) and r_1 relaxivity measurements. The T_1 relaxivity was measured at different field strengths. The r_1 ($mM^{-1}\,s^{-1}$) values were 102 (0.35 T), 143 (2.4 T), and 32 (9.4 T). In addition to published results, Figure 4 shows a plot of $1/T_1$ versus concentration of gadolinium contrast agent at 1.5 T, and the slope of the plot provides a measure of the relaxivity. As shown in the figure, the relaxivity ($r_1 = 85\ mM^{-1}\,s^{-1}$) of $Gd_3N@C_{80}$ (di-PEG5000)$(OH)_x$ is almost 24 times higher than that of commercial MRI contrast agents, Ominiscan and Magnevist.

High relaxivities were also observed with polyhydroxylated and polycarboxylated gadofullerenes. Mikawa et al. reported the r_1 ($mM^{-1}\,s^{-1}$) relaxivity for $Gd@C_{82}(OH)_{40}$ as: 67 (0.47 T), 81 (1 T), and 31 (4.7 T). Variable r_1 values have been

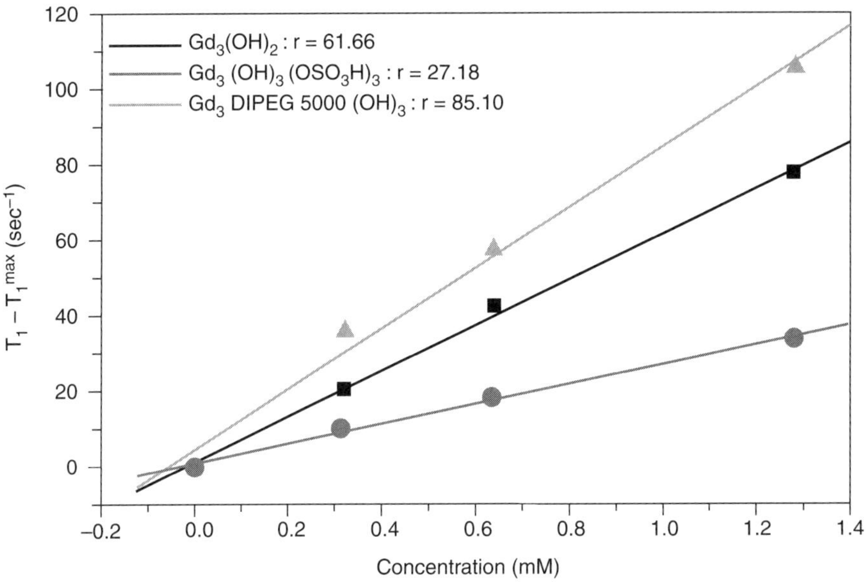

FIGURE 4 Plot of $1/T_1$ versus concentration of functionalized trimetasphere®.

reported by various investigators in the literature, which is because of the variation in the preparations (i.e., degree of hydroxylation, particle sizes, temperature, and field strength). For example, the relaxivity values, at 0.47 T, for the same material are reported as 20 and 47 by Wilson (33) and Zhang et al. (34), respectively. On the other hand, Laus et al. (24,25) prepared and reported r_1 (at 1.4 T) values as 83 and 98 $mM^{-1} s^{-1}$ for the former and 24 and 15 $mM^{-1} s^{-1}$ for the latter. More recently, the r_1 ($mM^{-1} s^{-1}$) values were reported as 9.1 (1.5 T) for mixed derivative $Gd@C_{82}O_6(OH)_{16}[NHCH_2CH_2COOH)_8]$ (30) and 37 (0.35 T), 38.9 (2.4 T), and 19.9 (9.4 T) for $Gd@C_{82}O_2(OH)_{16}[C(PO_3Et_2)_2]_{10}$ (31).

The high relaxivity of the gadofullerene derivatives is due to aggregate formation. The effect of pH on the relaxivity of derivatized gadofullerenes was studied by Toth et al. (26). The high relaxivity of polyhydroxylated and polycarboxylated gadofullerenes observed in water is substantially altered when solutes are present or pH is changed. The relaxivities, for both $Gd@C_{60}$ derivatives, increased considerably with decreasing pH until pH ~3, where precipitate was observed. For example, a factor of 2.6 for $Gd@C_{60}(OH)_x$ and 3.8 for $Gd@C_{60}[C(COOH)_2]_{10}$ were seen. Two possible explanations of the pH dependence of relaxivities were proposed. The pH might influence the proton exchange rate or the molecular rotation rate, the two main parameters that can limit proton relaxivity. To understand this, the temperature-dependent NMRD (nuclear magnetic resonance dispersion) and temperature-, concentration-, and pH-dependent dynamic and static light scattering studies were performed. On the basis of these studies, it was concluded that the decrease in relaxivity with an increase in pH is due to disaggregation of the gadofullerenes. For example, between pH 9 and 4, the average hydrodynamic diameters for aggregates were measured as follows: 70 and 700 nm for $Gd@C_{60}[C(COOH)_2]_{10}$, and 50 and 1200 nm for $Gd@C_{60}(OH)_x$. Gadofullerene derivatives with other type of functional groups and cage size also form nanocluster aggregates whose size and relaxivity varies with solution pH (29).

Decrease in relaxivity for Gd@C_{60}(OH) and Gd@C_{60}[C(COOH)$_2$]$_{10}$ in human serum albumin (HSA) was observed as opposed to predicted increase by the Solomon–Bloembergen–Morgan (SBM) theory (35,36). To understand these results, Laus et al. studied effects of sodium chloride and phosphate on the relaxivity. The increase in salt concentration decreased relaxivity ($mM^{-1}\,s^{-1}$), at 37°C and 60 MHz, significantly (83.2 in water vs. 31.6 in 150 mM NaCl for Gd@C_{60}(OH)$_x$ and 24.0 in water vs. 16 in 150 mM NaCl for Gd@C_{60}[C(COOH)$_2$]$_{10}$), suggesting disaggregation in the presence of salt. Disaggregation is more efficient in 10 mM phosphate than in 150 mM sodium chloride. The measured relaxivity values in the presence of 10 mM phosphate are 14.1 and 6.8 $mM^{-1}\,s^{-1}$ for Gd@C_{60}(OH)$_x$ and Gd@C_{60}[C(COOH)$_2$]$_{10}$, respectively. The specific phosphate effect might be related to the intercalation of $H_2PO_4^-$ and HPO_4^- ions into the hydrogen-bond network around the malonate or OH^- groups of gadofullerenes.

For metal chelates, the overall relaxivity has two components: outer sphere relaxivity and inner sphere relaxivity, which are explained by the SBM theory (35,36). In the case of gadofullerenes, there is no inner-sphere coordinated water because the metal(s) are entrapped in the cage, where bulk water cannot access for exchange. In an effort to understand the paramagnetic relaxation mechanism of gadofullerenes, Laus and coworkers (25) conducted temperature- and field-dependent ^{17}O and ^{1}H relaxation rates for aggregated and disaggregated forms of Gd@C_{60}(OH)$_x$ and Gd@C_{60}[C(COOH)$_2$]$_{10}$. For the aggregated gadofullerenes, the ^{17}O T_1 and T_2 values were different, and the authors concluded that the confinement of water molecules in the interstices of the aggregates is responsible for the high relaxivity. Rapid exchange of these water molecules with bulk contributes to the high relaxivity of aggregated gadofullerernes. The confinement of water molecules appears to be more important for the hydroxyl groups than for carboxylate groups. As mentioned above, the relaxivity reduces significantly upon disaggregation, and the temperature-dependent proton relaxivity of disaggregated gadofullerenes can be described as a sum of the outer-effect because of the translational diffusion of water molecules in the surroundings of the gadofullerenes and inner-sphere, from the proton exchange between the bulk and protonated hydroxyl and malonate groups contributions.

The Gd_3N@C_{80}(di-PEG5000)(OH)$_x$ was tested in vitro and in vivo imaging (32). For the purpose of understanding the nature of the flow during direct infusion of agents through the interstices of the brain's extracellular space, a model system is required that is free of physiologic effects but mimics the volume of distribution and pressure profiles observed during in vivo studies. For this study, a 0.6% agarose gel was used for convection-enhanced delivery (CED) in a process of slow and low pressure infusion into the brain parenchyma. The infusion of samples (0.5 μL/min for Gd_3N@C_{80}(di-PEG5000)(OH)$_x$ of 0.0261 mM and 0.2 μL/min Ominisan of 1 mM as a control) was applied for 120 minutes. The T_1-weighted imaging of vials containing samples was performed during infusion and 120 minutes postinfusion with a 2.4 T imager. The imaging results shown in Figure 5 demonstrate that the 30 times lower concentration of Gd_3N@C_{80}(di-PEG5000)(OH)$_x$ than Ominiscan gives similar image intensity.

For in vivo experiments, the contrast agent was infused directly into normal and tumor-bearing rat brain by means of concurrent bilateral infusion, and

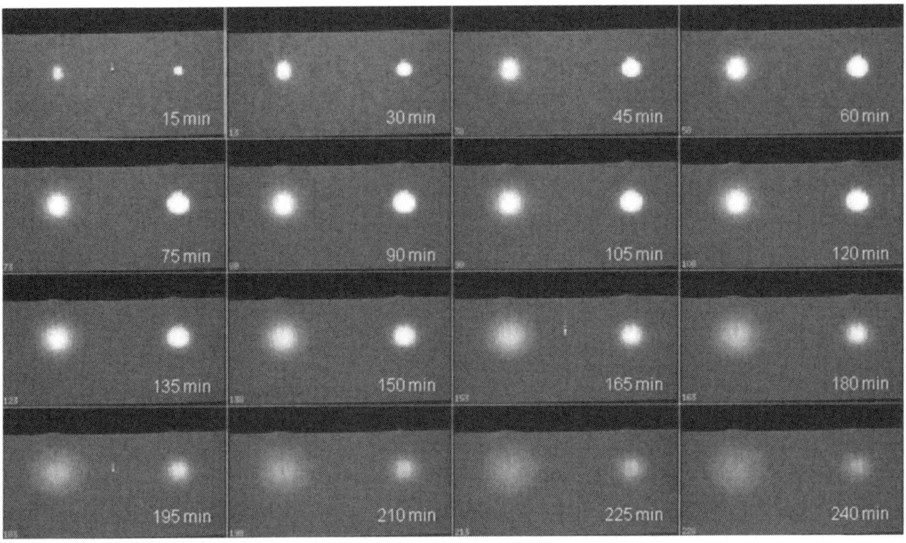

FIGURE 5 T_1-weighted MRI. T_1-weighted MRI of 0.6% agarose gel infused with 0.5 μL/min of 0.0261 $Gd_3N@C80$ (di-PEG5000)$(OH)_x$ (*right-side image*) and 0.2 μL/min of 1.0 mM ominiscan (*left-side image*). Displayed times are in minutes post-beginning of infusion. Infusion was applied for 120 minutes. Top two rows occur during the infusion process and bottom two during diffusion process.

T_1-weighted images were taken by using 2.4 T imager (32). Concentrations used were 0.013 mM $Gd_3N@C_{80}$(di-PEG5000)$(OH)_x$ and 0.5 mM Ominiscan. Visual inspection and signal intensity profiles through the infusion sites demonstrated that gadodiamide completely disappeared within three hours from the start of the infusion compared with the $Gd_3N@C_{80}$(di-PEG5000)$(OH)_x$, which diffuses very slowly. These results resemble the in vitro experiment given above. The dynamic study involving the tumor-bearing rat brain demonstrated a prolonged residence for $Gd_3N@C_{80}$(di-PEG5000)$(OH)_x$ within the tumor volume.

In vitro and in vivo MRI studies were also conducted with $Gd@C_{82}(OH)_{40}$. In vivo imaging studies were performed after iv administration of $Gd@C_{82}(OH)_{40}$ to CDF1 mice. A dose 20-times lower than that of $Gd(DTPA)^{2-}$ was used in these studies and gave strong signal enhancement in T_1-weighted MR images of lung, liver, spleen, and kidney. The biodistribution indicated that $Gd@C_{82}(OH)_{40}$ tends to be entrapped in the RES supposedly by forming large particles by aggregation or by the interaction with the plasma proteins. Another rat imaging study was conducted by Bolskar et al. (22) by using $Gd@C_{60}[C(COOH)_2]_{10}$. The study suggested that the material was as effective as an MRI contrast agent and excreted renally within one hour of injection.

HYDROCHALARONES—NEW BREAKTHROUGH IN MRI CONTRAST AGENTS

Endohedral metallofullerenes containing one Gd atom, derivatized either by polyhydroxylation or through addition of carboxyl groups to improve water solubility, exhibit high relaxivity, though the relaxivity has been attributed primarily to the formation of aggregates (as discussed above). From the ongoing discussions, it has been suggested that Fullerenes and its derivatives, including $Gd@C_{60}$ and $Gd@C_{80}$, form clusters in aqueous media. Additionally, biodistribution data of $Gd@C_{82}(OH)_{40}$ derivative suggested persistence of the material as clusters when delivered intravenously (27). The nanoclusters are known to cause thrombosis (29). The authors also suspected (27), on the basis of biodistribution studies of $Ho@C_{82}(OH)_n$, that these materials may accumulate in bone tissues and may not be well tolerated. Various strategies have been applied to eliminate large metallofullerene clusters including pH, buffer, cyclodextrin, protein binding, liposome formulations, and novel adducts.

Scientists from Luna Innovations have recently reported an alternative technology for developing a novel class of MRI contrast agents (37,38). The key improvements are (*i*) remarkably high relaxivity by adding groups onto fullerene cage of $Gd_3N@C_{80}$ (see later) that optimize magnetic coupling between the unpaired electrons on the enclosed paramagnetic gadolinium and water protons outside the cage; (*ii*) extremely high stability and kinetic inertia by trapping gadolinium inside the cage that is not available for chemical attack and unable to escape; and (*iii*) better control over the nanoparticle size; therefore, better control over clearance time and enhancement of MR images. These nanomaterials are so inert that high concentration of acids and high temperatures are required to pull gadolinium from the cage. Consequently, these materials are unlikely to release gadolinium in vitro and under in vivo conditions.

Additionally, this technology could serve as a platform upon which different moieties can be attached, without any toxicity related to free gadolinium, to direct the contrast agent to accumulate at specific targets and thereby improve diagnosis and management through imaging the anatomical distribution of those targets.

MacFarland and coworkers 33–34) synthesized a first series of novel water-soluble derivatives of $Gd_3N@C_{80}$, termed Hydrochalarones (derived from χαλαρω = relax) by addition of a series of glycol methyl ethers, ranging from monoethylene glycol to hexaethylene glycol (Fig. 6). Importantly, the polar ethylene glycols are

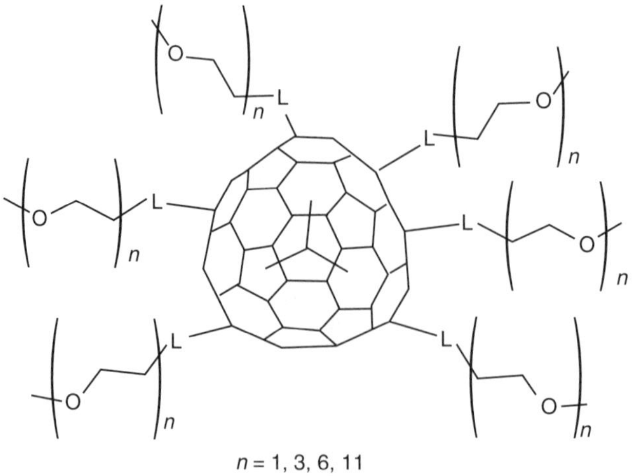

FIGURE 6 General structure of Hydrochalarones $Gd_3N@C_{80}$-R_x, where R = $-[N(OH)-(CH_2CH_2O)_nCH_3]_x$, where n = 1 to 6 and x is 10 to 22.

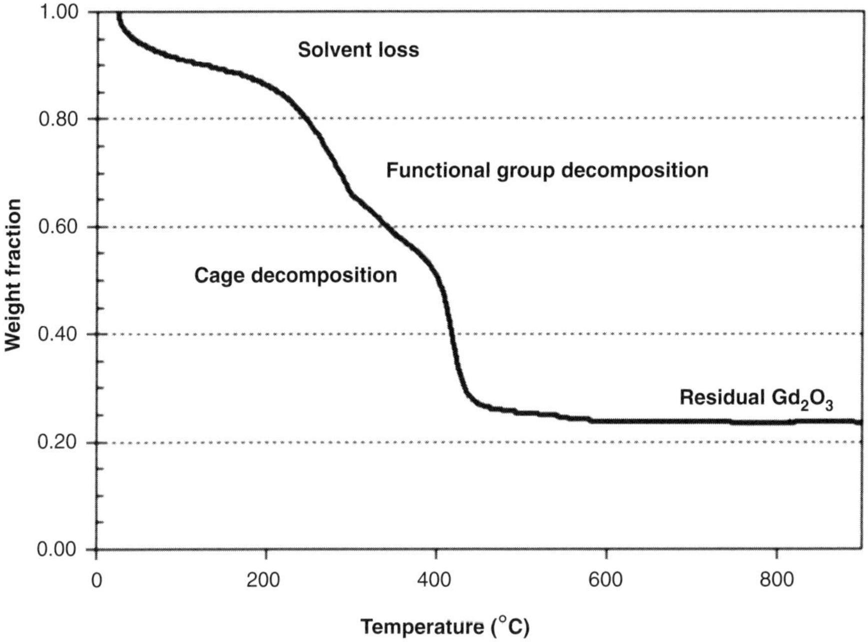

FIGURE 7 Thermogravimetric analysis of Hydrochalarones.

close to the fullerene cage, which increases its water solubility. The nomenclature, Hydrochalarone-*X*, for this new class of nanoparticles is derived from the length of oligo-ethylene glycol attached to the cage. In Hydrochalarone-*X*, *X* is the number of oxygens in the attached chain. This initial series of these nanomaterials were Cluster Hydrochalarones, typically 15 nm size--still much smaller than any reported nanoparticles so far and were designed to stay in the vasculature during imaging.

The structure elucidation and composition of these materials remain challenging, as the paramagnetic Gd atoms preclude using standard NMR and elemental analysis does not provide consistent results. However, the stability of the nanomaterials was demonstrated and the number of attached groups was inferred from thermogravimetric (TGA) analysis (Fig. 7), where oxidation of the side chains takes place at a lower temperature than that in the cage. For example, a known amount of dried Hydrochalarone sample was heated in the presence of air. The heating was ramped from room temperature to 300°C at 1°C/min, from 300°C to 900°C at 3°C/min, and then held at 900°C for 30 minutes. From room temperature to ~200°C, the weight loss is predominately due to solvent loss. The decomposition of the functional groups occurs from ~200°C to 400°C. (The inflection point at 300°C is due to the change in the temperature ramp.) The Fullerene cage oxidizes around 400°C to 450°C, resulting in residual gadolinium oxide.

It is known that a very stable species $Gd_3N@C_{80}$ is formed by combining two unstable species, Gd_3N and C_{80}. The Hydrochalarone derivatives are even more stable. For example, TGA shows $Gd_3N@C_{80}$ begins to oxidize at 180°C and disintegrates very rapidly thereafter. By contrast, the Hydrochalarone cages start to decompose at 400°C. Apparently the addition of polyethylene glycol groups stabilizes the carbon cage, presumably by relieving stress on the C=C bonds. Similarly, even heating in concentrated nitric acid at 80°C for four hours does not release the Gd atoms from the cage. Hydrochalarones are stable at room temperature for at least six months, monitored by relaxivity.

The molecular weights of Hydrochalarones are estimated from the ratio of the final residual oxide relative to the starting weight minus the solvent. The number of attached ethylene glycol groups was estimated from the difference of estimated molecular weights of the Hydrochalarones and the $Gd_3N@C_{80}$. The number of attached groups was calculated as: 20 to 22 for Hydrochalarone-1, 12 to 14 for Hydrochalarone-3, and 10 to 12 for Hydrochalarone-6.

Measurements of T_1 relaxivity, r_1, were performed on these nanomaterials in water by using a 20-MHz Maran Ultra relaxometer (Oxford Instruments, Oxfordshire, U.K.). The optimal member of this series, Hydrochalarone-6, has r_1 of 205 $mM^{-1}\,s^{-1}$ (68 $mM^{-1}\,s^{-1}$ Gd, compared with 3.8 $mM^{-1}\,s^{-1}$ for Gd DTPA, a commercial product). Chelate-based MRI contrast agents, with one or more coordinated water molecules that exchange with bulk water molecules, exhibit a relaxivity described by the SBM equation. In Hydrochalarone, the Gd atoms are entrapped in the cage and there is no possibility of water being coordinated to the paramagnetic metal centers. Thus, a different mechanism of spin-transfer between Gd and water must be at work. Previous work on $Gd@C_{60}(OH)_x$ and $Gd@C_{60}[C(COOH_yNa_{1-y})_2]_{10}$ implicated proton exchange from bulk water with hydroxyl and protonated malonate groups as integral to magnetic coupling. However, absence of –OH/–OD exchange in the FTIR experiment precluded this possibility.

The X-ray crystal structure shows that the $Gd_3N@C_{80}$ cage (39) is slightly distorted by the entrapped Gd_3N species and that the Gd atoms are closer to the carbon atoms of the cage. The unpaired f-orbital electrons of the Gd atoms may interact with electrons in the molecular orbitals on the six-carbon rings near the distortions. Water forms weak hydrogen bonds on benzene rings. Gel electrophoresis shows that Hydrochalarones bear a negative charge that could enhance hydrogen bonding. Unpaired f-orbital electrons of Gd may couple their spins through the electrons on the carbon molecular orbitals to water protons that are hydrogen bonded to the cage surface. This represents another example of "spin leakage" and supports the hypothesis that metallofullerenes behave as "superatoms." Hydrogen-bonded water molecules may act like inner-sphere water that rapidly exchange with many surrounding water molecules. Different length polyethylene glycol attachments may alter relaxivity by changing water diffusion rates through the hydration shells of the polyethylene glycol moieties.

In vivo MRI studies were performed at 4.7 T by using a small-animal scanner interfaced to a Varian INOVA console. Ten-times lower dose of Hydochalarone-6 (0.01 mmol/kg) compared to the dose of commercial agent (0.1 mmol/kg of Ominiscan) was administered intravenously to a BALB/c mouse. Figure 8 shows a coronal slice from a T_1-weighted whole-body image of a mouse before and 30 minutes post-Hydrochalarone administration. The persistent image enhancement of the aorta (*arrow*) indicates that Hydrochalarone-6 remains in the vasculature. Figure 9 shows signal intensity enhancement as a function of time for various organs, post–Hydrochalarone-6 administration, suggesting that the material persist in various organs up to 100 minutes. The new compound injected intravenously demonstrates enhanced contrast in a mouse MRI, showing the agent circulates for at least

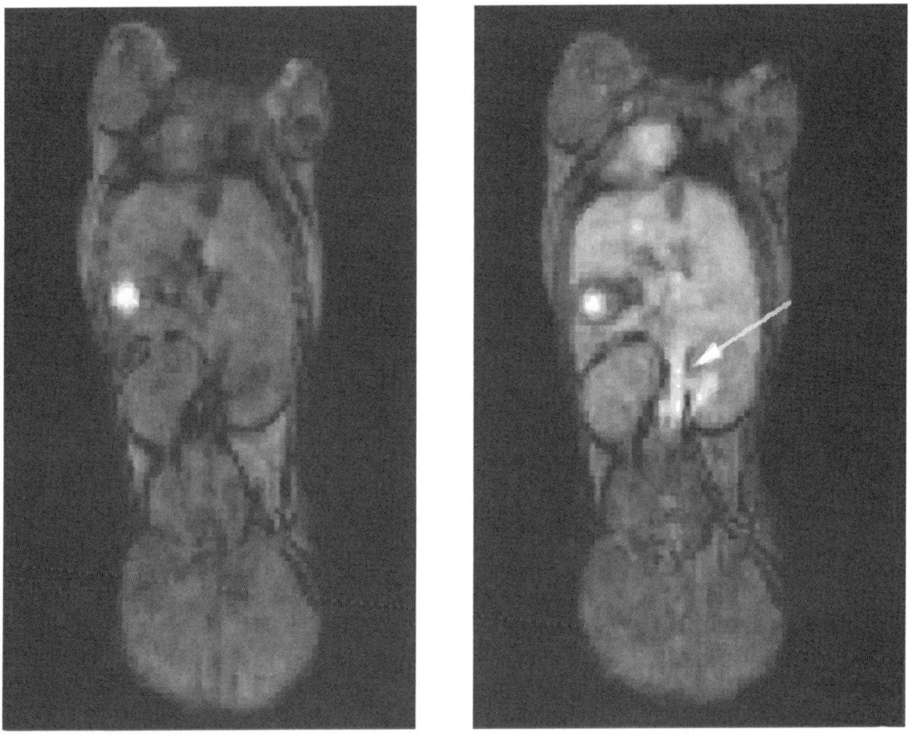

FIGURE 8 Coronal slice from a whole-body MRI of a mouse before (*left*) and 30 minutes after (*right*) administration of Hydrochalarone-6. The unidentified bright spot in both the pre- and post-contrast images is like an artifact.

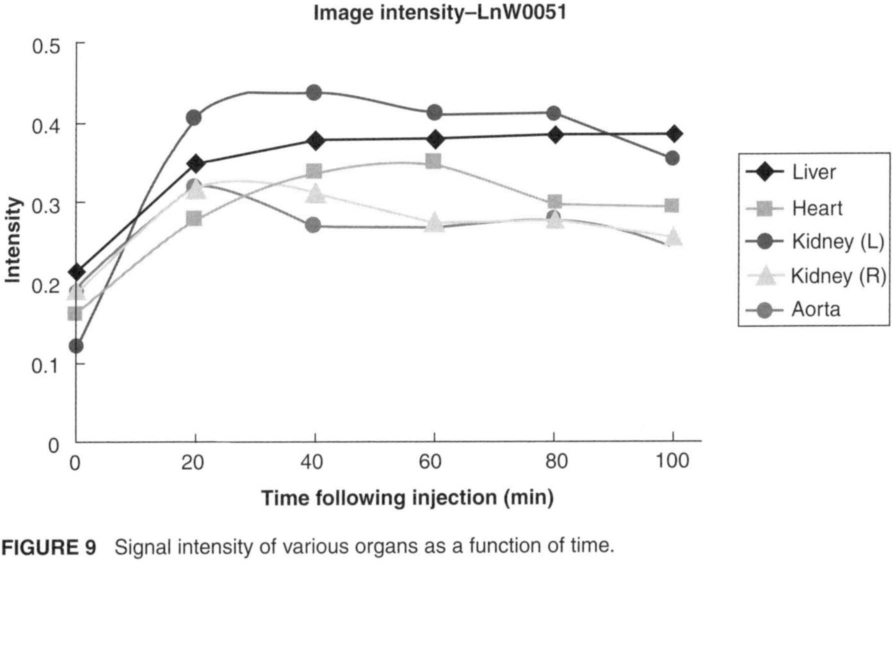

FIGURE 9 Signal intensity of various organs as a function of time.

30 minutes. This technology could serve as a platform for developing agents that can specifically target molecular entities uniquely expressed on certain cells, tissues, or organs thus providing caregivers improved ways to diagnose, treat, and monitor progression of a wide range of diseases.

More recently (40,41), Luna Innovations scientists designed smaller nanoparticles, typically 1 to 8 nm in size so that they extravasate readily. TGA was used to determine number of attachments on $Gd_3N@C_{80}$ in Hydrochalarone-1 and to demonstrate its stability. The molecular weight of Hydrochalarone-1 is estimated from the ratio of the final residual oxide relative to the starting weight minus the solvent. The molecular weight for Hydrochalarone-1 is estimated to be ~2040 ± 50, which would correspond to approximately 6 to 7 monoethylene glycol unit attachments.

The particle size of Hydrochalarones was measured by dynamic light scattering (DLS). Typically, the count rate is >25 kcps and the maximum correlation coefficient is <0.05 (occasionally <0.01), indicating a lack of particles of size >3 nm. This is consistent with results from dialysis where the majority of the Hydrochalarone-1 compound diffuses through 100 K MWCO dialysis tubing in a stagnant liquid within 20 hours.

The relaxivity measurements for these new materials were completed at 20 MHz in 30 M Tris Acetate buffer (pH 7.1). A plot of $1/T_1$ versus $[Gd_3N@C_{80}]$ is shown in Figure 10. The relaxivity ($mM^{-1}\,s^{-1}$) values were calculated as: 85 for Hydrochalarone-1, 130 for Hydrochalarone-3, and 110 for Hydrochalarone-6.

The promise of Hydrochalarones is demonstrated in small-animal MR images taken at high field strength. Figure 11 (*top*) shows whole-body three-dimensional gradient-echo MR images of Balb/C mice that were collected at 4.7 T. A precontrast image was acquired and Hydrochalarone-1 (100 μL of a 6.5 mM solution)

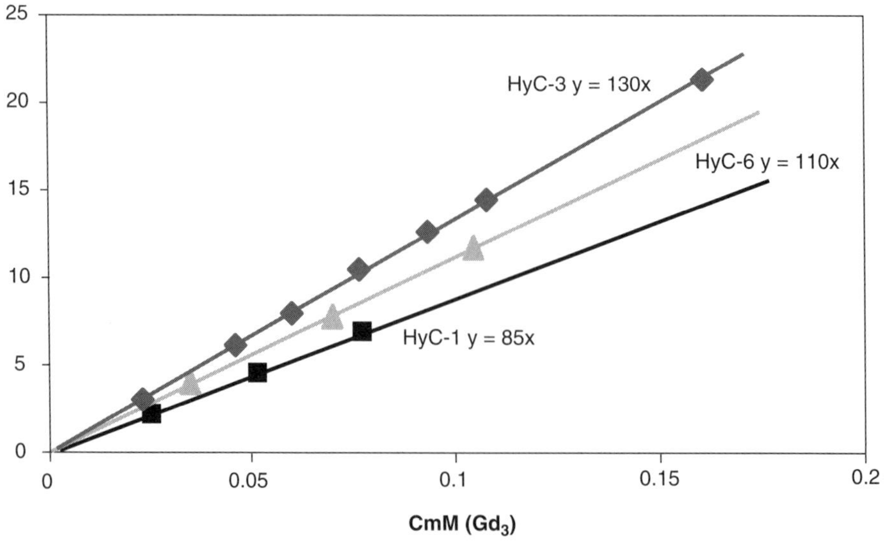

FIGURE 10 A plot of $1/T_1$ versus $[Gd_3N@C_{80}]$, mM. The slope of the plots is a measure of relaxivity (r_1) ($mM^{-1}\,s^{-1}$).

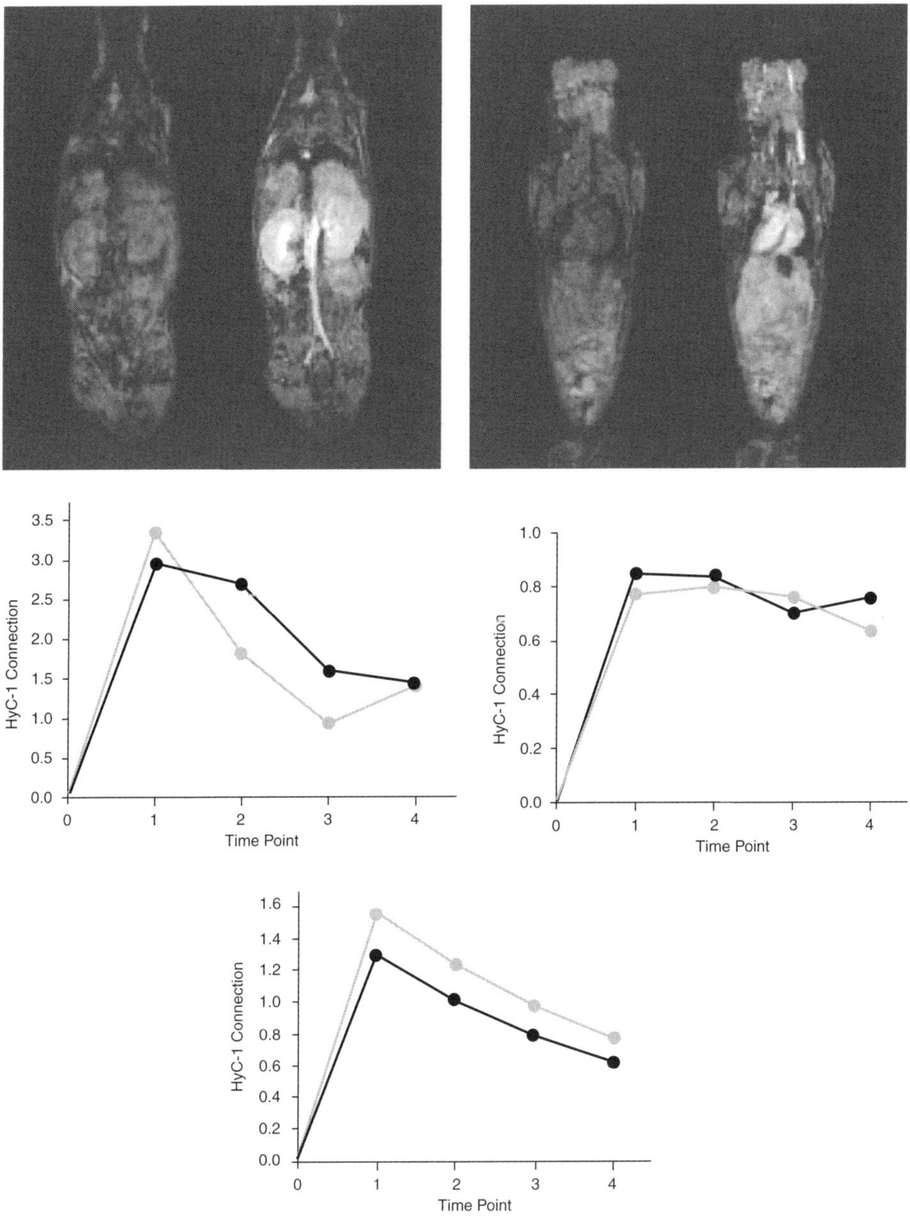

FIGURE 11 (*Top*) Whole-body 3D gradient-echo MR images of Balb/C mice collected at 4.7 T. A pre-contrast image was acquired and Hydrochalarone-1 (100 mL of a 6.5 mM solution) was injected via jugular catheter. Following injection of the agent, four additional whole body images, each requiring 20 minutes of data acquisition were acquired. Two coronal slices from images collected pre- and post-contrast are shown. (*Bottom*) Plots of Hydrochalarone-1 concentrations (arbitrary units), calculated from image intensity versus time curves for heart, liver, and kidney (reading from left to right) are shown below the images. Data collection time points on these graphs are separated from each other by approximately 20 minutes.

was injected via jugular catheter. Following injection of the agent, four additional whole-body images, each requiring 20 minutes of data acquisition, were collected. Two coronal slices from images collected pre- and post-contrast are shown. Plots of Hydrochalarone-1 concentrations (arbitrary units), calculated from image intensity versus time curves for liver, kidney, and heart are shown below the images [Fig. 11 (*bottom*)]. These experiments demonstrate that Hydrochalarones appear to have "stealth" properties in which rapid clearance mechanisms are not triggered. These are desirable properties, which provide a platform to develop targeting species for clinical diagnosis and disease management.

CONCLUSION

A summary of the recent work in the area of new developments in the MRI contrast agents is provided here. The commercial agents have demonstrated safety and efficacy in preclinical and clinical development; however, concerns over the newly identified gadolinium toxicity syndrome, NSF, remain to be a challenge in marketing these products. Data provided in the literature originally suggested good potential of gadofullerene derivatives as contrast agents. However, several issues were identified, including formation of aggregates causing concerns of thrombosis and RES uptake, and more importantly loss of relaxivity in vivo. Although, it was identified that the loss of relaxivity is slow and should not be a problem during MRI procedure. On the other hand, Hydrochalarones are stable, appear well tolerated in vivo, better control of particle size, provide excellent MRI enhancement, and have "stealth" properties, meaning that rapid clearance mechanisms seem not to be triggered. These are desirable properties for the development of Hydrochalarones as imaging agent platforms upon which can be attached a broad range of targeting species for improved clinical diagnosis and management of a number of diseases.

REFERENCES

1. Berrington de Gonzalez A, Darby S. Risk of cancer from diagnostics X-rays: Estimates for the UK and 14 other countries. Lancet 2004; 363:345–351.
2. Marckmann P, Skov L, Rossen K, et al. Nephrogenic systemic fibrosis: Suspected causative role of gadodiamide used for contrast-enhanced magnetic resonance imaging. J Am Soc Nephrol 2006; 17;2359–2362.
3. Thakral C, Alhariri J, Abraham JL. Long term retention of gadolinium in tissues from nephrogenic system fibrosis patient after multiple gadolinium-enhanced MRI scans: Case report and implications, contrast media and molecular imaging 2007; 2:199–205.
4. Public Health Advisory: Update on Magnetic Resonance Imaging (MRI) Contrast Agents Containing Gadolinium and Nephrogenic Fibrosing Dermopathy, December 22, 2006.
5. Wedeking P, Kumar K, Tweedle MF. Dissociation of gadolinium chelates in mice: Relationship to chemical characteristics. Mag Res Imag 1992; 10:641–648.
6. Lauffer RB. Paramagnetic metal complexes as water proton relaxation agents for NMR imaging: Theory and design. Chem Rev 1987; 87:901–927.
7. Kumar K. Macrocyclic polyamino carboxylate complexes of Gd(III) as magnetic resonance imaging contrast agents. J Alloys Comp 1997; 249:163–172.
8. Tweedle MF, Kumar K. Magnetic resonance imaging (MRI) contrasting agents. In: Clarke MJ, Alessio E, Sadler, PJ, eds. Metallopharmaceuticals II: Diagnosis and Therapy (Topics in Biological Inorganic Chemistry, 2), 1st ed. Emeryville, CA: Springer-Verlag Telos 1999, 1.
9. Kroto HW, Hearth JR, O'Brien SC, et al. Buckminsterfullerenes. Nature 1985; 318:162–163.

10. Jensen AW, Wilson SR, Schuster DI. Biological applications of fullerenes. Bioorg Med Chem 1996; 4:767–779.
11. Ross T, Prato M. Medicinal chemistry with fullerenes and fullerene derivatives. Chem Commun 1999; 8:663–669.
12. Sinohara H. Endohedral Metallofullerenes. Rep Prog Phys 2000; 63:843–862.
13. Nakamura E, Isobe H. Functionalized fullerenes in water. The first 10 years of their chemistry, biology, and nanoscience. Acc Chem Res 2003; 36:807–815.
14. Susanna B, Tatiana DR, Giampiero S, et al. Fullerene derivatives: An attractive tool for biological applications. Eur J Med Chem 2003; 38:913–923.
15. Ryan JJ, Bateman HR, Stover A, et al. Fullerene nanomaterials inhibit the allergic response. J Immunol 2007; 179:665–672.
16. Sijbesma R, Srdanov G, Wudl F, et al. Synthesis of a fullerene derivative for the inhibition of HIV enzymes. J Am Chem Soc 1993; 115:6510–6512.
17. Friedman SH, Decamp DL, Sijbesma RP. Inhibition of the HIV-1 protease by fullerene derivatives: Model building studies and experimental verification. J Am Chem Soc 1993; 115:6506–6509.
18. Dugan LL, Turetsky DM, Du C, et al. Carboxyfullerenes as neuroprotective agents. Proc Natl Acad Sci U S A 1997; 94:9434–9439.
19. Sharma P, Brown SC, Walter G, et al. Gd nanoparticulates: From magnetic resonance imaging to neutron capture therapy. Adv Powder Technol 2007; 18:663–698.
20. Bolskar RD. Gadofullerene MRI contrast agents. Nanomedicine 2008; 3:201–213.
21. Stevenson S, Rice G, Glass T, et al. Small-bandgap endohedral metallofullerenes in high yield and purity. Nature 1999; 401:55–57.
22. Bolskar RD, Benedetto AF, Husebo LO, et al. First soluble $M@C_{60}$ derivatives provide enhanced access to metallofullerenes and permit in vivo evaluation of $Gd@C_{60}[C(COOH)_2]_{10}$ as a MRI contrast agent. J Am Chem Soc 2003; 125:5471–5478.
23. Sitharaman B, Bolskar RD, Rusakova I, et al. $Gd@C_{60}[C(COOH)_2]_{10}$ and $Gd@C_{60}(OH)_x$: Nanoscale aggregation studies of two metallofullerene MRI contrast agents in aqueous solution. Nano Lett 2004; 4:2373–2378.
24. Laus S, Sitharaman B, Toth E, et al. Destroying gadofullerene aggregates by salt addition in aqueous solution of $Gd@C_{60}(OH)_x$ and $Gd@C_{60}[C(COOH)_2]_{10}$. J Am Chem Soc 2005; 127:9368–9369.
25. Laus S, Sitharaman B, Toth E, et al. Understanding paramagnetic relaxation phenomena for water-soluble gadofullerenes. J Phys Chem 2007; 111:5633–5639.
26. Toth E, Bolskar RD, Boret A, et al. Water-soluble gadofullerenes: Towards high relaxivity, pH-responsive MRI contrast agents. J Am Chem Soc 2005; 127:799–805.
27. Mikawa H, Kato H, Okumura M, et al. Paramagnetic water-soluble metallofullerenes having the highest relaxivity for MRI contrast agents. Bioconjug Chem 2001; 12:510–514.
28. Kato H, Kanazawa Y, Okumura M, et al. Lanthanoid endohedral metallofullerenols as MRI contrast agents. J Am Chem Soc 2003; 125:4391–4397.
29. Shu C-Y, Zhang E-Y, Xiang J-F, et al. Aggregation studies of water-soluble gadofullerene magnetic resonance imaging contrast agents: $[Gd@C_{82}O_6(OH)_{16}(NHCH_2CH_2COOH)_8]_x$. J Phys Chem 2006; 110:15597–15601.
30. Shu C-Y, Gan L-H, Wan C-R, et al. Synthesis and characterization of new water-soluble endohedral metallofullerene for MRI contrast agents. Carbon 2006; 44:496–500.
31. Shu C-Y, Wang C-R, Zhang J-F. Organophosphonate functionalized Gd@C82 as a magnetic resonance imaging. Contrast Agent Chem Mater 2008.
32. Fatouros PP, Corwin FD, Chen Z-J, et al. In vivo imaging studies of a new endohedral metallofullerene nanoparticles. Radiol Radiol 2006; 240:756–764.
33. Wilson LJ. Medical applications of fullerenes and metallofullerenes. In: The Electrochemical Society Interface, Winter, 1999. Pennington, NJ: The Electrochemical society.
34. Zhang S, Sun D, Li X, et al. Synthesis and solvent enhanced relaxation property of water-soluble endohedral metallofullerenes. Fullere Sci Tech 1997; 5:1635–1643.
35. Solomon I. Relaxation processes in a system of two spins. Phys Rev 1955; 99:559.
36. Bloembergen N, Morgan LO. Proton relaxation times in paramagnetic solutions: Effects of electron relaxation. J Chem Phys 1961; 34:842.

37. MacFarland DK, Walker KL, Lenk RP, et al. Hydrochalarones: A novel endohedral metallofullerene platform for enhancing magnetic resonance imaging contrast. J Med Chem 2008; 51:361–368.
38. MacFarland DK. Trimetasphere metallofullerene MRI contrast agents with high molecular relaxivity. In: D'Souza F, ed. ECS Transactions—Carbon Nanotubes and Nanostructures: Medicine and Biology, Vol. 13. Phoenix (In Press).
39. Stevenson S, Phillips JP, Reid JE, et al. Pyramidalization of Gd_3N inside a C80 cage. The synthesis and structure of $Gd_3N@C_{80}$. Chem Commun 2004; 2814–2815.
40. Garbow JR, Ackerman JJR, MaFarland D., et al. Abstracts of Papers, The World Molecular Imaging Congress, Nice, France, 2008.
41. Garbow JR, Ackerman JJR. A New Platform for the Development of Targeted, High Performance, MRI Contrast Agents: Hydrochalarones (Endohedral Metallofullerenes). US Radiology 2008 (In Press).

19 Semiconducting Quantum Dots for Bioimaging

Debasis Bera, Lei Qian, and Paul H. Holloway

Department of Materials Science and Engineering, University of Florida, Gainesville, Florida, U.S.A.

INTRODUCTION

There are several noninvasive imaging techniques available for molecular imaging purposes, such as fluorescence imaging, computed tomography (CT), magnetic resonance imaging (MRI), positron emission tomography (PET), single photon emission computed tomography (SPECT), ultrasonography, and many more (1). Across the electromagnetic spectrum, these techniques span from ultrasound to X-rays to gamma rays. Currently, MRI, optical imaging, and nuclear imaging are emerging as the key molecular imaging techniques (1). They differ in terms of sensitivity, resolution, complexity, acquisition time, and operational cost. However, these techniques are complementary to each other most of the time. There are several reviews on the physical basis of these techniques (1,2), instrumentation (3,4), and issues that affect their performance (5,6). Currently, a significant amount of research is aimed at using the unique optical properties of quantum dots (Qdots) in biological imaging. Much of optical bioimaging is based on traditional dyes (7,8), but there are several drawbacks associated with their use. It is well known that cell autofluorescence in the visible spectrum (9) leads to the following five effects: (*i*) The autofluorescence can mask signals from labeled organic dye molecules. (*ii*) Instability of organic dye under photoirradiation is well known in bioimaging, which results in only short observation times. (*iii*) In general, conventional dye molecules have a narrow excitation window, which makes simultaneous excitation of multiple dyes difficult. (*iv*) Dyes are sensitive to the environmental conditions, such as variation in pH. (*v*) Most of the organic dyes have a broad emission spectrum with a long tail at red wavelengths, which creates spectral crosstalk between different detection channels and makes it difficult to quantitate the amounts of different probes. Qdots, on the other hand, are of interest in biology for several reasons, including (*i*) higher extinction coefficients, (*ii*) higher quantum yields (QYs), (*iii*) less photobleaching, (*iv*) absorbance and emissions can be tuned with size, (*v*) generally broad excitation windows but narrow emission peaks, (*vi*) multiple Qdots can be used in the same assay with minimal interference with each other, (*vii*) toxicity may be less than conventional organic dyes, and (*viii*) the Qdots may be functionalized with different bioactive agents. In addition, near infrared (NIR) emitting Qdots can be used to avoid interference from the autofluorescence, because cell, hemoglobin, and water have lower absorption coefficient and scattering effects in the NIR region (650–900) (Fig. 1). Light is routinely used for intravital microscopy, but imaging of deeper tissue (500 μm–1 cm) requires the use of NIR light (10). Inorganic Qdots are more photostable under ultraviolet excitation than organic molecules, and their fluorescence is more saturated. In general, as-synthesized Qdots are very hydrophobic. Qdots have been synthesized by different bottom-up chemical methods, such as

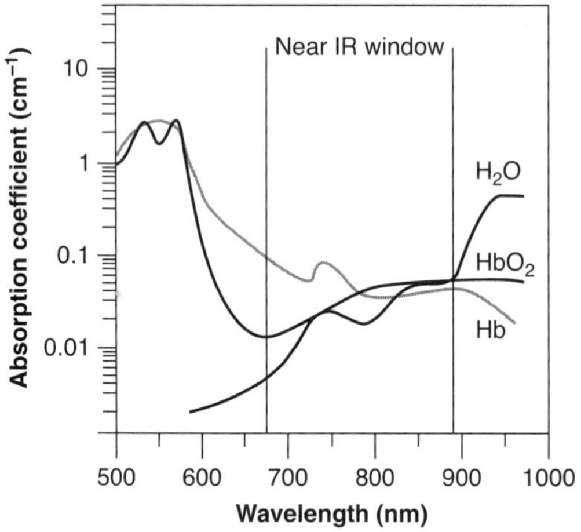

FIGURE 1 Absorption versus wavelength for hemoglobin and water, showing the NIR window for in vivo imaging to minimize absorption and scattering. *Abbreviations:* HbO_2, Oxygen attached hemoglobin; Hb, hemoglobin; H_2O, water. *Source*: From Ref. 10.

sol–gel (11,12), microemulsion (13,14), competitative reaction chemistry (15,16), hot solution decomposition method (17,18), microwave irradiation process (19,20), and hydrothermal synthesis procedure (21,22). For the production of highly crystalline, monodispersed Qdots, the hot solution decomposition method is the best method known to date. To convert Qdots from hydrophobic to hydrophilic, a silica shell is generally grown on the Qdots. Growth of silica shell can be achieved by microemulsion and/or sol–gel methods. Several review articles and book chapters (23–27) can be found with elaborate discussions on Qdots. Hence, the properties of Qdots are briefly overviewed in the following section.

SEMICONDUCTING QUANTUM DOTS

In semiconductors, the electronic ground state is commonly referred to as the valence band that is completely filled with electrons. The excited quantum states often lie in the conduction band, which is empty, or in the energy gap between the valence and conduction bands called the band gap. Therefore, unlike metallic materials, small continuous changes in electron energy within the semiconductor valence band are not possible. Instead a minimum energy is necessary to excite an electron in a semiconductor, and the energy released by de-excitation is often nearly equal to the band gap (28). When a semiconductor absorbs a photon, an electron may be excited to a higher energy quantum state. If the excited electron returns (relaxes) to a lower energy quantum state by radiating a photon, the process is called photoluminescence (PL) (29). Sometimes, one or more species are intentionally incorporated to the semiconductor. These impurities are called activators and they perturbed the band structure by creating local quantum states that may lie within the band gap (30). The predominant radiative mechanism in extrinsic luminescence is electron–hole recombination, which can occur via transitions between conduction band to acceptor state, donor state to valance band, or donor state to acceptor state.

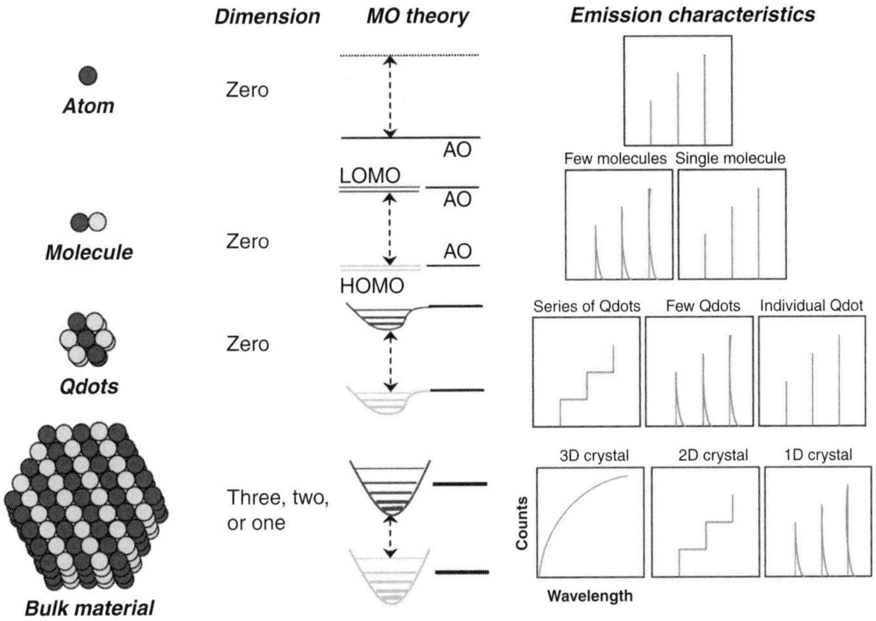

FIGURE 2 Schematic illustration of the changes of the density of quantum states with changes in the number of atoms in materials. *Abbreviations*: AO, atomic orbital; MO, molecular orbital; 3D, 2D, and 1D, three-, two-, and one-dimensional crystals. *Source*: From Ref. 23.

Nanostructured semiconductors Qdots have dimensions and numbers of atoms between the atomic-molecular level and bulk materials. Qdots have a band gap that depends on a complicated fashion upon a number of factors, including size of particle, bond type, and bond strength (23). Generally, a Qdot is composed of approximately 100 to 10,000 atoms (1–30 nm), and has optical properties distinct from its bulk counterpart (Fig. 2). These are often described as artificial atoms due to their δ-function–like density of states, which can lead to narrow optical line spectra with a very small Stoke's shift. This leads to the electronic states with wave functions that are more atomic-like. As the solutions for Schrödinger wave equation for Qdots are very similar to those for electrons bound to a nucleus, Qdots are called artificial atom. The typical intraband energy spacing for Qdots are in the range of 10 to 100 meV.

The most fascinating properties of Qdots are the drastic dependence in the optical absorption, exciton energies, and electron–hole pair recombination upon the size of Qdots. The dependence arises mainly from quantum confinement effect, a unique property of the Qdots (23). The quantum confinement effect modifies both energy and the density of states (DOS) near the band edges. Schematic diagrams of the DOS as a function of energy in Figure 2 show that the quantum states for Qdots lie between the discrete atomic levels and continuous bands for bulk materials. A blue shift (increase) of the band gap energy is observed when the Qdot diameter is reduced. This effect allows tuning of the energy gap by changing the size of the Qdots, while maintaining a narrow emission (full width half maximum of ~10–20 nm) (17). Figure 3 shows the emitted colors from Qdots with size. The band

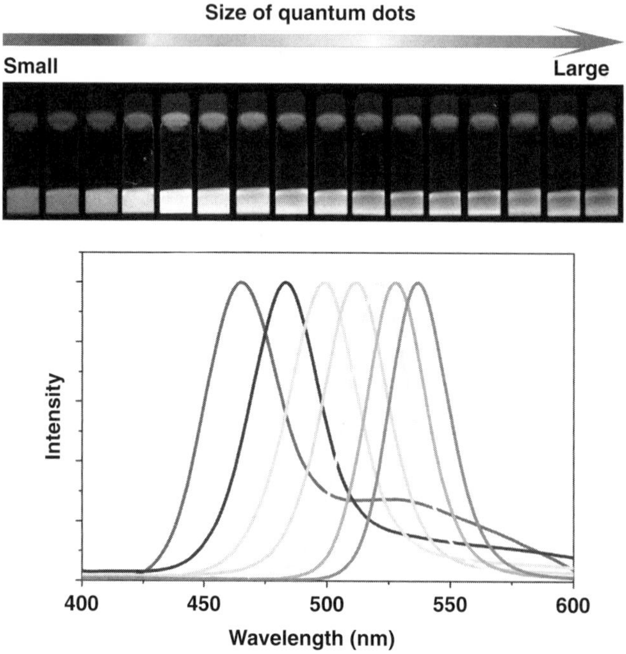

FIGURE 3 Colors and Qdots. (*Top*): Sixteen distinguishable emission colors from blue to red of CdSe Qdots excited by a near-UV lamp. (*Bottom*): Photoluminescence spectra of some of the different sized CdSe Qdots.

gap energy also depends on the composition of the semiconductor. To achieve emission of a particular color from a Qdot requires sufficient control during its synthesis because intrinsic properties are determined by factors such as size, shape, defect, impurities, and crystallinity.

Passivation

Surface defects, which are omnipresent in Qdots, act as "traps" for electrons or holes and lead to quenching of radiative recombination, thereby reducing the QYs. To achieve a stable emission with large QY, capping or passivation of the surface is crucial (31). The capping or passivation converts hydrophobic Qdots to hydrophilic. Therefore, surface modification of Qdots is very important for their biological application. Capping is generally carried out by incorporating an organic or inorganic layer onto the Qdots. Only a partial passivation of the Qdot surface can be achieved by incorporating select polymers onto the Qdots. Some advantages of organic capping layers include simultaneous achievement of colloidal suspension capping and the ability to bio-conjugate the Qdots. Figure 4(A) illustrates a Qdot passivated with organic molecules. Chemically reduced bovine serum albumin (BSA) has been used simultaneously to passivate and functionalize the surface of CdTe Qdots to make them water soluble (32). Denatured BSA (dBSA) conjugated to the CdTe Qdots surface improved the chemical stability and the PL quantum yield (32). This study showed that over a pH range of 6 to 9, the solution of dBSA-coated CdTe Qdots were stable and bright, but higher and lower pH values led to dramatic decreases

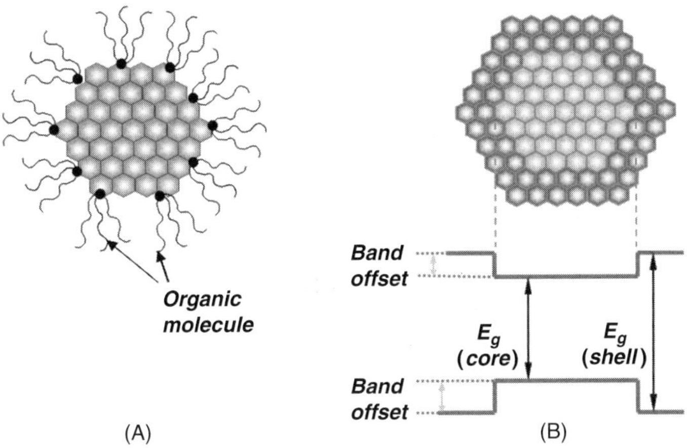

FIGURE 4 Schematic illustration of (**A**) an organically capped Qdot and (**B**) an inorganically passivated Qdot (core/shell structure of Qdot). An energy diagram shows the band-gap difference (E_g) of core and shell of inorganically passivated Qdots. *Source*: From Ref. 31.

in PL intensity and chemical stability. Similarly, concentrations of dBSA that were too high or too low in the Qdots solution resulted in a decreased PL quantum yield. Recently, DNA-passivated CdS Qdots were reported to be stable in and nontoxic to biological systems (33).

For inorganic passivation, a material with a larger band gap is grown either epitaxially (crystalline) or as a nonepitaxial crystalline or amorphous layer on the core [Fig. 4(B)] (34). Generally, a significant QY improvement was observed by introducing an inorganic shell layer. Recently, multiple layers on the core have been used to circumvent several issues such as (*i*) to make hydrophilic Qdots (35), (*ii*) to functionalize or derivatize with biological agents (36), (*iii*) to introduce multimodal imaging capabilities for biological applications (37), and (*iv*) to enhance the fluorescence QY by minimizing Auger recombination (13).

Silica-Coated Qdots

As mentioned earlier, Qdots are often synthesized in nonpolar, nonaqueous solvents, leaving them hydrophobic. In addition, except for some oxide-based Qdots, which are assumed to be lesser toxic, most of the Qdots contain toxic ions [e.g., cadmium (Cd), Lead (Pb), arsenic (As), etc.]. Furthermore, functionalization of Qdots is very important for biological application. To address these issues, a silica shell may be grown on Qdots [Fig. 5(A)]. Aqueous-based synthesis methods generally are used to produce silica-capped Qdots (35). Recently, it is shown that a silica shell can prevent the leakage of toxic Cd^{2+} from IR-emitting CdTe Qdots. Cytotoxicity and the potential interference of Qdots with cellular processes are the subject of intensive studies (38,39). The silica shell also allows easy functionalization with biomolecules such as proteins (37,40) and results in greater photostability. The luminescent properties of silica-coated Qdots depend on the charge trapped on the surface as well as on the local electric field. The field-dependent emission from Qdots is called quantum-confined Stark effect (41). External electric field or internal local field results in shifts of both emission wavelength and intensity. By neutralizing a

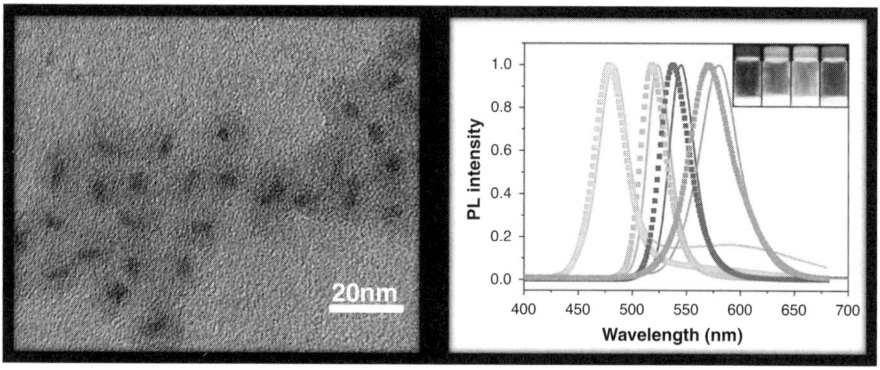

(A) (B)

FIGURE 5 (**A**) High-resolution transmission electron microscopy shows silica-coated CdSe Qdots. (**B**) Photoluminescence spectra from CdSe (*solid line*) and CdSe/SiO_2 core/shell (*dotted line*) Qdots with different sizes. *Inset*: photograph of core/shell Qdots excited by a hand-held UV lamp.

surface positive charge, we recently found that the emission from CdSe Qdots was blue shifted, and the quantum yield increased dramatically [Fig. 5(B)]. Although there are no attempts found in the literature, the electric field induced change of emission from Qdots can be potentially useful for biological imaging and sensing.

BIOIMAGING

Nanomaterials-based bioimaging is an emerging technology in medical science, which has the potential to revolutionalize the diagnosis and treatment of disease in near future (6,42–45). Among nanostructured materials, Qdot-based size-tuned emission color offers the potential to develop a multicolor optical coding technique (e.g., by functionalizing different sized CdSe Qdots with different molecules). Researchers have used Qdots for in vivo and in vitro imaging and diagnostic of live cell as a complement to or replacement of conventional organic dyes (46–48). Some of the early-stage studies are tabulated in Table 1. As mentioned above, Qdots

TABLE 1 Some In Vitro and In Vivo Bioimaging Studies Using Qdots

Qdots	Purpose	Imaging techniques	Emission/ size of Qdots	References
CdSe/CdS/SiO_2	Mouse fibroblast cell imaging	In vitro fluorescence	550 nm and 630 nm	(49)
CdSe/ZnS	Biological detection/sensing	In vitro fluorescence	1–4 nm	(50)
CdSe/ZnS/SiO_2	Phagokenetic track imaging	In vitro fluorescence	554 nm and 626 nm	(51)
CdSe/ZnS	Tumor vasculature and lung endothelium imaging	In vitro and in vivo fluorescence	$<$10 nm	(52)
CdTe/CdSe	Cancer cell lymph nodes imaging	In vivo fluorescence	NIR	(53)
CdSe/ZnS	Maltose-binding protein	In vitro FRET	560 nm	(54)

are promising candidate for bioimaging applications due to their tunable color and narrow emission spectra for multiple imaging, broad absorption bands, and photostability. Exploiting some properties of Qdots, such as (*i*) sharp and UV–NIR tunable fluorescence, (*ii*) charge transfer through fluorescence resonance energy transfer (FRET), (*iii*) surface-enhanced Raman spectroscopy (SERS), (*iv*) radio-opacity and paramagnetic properties, and (*v*) magnetic resonance imaging (MRI) contrast agent, it has been shown that Qdots can be used for bioimaging purpose. In the following section, we will briefly discuss these imaging techniques.

Fluorescence for Bioimaging

Qdots fluorescence-based bioimaging (55–57) can be broadly classified into four types of modes: intensity, spectrum, lifetime, and time-gated. All of these modes can be used at the same time for multimodality imaging. Generally, a high QY from Qdots is required for intensity-based imaging. On the other hand, narrow emitting spectra make Qdots suitable for multiple colors imaging. The longer fluorescence lifetime of Qdots compared with that of tissue avoid the noise from autofluorescence of tissues. Therefore, there is an advantage to use both lifetime and time-gated modes simultaneously (Fig. 6). Photoluminescence from Qdots has been a widely

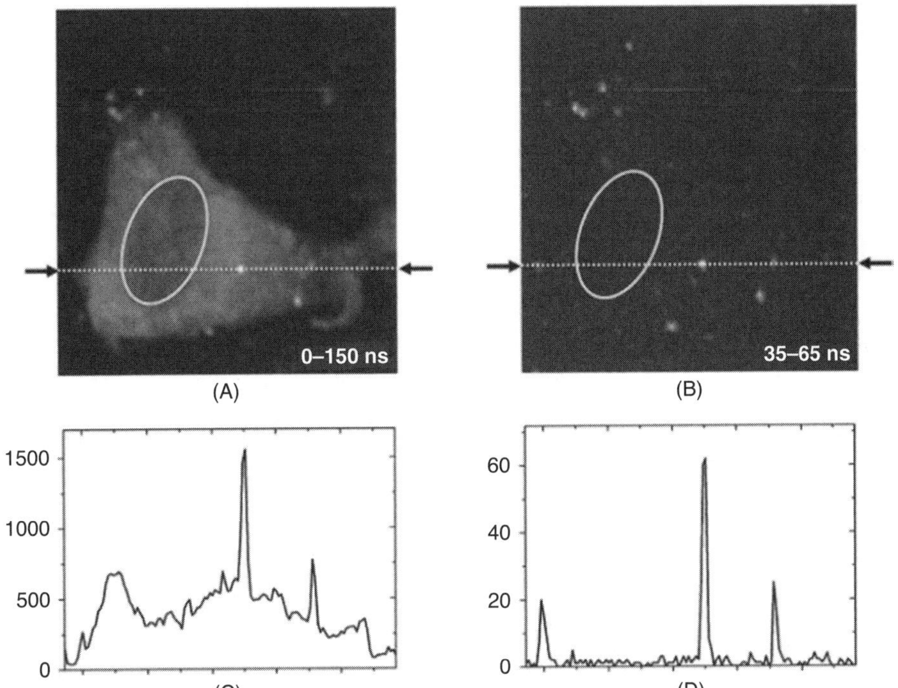

FIGURE 6 Time-resolved confocal images of a fixed 3T3 cell: (**A**) A micrograph acquired from all the detected photons and (**B**) Time-gated micrograph constructed from only photons that arrived 35 to 65 ns after the laser pulse (laser intensity: 0.1 kW/cm^{-2}; integration time per pixel: 25 ms; scale bar: 5 μm). Inset figures show cross sections along the same horizontal line (indicated by the *black arrows*) for (**A**) and (**B**). *Source*: From Ref. 59.

used tool in biology. In 1998, Bruchez and his group showed that Qdots were potential candidates for biological applications (49). To establish the use of Qdots, biotin was covalently bound to the Qdot surface and used to label fibroblasts, which was incubated in phalloidin-biotin and streptavidin. For biological and medical applications, it is of importance to study the photophysical properties of Qdots in living cells (58), particularly photo-induced optical properties of the intracellular Qdots. After injecting thiol-capped CdTe Qdots into living cells, the PL intensity increased with time and the emission peak blue-shifted (58). De-oxygenation prevented the PL blue shift, suggesting that photoactivated oxygen was responsible. The activated oxygen is presumably formed from the oxygen that intercalates the thiol layer at the Qdot core surface. When Qdots are used as fluorescence probes for cellular imaging, the effects of the PL blue shift and photobleaching must be considered.

Spectral encoding Qdot technology (60,61) is expected to open new opportunities in gene expression studies, high-throughput screening, and medical diagnostics. The broad absorption spectra of the Qdots allow single wavelength excitation of emission from different-sized Qdots. Multicolor optical coding for biological assays has been achieved by using different sizes of CdSe Qdots with precisely controlled ratios. The use of 10 intensity levels and 6 colors could theoretically encode one million nucleic acid or protein sequences. Han et al. embedded different-sized Qdots into highly uniform and reproducible polymeric microbeads, which yielded bead identification accuracies as high as 99.99% (62).

The luminescent lifetime of CdSe Qdots (several tens of nanoseconds) is longer than that of cell autofluorescence (~1 ns), which permits measurement of marker spectra and location without high backgrounds through the use of time-gated fluorescent spectroscopy and/or microscopy. In addition, the photostability of CdSe is much better than that of conventional organic dyes (63), allowing data acquisition over long times with continuous excitation. Compared with conventional organic dyes, Qdots have longer lifetime allowing acquisition of low background PL images by using time-gated fluorescent microscopy, as shown in Figure 6 (59). Figure 7 shows (64) micrographs from CdSe Qdots–based deep tissue imaging of the vasculature system highlighting various structures. The mice used in this study showed no ill effects from the Cd-containing labels.

In order to enhance the lifetime of the emission, some transition or rare earth elements are intentionally incorporated into the Qdots. These activators create local quantum states that lie within the band gap and provide states for excited electrons or traps for charge carriers and result in radiative relaxations towards the ground state. For transition metal ions such as Mn^{2+}, the lifetime of the luminescence (34,65,66) is of the order of milliseconds due to the forbidden d–d transition. Santra et al. (36) and Holloway et al. (67,68) demonstrated in vivo bioimaging capability by using amine-modified Mn-doped CdS/ZnS core/shell Qdots conjugated to a TAT peptide. After administering the Qdots through the right common carotid artery that supplies blood only to the right side of a rat's brain, Mn-doped Qdots loaded brain was sliced for histological analysis. Transmission optical and fluorescence micrographs [Figs. 8(A) and 8(B)] of a cross section of fixed brain tissue clearly showed the blood capillaries [broken white circle in Fig. 8(A)] and surrounding brain cells. It was also shown that the TAT-conjugated Qdots reached the nucleus of the brain cells [green-circled brown spots in Fig. 8(A)]. It is well known that the TAT peptide can rapidly translocate through the plasma membrane and accumulate in the cell nucleus (36). The histological analysis of the brain tissue

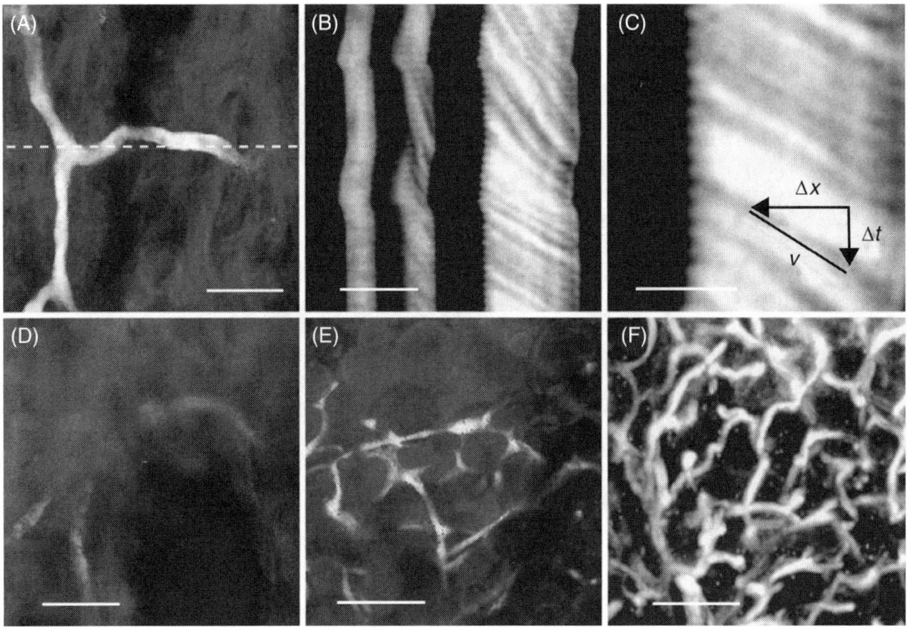

FIGURE 7 In vivo fluorescence imaging of vasculature labeled by a tail vein injection of CdSe Qdots (4.3 nm). (**A**) Fluorescent capillaries containing 1 μM Qdot. Dashed line indicates position of line scan shown in part (**B**). (**C**) Zoom of section in part (**B**), showing undulations in capillary due to heartbeat. (**D**) Comparison image at the same depth as in part (**A**), acquired by injecting FITC-dextran at its solubility limit. (**E**) Image of the surface of adipose tissue surrounding the ovary (blue, autofluorescence; yellow, 20 nM Qdots in blood). (**F**) Projection of capillary structure through 250 μm of adipose tissue. Excitation was at 880 nm for parts (**A**) to (**C**), 780 nm for parts (**D**) to (**F**). Scale bars: 20 μm [(**A**), (**B**), and (**D**)], 10 μm (C), 50 μm [(**E**) and (**F**)]. *Source*: From Ref. 64.

supports the fact that TAT-conjugated Qdots crossed the blood–brain barrier, migrated to brain parenchyma, and reached the cell nuclei. Endothelial cells in the blood capillaries were found heavily loaded with CdS:Mn/ZnS Qdots and appeared as bright yellow lines in Figure 8(B).

The use of NIR or IR photons is promising for biomedical imaging in living tissue owing to longer attenuation distances and lack of autofluorescence in the IR region. This technology often requires exogenous contrast agents with combinations of hydrodynamic diameter, absorption, quantum yield, and stability that are not possible with conventional organic dyes (53). Qdots-based contrast agent offers these properties. In addition, the emission can be tuned to the NIR window (Fig. 1) either by controlling the size of the Qdots or by incorporating rare-earth activators. The emission of CdTe/CdSe Qdots can be tuned into the NIR, while preserving the absorption cross section. It was shown that a polydentate phosphine coating onto the Qdots made the Qdots water soluble, allowing them to be dispersed in serum. Injection of only 400 pmol of NIR emitting Qdots permitted real-time imaging of sentinel lymph nodes that were 1 cm below the surface using an excitation power density of only 5 mW/cm^2 (53).

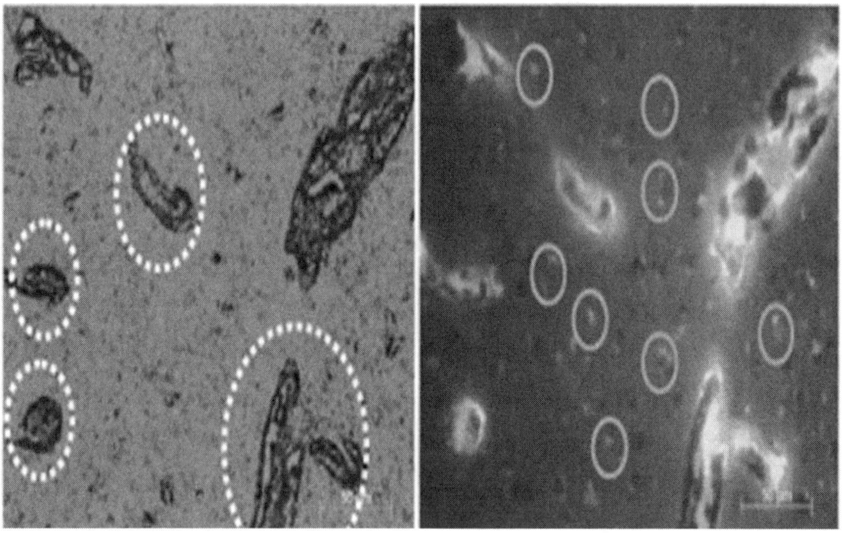

FIGURE 8 (**A**) Transmission optical and (**B**) fluorescence microscopy images (40×) of cross-section of fixed brain tissue showing luminescence in part (**B**) from CdS:Mn/ZnS core/shell Qdots. *Source*: From Ref. 36.

Use of Fluorescence Resonance Energy Transfer (FRET) in Bioimaging

Fluorescence resonance energy transfer (FRET) is a phenomenon in which photoexcitation energy is transferred from a donor fluorophor to an acceptor molecule. Based on Förster theory, the rate of this energy transfer depends on the spectral overlap of donor emission and acceptor absorption and the donor–acceptor spatial arrangement (69). The ability of Qdots to participate in FRET provides a mechanism for signal transduction in optical sensing schemes. CdSe Qdots can be used to build on–off switches by utilizing Förster resonance energy transfer between the Qdot donor and an organic acceptor. In such this optical sensing scheme, Qdots could act both as donor and as acceptor. Tran et al. and Clapp et al. (70,71) have studied such an energy transfer between donor Qdots and acceptor dye molecules. In the blend of water-soluble CdSe/ZnS Qdots and maltose-binding protein (MBP), MBP was assembled onto the surface of Qdots by both electrostatic self-assembly and metal affinity coordination (Fig. 9). With increased fraction of MBPs, emission from the dye increased while that from Qdots decreased. In addition, the emission intensity from dye-labeled MBPs was dependent on the emitted color from different sized Qdots upon the spectral overlap (as expected from Förster theory). Clapp et al. also investigated Qdots as an acceptor in FRET process (72). This on–off switch has the potential to be used as a sensor in many important applications, including healthcare, environmental monitoring, and biodefense systems.

All of these experiments confirmed that water-soluble Qdots have potential applications in biosensor or bioimaging. FRET has been utilizing as the way to probe biological activity. Patolsky et al. reported that telomerization and DNA replication can be monitored with CdSe/ZnS Qdots (73). As telomerization proceeded, the emission from Texas-Red–labeled dUTP increased while that from Qdots decreased

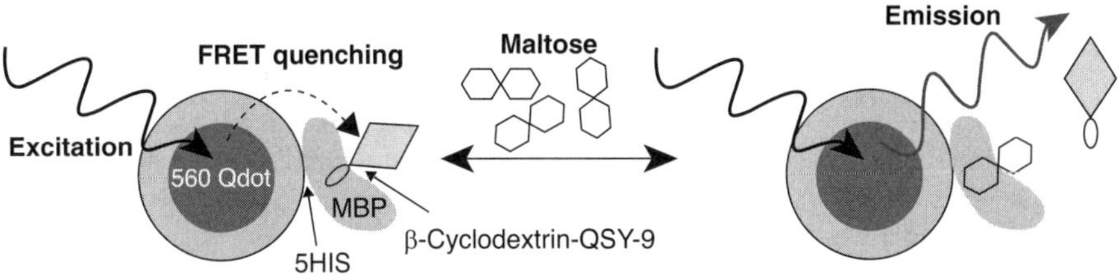

FIGURE 9 A schematic showing on the function of 560 nm Qdot-MBP nanosensor. Each 560 nm emitting Qdot is surrounded by an average of ~10 MBP moieties (a single MBP is shown for simplicity). Formation of Qdot-MBP-β-CD-QSY9 (maximum absorption ~565 nm) results in quenching of Qdot emission. Added maltose displaces β-CD-QSY9 from the sensor assembly, resulting in an increase in direct Qdot emission. *Source*: From Ref. 54.

via FRET. In replication studies, the Texas-Red dUTP was brought into proximity with the nanocrystal, resulting in FRET from Qdots to organic dye. These results suggested the possibility of using Qdots in the detection of cancer cells or in amplification of DNA on chip arrays.

FRET coupled to quenching provided alternative path for sensing. The luminescence via FRET is turned on by the appearance of analyte, which displaces a quencher or a terminal energy acceptor. Medintz et al. have developed a sensor for maltose by adapting their CdSe–MBP conjugates for analyte displacement strategies (54). First, a β-cyclodextrin conjugated to a nonfluorescent QSY9 quencher dye was docked to the MBP saccharide-binding site of the CdSe/ZnS–MBP. Second, maltose displaced the β-cyclodextrin–QSY9 conjugate to restore Qdots emission. This approach is general and the concept of an antibody fragment bound to a Qdot surface through noncovalent self-assembly should find wider use for other analytes of interest.

Surface-Enhanced Raman Spectroscopy

Surface-enhanced Raman Spectroscopy (SERS) is a near-field probe, which is sensitive to local environment conditions. Qdot-based SERS can be used in two ways in biomedical application. The first approach is as single molecule-based SERS that measures the unique fingerprint spectra of pure analytes on a Qdot. In the second approach, Qdots are covered by a monolayer of analyte. Therefore, SERS spectra are obtained from an ensemble of nanoparticles. In this case, population-average data are determined and one can get robust data from complex milieu or whole blood circulation.

By using single molecular SERS technique, along with Au/silica or core/shell Qdots, biomarker can be very sensitive. Currently, bioconjugated SERS have also been developed to identify protein biomakers on the surfaces of living cancer cells. For example, targeted gold nanoparticles are prepared by using a mixture of thiol-PEG and a heterofunctional PEG for live cancer cell detection, which binds to the epidermal growth factor receptor (EGFR) with high specificity and affinity (74). Human head and neck carcinoma cells are EGFR-positive and can give strong SERS signals (75).

In contrast, the human, non–small cell lung carcinoma does not express EGFR receptors, showing little or no SERS signals. Single cell profiling studies are of great clinical significance because EGFR is a validated protein target for monoclonal antibody and protein-kinase–based therapies. In addition, Qdots-based SERS can be used for in vivo tumor targeting and detection. Qian et al. reported that they injected a small dosage of nanoparticles into subcutaneous and deep muscular sited in live animals and highly resolved SERS signals were obtained (76). It is estimated that the achievable penetration depth is about 1 to 2 cm for in vivo SERS tumor detection.

Radio-Opacity and Paramagnetic Properties

CdS:Mn/ZnS core/shell Qdots were characterized for radio-opacity and magnetic hysteresis (67,68,77) for possible use as contrast agent in CT and MRI due to electron dense Cd and paramagnetic Mn, respectively (77). For radio-opacity, the Qdots sample was compared with a conventional radio-opaque dye, *Omnipaque*, used for CT scans and angiography. From Figure 10(A), it was determined that the X-ray absorption of Qdots was less than that of *Omnipaque*. In this respect, Qdot may not provide sufficient contrast for current radiographic practice. A superconductor quantum interface device (SQUID) magnetometer was used to measure the magnetization of CdS:Mn/ZnS Qdots. A typical room temperature hysteresis curve for paramagnetic CdS:Mn is shown in Figure 10(B). Hysteresis was observed, but it is too small for MRI imaging.

Magnetic Resonance–Based Bioimaging

Magnetic resonance imaging (MRI) is essentially proton nuclear magnetic resonance (NMR) (6,78). Protons are excited with short pulses of radio frequency radiation, and the free induction decay as they relax is measured and deconvoluted by a Fourier transform, which provides an image of the tissue. Areas of high proton density (e.g., water or lipid molecules) have a strong signal and appear bright. Areas of bone or tendon, which have a low proton density, have a weak signal and appear dark. A major limitation of MRI is its inability to distinguish differences in various types of soft tissues where the relative proton densities can be very similar. Regions having air pockets and fecal matter, such as the bowel, are hard to image because

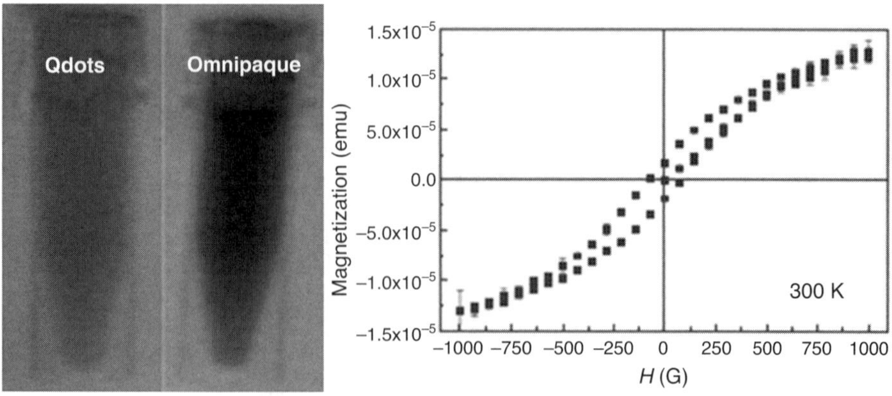

(A) (B)

FIGURE 10 (**A**) Fluoroscopy images of CdS:Mn/ZnS Qdot (*left*) and Omnipaque (*right*) of equal concentration under same magnification.(**B**) Magnetization curve for CdS:Mn/ZnS Qdots. *Source*: From Ref. 67.

of inconsistent proton density. Therefore, various contrast agents such as perfluorochemicals, oils, fats, and nanomaterials, are studied to circumvent these imaging problems. Unlike organic molecules, nanomaterials-based contrast agents are miscible in aqueous systems that allow them to be used intravenously. Therefore, they are well suited for in vivo applications such as tracking blood flow in the brain. The advantage of Qdots over other nanoparticles is that they offer multimodal imaging capabilities (37). However, appropriate functionalization of the Qdots is needed in order to make the Qdot a suitable contrast agent for MRI.

Qdot-based contrast agents change the strength of the MRI signal at a desired location. For example, paramagnetic contrast agents change the rate at which protons decay from their excited state to the ground state, allowing more rapid decay through energy transfer to a neighboring nucleus (78). As a result, regions containing the paramagnetic contrast agent appear darker in an MRI than regions without the agent. When paramagnetic Qdots are delivered to the liver, the uptake rate of Qdots by healthy liver cells is much higher than that by diseased cells. Consequently, the healthy regions are darker than the diseased regions. Several experimental reports in the literature demonstrate the usefulness and multimodal use of Qdots in MRI applications (79–83). In these reports, the Qdots are often coated with a water-soluble paramagnetic coating to enhance contrast.

Yang et al. (81) synthesized a water soluble Gd-functionalized silica-coated CdS:Mn/ZnS Qdots and studied them for MRI contrast agent. Longitudinal (T_1) and traverse (T_2) proton relaxation times were measured with a single slice, spin–echo image sequence at 4.7 T. Increased MR signal intensity, as shown in Figure 11(A), was observed (67,68,81) with increasing Gd concentrations due to the shorter water relaxation time T_1. In T_2-weighted images, the MR signal intensity was substantially decreased by the effects of increased Gd on the T_2 of water [Fig. 11(B)]. For control experiments, T_1- and T_2-weighted images of serial dilutions of Qdots without Gd^{III} ions were recorded and could not be distinguished from those of deionized (DI) water.

Normalized T_1- and T_2-weighted intensities versus repetition time (T_R) and echo time (T_E), respectively, for DI water and a series of diluted Gd-Qdots (from 0.36 to 0.0012 mM of Gd) showed increasingly faster recovery of longitudinal magnetization and faster decay of transverse magnetization for increased Gd concentrations [67,68,81; as shown in Fig. 11(C) and 11(D)]. The efficacy of a contrast agent is generally expressed by its relaxivity (R_i), that is defined by $1/T_i = 1/T^0 + R_i[\mathrm{Gd}]$ (84), where T_i is the relaxation time for a contrast agent solution concentration of [Gd], and T^0 is the relaxation time in the absence of a contrast agent. The relaxivities R_1 and R_2 were found to be 20.5 and 151 $\mathrm{mM}^{-1}\ \mathrm{s}^{-1}$, respectively. Compared to commercially available contrast agents, Gd-Qdots exhibited higher R_1 and R_2 values under the same magnetic field strength of 4.7 T (85). High relaxivities were attributed to a reduced tumbling rate of the Gd^{3+}-based contrast agents by grafting the contrast agent to rigid macromolecules and avoiding free rotation of the chelate (86,87). Although the Gd-Qdots can serve as either a T_1 or T_2 contrast agent, the R_2/R_1 ratio of ~7.4 indicates that they may be most effective as a T_2 contrast agent.

CONCLUDING REMARK

Recent advance in precisely controlled synthesis of Qdots, their optical and structural characteristics, and their functionalization by addition of inorganic, organic,

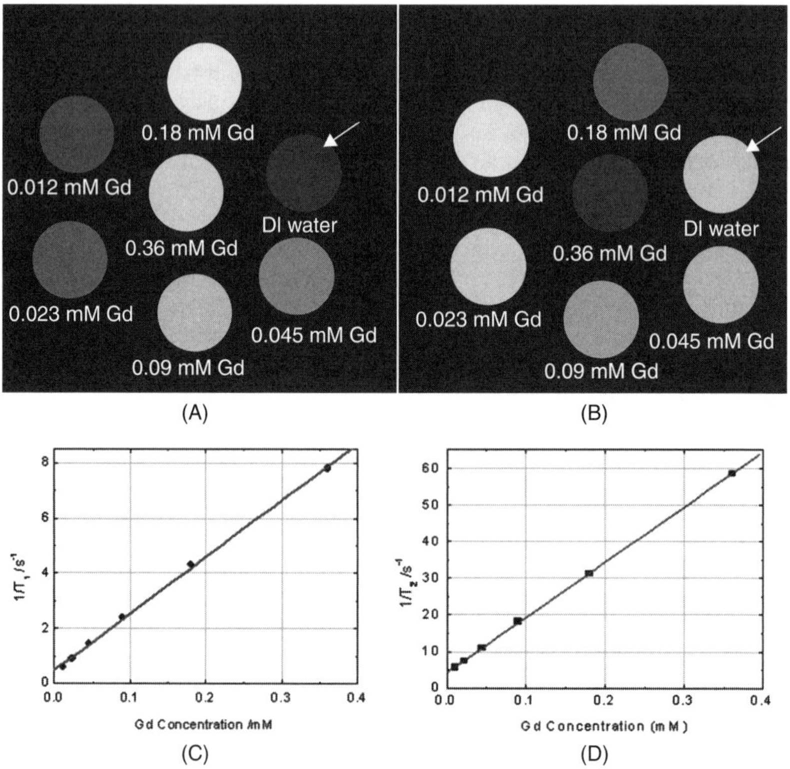

FIGURE 11 Magnetic resonance data from Gd-functionalized silica-coated CdS:Mn/ZnS Qdots: (**A**) T_1-weighted [repetition time (T_R) = 11,000 ms, echo time (T_E) = 4.2 ms) and (**B**) T_2-weighted (T_R = 11,000 ms, T_E = 24 ms) images of DI water and serial dilutions of Gd-Qdots (0.36, 0.18, 0.09, 0.045, 0.023, 0.012 mM of Gd). Linear plots of Gd concentration versus $1/T_1$ (**C**) and $1/T_2$ (**D**) to obtain ionic relaxivities of R_1 and R_2 of Gd-Qdots (81).

and/or biological molecules were discussed. These studies support the promise of a quantum leap in the extensive use of Qdots in future biological applications. It is predicted that Qdots will provide unprecedented sensitivity and selectivity over the traditional practices on molecular imaging. The use of Qdots emitting in near-infrared region will provide greater sensitivity and the longer lifetime of their excited states as compared to organic fluorophores and proteins for improved bioimaging. Despite the advantages for Qdots-based bioimaging, few issues related to Qdots need to be addressed before in vivo use, especially their toxicity. Son et al. (88) reported an ion exchange at the surface of CdSe Qdots that suppressed their PL intensity and led to the release of Cd^{2+}, which in known to be toxic to human. The only way to partially recover the PL emission was to add excess Cd^{2+}, which is unacceptable for biological application. In addition, bare Qdots were reported to be cytotoxic (89). Some of the Qdots properties are limiting, such as the fact that their typical size is a few times larger than that of the traditional organic marker dyes. As research on nanoparticles with novel properties continues, it should be possible to overcome these drawbacks and to develop multifunctional, multimodal Qdot-based systems for better biological imaging within a few years.

ACKNOWLEDGEMENT

The work was supported by ARO Grant W911NF-07-1-0545.

REFERENCES

1. Cassidy PJ, Radda GK. Molecular imaging perspectives. J R Soc Interface 2005; 2(3):133–144.
2. Schillaci O, Danieli R, Padovano F, et al. Molecular imaging of atheroslerotic plaque with nuclear medicine techniques (Review). Int J Mol Med 2008; 22(1):3–7.
3. Lecchi M, Ottobrini L, Martelli, C, et al. Instrumentation and probes for molecular and cellular imaging. Q J Nucl Med Mol Imaging 2007; 51(2):111–126.
4. Pichler BJ, Wehrl HF, Judenhofer MS. Latest advances in molecular imaging instrumentation. J Nucl Med 2008; 49:5S–23S.
5. Levenson RM, Lynch DT, Kobayashi H, et al. Multiplexing with multispectral imaging: From mice to microscopy. ILAR J 2008; 49(1):78–88.
6. Sharma P, Brown S, Walter G, et al. Nanoparticles for bioimaging. Adv Colloid Interface Sci 2006; 123:471–485.
7. Hillman EMC. Optical brain imaging in vivo: Techniques and applications from animal to man. J Biomed Opt 2007; 12(5):051402 1–4.
8. Luker GD, Luker KE. Optical imaging: Current applications and future directions. J Nucl Med 2008; 49(1):1–4.
9. Wu WW, Li AD. Optically switchable nanoparticles for biological imaging. Nanomedicine 2007; 2(4):523–531.
10. Weissleder R. A clearer vision for in vivo imaging. Nat Biotechnol 2001; 19:316–317.
11. Bera D, Qian L, Sabui S, et al. Photoluminescence of ZnO quantum dots produced by a sol–gel process. Opt Mater 2008; 30(8):1233–1239.
12. Bera D, Qian L, Holloway PH. Time-evolution of photoluminescence properties of ZnO/MgO core/shell quantum dots. J Phys D Appl Phys 2008; 41:182002 1–4.
13. Qian L, Bera D, Holloway P. Photoluminescence from ZnS/CdS: Mn/ZnS quantum well quantum dots. Appl Phys Lett 2008; 92(8):093103 1–3.
14. Yang H, Holloway PH. Enhanced photoluminescence from CdS: Mn/ZnS core/shell quantum dots. Appl Phys Lett 2003; 82(12):1965–1967.
15. Yang HS, Holloway PH, Ratna BB. Photoluminescent and electroluminescent properties of Mn-doped ZnS nanocrystals. J Appl Phys 2003; 93(1):586–592.
16. Yang H. Syntheses and applications of Mn-doped II–VI Semiconductor Nanocrystals. In: Materials Science and Engineering [Ph.D. Dissertation]. Gainesville, FL: University of Florida, 2003:129.
17. Qian L, Bera D, Holloway PH. Temporal evolution of white light emission from CdSe quantum dots. Nanotechnology 2008; 19(28):285702 1–4.
18. Lee H, Holloway PH, Yang H. Synthesis and characterization of colloidal ternary ZnCdSe semiconductor nanorods. J Chem Phys, 2006; 125(16):164711 1–7.
19. Qian HF, Li L, Ren JC. One-step and rapid synthesis of high quality alloyed quantum dots (CdSe–CdS) in aqueous phase by microwave irradiation with controllable temperature. Mater Res Bull 2005; 40(10):1726–1736.
20. Cheng WQ, Liu D, Yan ZY. Spectroscopic study on CdS nanoparticles prepared by microwave irradiation. Spectrosc Spectral Anal 2008; 28(6):1348–1352.
21. Yang WH, Li WW, Dou HJ, et al. Hydrothermal synthesis for high-quality CdTe quantum dots capped by cysteamine. Mater Lett 2008; 62(17–18):2564–2566.
22. Yang HQ, Yin WY, Zhao H, et al. A complexant-assisted hydrothermal procedure for growing well-dispersed InP nanocrystals. J Phys Chem Solids 2008; 69(4):1017–1022.

23. Bera D, Qian L, Holloway PH. Phosphor quantum dots. In: Kitai A, ed. Luminescent Materials and Application, West Sussex, England: John Wiley & Sons, 2008:19.
24. Yoff AD. Low-dimensional systems—Quantum-size effects and electronic-properties of semiconductor microcrystallites (zero-dimensional systems) and some quasi-2-dimensional systems. Adv Phys 1993; 42(2):173–266.
25. Yoffe AD. Semiconductor quantum dots and related systems: Electronic, optical, luminescence and related properties of low dimensional systems. Adv Phys 2001; 50(1):1–208.
26. Alivisatos AP. Perspectives on the physical chemistry of semiconductor nanocrystals. J Phys Chem 1996; 100(31):13226–13239.
27. Bukowski TJ, Simmons JH. Quantum dot research: Current state and future prospects. Crit Rev Solid State Mater Sci 2002; 27(3–4):119–142.
28. Bajaj KK. Use of excitons in materials characterization of semiconductor system. Mater Sci Eng R Rep 2001; 34(2):59–120.
29. Gfroerer TH. Photoluminescence in analysis of surface and interfaces. In: Meyers RA, ed. Encyclopedia of Analytical Chemisty. Chichster, England: John Wiley & Sons Ltd., 2000:9209–9231.
30. Shionoya S, Yen WM. Phophor Handbook. New York, NY: CRC Press, 1998:608.
31. Yang H, Holloway PH. Efficient and photostable ZnS-passivated CdS:Mn luminescent nanocrystals. Adv Funct Mater 2004; 14(2):152–156.
32. Wang Q, Kuo YC, Wang YW, et al. Luminescent properties of water-soluble denatured bovine serum albumin-coated CdTe quantum dots. J Phys Chem B 2006; 110(34):16860–16866.
33. Ma N, Yang J, Stewart KM, et al. DNA-passivated CdS nanocrystals: Luminescence, bioimaging, and toxicity profiles. Langmuir 2007; 23(26):12783–12787.
34. Yang H, Holloway PH, Cunningham G, et al. CdS:Mn nanocrystals passivated by ZnS: Synthesis and luminescent properties. J Chem Phys 2004; 121(20):10233–10240.
35. Yang HS, Holloway PH, Santra S. Water-soluble silica-overcoated CdS: Mn/ZnS semiconductor quantum dots. J Chem Phys 2004; 121(15):7421–7426.
36. Santra S, Yang H, Stanley JT, et al. Rapid and effective labeling of brain tissue using TAT-conjugated CdS: Mn/ZnS quantum dots. Chem Commun 2005; 25:3144–3146.
37. Santra S, Bagwe RP, Dutta D, et al. Synthesis and characterization of fluorescent, radio-opaque, and paramagnetic silica nanoparticles for multimodal bioimaging applications. Adv Mater 2005; 17(18):2165–2169.
38. Boldt K, Bruns OT, Gaponik N, et al. Comparative examination of the stability of semiconductor quantum dots in various biochemical buffers. J Phys Chem B 2006; 110(5):1959–1963.
39. Wolcott A, Gerion D, Visconte M, et al. Silica-coated CdTe quantum dots functionalized with thiols for bioconjugation to IgG proteins. J Phys Chem B 2006; 110(11):5779–5789.
40. Santra S, Dutta D, Moudgil BM. Functional dye-doped silica nanoparticles for bioimaging, diagnostics and therapeutics. Food and Bioproducts Process 2005; 83(C2):136–140.
41. Empedocles SA, Bawendi MG. Quantum-confined stark effect in single CdSe nanocrystallite quantum dots. Science 1997; 278(5346):2114–2117.
42. Santra S, Xu JS, Wang KM, et al. Luminescent nanoparticle probes for bioimaging. J Nanosci Nanotechnol 2004; 4(6):590–599.
43. Wang Y, Deng YL, Qing H, et al. Advance in real-time and dynamic biotracking and bioimaging based on quantum dots. Chem J Chin Univ—Chin 2008; 29(4):661–668.
44. Parak WJ, Pellegrino T, Plank C. Labeling of cells with quantum dots. Nanotechnology 2005; 16(2):R9–R25.
45. Li B, Cai W, Chen XY. Semiconductor quantum dots for in vivo imaging. J Nanosci Nanotechnol 2007; 7(8):2567–2581.
46. Sapsford KE, Pons T, Medintz IL, et al. Biosensing with luminescent semiconductor quantum dots. Sensors 2006; 6(8):925–953.
47. Sukhanova A, Devy M, Venteo L, et al. Biocompatible fluorescent nanocrystals for immunolabeling of membrane proteins and cells. Anal Biochem 2004; 324(1):60–67.
48. Dahan M, Levi S, Luccardini C, et al. Diffusion dynamics of glycine receptors revealed by single-quantum dot tracking. Science 2003; 302(5644):442–445.

49. Bruchez M, Moronne M, Gin P, et al. Semiconductor nanocrystals as fluorescent biological labels. Science 1998; 281(5385):2013–2016.
50. Chan WCW, Nie SM. Quantum dot bioconjugates for ultrasensitive nonisotopic detection. Science 1998; 281(5385):2016–2018.
51. Parak WJ, Boudreau R, Le Gros M, et al. Cell motility and metastatic potential studies based on quantum dot imaging of phagokinetic tracks. Adv Mater 2002; 14(12):882–885.
52. Akerman ME, Chan WCW, Laakkonen P, et al. Nanocrystal targeting in vivo. Proc Natl Acad Sci U S A 2002; 99(20):12617–12621.
53. Kim S, Lim YT, Soltesz EG, et al. Near-infrared fluorescent type II quantum dots for sentinel lymph node mapping. Nat Biotechnol 2004; 22(1):93–97.
54. Medintz IL, Clapp AR, Mattoussi H, et al. Self-assembled nanoscale biosensors based on quantum dot FRET donors. Nat Mater 2003; 2(9):630–638.
55. Michalet X, Pinaud FF, Bentolila LA, et al. Quantum dots for live cells, in vivo imaging, and diagnostics. Science 2005; 307(5709):538–544.
56. Hassan M, Klaunberg BA. Biomedical applications of fluorescence imaging in vivo. Comp Med 2004; 54(6):635–644.
57. Medintz IL, Uyeda HT, Goldman ER, et al. Quantum dot bioconjugates for imaging, labelling and sensing. Nat Mater 2005; 4(6):435–446.
58. Zhang Y, He J, Wang PN, et al. Time-dependent photoluminescence blue shift of the quantum dots in living cells: Effect of oxidation by singlet oxygen. J Am Chem Soc 2006; 128(41):13396–13401.
59. Dahan M, Laurence T, Pinaud F, et al. Time-gated biological imaging by use of colloidal quantum dots. Opt Lett 2001; 26(11):825–827.
60. Wu CL, Hong JQ, Guo XQ, et al. Fluorescent core-shell silica nanoparticles as tunable precursors: Towards encoding and multifunctional nano-probes. Chem Commun 2008; 6:750–752.
61. Gao XH, Yang LL, Petros JA, et al. In vivo molecular and cellular imaging with quantum dots. Curr Opin Biotechnol 2005; 16(1):63–72.
62. Han MY, Gao XH, Su JZ, et al. Quantum-dot-tagged microbeads for multiplexed optical coding of biomolecules. Nat Biotechnol 2001; 19(7):631–635.
63. Dubertret B, Skourides P, Norris DJ, et al. In vivo imaging of quantum dots encapsulated in phospholipid micelles. Science 2002; 298(5599):1759–1762.
64. Larson DR, Zipfel WR, Williams RM, et al. Water-soluble quantum dots for multiphoton fluorescence imaging in vivo. Science 2003; 300(5624):1434–1436.
65. Bera D, Holloway PH, Yang H. Efficient and Photostable ZnS-passivated CdS: Mn Quantum dots. In: Joint International Meeting of Electrochemical Society. Cancun, Mexico: The Electrochemical Society, 2006.
66. Yang HS, Santra S, Holloway PH. Syntheses and applications of Mn-doped II–VI semiconductor nanocrystals. J Nanosci Nanotechnol 2005; 5(9):1364–1375.
67. Holloway PH, Yang H, Lee H, et al. Nanophosphor: PL, EL, and biological markers. Proceeding of 13th International Display Workshops, IDW'06.2006, Otsu, Japan.
68. Holloway PH, Yang H, Lee H, et al. Nanophosphor: PL, EL, and biological markers. Proceeding of 3rd International Symposium on Display & Lighting Phosphor Materials, ISDLPMW'06, 2006, Yokohama, Japan.
69. Lakowicz JR. Principles of Fluorescence Spectroscopy, 2nd ed. New York, NY: Academic Publishers, 1999.
70. Tran PT, Goldman ER, Anderson GP, et al. Use of luminescent CdSe–ZnS nanocrystal bioconjugates in quantum dot-based nanosensors. Phys Status Solidi B—Basic Res 2002; 229(1):427–432.
71. Clapp AR, Medintz IL, Mauro JM, et al. Fluorescence resonance energy transfer between quantum dot donors and dye-labeled protein acceptors. J Am Chem Soc 2004; 126(1):301–310.
72. Clapp AR, Medintz IL, Fisher BR, et al. Can luminescent quantum dots be efficient energy acceptors with organic dye donors? J Am Chem Soc 2005; 127(4):1242–1250.
73. Patolsky F, Gill R, Weizmann Y, et al. Lighting-up the dynamics of telomerization and DNA replication by CdSe–ZnS quantum dots. J Am Chem Soc 2003; 125(46):13918–13919.

74. Herbst RS, Shin DM. Monoclonal antibodies to target epidermal growth factor receptor-positive tumors—A new paradigm for cancer therapy. Cancer 2002; 94(5):1593–1611.
75. Reuter CWM, Morgan MA, Eckardt A. Targeting EGF-receptor-signalling in squamous cell carcinomas of the head and neck. Br J Cancer 2007; 96(3):408–416.
76. Qian X, Peng X-H, Ansari DO, et al. In vivo tumor targeting and spectroscopic detection with surface-enhanced Raman nanoparticle tags. Nat Biotech 2008; 26(1):83–90.
77. Santra S, Yang HS, Holloway PH, et al. Synthesis of water-dispersible fluorescent, radio-opaque, and paramagnetic CdS : Mn/ZnS quantum dots: A multifunctional probe for bioimaging. J Am Chem Soc 2005; 127(6):1656–1657.
78. Martin CR, Mitchell DT. Nanomaterials in analytical chemistry. Anal Chem 1998; 70(9):322A–327A.
79. Tan WB, Zhang Y. Multi-functional chitosan nanoparticles encapsulating quantum dots and Gd-DTPA as imaging probes for bio-applications. J Nanosci Nanotechnol 2007; 7(7):2389–2393.
80. Mulder WJM, Koole R, Brandwijk RJ, et al. Quantum dots with a paramagnetic coating as a bimodal molecular imaging probe. Nano Lett 2006; 6(1):1–6.
81. Yang HS, Santra S, Walter GA, et al. Gd-III-functionalized fluorescent quantum dots as multimodal imaging probes. Adv Mater 2006; 18(21):2890.
82. van Tilborg GAF, Mulder WJM, Chin PTK, et al. Annexin A5-conjugated quantum dots with a paramagnetic lipidic coating for the multimodal detection of apoptotic cells. Bioconjug Chem 2006; 17(4):865–868.
83. Bakalova R, Zhelev Z, Aoki I, et al. Silica-shelled single quantum dot micelles as imaging probes with dual or multimodality. Anal Chem 2006; 78(16):5925–5932.
84. Toth E, Bolskar, RD, Borel A, et al. Water-soluble gadofullerenes: Toward high-relaxivity, pH-responsive MRI contrast agents. J Am Chem Soc 2005; 127(2):799–805.
85. Rohrer M, Bauer H, Mintorovitch J, et al. Comparison of magnetic properties of MRI contrast media solutions at different magnetic field strengths. Invest Radiol 2005; 40(11):715–724.
86. Pierre VC, Botta M, Raymond KN. Dendrimeric gadolinium chelate with fast water exchange and high relaxivity at high magnetic field strength. J Am Chem Soc 2005; 127(2):504–505.
87. Tan WB, Zhang Y. Multifunctional quantum-dot-based magnetic chitosan nanobeads. Adv Mater 2005; 17(19):2375.
88. Son DH, Hughes SM, Yin YD, et al. Cation exchange reactions-in ionic nanocrystals. Science 2004; 306(5698):1009–1012.
89. Green M. Semiconductor quantum dots as biological imaging agents. Angew Chem Int Ed 2004; 43(32):4129–4131.

20 Application of Near Infrared Fluorescence Bioimaging in Nanosystems

Eunah Kang, Ick Chan Kwon, and Kwangmeyung Kim

Biomedical Research Center, Korea Institute of Science and Technology, Seoul, South Korea

INTRODUCTION TO NEAR INFRARED FLUORESCENCE BIOIMAGING

Near infrared fluorescence (NIRF) bioimaging, capable of imaging the whole body of small animals in vivo, now reveals the interrelatedness of numerous nanosystems with internal biological environments in the fields of drug delivery and molecular imaging (1,2). Traditional drug delivery nanosystems are coordinated with numerous active moieties including drugs, imaging probes, targeting moieties, antibodies, glycoproteins, peptides, receptor-binding ligands, and aptamers, etc. (2,3). Therefore, studying the in vivo characteristics is very important for understanding the interaction between nanosystems and biological environments. Direct visualization by NIRF bioimaging affords noninvasive real-time monitoring of preferential localization resulting from the interaction of these targeting moieties. Recently, the development of new bioimaging systems and imaging probes allows NIRF bioimaging to detect molecular events, actions of specific molecules, and protein recognition down to the molecular level. These specific identifications can be displayed as an amplified signal at localized body sites, presenting the functional status of target diseases. NIRF bioimaging with spatial and temporal resolution in vivo allows for the assessment of biodistribution, the targeting efficacy of drug carriers and imaging probes via the guidance of passive, active, and activatable molecular targeting. In this chapter, numerous applications of NIRF bioimaging characterizing nanosystems in drug delivery and molecular imaging are reviewed, and strategies of nanosystems are classified for NIRF bioimaging.

Bioimaging Technology in Drug Delivery Systems and Molecular Imaging

Nanotechnology offers many advantages to drug delivery systems and the molecular imaging field as well as has the potential to literally revolutionalize both of these fields. In terms of drug delivery systems, liposomes, micelles, dendrimers, and metal colloidals (diameters less than 100 nm) have been extensively studied to enhance the efficacy of therapeutic agents (4–8). Owing to their small size and excellent biocompatibility, nanosized drug carriers can circulate in the bloodstream for a long period of time, enabling them to reach a target site and effectively deliver therapeutic agents, all the while minimizing the inefficiency and side effects of free drugs. In addition, nanosystem-based imaging probes (nanoprobes) have yielded new strategies for designing imaging probes that efficiently detect target biomolecules or diagnose diseases. These nanoprobes have large surface areas (ideal for efficient modification with a wide range of imaging moieties), prolonged plasma half-life, enhanced stability, improved targeting, and reduced nonspecific binding, etc. These nanosystems consist of a variety of materials (e.g., peptides, DNA, proteins, antibodies, liposomes, dendrimers, micelles, nanoparticles, metal colloids, etc.) that are all intended to act as therapeutic agents and/or imaging probes.

In vitro physicochemical properties such as a particle's size, surface chemistry, and surface charge are generally characterized by traditional techniques (electron microscopy, dynamic light scattering, energy dispersive X-ray, zeta potential, etc.). The physicochemical properties affecting a nanoparticle's behavior in a biological system might, in part, determine the biodistribution, safety, targeting efficacy, and multifunctional efficacy in drug delivery systems and molecular imaging. The in vitro physicochemical properties of nanosystems, however, do not always reflect their in vivo behaviors, because these properties are probably dependent on environmental conditions and must therefore be assayed not only in vitro but also under in vivo conditions. Unfortunately, to date, only limited information is available on the interaction between nanosystems and biological environments. The optimum physicochemical characteristics of drug carriers and imaging probes that can efficiently deliver and image target biological molecules have not been fully characterized. Despite the benefits that nanosystems have contributed to medicine, some applications remain to be improved, for example, specific targeting to the acting site, efficient drug delivery inside the target cells or tissues, and early-stage diagnosis, etc. Therefore, well-established methodologies for the in vivo characterization of nanosystems are urgently required for improving drug delivery systems and molecular imaging. An effective approach for achieving efficient drug delivery and molecular imaging would be to rationally develop nanosystems based on the understanding of their interactions with the biological environment and molecular mechanisms in vivo. Furthermore, underlying mechanisms need to be understood in order to enhance the efficacy of the encapsulated therapeutic agents, with respect to the targeting of biomolecules, target tissue uptake, real-time trafficking, and accumulation.

NIRF Bioimaging in Nanosystems

Currently, in vivo bioimaging techniques, including magnetic resonance imaging (MRI) (9,10), positron emission tomography (PET) (11,12), computed tomography (CT) (13), and near infrared (NIR) fluorescence imaging (14,15), are extensively used to characterize the in vivo behaviors of nanosized drug carriers and imaging probes. Imaging probe–labeled nanosystems can be monitored in real-time and visualized in a noninvasive way, allowing for clinical uses in animals and humans. With the help of various imaging systems, the most important characteristics of nanosystems in vivo were studied, including in vivo ADME (absorption, distribution, metabolism, and exertion), targeting efficacy, their multifunctional properties as drug carriers and imaging probes, and the efficacy of nanosystems in drug delivery systems and molecular imaging. Also, in vivo experimental uncertainties arising from inter-animal variations are greatly reduced, because each animal serves as its own "control" for consecutive analyses at the same condition. With the help of bioimaging technique, consecutive bioimaging experiments can need fewer animals wherein the same animal is repetitively and reproducibly assayed without any sacrifice time point experiment.

NIRF bioimaging is particularly well suited for bioimaging, because NIR fluorescent probes are safe, sensitive, and labeled specifically to target small molecules, proteins, and nanosystems. The fundamental barriers to the optical imaging of tissues are light scattering, autofluorescence, and absorption by tissues in the mid-visible range (14). NIR light in the region between 650 and 900 nm avoids some of these limitations because of a higher tissue penetration and minimal absorbance by the surface tissue. This is because hemoglobin, the main absorber of visible light, and water and lipids, the primary absorbers of infrared light, have lower absorptions in the NIRF range (Fig. 1) (14–17). With these benefits of using the NIR region, a large number of organic NIR fluorescent agents have been developed to examine fundamental processes at the organ, tissue, cellular, and molecular levels. The ideal characteristics of NIR fluorescent agents for optical imaging possess peak fluorescence in ranges from 700 to 900 nm, a high quantum yield, a

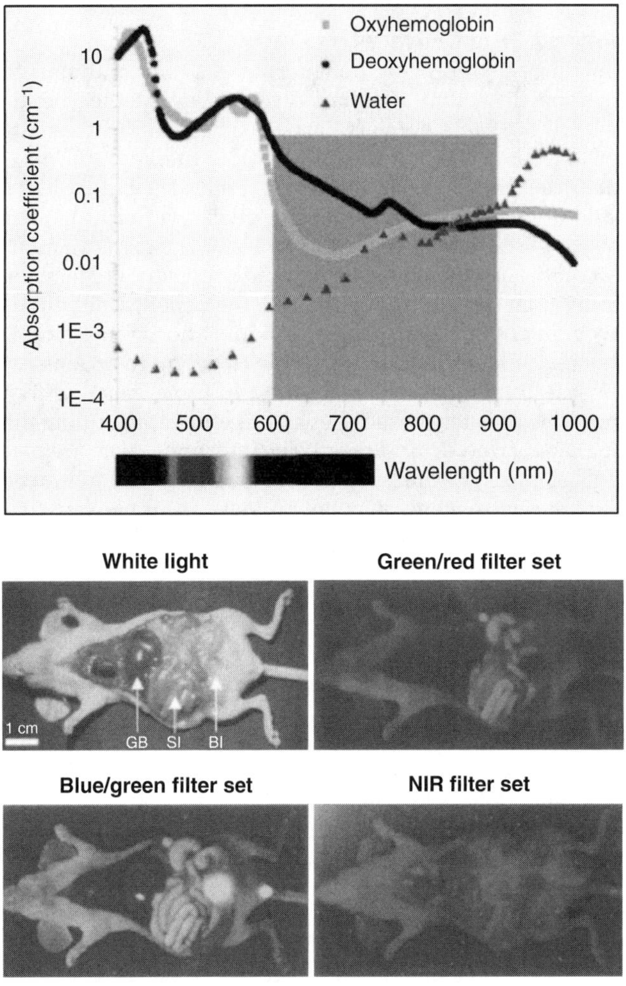

FIGURE 1 Characteristics of NIR region. (**A**) Low light absorption of oxyhemoglobin and deoxyhemoglobin and insensitive absorption of water in NIR region (600–900), (**B**) low autofluorescence in NIR region compared with white, green, and red emission region. *Source*: From Refs. 16 and 18.

narrow excitation/emission spectrum, and functional groups for chemical conjugations. These optical characteristics of the NIR region allow for deep tissue imaging with reduced autofluorescence and scattering in vivo (18). In total, the NIRF imaging system allows relatively long penetration depth (<10–15 cm), and thus is useful for extensive small animal studies. Furthermore, NIRF bioimaging has advanced with the development of fluorescent probes that are physicochemically stable, nontoxic, highly biocompatible, biodegradable, and extractable (19). Inorganic fluorescent agents such as quantum dots (QDs), when used as targeting biomarkers, also provide certain assurances for bioimaging, as they are resistant to photobleaching, allow for the targeting of dozens of molecules, and give higher quantum yields (18). For these listed reasons, they are not susceptible to many of the common problems of other organic fluorescent agents.

NIRF Bioimaging for Nanosized Drug Delivery Systems

Recently, drug delivery system and molecular imaging have been combined to generate multifunctional nanoparticles that simultaneously delivered therapeutic agents and imaging probes with the same delivery carriers. These multifunctional drug carriers show great promise in the emerging field of new therapy and diagnosis, because they allow detection as well as monitoring of an individual patient's diseases at an early stage and delivering disease-specific agents over an extended period for enhanced therapeutic efficacy. Moreover, real-time and noninvasive monitoring of the drug carriers could enable a clinician to rapidly decide whether the regimen is effective in an individual patient or not. Despite the benefits that multifunctional drug carriers have rendered to medicine, some applications remain challenging, for example, in vivo real-time monitoring, specific targeting to action site, or efficient drug delivery inside the target cells or tissues. Improvements within the field of NIRF bioimaging provide extensive studies useful for characterizing the in vivo fate of multifunctional nanosized drug carriers. The information on circulating, or localized, nanosized drug carriers from NIRF bioimaging has close relevance with the in vivo fate of different nanosystems in small animals or humans, in that drug carriers are passively targeted, actively targeted, or nonspecifically localized. In this chapter, we highlight a few bioimaging methods that are useful for the in vivo characterization of different nanosized drug carriers.

NIRF Bioimaging for Nanosized Drug Carriers

Researchers are actively exploring nanosystems made of lipid-based micelles, natural/synthetic nanoparticles, and inorganic particles (such as quantum dots, gold particles, paramagnetic particles, and carbon nanotubes) in drug delivery systems. As "seeing is believing," NIRF bioimaging opens a new era of nanosized drug carriers capable of providing visual proof. Noninvasive NIRF bioimaging is a developing field for preclinical animal models in vivo. Simple NIRF fluorophores were conjugated with drug carriers, allowing for NIRF bioimaging in vivo, and then were resolved with respect to both time and space so that imaging and drug delivery can potentially be carried out at the same time (20). Combining therapeutics with imaging diagnosis, bioimaging captures information that is significant to aspects of biodistribution and delivery efficacy by using a quantified signal. The quantitative assessment of the generated signal and the activity at the molecular level are keys to success in bioimaging.

NIR fluorophores–labeled drug carriers allow for the capturing of images characterizing biodistribution and in vivo kinetics. The biodistribution of NIRF imaging provides direct evaluation of a drug carrier such as (*i*) take-up within target organs, (*ii*) undesirable rapid liver uptake, (*iii*) tissue distribution, and (*iv*) nonselective accumulation into undesired organs. The dynamic equilibrium determined by equilibrated contrast can help suggest the optimal period of multiple doses. The contrast enhancement of each organ over time was measurable as a quantitative value by repeated scanning of the whole body. Visualizing the quantitative fluorescence signal with temporal and spatial resolution offers direct understanding of physiological conditions as drug carriers are administered. Notable studies in drug delivery and bioimaging are reviewed.

Park et al. (21), Hwang et al. (22), and Min et al. (23) performed diverse studies into the in vivo biodistribution of chitosan-based nanoparticles by using NIRF bioimaging (Fig. 2) (21–23). The in vivo biodistribution of the NIR fluorophores, Cy5.5-labeled chitosan nanoparticles, were clearly shown via NIRF bioimaging in tumor-bearing mice. These studies showed that high fluorescence intensities in inoculated tumor tissues were easily distinguished from the background tissue signal, indicating that the chitosan nanoparticles being used as anticancer drug carriers were passively localized in tumor. This is because nanosized particles or micelles are more permeable and are retained for longer period than traditional vectors (the so-called enhanced and retention effect; EPR effect), resulting in efficient passive accumulation in solid tumor tissues. This unique biodistribution in a whole body presents information concerning the drug efficacy, that is, how much of a drug is efficiently reaching a target tumor in real time and in a noninvasive way in live animals. This information is simply generated by quantifying the fluorescence signal ratio of a tumor to the background tissue signal. Ex vivo study also showed that chitosan nanoparticles were mainly taken into a tumor, compared to other organs. The estimated quantitative biodistribution of chitosan nanoparticles in each organ was presented as fluorescence intensity over time. The NIRF bioimaging method provided quantitative analysis with temporal and spatial resolution in the whole body of small animals, including tumor targeting efficacy, in vivo biodistribution, and therapeutic efficacy of nanosized drug carriers.

At the molecular level, NIRF bioimaging also provided in vivo biodistribution of nonviral polymeric carriers in gene therapy. For example, NIR indocyanine dye (IR820)–labeled polyethyleneimine (IR820-PEI) was used as nanosized gene carriers, and this IR820-PEI was monitored in real time and in a noninvasive manner in live animals (24). The IR820-PEI presented high chemical stability and good optical properties. The IR820-PEI is able to bind to DNA and the delivery process can be monitored in vivo by using noninvasive NIRF bioimaging (Fig. 3). NIRF bioimaging data suggest that intravenously injected IR820-PEG/DNA complexes predominantly accumulated in liver tissue. One hour after injection, the gene carriers began to accumulate in the liver, a typical store organ, and importantly even 24 hours after the injection, the bright NIRF signals was observed in liver tissue. After three hours, the NIRF signal intensity increased. This in vivo biodistribution of PEI carriers is in good agreement with published works on other systems. With the help of NIRF bioimaging, the in vivo biodistribution of gene carriers was successfully visualized and quantified in live animals.

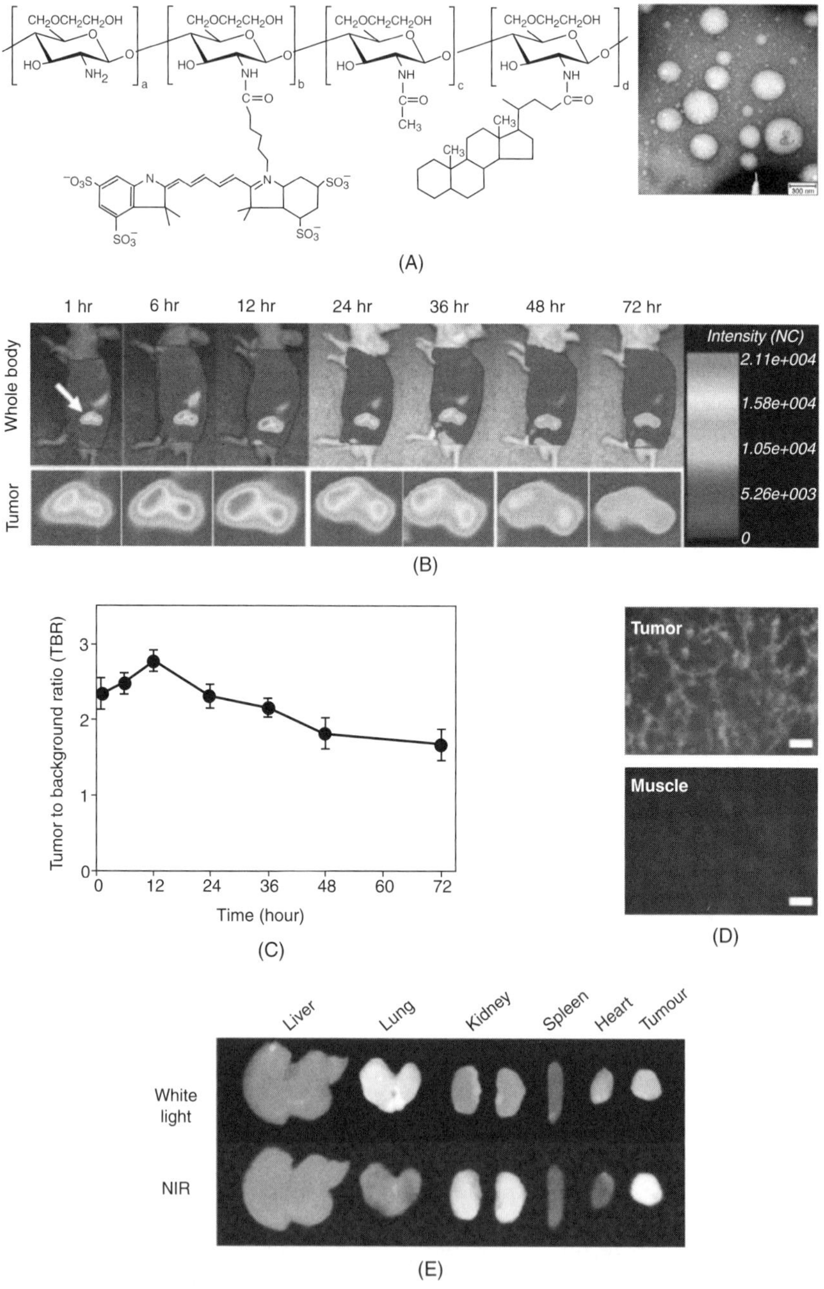

FIGURE 2 Tumor accumulation and tissue distribution of docetaxel-loaded chitosan nanoparticles in tumor bearing mice. (**A**) Chemical structure of Cy5.5 labeled and cholanic acid–modified glycol chitosan conjugate and the amphiphilic conjugate–formed polymeric nanoparticles in aqueous condition with an average diameter of 250 nm. (**B**) Real-time tumor accumulation of docetaxel-loaded HGC. NIRF images were taken by eXplore Optics system after intravenous injection. (**C**) Quantitative signal ratio of tissue to background. (**D**) Fluorescence images of sliced tumor tissue and normal muscle tissue. (**E**) Ex vivo NIR fluorescence images of excised organ after 72 hours of intravenous injection of docetaxel-loaded HGC. Images were taken by a Kodak image Station 4000 MM. *Source*: From Ref. 22.

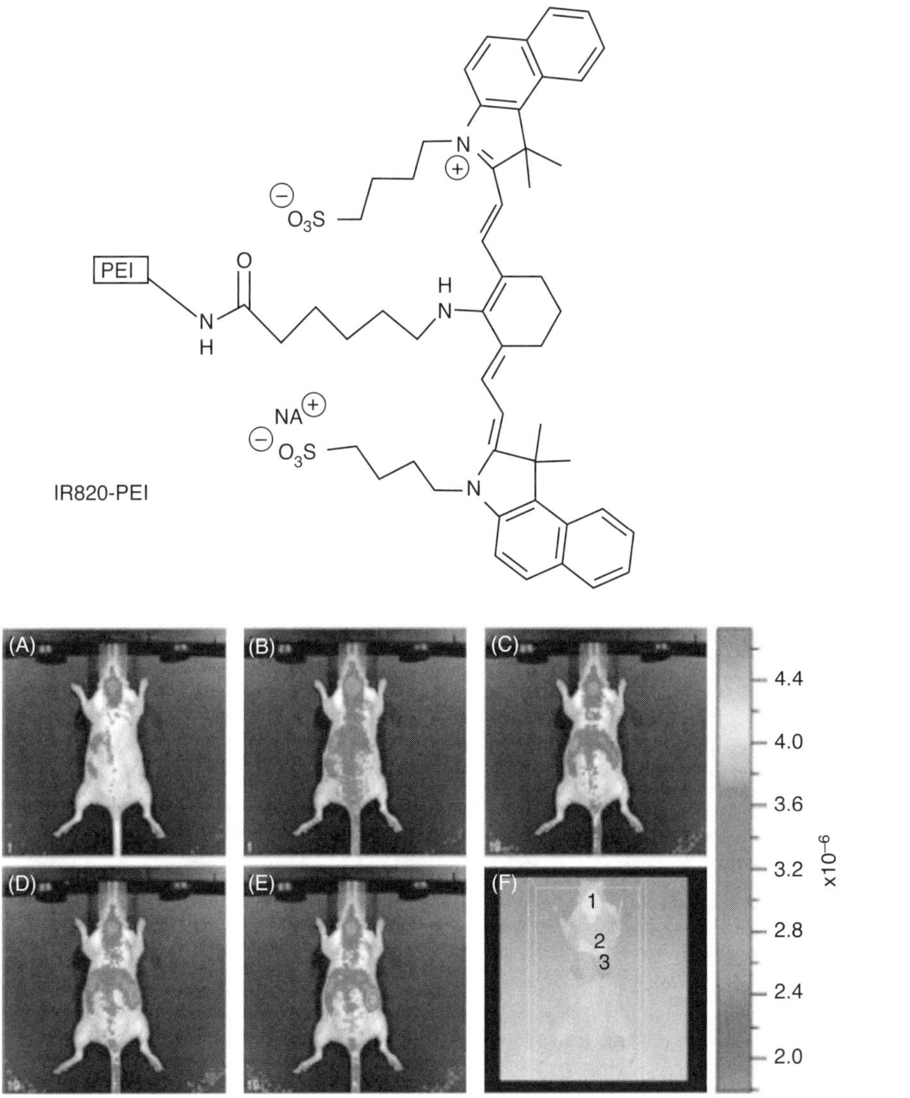

FIGURE 3 In vivo NIRF imaging of gene expression: (**A**) chemical scheme of PEI-conjugated NIR probe (IR820-PEI) and (**B**) in vivo NIRF imaging of nude mice after 2 mg/mL of intravenous injection of IR820-PEI-DNA. The images were taken over time of before, one minute, one hour, two hours, and three hours. *Source*: From Ref. 24.

NIRF Bioimaging for Active Targetable Drug Carriers

For the active drug targeting system, researchers have spent a great deal of efforts aimed at developing methods of efficiently delivering a drug to target cells or tissues through active targeting. Active drug targeting is usually achieved by chemical attachment to a targeting component that strongly interacts with antigens (or receptors) displayed on the target tissue, leading to preferential accumulation of the drug in the target organ, tissue, or cell. In the active drug targeting system, various targeting moieties, antibodies, glycoproteins, peptides, receptor-binding ligands, or aptamers are coordinated on the surface of drug delivery system. The targeting efficacy of various active targeting drug delivery systems could be easily monitored by using the NIRF imaging system.

For example, the active targeting efficacy of chitosan nanoparticles modified with atherosclerosis homing peptide (CRKRLDRNC; termed AP peptide) was clearly visualized by using the NIRF bioimaging system (25,26). As an atherosclerotic lesion–specific drug carrier, AP-conjugated chitosan nanoparticles labeled with the NIRF dye, Cy5.5, showed higher binding affinity to TNF-α–stimulated bovine aortic endothelial cells (BAECs), which show a 2–3-fold increased in IL-4R expression (27). Through these assays and by using confocal laser scanning microscopy, fluorescence spectrophotometer, and fluorescence microscopy, we confirmed that AP-conjugated chitosan nanoparticles exhibited a much higher binding affinity to TNF-α–stimulated BAECs compared to normal BAECs or to that of untargeted chitosan nanoparticles in static and flow conditions. To determine the targeting ability of AP-conjugated chitosan nanoparticles to atherosclerotic plaques in vivo, Park et al. performed NIRF bioimaging on a low-density lipoprotein receptor-deficient ($Ldlr^{-/-}$) atherosclerosis animal model (25). As shown in Figure 4, the fluorescence photon counts from the atherosclerotic aortic arch were significantly higher than those of the normal aortic arch. The fluorescence microscopy study supported the notion that AP-conjugated chitosan nanoparticles bound to the luminal surfaces of atherosclerotic lesions and also found inside the lesions were composed mainly of smooth muscle cells and macrophages, whereas untargeted chitosan nanoparticle binding to normal aortic tissue was negligible.

The acidic extracellular pH of tumor tissues allows for a cancer treatment strategy by constructing pH-sensitive polymeric micelles. Extracellular pH that ranges from 7.0 to 6.5 in animal tumor models and clinical tumors is mainly due to the anaerobic respiration and subsequent glycolysis (28–30). Based on the acidic tumor environments, Lee et al. developed a novel pH-responsive pop-up polymeric micelles that presents a nonspecific cell penetrating (TAT peptide) on its surface in response to small change in pH (31). Above pH 7.2, TAT was anchored on the inside of the polymeric micelle via a pH-sensitive molecular chain actuator—a short poly(L-histidine), thus being shielded by the PEG shell of micelle. The core part of the micelle was constructed for disintegration in the early endosomal pH (pH < 6.5) of tumor cells. In a lower pH (pH < 6.5), the pH-responsive pop-up polymeric micelles start to physically dissociate, and quickly release doxorubicin (DOX), leading to increased DOX activity in various wild and multidrug resistant cell lines. A dorsal skin-fold window chamber model allows in situ monitoring of administered drug formulations on vascularized tumors. The pH-responsive pop-up polymeric micelles that encapsulated fluorescein-labeled lipid DHPE allow for viewing the tumor-specific accumulation of the micelles. Sixty minutes postinjection of micelles, the intensity within the tumor was significant, suggesting rapid entry of the pH-responsive pop-up polymeric micelles. Using the NIRF imaging system, it was supported that pH-responsive pop-up polymeric micelles labeled with Cy5.5 have high tumor-targeting efficiency.

In addition, thermally sensitive macromolecular drug carrier was targeted in a solid tumor by the method of hypothermia treatment. The targeting of thermally sensitive drug carrier was visualized by vasculature window of tumor and by the

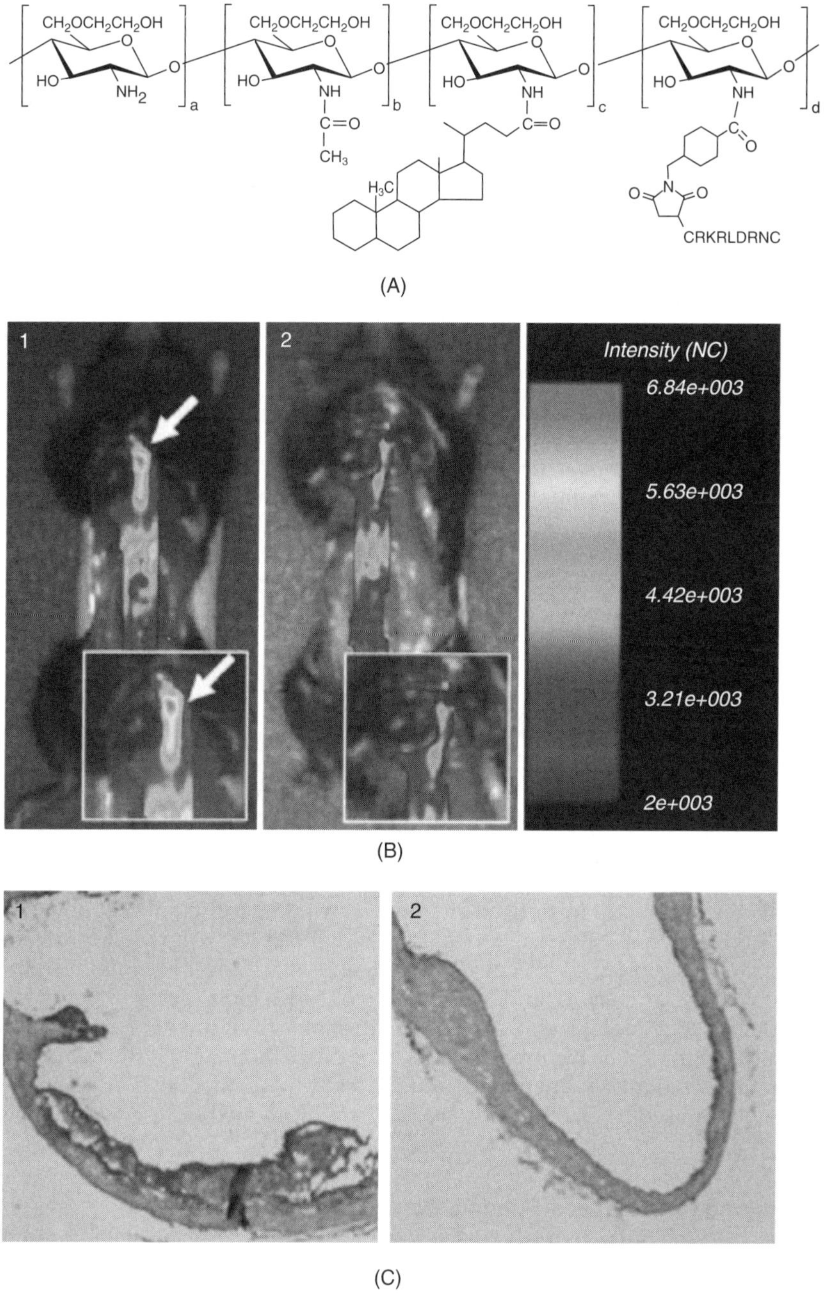

FIGURE 4 In vivo NIRF imaging of atherosclerotic lesion that was targeted by chitosan nanoparticles tagged with atherosclerotic homing peptide (AP). (**A**) Chemical structure of atherosclerotic homing peptide-modified glycol chitosan polymer. (**B**) NIRF imaging of aorta in atherosclerosis-induced *Ldlr*$^{-/-}$ mouse (1) and in normal mouse (2). (**C**) Oil Red O lipid stained aortas in atherosclersosis-induced *Ldlr*$^{-/-}$ mouse (1) and in normal mouse (2). *Source*: From Ref. 25.

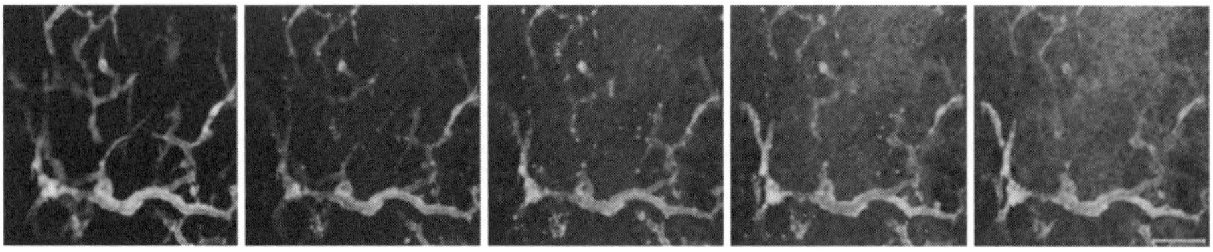

FIGURE 5 Macromolecular drug carrier in a solid tumor. The visualized activity and targeting of thermally sensitive macromolecular drug carrier in a solid tumor, thermally sensitive elastin like polypeptide 1 (Alexa 488 green labeled), and thermally insensitive elastin like 2 (Alexa 546 red labeled) in a tumor before and during hyperthermia treatment. Sequentially, tumor was heated to 41.5°C and then cooled to 37°C. The subsequential heat treatment activated and localized thermally sensitive elastin-like polypeptide in a target site. *Source*: From Ref. 32.

activation of fluorescence conjugated drug carrier (Fig. 5) (32). These NIRF imaging tools provided quantitative analysis with temporal and spatial resolution in the whole body of a small animal, including tumor accumulation of drug carrier, tissue distribution, and efficacy of drug carrier for treatment. Pharmacokinetics and biodistribution during molecular events associated with nanosized drug carriers became possible in the whole body. This new bioimaging technique will allow many researchers to visualize the in vivo fate of different nanosystems in drug delivery systems.

NIRF BIOIMAGING FOR BIOLOGICAL PROCESS

Progress in the field of molecular and cell biology has helped us to understand a number of biological disorders and has ultimately led to the development of novel therapeutics. To better understand human disorders, it is critical to identify which biological processes occur where, when, and under what physiological conditions. Among the various bioimaging modalities, fluorescence optical imaging technologies are powerful analytical methods not only in vitro but also in vivo. With the aid of recent developments in optical imaging instruments in combination with sophisticated NIRF nanoprobes, the technology now has been expanded to detect and monitor various biological processes in cells and in vivo in real-time. Herein, the unique concepts, properties, and applications of NIRF bioimaging systems for the imaging of various biological processes in vitro and in vivo will be discussed. Targeting and recognition at the molecular level of NIRF bioimaging: the protease-mediated, protein recognition–mediated, and protein kinase–mediated bioimaging approaches will be used as examples. Aspects that require further research will be highlighted.

NIRF Bioimaging of Proteases-Specific Nanoprobes

Proteases are enzymes that hydrolyze specific peptide bonds of proteins known to be overactive in a number of pathologies. The main diseases in which proteases or their inhibitors are involved include cancer, inflammation, diseases of the vasculature, Alzheimer's disease as well as infectious diseases. Therefore it is important to know which protease degrades where, when, and under what physiological conditions to better understand the onset and progression of these diseases. Accurate protease detection systems constitute crucial tools not only for drug screening systems used to identify drugs that target proteases, but also for the early diagnosis of diseases such as cancer, in order to enable the successful treatment of patients. Many approaches have been developed to visualize protease activities utilizing peptide chemistry. The most common detection method for protease activity is the use of peptide protease substrates containing chromophores at their termini. Cleavage between the peptide substrates and chromophores by activated proteases results in significant absorbance changes. Although this system is sensitive, its application is limited due to modest fluorescent changes that are too weak for using bioimaging systems. Therefore, nanoprobes with high fluorescent signal amplification strategies and specific recognition properties by target proteases are essential for the highly sensitive NIRF-based bioimaging systems. To improve the efficiency of NIRF bioimaging techniques, peptide cleavage mediated by its target protease was employed so that a signal can be associated with a specific disease. This activatable probe possessing the cleavable peptide linkage is optically silent in its quenched state and becomes highly fluorescent after the proteolysis of protease substrate linkers by the target protease. The peptide linkers therefore were chosen from families of possible protease enzyme substrate. Representative platform study was presented in Figure 6 (33). Using this platform, specific molecular events in vivo have been imaged for specific diseases and processes including breast cancer (34), E-selectin as a proinflammatory marker (35), atherosclerosis (36), thrombin activity (37), etc. Detailed characteristics and properties of various activatable nanoprobes will not be discussed herein, because they have been extensively reviewed elsewhere (19,38–40).

NIRF bioimaging system could detect MMPs activity directly within hours after treatment with MMPs activatable nanoprobes. The most widely applied NIRF bioimaging systems for MMPs detection in vivo were developed by Weissleder and colleagues (Fig. 6). They first utilized NIRF bioimaging systems for the detection of MMPs in vivo by using polymer-based MMPs activatable nanoprobes. The probe contains methoxy PEG-protected poly-L-lysine (PLL) co-polymer (PGC) as a backbone to which the Cy5.5 dyes are conjugated through a MMPs cleavable peptide substrate (41,42). The self-quenched probe showed 15-fold lower NIRF signals compared with free Cy dye at equimolar concentrations and 12-fold increased fluorescence signals in the presence of MMPs. Serial fluorescent images in MMPs-positive tumor-bearing mice showed that this NIRF imaging method allowed for the detection of tumors (42). A similar strategy utilized Cy dye conjugated to an enzymatically cleavable polymer backbone demonstrated that NIRF bioimaging could detect cathepsins B overexpressing diseases, such as breast cancer (43), rheumatoid arthritis (44), and atherosclerosis (45). In contrast to polymer-based NIRF nanoprobes, Lee et al. recently created an alternative, simple, and one-step NIRF bioimaging system utilizing gold nanoparticles (AuNP)–based MMPs activatable nanoprobes for use in protease inhibitor drug screening and early diagnosis of cancer in vivo (Fig. 7) (46). Biocompatible AuNPs have a considerable advantage in obtaining NIRF images because of NIRF quenching properties (47). This AuNP probe quenches conjugated NIRF dyes with high efficiency and is specifically activated by the target MMPs, which makes it possible to detect nanomolar amounts of protease, both in vitro and in an animal model. In vivo NIRF bioimaging experiments show an apparent positive contrast in MMPs-positive tumor-bearing mice.

Apoptosis, a programmed cell death process in multicellular organisms, plays a key role in the pathogenesis of many disorders, such as autoimmune and

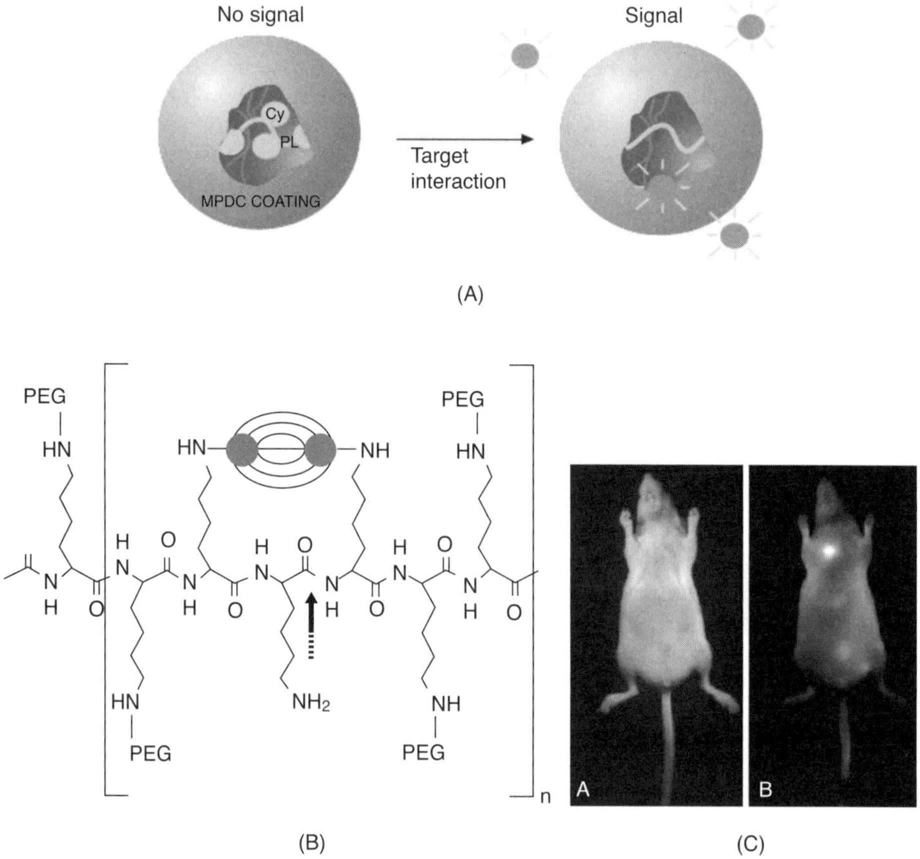

FIGURE 6 Activatabel NIRF imaging probe using protease-mediated cleavage site. (**A**) Schematic picture of activatabel probe, (**B**) poly(L-lysine) conjugated with peptide cleavage site and NIRF fluorophore, (**C**) in vivo NIRF signal activated in only local tumor in tumor implanted nude mouse. *Source*: From Ref. 33.

neurodegenerative disorders, cardiovascular disease, and tumor responses to chemotherapy or radiotherapy (48). The majority of effective anticancer therapies including most anticancer drugs and gamma-irradiation exert their lethal effect by inducing apoptosis. Therefore a defective apoptotic pathway in cancer cells often leads to treatment failure. Given the central role of apoptosis, it would be desirable to have a noninvasive imaging method to monitor this process in cancer patients undergoing chemotherapy and radiation treatments as well as for the development of apoptosis-related new drugs (49,50). Bullok et al. developed a peptide-based, cell-permeable, caspase-activatable NIRF nanoprobe, TcapQ647, and successfully detected apoptosis in cell culture and in vivo by using NIRF bioimaging (51,52). TcapQ647 comprised of D-amino acid TAT-peptide–based permeation peptide sequence to allow cell-permeation, and an L-amino acid, Asp–Glu–Val–Asp, as a caspase recognition substrate. Cleavable caspase-specific substrate was strongly quenched by addition of the far-red quencher, QSY 21, and the fluorophore,

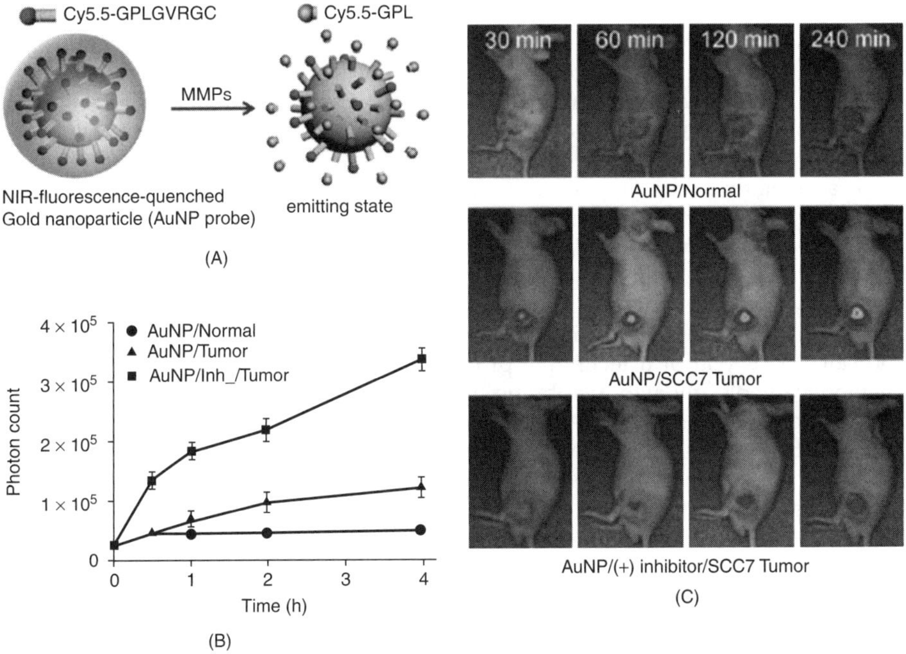

FIGURE 7 In vivo NIRF imaging detecting the activity of MMP. (**A**) Schematic picture of an AuNP probe. Cy 5.5/MMP-active peptide cleavage site was conjugated on AuNP. (**B**) Quantitative NIRF intensity analysis from MMP positive tumor bearing mice. (**C**) In vivo NIRF images of normal and subcutaneous MMP-overexpressed SCC7-tumor–bearing mice after injection of the AuNP. *Source*: From Ref. 46.

Alexa Fluor 647. Recombinant enzyme assays indicated that the activity of effector caspases, caspase-3 and 7, in living cells could be monitored by using TcapQ647-assisted NIRF bioimaging. Furthermore, in vivo experiments demonstrated that NIRF imaging allowed for the detection of parasite-induced apoptosis in human colon xenograft and liver abscess mouse models. Such a NIRF-based apoptosis-detecting process could simultaneously monitor delivery and the apoptotic potential of a new drug. Overall, the enhanced NIRF bioimaging methods highlight its potential in protease-related drug screening and NIRF tomographic imaging in vivo.

NIRF Bioimaging of Target Protein Recognition

The advanced techniques of cell biology permitted NIRF imaging to target protein recognition. Proteins in cellular systems that detect specific moieties or recognize specific local environments have been employed for bioimaging in the following manners: (*i*) an unaltered protein itself has a specific interaction with a local site (53–55), (*ii*) genetically engineered protein are generated expressing a specific receptor (56–57), and (*iii*) generation of monoclonal antibodies (58). Annexin V or C2-domain of synaptotagmin I binds to phophatidylserine (PS) that was externalized on the membrane surface of the apoptotic cells (52–55). As protein indicator, Annexin V, C2-domain of synaptotagmin I, has been derived to detect apoptotic cell as a strategy for molecular imaging. Simple conjugation of NIR fluorophore to annexin V (53,54), or a biotinylated C2-domain of synaptotagmin (55) were examples of an NIRF imaging indicator.

The use of genetically engineered proteins has been established as an imaging indicator by Tannous and colleagues (56). It has been shown that biotinylated transmembrane proteins on the cell surface was expressed by the reporter gene and biotin acceptor peptide (BAP), and that cells bearing this transmembrane protein reporter gene and biotin on the cell membrane could be imaged by using streptavidin-mediated fluorophore in vivo (Fig. 8) (56). BAP was incorporated between the N terminus of the transmembrane domain. Endogenous biotin ligases covalently linked a single biotin moiety to the BAP fusion protein. Biotinylated fusion protein was imaged using streptavidin-mediated fluorophore as an imaging indicator. This platform technique provided imaging tools of tumors, expressing metabolically biotinylated membrane surface receptor.

Recently, various biomarkers-modified nanosystem-based imaging probes (nanoprobes) have been extensively studied in molecular imaging field. Nanoprobes have yielded new strategies for designing imaging probes that efficiently detect target biomolecules or diagnose diseases. These nanoprobes have large surface, prolonged plasma half-life, enhanced stability, improved targeting, and reduced nonspecific binding, etc. Therefore, various biomarkers, such as peptides, proteins, antibodies, and aptomers, etc., were chemically attached to polymeric nanoparticles and QD. A wide variety of affinity biomarkers-conjugated nanosystems have been easily synthesized, and their in vivo unique ability to recognize target proteins could be visualized by using NIRF bioimaging in live animals. Nanosystem-based new imaging probes provide some advantages, including (*i*) a long circulation in the bloodstream, (*ii*) the ability to attach a high number of biomarkers to the polymer, (*iii*) a lack of immunogenicity and toxicity, and (*iv*) the ability to cross leaky endothelial barriers in tumors (59). For example, PEG-modified (PEGylated) PLL conjugated with a monoclonal antibody fragment and an E-selectin–binding peptide coupled to an NIR fluorophore showed a high binding affinity for E-selectin (CD62E), a proinflammatory marker that is upregulated in many tumors and is frequently localized in dividing microvascular endothelial cells and tissues undergoing active angiogenesis. This polymer-based targeted agent has a high binding specificity (20–30-fold over nonspecific uptake) and has been used to image E-selectin expression on human endothelial cells (60). Also, a quantum dot with emission spectra in NIRF region was conjugated with epidermal growth factor (Fig. 9) (61). EGF conjugated QD was selectively accumulated in EGF receptor over expressed tumor, compared with the administered bare QD. In vivo NIRF imaging showed quantitative analysis of tumor accumulation of protein reporter, tissue distribution, and clearance time. Ex vivo NIR imaging of organs displayed quantitative biodistribution of each organ as well as targeted tumor. Proteins capable of acting as a reporter are useful to build NIRF imaging as well as for exploiting underlying cellular mechanisms. Understanding cellular and molecular events in depth is necessary to propose new reporter systems for NIRF bioimaging.

CONCLUSION

Despite technical advances in many areas of analytic systems, the detection and imaging of molecular events and biological process in vivo remains poor. A meaningful impact on early disease screening, staging, and treatment is unlikely to occur

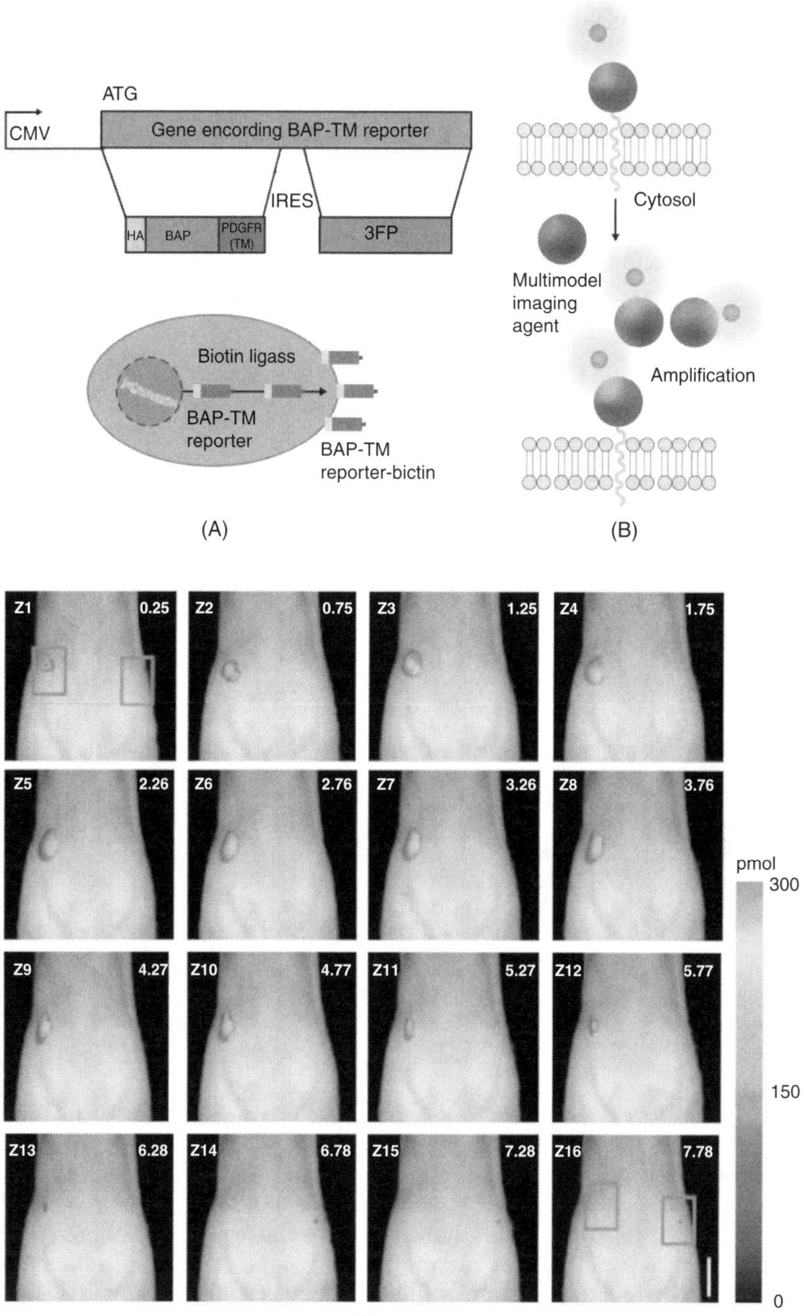

FIGURE 8 The protein imaging reporter for monitoring cellular events. (**A**) Biotinylated transmembrane proteins were genetically coded. Biotin-bearing cells and biotinylated cell membrane were imaged by using fluorescently labeled streptavidin. (**B**) Quantitative fluorescence intensity in tumor. (**C**) In vivo fluorescence image of glioma tumors expressing biotinylated transmembrane protein (*left*) and control tumors (*right*) after intravenous streptavidin-Alexa 690. Images were presented at different Z depth. *Source*: From Ref. 56.

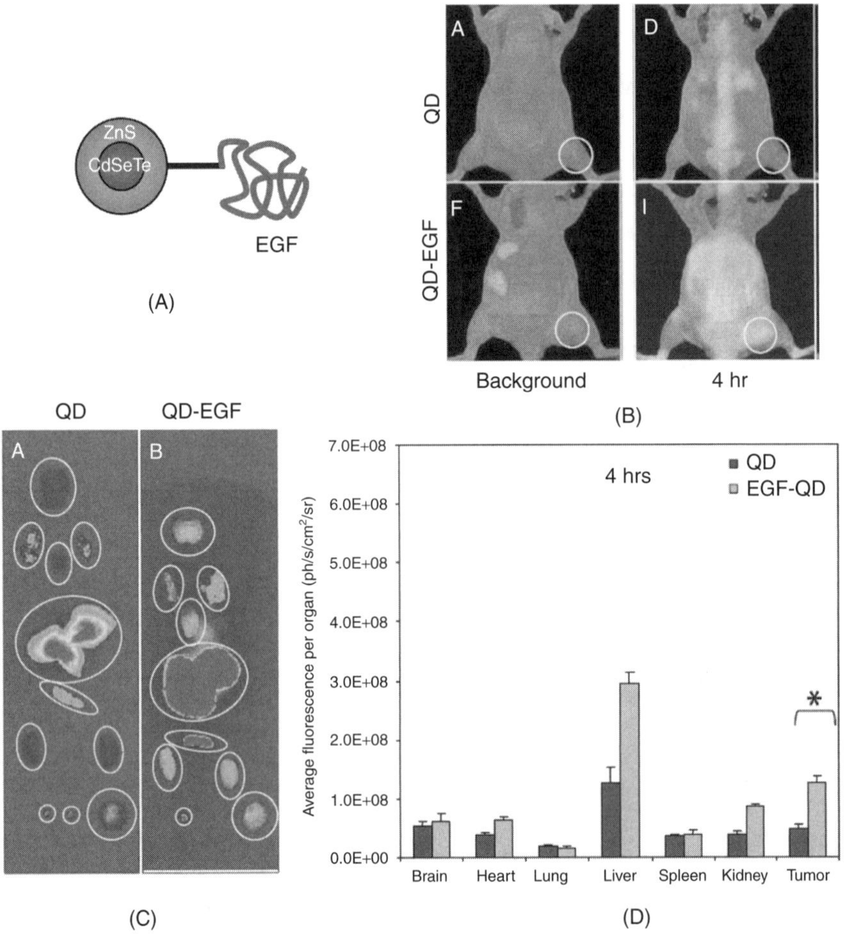

FIGURE 9 Imaging of protein–protein interaction: (**A**) schematic picture of epidermal growth factor conjugated quantum dot, (**B**) in vivo NIRF imaging of EGF receptor overexpressed tumor bearing mice after four hours of intravenous injection of EGF conjugated QD, (**C**) ex vivo NIRF image for each organ, and (**D**) quantitative analysis of organs. *Source*: From Ref. 59.

until target sensitivity and resolution in vivo improves significantly. Interdisciplinary research that combines optical imaging science and NIRF nanoprobes has generated novel NIRF bioimaging tools that allow researchers to detect and monitor various nanosystems in drug delivery and molecular imaging. Dynamics and kinetics resulting from a biological response at the molecular level is possibly assessed by NIRF bioimaging in vivo. These techniques can trace the fate of biomolecules or drugs directly, and noninvasively. NIRF bioimaging provides unique data that can aid in selecting the best drug candidates, determining optimal dosing regimens, and lowering risks of failure. Moreover, NIRF bioimaging systems do not require expensive instrumentation and are able to be analyzed in a rapid and efficient fashion. NIRF bioimaging will offer the unique opportunity to study and diagnose various diseases at the molecular and cellular level. The application of NIRF bioimaging system is just the beginning, and much work remains to be done, including standardization, improvements in image analysis, and characterization of the accuracy or reproducibility of imaging techniques. When NIRF bioimaging is fully integrated into the field of medical imaging, the field of diagnostics will be revolutionized.

REFERENCES

1. Dobrovolskaia MA, McNeil SE. Immunological properties of engineered nanomaterials. Nat Nanotechnol 2007; 2:469–478.
2. Nie S, Xing Y, Kim GJ, et al. Nanotechnology applications in cancer. Annu Rev Biomed Eng 2007; 9:257–288.
3. Alexis F, Pridgen E, Molnar LK, et al. Factors affecting the clearance and biodistribution of polymeric nanoparticles. Mol Pharm 2008; 5:505–515.
4. Tallury P, Payton K, Santra S. Silica-based multimodal/multifunctional nanoparticles for bioimaging and biosensing applications. Nanomed 2008; 3:579–592.
5. Cho K, Wang X, Nie S, et al. Therapeutic nanoparticles for drug delivery in cancer. Clin Cancer Res 2008; 14:1310–1316.
6. Kaur IP, Bhandari R, Bhandari S, et al. Potential of solid lipid nanoparticles in brain targeting. J Control Release 2008; 127:97–109.
7. Parveen S, Sahoo SK. Polymeric nanoparticles for cancer therapy. J Drug Target 2008; 16:103–123.
8. Agarwal A, Saraf S, Asthana A, et al. Ligand based dendritic systems for tumor targeting. Int J Pharm 2008; 350:3–13.
9. Sun C, Lee JS, Zhang M. Magnetic nanoparticles in MR imaging and drug delivery. Adv Drug Deliv Rev 2008; 60:1252–1265.
10. Perez JM, Josephson L, O'Loughlin T, et al. Magnetic relaxation switches capable of sensing molecular interactions. Nat Biotechnol 2002; 20:816–820.
11. Nahrendorf M, Zhang H, Hembrador S, et al. Nanoparticle PET-CT imaging of macrophages in inflammatory atherosclerosis. Circulation 2008; 117:379–387.
12. Burton JB, Johnson M, Sato M, et al. Adenovirus-mediated gene expression imaging to directly detect sentinel lymph node metastasis of prostate cancer. Nat Med 2008; 14:882–888.
13. Segal E, Sirlin CB, Ooi C, et al. Decoding global gene expression programs in liver cancer by noninvasive imaging. Nat Biotechnol 2007; 25:675–680.
14. Licha K, Hessenius C, Becker A, et al. Synthesis, characterization, and biological properties of cyanine-labeled somatostatin analogues as receptor-targeted fluorescent probes. Bioconjug Chem 2001; 12(1):44–50.
15. Klohs J, Wunder A, Licha K. Near-infrared fluorescent probes for imaging vascular pathophysiology. Basic Res Cardiol 2008; 103(2):144–151.
16. Mahmood U. Near infrared optical applications in molecular imaging. Earlier, more accurate assessment of disease presence, disease course, and efficacy of disease treatment. IEEE Eng Med Biol Mag 2004; 23(4):58–66.
17. Weissleder R. A clearer vision for in vivo imaging. Nat Biotechnol 2001; 19(4):316–317.
18. Frangioni JV. In vivo near-infrared fluorescence imaging. Curr Opin Chem Biol 2003; 7(5):626–634.
19. Funovics M, Weissleder R, Tung CH. Protease sensors for bioimaging. Anal Bioanal Chem 2003; 377(6):956–963.
20. Licha K, Olbrich C. Optical imaging in drug discovery and diagnostic applications. Adv Drug Deliv Rev 2005; 57(8):1087–1108.
21. Park K, Kim JH, Nam YS, et al. Effect of polymer molecular weight on the tumor targeting characteristics of self-assembled glycol chitosan nanoparticles. J Control Release 2007; 122(3):305–314.

22. Hwang HY, Kim IS, Kwon IC, et al. Tumor targetability and antitumor effect of docetaxel-loaded hydrophobically modified glycol chitosan nanoparticles. J Control Release 2008; 128(1):23–31.
23. Min KH, Park K, Kim YS, et al. Hydrophobically modified glycol chitosan nanoparticles-encapsulated camptothecin enhance the drug stability and tumor targeting in cancer therapy. J Control Release 2008; 127(3):208–218.
24. Masotti A, Vicennati P, Boschi F, et al. A novel near-infrared indocyanine dye-polyethylenimine conjugate allows DNA delivery imaging in vivo. Bioconjug Chem 2008; 19(5):983–987.
25. Park K, Hong HY, Moon HJ, et al. A new atherosclerotic lesion probe based on hydrophobically modified chitosan nanoparticles functionalized by the atherosclerotic plaque targeted peptides. J Control Release 2008; 128(3):217–223.
26. Hong HY, Lee HY, Kwak W, et al. Phage display selection of peptides that home to atherosclerotic plaques: IL-4 receptor as a candidate target in atherosclerosis. J Cell Mol Med 2007, postprint;10.1111/j.1582–4934.2007.00189.x.
27. Lugli SM, Feng N, Heim MH, et al. Tumor necrosis factor alpha enhances the expression of the interleukin (IL)-4 receptor alpha-chain on endothelial cells increasing IL-4 or IL-13-induced Stat6 activation. J Biol Chem 1997; 272(9):5487–5494.
28. Van Sluis R, Bhujwalla ZM, Raghunand N, et al. In vivo imaging of extracellular pH using 1H MRSI. Magn Reson Med 1999; 41(4):743–750.
29. Lee ES, Na K, Bae YH. Super pH-sensitive multifunctional polymeric micelle. Nano Lett 2005; 5(2):325–329.
30. Yin H, Lee ES, Kim D, et al. Physicochemical characteristics of pH-sensitive poly(L-histidine)-b-poly(ethylene glycol)/poly(L-lactide)-b-poly(ethylene glycol) mixed micelles. J Control Release 2008; 126(2):130–138.
31. Lee ES, Gao Z, Kim D, et al. Super pH-sensitive multifunctional polymeric micelle for tumor pH(e) specific TAT exposure and multidrug resistance. J Control Release 2008; 129(3):228–236.
32. Dreher MR, Liu W, Michelich CR, et al. Thermal cycling enhances the accumulation of a temperature-sensitive biopolymer in solid tumors. Cancer Res 2007; 67(9):4418–4424.
33. Weissleder R, Tung CH, Mahmood U, et al. In vivo imaging of tumors with protease-activated near-infrared fluorescent probes. Nat Biotechnol 1999; 17(4):375–378.
34. Law B, Curino A, Bugge TH, et al. Design, synthesis, and characterization of urokinase plasminogen-activator-sensitive near-infrared reporter. Chem Biol 2004; 11(1):99–106.
35. Kang HW, Weissleder R, Bogdanov A Jr. Targeting of MPEG-protected polyamino acid carrier to human E-selectin in vitro. Amino Acids 2002; 23(1–3):301–308.
36. Jaffer FA, Kim DE, Quinti L, et al. Optical visualization of cathepsin K activity in atherosclerosis with a novel, protease-activatable fluorescence sensor. Circulation 2007; 115:2292–2298.
37. Jaffer FA, Tung CH, Gerszten RE, et al. In vivo imaging of thrombin activity in experimental thrombi with thrombin-sensitive near-infrared molecular probe. Arterioscler Thromb Vasc Biol 2002; 22:1929–1935.
38. Rao J, Dragulescu-Andrasi A, Yao H. Fluorescence imaging in vivo: Recent advances. Curr Opin Biotechnol 2007; 18(1):17–25.
39. Ntziachristos V, Bremer C, Weissleder R. Fluorescence imaging with near-infrared light: New technological advances that enable in vivo molecular imaging. Eur Radiol 2003; 13(1):195–208.
40. Sevick-Muraca EM, Houston JP, Gurfinkel M. Fluorescence-enhanced, near infrared diagnostic imaging with contrast agents. Curr Opin Chem Biol 2002; 6(5):642–650.
41. Bremer C, Bredow S, Mahmood U, et al. Optical imaging of matrix metalloproteinase–2 activity in tumors: Feasibility study in a mouse model 1. Radiology 2001; 221:523–529.
42. Bremer C, Tung CH, Weissleder R. In vivo molecular target assessment of matrix metalloproteinase inhibition. Nat Med 2001; 7(6):743–748.
43. Bremer C, Tung CH, Bogdanov A Jr, et al. Imaging of differential protease expression in breast cancers for detection of aggressive tumor phenotypes. Radiology 2002; 222(3):814–818.

44. Ji H, Ohmura K, Mahmood U, et al. Arthritis critically dependent on innate immune system players. Immunity 2002; 16(2):157–168.
45. Wunder A, Tung CH, Muller-Ladner U, et al. In vivo imaging of protease activity in arthritis: A novel approach for monitoring treatment response. Arthritis Rheum 2004; 50(8):2459–2465.
46. Lee S, Cha EJ, Park K, et al. A near-infrared-fluorescence-quenched gold-nanoparticle imaging probe for in vivo drug screening and protease activity determination. Angew Chem Int Ed Engl 2008; 47(15):2804–2807.
47. Dubertret B, Calame M, Libchaber AJ. Single-mismatch detection using gold-quenched fluorescent oligonucleotides. Nat Biotechnol 2001; 19(4):365–370.
48. Vaux DL, Korsmeyer SJ. Cell death in development. Cell 1999; 96(2):245–254.
49. Fischer U, Schulze-Osthoff K. New approaches and therapeutics targeting apoptosis in disease. Pharmacol Rev 2005; 57(2):187–215.
50. Kim K, Lee M, Park H, et al. Cell-permeable and biocompatible polymeric nanoparticles for apoptosis imaging. J Am Chem Soc 2006; 128:3490–3491.
51. Bullok K, Piwnica-Worms D. Synthesis and characterization of a small, membrane-permeant, caspase-activatable far-red fluorescent peptide for imaging apoptosis. J Med Chem 2005; 48(17):5404–5407.
52. Bullok KE, Maxwell D, Kesarwala AH, et al. Biochemical and in vivo characterization of a small, membrane-permeant, caspase-activatable far-red fluorescent peptide for imaging apoptosis. Biochemistry 2007; 46:4055–4065.
53. Schellenberger EA, Sosnovik D, Weissleder R, et al. Magneto/optical annexin V, a multimodal protein. Bioconjug Chem 2004; 15:1062–1067.
54. Ntziachristos V, Schellenberger EA, Ripoll J, et al. Visualization of antitumor treatment by means of fluorescence molecular tomography with an annexin V-Cy5.5 conjugate. Proc Natl Acad Sci U S A 2004; 101:12294–12299.
55. Jung H, Kettunen M, Davletov B, et al. Detection of apoptosis using the C2A domain of synaptotagmin I. Bioconjug Chem 2004; 15:983–987.
56. Tannous BA, Grimm J, Perry KF, et al. Metabolic biotinylation of cell surface receptors for in vivo imaging. Nat Methods 2006; 3:391–396.
57. Ward MW, Rehm M, Duessmann H, et al. Real time single cell analysis of Bid cleavage and Bid translocation during caspase-dependent and neuronal caspase-independent apoptosis. J Biol Chem 2006; 281(9):5837–5844.
58. Xu H, Baidoo K, Gunn AJ, et al. Design, synthesis, and characterization of a dual modality positron emission tomography and fluorescence imaging agent for monoclonal antibody tumor-targeted imaging. J Med Chem 2007; 50:4759–4765.
59. Park JH, Kwon S, Nam JO, et al. Self-assembled nanoparticles based on glycol chitosan bearing 5beta-cholanic acid for RGD peptide delivery. J Control Release 2004; 95:579–588.
60. Yokoyama M, Okano T, Sakurai Y, et al. Toxicity and antitumor activity against solid tumors of micelle-forming polymeric anticancer drug and its extremely long circulation in blood. Cancer Res 1991; 51:3229–3236.
61. Diagaradjane P, Orenstein-Cardona JM, Colon-Casasnovas NE, et al. Imaging epidermal growth factor receptor expression in vivo: Pharmacokinetic and biodistribution characterization of a bioconjugated quantum dot nanoprobe. Clin Cancer Res 2008; 14(3):731–741.

Index

ADC. *See* Analog-to-digital converter
α-Fetoprotein (AFP), 101
Analog-to-digital converter (ADC), 169, 171
Antioxidant enzyme catalase (CAT), 38, 45
Apolipoprotein E (Apo E), 219
Apomorphine brain concentrations, 229, 230f
Area under the curve (AUC), 220, 221, 223, 224, 228, 229, 234
Assembled biocapsule consisting of two micromachined membranes, 120, 121f
Atom transfer radical polymerization (ATRP), 94, 96–97, 97f
Atomic force microscopy (AFM), 315–319, 321–324, 327, 328–329
Atomic resolution image of an Au island on an amorphous carbon substrate, 265f
ATRP. *See* Atom transfer radical polymerization

BCS. *See* Biopharmaceutic Classification System
β–Lactoglobulin (BLG), 72, 78, 83
Biodegradable polymers, 18–21
  lactide/glycolide copolymers, 18–19
  poly(ε–caprolactones), 19–20
  polyanhydrides, 20–21
  polyorthoesters, 21
Biological micro electro mechanical systems (BioMEMS), 120
Biological response modifier (BRM), 57
BioMEMS. *See* Biological micro electro mechanical systems
Biopharmaceutic classification and drug molecules, 17
Biopharmaceutic Classification System (BCS), 17, 194, 206, 207
Blood-brain barrier (BBB), 4, 219
Bonded together to form a cell-containing cavity bounded by membranes, 121f
Bone marrow (BM), 57, 143, 199
Bovine serum albumin (BSA), 8, 71, 177
BRM. *See* Biological response modifier
Brucella abortus, 82
Calcium phosphate-PEG-insulin-casein oral insulin delivery system, 119f
Calculated X-ray powder diffractogram of mineral pseudo-brookite, 273f
Cancer treatment
  chemotherapeutic agents in, 58–59
    chemical damage of DNA in cell nuclei, 58–59
    of synthesis of mitotic spindles, 59
    of synthesis of pre-DNA molecule, prevention, 58
  nanoparticulate-based drug delivery in
    GNP for anticarcinogenic drug delivery, 60–61
    liposomes in, 61
    magnetic nanoparticles, 61–63
    nanoparticle-based delivery of, 64–65
    in photodynamic therapy, 64
    polymers as drug delivery systems, 63–64
  techniques of, 57–58
Carbon magnetic nanoparticles (CMNPs), 63
Carbon nanotube (CNT), 101, 316–321
CAT. *See* Antioxidant enzyme catalase
Central nervous system (CNS), 220, 223, 226, 332t
CFU-GM. *See* Colony-forming unit-granulocyte macrophage
CLM. *See* Coupled luminescent method
CLSM. *See* Confocal laser scanning microscopy
CMNPs. *See* Carbon magnetic nanoparticles
CNT. *See* Conjugated carbon nanotube
Colony-forming unit-granulocyte macrophage (CFU-GM), 190, 199–200
  colony of, 199f
Complement activation pathways, 197, 198f
Confocal laser scanning microscopy (CLSM/LSCM), 172–173
Conjugated carbon nanotube (CNT), 101, 316–321

Coronal slice from a whole-body MRI of mouse, 343f
Coupled luminescent method (CLM), 206
CPEDR. *See* Crystal powder electron diffraction ring
Crystal powder electron diffraction ring (CPEDR), 270, 283–287

Dehydration-rehydration vesicle (DRV), 38, 40
Dendrimers, 10–11, 55, 101, 138–140, 144, 178, 294, 212, 367–368
Dendrimer synthesis, 139f
Density of states (DOS), 351, 352
Dermal/transdermal drug delivery
  applications, 140–145
    cosmetics/sunscreens/skin protectants, 140–141
    drug molecules, 141–142
    gene delivery, 143–145
    protein delivery, 142–143
  lipid-based nanosystems, 130–135
    deformable liposomes, 132–133
    ethosomes, 133–134
    lipid nanoparticles, 134–135
    liposomes, 130–132
  polymer-based nanosystems, 137–140
    dendrimers, 138–140
    polymeric nanocapsules/nanoparticles, 137–138
  surfactant-based nanosystems, 135–137
    nanoemulsion, 137
    niosomes, 135–137
Dichlorodihydrofluoroscein (DCFH), 211
DLS. *See* Dynamic light scattering
Drug-loaded poly(D,L-lactic-co-glycolic acid) nanoparticles, 23f
DRV. *See* Dehydration-rehydration vesicle
Dynamic light scattering (DLS), 344

EBID. *See* Electron beam-induced deposited
ECM. *See* Extracellular matrix
EDX quantification, 260–263
EDX spectrum, 259
EGFR. *See* Epidermal growth factor receptor
Electron beam-induced deposited (EBID), 328
Electron diffraction patterns of silicon crystal, 287f
Electron energy loss spectroscopy, 260–263
Electron Guns, 256t
Electron microscopy
  basic theory of, 253–255
  EDX quantification in, 260–263
  electron energy loss spectroscopy in, 260–263
  instrumentation of, 255–257
  in nanoparticle characterization, 263–266
  stem-related techniques in, 257–260
    energy-dispersive X-ray spectroscopy, 257–260
    Z-contrast imaging, 257
ELNES. *See* Energy loss near-edge fine structure
Emulsification solvent diffusion method, 24, 25, 54, 25f
Energy loss near-edge fine structure (ELNES), 263
Energy-dispersive X-ray spectrum, 259f
Enzyme and Ac-enzyme, 40f
Epidermal growth factor receptor (EGFR), 60, 360
Ethosomal skin penetration, 133f
Extracellular matrix (ECM), 184, 186

Fc receptor (FcR), 206
FCM. *See* Fluorescence confocal microscopy
Field emission gun (FEG), 252, 256, 260, 264
FEG. *See* Field emission gun
Fixed brain tissue, 356, 358f
Fibroblast cells, 9, 105, 106f
Fluorescence confocal microscopy (FCM), 173, 186
Fluorescence microscopy pictures showing cells, 183f
Fluorescence resonance energy transfer (FRET), 176, 355, 358–359
Fluorescence spectra, 170f
Fluoroscopy images
  of CdS:Mn/ZnS Qdot, 360f
  Omnipaque, 360f
FRET. *See* Fluorescence resonance energy transfer
Fullerene-based nanostructures
  contrast agents in MRI, 330–331
  current gadolinium chelate-based, imitations of, 333
  EW frontiers in MRI, 334
  gadolinium chelates-based MRI, 331–333
  hydrochalarones-new breakthrough in, 339–346

nanomaterial for biomedical applications, 333–334
trimetaspheres®–based MRI contrast agents, 334–339

GaN nanowires, 323–327, 325f
Generally regarded as safe (GRAS), 71, 73
Gold nanoparticles (GNPs)
biomolecular delivery of, 105–107
diagnostic applications of, 98–102
composite biosensors of, 101
electrochemical biosensors in, 99–101
optical biosensors of, 98–99
polymer-hybridized biosensors of, 102
drug delivery, 105
function of, 92–98, 94t
"grafting to", 93–94
"grafting from", 94–98
hyperthermia using, 107–108
therapeutic applications of, 102–105
therapy via active and passive in vivo targeting, 102–105
GRAS. *See* Generally regarded as safe

HAADF detectors, 255f
HCS. *See* High content screening
HEK. *See* Human embryonic kidney
High content screening (HCS), 172, 186
High-pressure homogenization (HPH), 38, 42
High-resolution transmission electron microscopy (HRTEM), 270–272, 275, 282, 293–294, 297, 299, 301, 302, 305, 317
HPH. *See* High-pressure homogenization
Human embryonic kidney (HEK), 177, 178, 210
Human serum albumin (HSA), 71–73, 78–79, 80, 83, 86, 192–193, 338
HRTEM. *See* High-resolution transmission electron microscopy
HRTEM image resolution
and visibility of zone axes, 302t
Hyperthermia, 58

In vitro blood interaction
blood coagulation, 196–197
experimental procedure, 196
nanoparticles on coagulatory changes, 196–197
cell-based, 199–200
complement activation, 197–199
cytotoxic activity of NK cells, 205–207
activating and inhibitory receptors, 206–207
cytokines, 206
FC receptor, 206
hemolysis, 193–194
leukocyte proliferation, 200–201
macrophage/neutrophil function, 201–205
chemotaxis, 201–202
cytokine induction, 203–204
macrophage chemotaxis, 202–203
oxidative burst, 205
phagocytosis, 203
pharmacological/toxicological assessments, 207–208
plasma protein binding, 190–193
isothermal titration calorimetry, 192
size-exclusion chromatography, 192
surface plasmon resonance, 193
platelet aggregation induced by nanoparticles, 194–196
flow cytometry, 195
vascular thrombosis, 196
zymography, 195
special considerations, 208
apoptosis and mitochondrial dysfunction, 211–213
cell cycle analysis, 210
cell viability using MTT, 209–210
cytotoxicity, 208
loss of monolayer adherence test, 210
membrane integrity, 208–209
oxidative stress, 211
target organ toxicity, 210–211
In vitro characterization (nanoparticle cellular interaction)
confocal laser scanning microscopy in, 172–173
fluorescence confocal microscopy, 173
high content screening in, 172
histogram files, 171
laser capture microdissection, 173–174
laser scanning cytometry in, 171
sterility of, 174
detection of endotoxin, 174–175
detection of mycoplasma, 175–176
microbial contamination, 175
targeting studies in
β–galactosidase, 186
cell adhesion study, 184–186
cell binding and transfection studies, 176–179

In vitro characterization (nanoparticle cellular interaction) (*Continued*)
cell viability (MTS), 183
cellular uptake studies, 179–183
colony in soft agar, 184
electrophoretic mobility shift, 184
In vitro NPDDS
boundary layer effects, 164
drug release from particulate drug carriers, 156
empiric models of drug release, 159–160
erodible and biodegradable systems, 164–165
factors influencing drug release, 157–159
kinetics of drug release from micro/nanoparticles, 156–157
methods of measurement of drug release, 165–166
monolithic devices (microparticles), 161–163
pore effects, 163–164
reservoir-type devices (microcapsules), 160–161
swelling controlled system, 165
In vivo NIRF imaging
detecting activity of MMP, 379f
of gene expression, 373f
In vivo SLN, 219–232
administered intravenously, 223–226
by intraperitoneal administration, 226–228
as delivery systems by duodenal route, 228–232
as ocular drug delivery systems, 223
as potential MRI diagnostics, 222–223
microemulsions, 232–235
microemulsions for transdermal application of Apomorphin, 233–235
prepared from warm microemulsions, 221–222
unloaded biodistribution in, 222
Inhibition in pancreatic cancer cell lines, 185f
Isothermal titration calorimetry (ITC), 191–192

Lactate dehydrogenase (LDH), 208–209
LAL. *See* Limulus amebocyte lysate
Large unilamellar vesicles (LUVs), 130f, 131
Laser capture microdissection (LCM), 173–174, 186
Laser scanning cytometry (LSC), 171, 172, 186
Layer-by-layer (LBL), 101, 102
LCM. *See* Laser capture microdissection
Light for releasing encapsulated materials, 109f
Limulus amebocyte lysate (LAL), 174
Liposomes, 7, 16, 36, 38, 39, 61, 62f131t
Living radical polymerization (LRP), 96
Low-density lipoprotein (LDL), 7, 219
Lower critical solution temperature (LCST), 108
LRP. *See* Living radical polymerization
Luteinizing hormone-releasing hormone (LHRH), 63

Macromolecular drug carrier, 376f
Magnetic resonance imaging (MRI), 9, 222
Maltose-binding protein (MBP), 358–359, 359F
Matrix metalloproteinase (MMP), 195–196, 377, 379f
Mean residence time (MRT), 220, 221
Membrane integrity assays (MIAs), 208
Methotrexate (MTX), 104
Micrococcus lysodeikticus, 108
Microscopic/spectroscopic characterization
electron microscopy in, 239– 242
scanning electron microscopy, 241–242
transmission electron microscopy, 240–241
optical absorption, 242–245
outlook of, 248
Raman spectroscopy, 245–248
MLVs. *See* Multilamellar vesicles
MRI. *See* Magnetic resonance imaging
MRI contrast agents
commercially available gadolinium-based, 332t
EW frontiers in, 334
hydrochalarones-new breakthrough in, 339–346
structures of actives in, 331f
structures of actives in commercial, 331f
trimetaspheres®-based, 334–339
Multichannel analyzer (MCA), 258, 258
Multilamellar vesicles (MLVs), 130f, 131, 132
Mycoplasma, 174, 176

Nanocarriers
of cosmetics/skin protectants delivered by, 140t
for enzymes, 36
of macromolecules, 142t
of small drugs, 142t
Nano-Crystallography Database (NCD), 304
Nanoparticle-in-microsphere oral system (NiMOS), 86, 87, 87f

Nanoparticles
average number of covalent bonds, 244f
biodegradable polymers used in, 17–18
conformational change-induced for DNA detection, 100f
edge energy values obtained for, 243f
of DNA-loaded protein, 80f
EDX line profile, 268f
in vitro experiments
apoptosis/mitochondrial dysfunction, 211–213
cell cycle analysis, 210
cell viability using MTT, 209–210
cytotoxicity, 208
loss of monolayer adherence test, 210
membrane integrity, 208–209
oxidative stress, 211
target organ toxicity, 210–211
in vivo NIRF imaging of, 375f
in vivo fate of, 81f
NiMOS through oral route, 87f
limitations of, 123
marketed or scientifically explored, 2t
of protein by phase separation method, 77f
of protein, by emulsion/solvent extraction method, 79f
particle diameter for (Au), 245f
passive targeting by enhanced permeability/retention effect, 84f
pH-sensitive molecular gates, 118f
platelet aggregation induced
flow cytometry, 195
vascular thrombosis, 196
zymography, 195
scanning transmission, 267f
SPR, 191–193, 193f
surface enhanced Raman spectra of, 247f
surface modifications of, 76f
therapeutic applications of, 103f
transmission electron microscopy image of Au, 264f
tumor accumulation/tissue distribution, 372f
voltametric identification of, 101f
Nanoparticulate drug delivery systems (NPDDS)
antibody targeting of, 7–8
by freeze-drying, 28f
blood-brain barrier cancer treatment, 4–7
in diagnostic medicine, 9
features of 1–3
hydrogel nanoparticles in, 9
lipid nanoparticles/nanostructured lipid carriers, 8
major applications of, 3–4
mucoadhesive/gastrointestinal tract absorption, 8–9
of protein polymers used for, 85t
research reports covering various applications of, 5–6t
terminologies used for, 3t
vaccine delivery for, 8
Nanophotothermolysis of cancer cells, 60f
Nanostructures
comparative account of, 147t
mechanical properties of
atomic force microscopy for bend test, 315–317
of carbon nanotubes, 317–321
of nanosprings, 327–328
of nanowires, 321–327
Nanostructured lipid carriers (NLCs), 134
Nanosystems used for skin applications, 129f
Natural killer (NK), 190
NCD. *See* Nano-Crystallography Database
Near infrared fluorescence (NIRF), 367
Near infrared region (NIR), 108
NIR region, characteristics of, 369f
NPDDS, for gene delivery
block inonomer complexes in, 54
dendrimers in, 55
gene gun or ballistic particle-mediated gene delivery in, 52
nanogels in, 54
nanoparticle-mediated transport of, 52–53
nanospheres/nanocapsules/aquasomes, 55
polymeric micelles, 53–54
self-assembling gene delivery systems, 53
NPDDS, for macromolecules
dermal and transdermal delivery, 44–46
enzymes with therapeutic activity, 36
lipidic nanoparticulate drug delivery systems, 36–42
acylation of enzymes to promote hydrophobic interaction, 39–41
chemical link of enzymes directly to liposome surface, 41
liposomes, 36–39
solid lipid nanoparticles, 42
pharmaceutical nanocarriers for enzymes, 36
polymeric, 43–44

NPDDS, for treatment of diabetes, 117–123
 biomems for insulin delivery, 120
 for insulin delivery, 120–123
 insulin delivery, 119–120
 limitations of, 123
 oral insulin administration, 118–119
 polymeric nanoparticles, 117–118
NRF bioimaging
 for active targetable drug carriers, 373–376
 for biological process, 376–380
  of proteases-specific nanoprobes, 376–379
  of target protein recognition, 379–380
 introduction to, 367–376
 for nanosized drug carriers, 370–373
 for nanosized drug delivery systems, 370
 in nanosystems, 368–370
 technology in, 367–368
Nuclear magnetic resonance (NMR), 360

Omnipaque, 360
Optical density (OD), 194

PAGE. *See* Polyacrylamide gel electrophoresis
Passive Skin Permeation, 129t
PBS. *See* Phosphate-buffered saline
Percentage reduction in blood glucose level after nasal administration, 107f
Percentage reduction in blood glucose level after oral administration, 107f
PET. *See* Positron emission tomography
PFH. *See* Plasma free hemoglobin
PHAM. *See* Phytohemaglutinin
Phosphate-buffered saline (PBS), 169
Photoluminescence (PL), 350
Photomultiplier tube (PMT), 169, 171, 173
Phytohemaglutinin (PHAM), 182, 200
Pinhole, 172
Plasma free hemoglobin (PFH), 194
Platelet-rich plasma (PRP), 194
PMT. *See* Photomultiplier tube
Plot of 1/T 1 versus concentration of functionalized trimetasphere, 337f
Plots of intensity of surface enhanced Raman spectra, 248f
Poly(ε–caprolactones), 19–20
Poly(alkyl cyanoacrylate) nanoparticles by anionic polymerization, 26f
Polyacrylamide gel electrophoresis (PAGE), 190
Polyamidoamine dendrimer of varying surface groups, 139f
Polyanhydrides, 20–21
Polycaprolactone (PCL), 17, 19, 86, 87f
Polymer material, responses to, 19, 19t
Polymerase chain reaction (PCR), 175–176
Polymerization in living radical, 95f
Polymer nanoparticles, production of, 21–29
 colloidal drug carriers, 29–31
 emulsification solvent in, 25
 emulsion polymerization, 25–27
 emulsion-solvent evaporation/extraction, 22–24
 laboratory-scale production of, 22
 large-scale pilot production of (drug-loaded), 27–29
 salting out, 24
Polyorthoesters, 21–22
Positron emission tomography (PET), 330, 349, 348
Precession electron diffraction (PED), 270, 275, 282, 285, 287–290, 293, 294, 297–301, 303, 311–313
Probe D-289, 134f
Protein-based NPDDS
 biological characteristics of protein nanoparticles, 81–87
  biocompatibility/immunogenicity, 81–83
  biodegradation, 83–84
  biodistribution and applications, 84–87
 characteristics of protein polymers, 70–73
  animal/insect proteins, 70–72
  plant proteins, 73
 factors influence characteristics of, 73–77, 74f
  drug properties, 76–77
  protein composition, 73–74
  protein solubility, 74–75
  surface properties, 75–76
 preparation methods, 77–81
  coacervation/desolvation, 77–79
  complex coacervation, 80–81
  emulsion/solvent extraction, 79–80
Protein imaging reporter, 381f
Protein-protein interaction, 382f
Protein polymers, 70t
PRP. *See* Platelet-rich plasma

Qdot
 colors and, 352f
 fluoroscopy images of, 360f
 high-resolution transmission electron microscopy, 354f

in vitro/in vivo bioimaging studies, 354t
in vivo fluorescence imaging of, 357f
MRI data from, 362f
schematic illustration of, 353f
schematic showing function of, 359f
silica-coated, 353–354

RAFT. *See* Reversible addition-fragmentation chain transfer
Raman spectroscopy, 245–248
Reciprocal lattice geometry
2D symmetries and, 299–301
advanced instrumentation, 302–303
Blackman corrections of electron diffraction intensities, 283–285
by powder X-ray diffractometry, 272–274
crystallography databases for, 303–304
from fine-grained CPEDR patterns, 285–287
from PED patterns, 287–293
in lattice-fringe fingerprint plots, 293–299
kinematic/quasi-kinematic approximations, 275–282
limitations to structural fingerprinting, 303
usage of crystallographic reliability (r) values, 301–302
utilizing fast electrons for structural fingerprinting in, 274–275
Recombinant DNA (rDNA), 69
Red blood cells (RBCs), 194
RESS processes, schematic diagram, 29f
Restriction fragment length polymorphism (RFLP), 175
Reticuloendothelial system (RES), 63, 81, 220
Reversible addition-fragmentation chain transfer (RAFT), 95f, 96

SAS, schematic diagram of, 29f
Scanning electron micrographs, 121f
Scanning electron microscopy (SEM), 53, 239, 241–242
Scanning transmission electron microscopy (STEM), 252–257, 260, 263, 264f, 265–266, 311
Scanometric DNA, 99f
Selected area electron diffraction (SAED), 289–290, 293
Self-assembled monolayer (SAM), 96, 97, 100
Self-nanoemulsified drug delivery system (SNEDDS), 8, 9
SEM. *See* Scanning electron microscopy
Semiconducting quantum dots
description of, 350–354
passivation, 352–353
silica-coated Qdots, 353–354
bioimaging of
fluorescence for, 355–358
magnetic resonance-based, 360–361
of FRET in, 358–359
radio-opacity and paramagnetic properties, 360
surface-enhanced Raman spectroscopy in, 359–360
SEM image, 64f
Short interfering RNA (siRNA), 52–53, 54, 177
Signal intensity, of various organs as a function of time, 343f
Silver nanowires, 324, 326f
Single photon emission computed tomography (SPECT), 330, 349
Size-exclusion chromatography (SEC), 191, 192
Sketches to illustrate precession electron diffraction (PED) geometry, 270, 275, 283–287, 293, 294, 297, 299–301, 305, 311–313
Skin, structure of, 127f
Skin penetration, of deformable vesicles, 132f
SLNs. *See* Solid lipid nanoparticles
Small unilamellar vesicles (SUVs), 131
SOD. *See* Superoxide dismutase
Solid lipid nanoparticles (SLNs), 42, 134, 219
Spatial resolution in thin specimen, 260f
SPECT. *See* Single photon emission computed tomography
SQUID. *See* Superconductor quantum interface device
Superconductor quantum interface device (SQUID), 336, 360
Superoxide dismutase (SOD), 38, 41, 43, 45, 46, 87
Superparamagnetic iron oxide nanoparticles (SPION), 63
Surface plasmon resonance (SPR), 191, 192, 193f, 310
Surface-enhanced Raman scattering (SERS), 99, 246, 355, 359
Surface-initiated free radical polymerization (FRP), 95
Surface initiated polymerization (SIP), 94, 96, 98, 108
SUVs. *See* Small unilamellar vesicles

TBARS. *See* Thiobarbituric acid reactive substances
T 1–weighted MRI, 339f
TCI. *See* Transcutaneous immunization
TEM. *See* Transmission electron microscope
Therapeutic enzymes, 37t
Thermogravimetric analysis, 341f
Thiobarbituric acid reactive substances (TBARS), 211
Three-dimensional atomic force microscopic images, 98f
Time-resolved confocal images of a fixed 3T3 cell, 355f
Toxoplasma gondii, 219
Toxoplasmic encephalitis, 219
Transcutaneous immunization (TCI), 143
Transmission electron microscope (TEM), 239–242, 256, 257, 270, 272, 275, 277, 285, 288, 293, 299, 302, 303, 318, 321
Transmission electron microscopy, 240–241
Trimetasphere, 334–339, 336f
Tumor necrosis factor (TNF), 60, 105

ULVs. *See* Unilamellar vesicles
Unilamellar vesicles (ULVs), 131, 132
United States Pharmacopoeia (USP), 165
USP. *See* United States Pharmacopoeia

Various nanosystems, 130f
Vascular endothelial growth factor (VEGF), 225–226
VEGF. *See* Vascular endothelial growth factor

Z-contrast imaging, 257